STRENGTH OF MATERIALS

A.K. UPADHYAY
B.E. (Civil), M.E.

S.K. KATARIA & SONS®
Publisher of Engineering Books
4885/109, Prakash Mahal, Dr. Subhash Bhargav Lane,
Opp. Delhi Medical Association, Daryaganj, New Delhi–110002 (INDIA)
Phone: +91-11-43551243, *Mobile:* +91-9871775858, 7678425209
e-mail: katariabook@gmail.com, katariabooks@yahoo.com
Website: www.skkatariaandsons.com

Published by:

S.K. Kataria & Sons®
4885/109, Prakash Mahal, Dr. Subhash Bhargav Lane,
Opposite Delhi Medical Association, Daryaganj, New Delhi-110002 (INDIA)
Phone: +91-11-43551243, *Mobile:* +91-9871775858, 7678425209
e-mail: katariabook@gmail.com, katariabooks@yahoo.com
Website: www.skkatariaandsons.com

Send all correspondence to : **M/s S.K. Kataria & Sons,** New Delhi
Email: katariabook@gmail.com

First Edition: January, 2011; Reprint: 2026

Laser Typesetting : Plus Computers, Delhi

Printed at: Arora Printing Press, Delhi

Preface

I take the pleasure of introducing this book entitled "Strenth of Materials" for the diploma and degree classes. In this book I have tried to present the subject matter in a very simple manner keeping in mind the needs and requirement of all levels of students.

This book contains total of eleven chapters covering the entire course contents of BTE of various states. For each topic sufficient number of numericals are solved. Numerical from question papers of various BTE are solved and added.

At the end of each chapters objective type questions and theoretical questions have been provided which will help the students for the preparation of various examinations. Laboratory practicals have been presented at the end of the book.

Constructive criticism from colleagues and students of the various institutes will be highly appreciated.

January, 2011 **A.K. Upadhyay**

An investment in knowledge yields the best results.

—Benjamin Franklin

Contents

CHAPTER 1

Properties of Materials

1.1 CLASSIFICATION OF MATERIALS

Chapter Outline

- Classification of Materials
- Mechanical Properties of Materials
- Test for the Mechanical Properties of Materials

Materials are classified into following categories from engineering point of view:

(*a*) Elastic Material
(*b*) Plastic Material
(*c*) Elasto-plastic Material
(*d*) Ductile Material
(*e*) Brittle Material

(*a*) Elastic Material

A material is said to elastic material if it undergoes deformation on external loading and the deformation disappear on removal of the load. Example— steel and rubber.

(*b*) Plastic Material

A material is said to be plastic if it undergoes deformation on external loading and the deformation does not disappear at all on removal of load. Example— lead and gold.

DO YOU KNOW?

A material is said to be isotropic if it is equally elastic in all directions.

(*c*) Elasto-plastic Material

A material is said to be elasto-plastic if it undergoes deformation on external loading and the deformation disappears partially on removal of the load. Hence, a permanent deformation remains even after removal of the load in elasto-plastic material.

(*d*) Ductile Material

Ductile material is a material which can undergo large deformation before failure. In other words, a material which can be drawn into wires is known as ductile material. Example— Aluminium, copper, magnesium, nickel, etc.

DO YOU KNOW?

Most of the materials become plastic under the action of large forces and behave in a manner similar to a viscous fluid.

An advantage of ductility is that visible distortions may occur if the loads become too large, thus providing an opportunity to take remedial action before an actual fracture occurs.

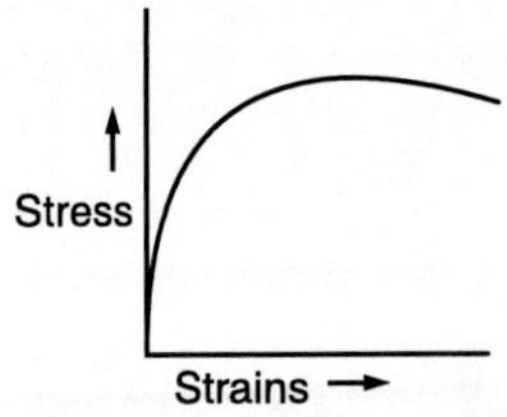

Fig.1.1 Typical stress-strain curve for ductile material

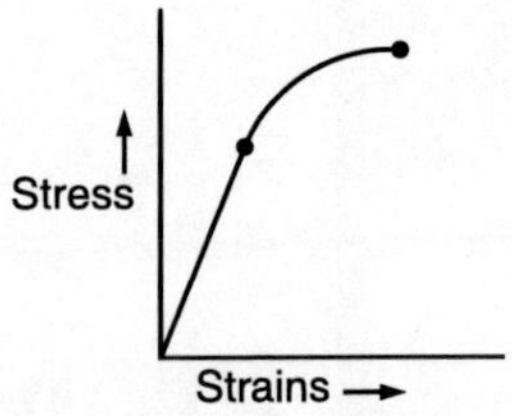

Fig. 1.2 Typical stress-strain curve for brittle material

(e) Brittle Material

A brittle material is a material in which no deformation takes place when external load is applied and it fails by rupture. In other words, materials that fail in tension at relatively low values of strain are known as brittle materials. Example— concrete, cast-iron, glass, ceramic materials.

> **DO YOU KNOW?**
> Ordinary glass is nearly ideal brittle material.

1.2 MECHANICAL PROPERTIES OF MATERIALS

Some of the important mechanical properties of materials are as follows :

(i) **Hardness :** The hardness of a material is a measure of its resistance to penetration by an identer. It is also a measure of strength and often has the units of stress. The identer is fabricated to a hard material and the shape of tip may be conical, spherical or pyramidal.

(ii) **Elasticity :** Elasticity is the property of a material by virtue of which it returns to its original dimension during unloading.

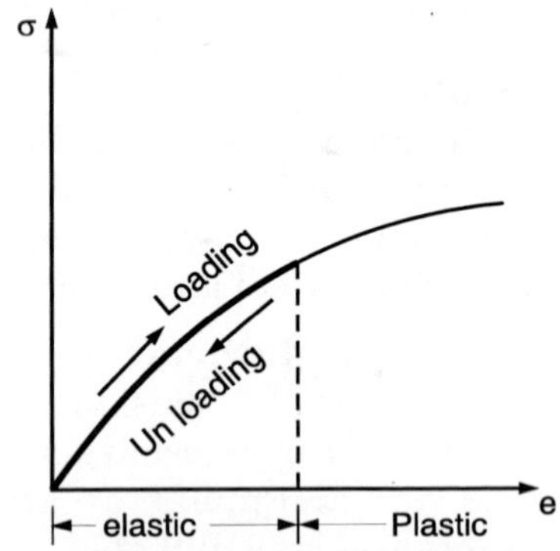

Fig. 1.3 Elastic behaviour

(iii) **Plasticity :** Plasticity is the property of a material by virtue of which permanent deformation of the material takes place after removal of load.

(iv) **Ductility :** Ductility is the property of a material by virtue of which it can be drawn into wires before rupture. It is measured by the percentage of elongation and percentage of reduction of area of a test piece before rupture.

> **DO YOU KNOW?**
> Ductile materials are capable of absorbing large amount of energy prior to fracture.

(v) **Brittleness :** Lack of ductility is brittleness.

(vi) **Toughness :** The property of a metal by virtue of which it can be bent, twisted or stretched under a high stress before rupture is called toughness.

> **DO YOU KNOW?**
> With increasing carbon content, steel becomes less ductile but has a higher yield stress and higher ultimate stress.

(vii) **Malleability :** The property of a metal by virtue of which it can be beaten or rolled into their sheets without rupture is called malleability.

(viii) **Strength :** It is the ability of a material to resist the action of external force without breaking. The tensile strength, compressive strength and shear strength represents the strength of a material.

1.3 TEST FOR THE MECHANICAL PROPERTIES OF MATERIALS

To know the strength of materials, the following tests are carried out in the laboratory :

(*i*) Tension test
(*ii*) Compression test
- Torsion test

(*iii*) Impact test
(*iv*) Fatigue test

(*i*) Tension Test

The most common material test is the tension test. The test is generally carried out in universal testing machine. The test involves taking a small sample with fixed cross-sectional area and then pulling it with a controlled, gradually increasing force until the sample breaks. The stress-strain curve is then drawn and ultimate strength is then found out.

Objectives : The objective is to determine the following properties :

(*i*) Yield stress (*ii*) Proportional limit (*iii*) Ultimate stress
(*iv*) Modulus of elasticity (*v*) Ductility (*vi*) Reduction in area

Specimen : A round specimen is generally used for tension test. As per Is : 1608, the standard dimensions of the specimen are :

(*i*) The diameter of central portion should not be less than 4 mm.

(*ii*) Gauge length $L_0 = 5.65 \sqrt{A_0}$; where A_0 is the cross-sectional area of central portion.

Procedures :

(*i*) Measure the central diameter of the specimen by a screw micrometer and mark the gauge length.

(*ii*) The test machine is set up and insert the specimen into grips of the machine.

(*iii*) Fix the extensometer and vernier scale and set the readings to zero position.

(*iv*) Start the machine and take readings on extensometer at regular intervals.
The rate of loading should be so selected that the reading of the extensometer can be taken accurately.

(*v*) As soon as the extension becomes unproportionaly large, remove the extensometer and start taking readings on the vernier scale.

(*vi*) Apply the load till the specimen breaks and then stop the machine.

(*vii*) Release the specimen from the machine and then by joining the broken pieces together, measure the final length and diameter of the specimen.

(*viii*) Plot the stress strain curve and find out the various stresses.

(*ii*) Compression Test

Compression test is opposite to tension test. This test is performed for testing brittle materials like cast iron, concrete, etc. A universal testing machine or simple compression testing machine is used for compression test. The specimens used in this test are cylindrical or cubical.

Procedures :

(*i*) Measure the dimensions of the test specimen.

(*ii*) Place the specimen between the upper and lower platens care should be taken that the centre lines of the platens and the specimen coincides as far as possible.

(*iii*) Fix the compresometer to the specimen to measure the decrease in length.

(*iv*) Apply the load at a specified rate. If a brittle material is to be tested, only the load at failure is recorded. When a ductile material is to be tested, the load at different values of compression is recorded.

(*v*) Stop applying load when the specimen fails. Note down the load at which the specimen fails.

(*vi*) Calculate the compressive stress by dividing the failure load (P) with original cross-sectional area (A) of specimen *i.e.*,

$$\sigma_c = \frac{P}{A}$$

Type of failure : In a ductile material, no fracture takes place and definite failure plane is not developed. The specimen bulges laterally and take a barrel like shape.

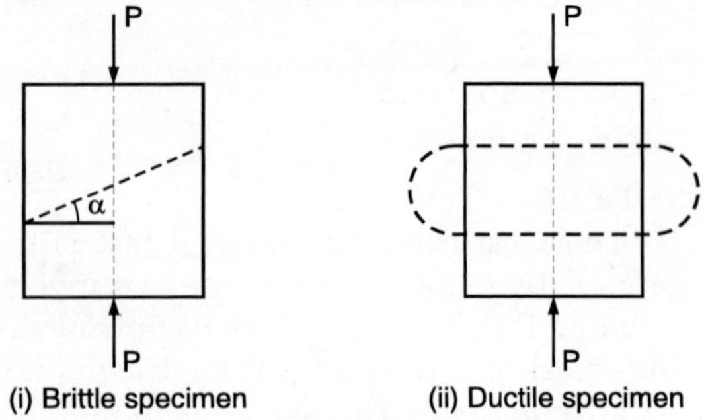

Fig. 1.4

In brittle material, specimen fails diagonally as shown in figure 1.4 (*i*). Theoretically the angle of failure plane with horizontal should be 45°, but due to friction the angle varies between 50 to 60°.

(*iii*) Impact Test

Impact strength of a material is its capacity to absorb shock energy before it fracture. The amount of energy absorbed by a material during fracture is a measure of a given materials toughness. Thus, this test is used to measure the toughness of a material under shock loading conditions. There are two types of impact tests :

(*a*) Izod impact test

(*b*) Charpy impact test

(*a*) Izod impact test : Izod impact test was invented in early 1900's by a metallurgist named Izod. It consists of a pendulum with a determined weight at the end of its arm. The pendulum swings down and striking the specimen while it is held securely in a vertical position. The impact strength is determined by the loss of energy of the pendulum as obtained by precisely measuring the loss of height in the pendulum's swing.

Specimen : The test specimen is generally 75 mm × 10 mm × 10 mm in size and consists of a 45° V notch on the surface which is 2 m deep.

Procedures :

(*i*) Keep the pendulum in its highest position and note down the reading of the pointer for zero error.

(*ii*) Release the pendulum from its highest position and note down the reading when it comes back. The pointer should indicate the same reading. Any deviation should be noted.

(*iii*) Fix the specimen properly in the gripping dies.

(*iv*) Held the pendulum at its highest position and release it. Note down the reading when the pendulum comes back.

(*v*) Find the difference between the two readings. This difference will indicate the energy absorbed for breaking the specimen.

Calculation : The energy absorbed is calculated as follows : Refer figure 1.6.

Let W = Weight of the pendulum

H = Height of fall of pendulum

h = Height of rise of pendulum after striking

x = Distance of C.G. of pendulum from the axis of rotation

θ = Angle of fall
α = Angle of rise

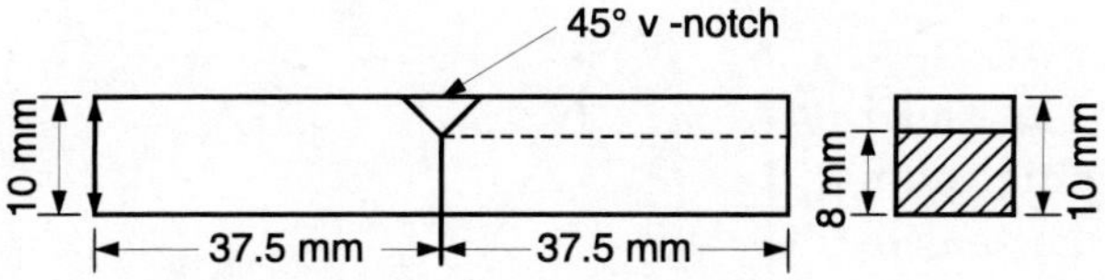

Fig. 1.5 V-notched test piece

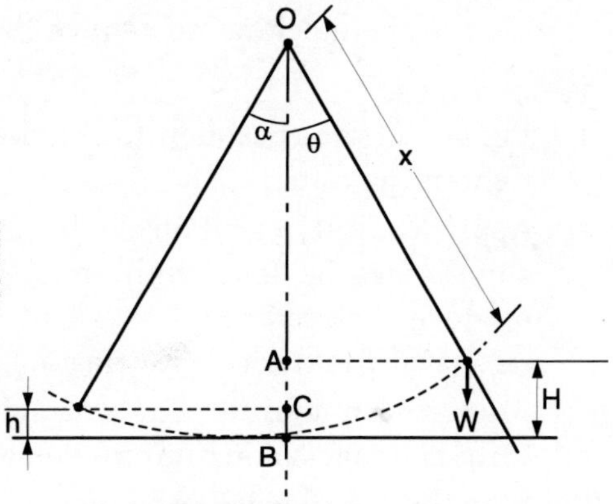

Fig. 1.6

$$\text{Initial energy} = WH \qquad [\because OB = x \text{ and } OA = x\cos\theta]$$
$$= W \times AB = W(x - x\cos\theta)$$
$$= Wx(1 - \cos\theta)$$
$$\text{Final energy} = Wh = W \times CB$$
$$= Wx(OB - OC) \qquad [\because OB = x \text{ and } OC = x\cos\alpha]$$
$$= W(x - x\cos\alpha)$$
$$= Wx(1 - \cos\alpha)$$
$$\text{Energy absorbed} = \text{Initial energy} - \text{Final energy}$$
$$= Wx(1 - \cos\theta) - Wx(1 - \cos\alpha)$$
$$= Wx(1 - \cos\theta - 1 + \cos\alpha)$$
$$= Wx(\cos\alpha - \cos\theta).$$

(*b*) Charpy impact test : Charpy impact test is an improved vision of Izod test and is more common than Izod test. A simply supported test piece is used in this method. The size of the specimen is 55 mm × 10 mm × 10 mm with a 45° – V notch, 2 mm deep. The principle involved is similar to Izod test. The specimen is broken by using swinging pendulum falling from a certain height. The energy absorbed in breaking the specimen is calculated in similar way as in Izod test.

(*iv*) Fatigue test : Fatigue is a phenomenon leading to fracture under repeated applied stress. Fatigue strength of the material is the limiting stress upto which the material can withstand without failure a specified number of cycles of stress. The maximum stress below which a material can withstand infinite number of cycles of stress is called the *endurance limit* of the material.

The various types of fatigue test are:

(*i*) Axial stress cyclic test
(*ii*) Bending stress cyclic test
(*iii*) Torsional stress cyclic test
(*iv*) Combination of axial, bending and torsional stress cyclic test.

The bending stress cyclic test is the most common one. In this a number of identical test specimens (usually 8 to 10) are made from the material. The specimen is fixed at one end and free at other of the machine, to behave like a cantilever.

Procedures:

(*i*) Fix one of the specimen to the machine and tighten the nuts.

(*ii*) Apply a suitable load on the beam. The load should not be less than that, which can produce a stress equal to 3/4 of the tensile strength of material upto elastic limit.

(*iii*) Start the machine. The specimen passes through tension and compression alternately.

(*iv*) Stop the machine when the specimen breaks.

(*v*) Note down the maximum bending stress (σ_b) and number of cycles (N) to cause failure.

(*vi*) Fix another specimen and after changing the applied load repeat steps (*iii*) to (*v*).

(*vii*) Continue this process till we get considerable set of values of σ_b and N.

(*viii*) Plot the graph between σ_b and N in log-log scale and find the endurance limit as shown in figure 1.7.

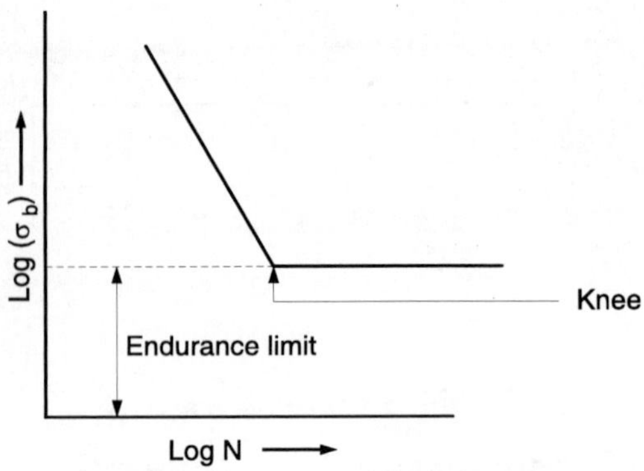

Fig. 1.7 σ_b vs N curve of mild steel.

SUMMARY

1. **Elastic material :** If a material undergoes deformation on loading and the deformation disappears on removal of load, then the material is known as elastic material.
2. **Plastic material :** If a material undergoes deformation on loading and the deformation does not disappear on removal of load, then the material is said to be plastic material.
3. **Ductile material :** A material is said to be ductile material if it can be drawn into wires.
4. **Brittle material :** A material is known as brittle material in which no deformation takes place an loading but fails by rupture.
5. **Toughness :** It is the property of a material by virtue of which it can be twisted, bent or stretched undue high stress.
6. **Malleability :** It is the property of a material by virtue of which it can be beaten or rolled into thin sheets without rupture.
7. **Endurance Limit :** Endurance limit is the maximum stress below which a material can withstand infinite number of cycles of stress.

OBJECTIVE TYPE QUESTIONS

1. Fill in the Blanks :

(*i*) The property which enables a material to come back to its original shape and size after removal of load is called ....

(*ii*) The property of a material by virtue to which it can be drawn into thin wire is known as ....

(*iii*) The property of a material by virtue of which it can be beaten or rolled into plate is known as ....

(*iv*) Cast iron is a .... material.

(*v*) Fatigue of metal is caused by ....

Answers

(*i*) Elasticity (*ii*) ductility (*iii*) malleability (*iv*) brittle
(*v*) varying stresses

THEORETICAL QUESTIONS

1. Define the following:
   (*a*) Elasticity (*b*) Plasticity
   (*c*) Ductility (*d*) Toughness
   (*e*) Malleability
2. Explain the following:
   (*a*) Elastic material (*b*) Plastic material
   (*c*) Ductile material (*d*) Brittle material
3. Discuss the mechanical properties of materials.
4. Name the various tests used to know the mechanical properties of materials.
5. Discuss briefly tensile test.
6. What do you mean by endurance limit of a material? How endurance limit is find out for mild steel?
7. Why impact test is carried out on a material? Describe briefly Izod impact test.

❑❑❑

CHAPTER 2

Stresses and Strains

2.0 LOAD

The external force acting on a body is called the load.

2.1 UNITS OF LOAD

In MKS : Unit of load is kg.

In SI : Unit of load is N (Newton).

2.2 CLASSIFICATION OF LOADS

The loads may be classified as:

(A) According to the Manner of Application on the Body:

(*i*) **Point Load:** Point load is also known as concentrated load. The load which acts at a point is called a point load.

(*ii*) **Distributed Load:** The load which is distributed in some manner over the length of a beam is known as distributed load.

If the distribution is uniform over the length of the beam. Then the load is called uniformly distributed load or U.D.L.

If the load varies uniformly from one point to another along the length of the beam, then the load is called uniformly varying load. Triangular or Trapezoidal type of loading falls under this category, as shown in figures 2.3 and 2.4.

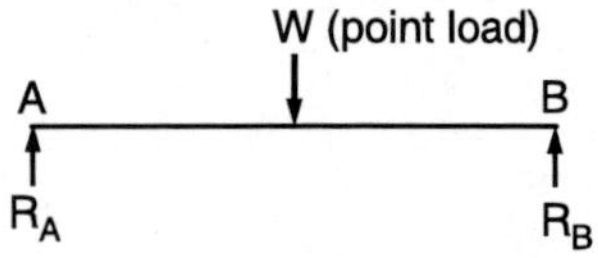

Fig. 2.1.

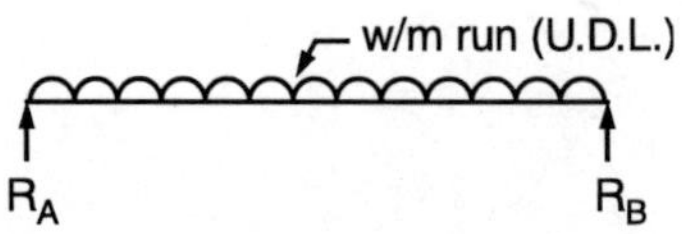

Fig. 2.2.

Chapter Outline

- Load
- Units of Load
- Classification of Loads
- Stress
- Units of Stress
- Types of Stresses
- Strain
- Types of Strains
- Elasticity
- Elastic Limit
- Hooke's Law
- Principle of Superposition
- Young's Modulus of Elasticity (E)
- Modulus of Rigidity (G)
- Bulk Modulus (K)
- Longitudinal and Lateral Strain
- Poisson's Ratio (M)
- Relations Between E, G and K
- Stress-strain Curve
- Maximum or Ultimate Stress
- Working Stress
- Factor of Safety
- Breaking Stress or Rupture Strength
- Proof Stress
- Deformation of a Body Subjected to Axial Load
- Bars of Varying Sections
- Extension of a Tapering Rod
- Bars of Composite Section
- Thermal Stress and Strain

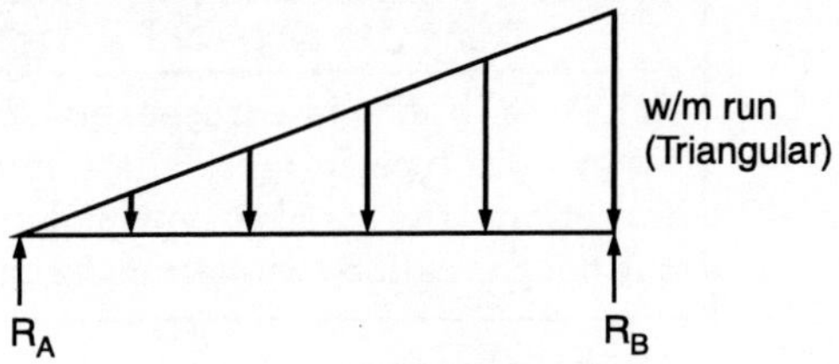

Fig. 2.3.

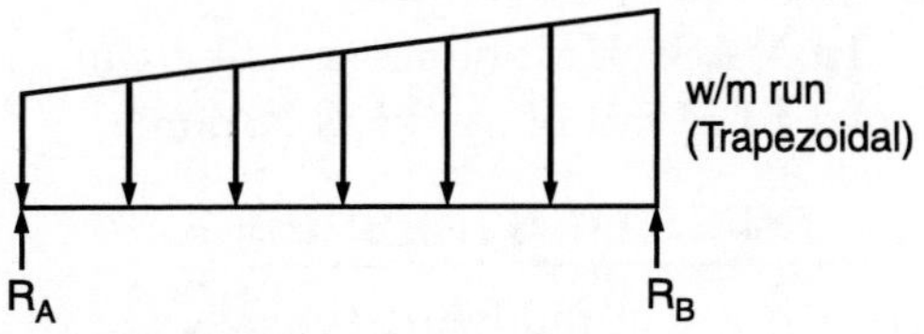

Fig. 2.4.

(B) According to the Effects Produced on the Body:

(*i*) **Tensile loads:** The loads which tend to increase the length of the body are known as tensile loads.

(*ii*) **Compressive loads:** The loads which tend to decrease the length of the body are known as compressive loads.

(*iii*) **Shearing loads:** The loads which tend to slide a part of a body over the other across any section are known as shearing loads.

(*iv*) **Twisting or torsional loads:** The loads which tend to rotate one end of the body relative to the other about its longitudinal axis are known as twisting or torsional loads.

(*v*) **Bending loads:** The loads which tend to produce curvature or bending in the body are known as bending loads.

2.3 STRESS

Whenever a body is subjected to an external force, it tends to deform. The body resists deformation due to the internal resistance developed within the body and this resistance is equal to the applied force.

Consider a bar of uniform cross-sectional area 'A'. The bar is subjected to tensile force P. Consider a section $X-X$ of the bar.

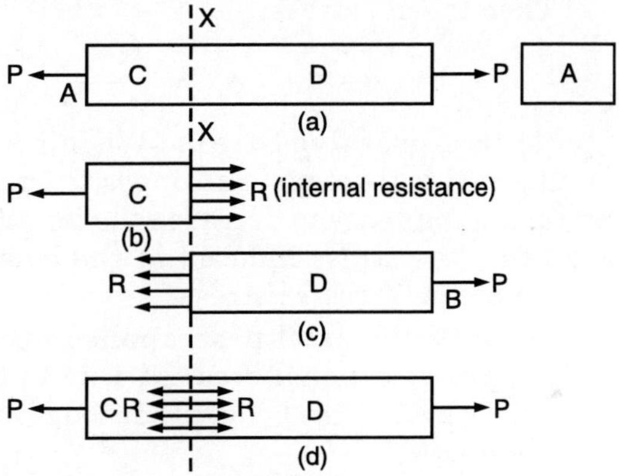

Fig. 2.5.

The part 'C' of the bar will be in equilibrium if $P = R$ (Resistance). Similarly, part 'D' of the bar will be in equilibrium if $P = R$. This internal resistance against deformation is the stress.

Thus, the force of resistance offered by a body against deformation is known as *stress.*

The internal force of resistance to deformation per unit cross-sectional area of the body is called *intensity of stress* or *stress.*

Mathematically, stress may be defined as the force per unit area. It is denoted by σ (sigma).

Stress, $$\sigma = \frac{P}{A}$$

where P = Applied force or load on the body

A = Cross-sectional area of the body.

Note: In framed structures, all the individual members are of definite length as required by the geometry of the structure. If any member is not of correct length, stress in other members will be set up when that member is fitted. Such stress is known as **stress due to lack of fit**.

2.4 UNITS OF STRESS

In MKS : Unit of stress is kg/cm^2.
In SI : Unit of stress is N/mm^2.

> **DO YOU KNOW?**
> When a body is subjected to external forces in more than one directions, the stress developed in the body is called *compound stress.*

2.5 TYPES OF STRESSES

Stresses are of following types:

(*i*) Direct stress (*ii*) Shear stress (*iii*) Bending stress
(*iv*) Torsional stress (*v*) Thermal stress.

(*i*) Direct or Normal Stress: The stresses which are normal to the plane on which they act are called direct or normal stresses.

Direct stresses are either *tensile* or *compressive.*

(*a*) **Tensile Stress :** When a body is subjected to two equal and opposite pulls and the body has a tendency to elongate, the stress induced at any cross-section of the body is termed as tensile stress.

Consider a body of length '*l*' and cross-sectional area '*A*' subjected to two equal and opposite pulls '*P*'.

Then tensile stress,

$$\sigma_t = \frac{P}{A}$$

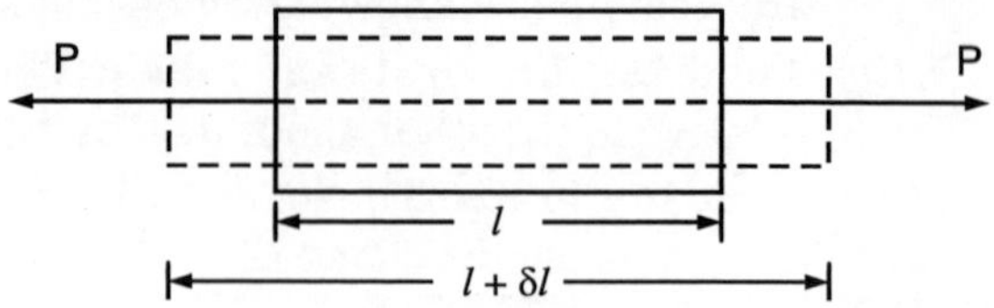

Fig. 2.6.

(*b*) **Compressive Stress :** When a body is subjected to two equal and opposite pushes and has a tendency to decrease the length of the body, the stress induced in the body is known as compressive stress.

Let P be the axial push applied on the body of cross-sectional area '*A*'. Due to this push the length of the body is reduced from l to $(l - \delta l)$.

Then compressive stress,

$$\sigma_c = \frac{P}{A}.$$

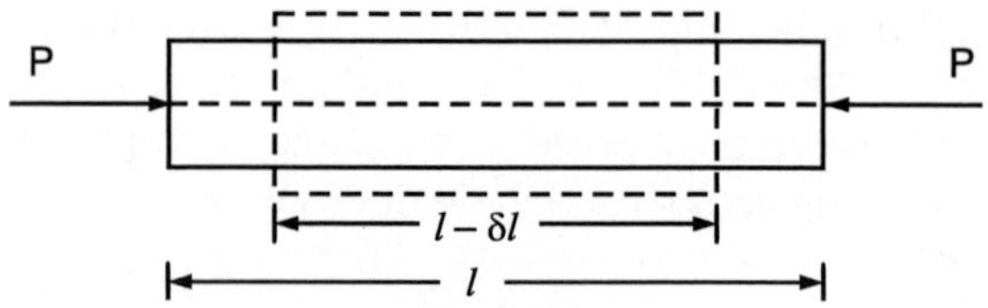

Fig. 2.7.

(*c*) **Shear Stress :** If two equal and opposite parallel forces, whose line of action are not the same, act on a body, then there is a tendency of one part of the body to slide over the other part. The stress induced in the body is known as *shear stress.* shear stress is tangential to the area over which it acts.

Area of the body parallel to P, $A = l \times b$ (shear area)

$$\therefore \quad \text{Shear stress, } \tau = \frac{\text{Shear force}}{\text{Shear area}}$$

$$= \frac{P}{A} = \frac{P}{(l \times b)}.$$

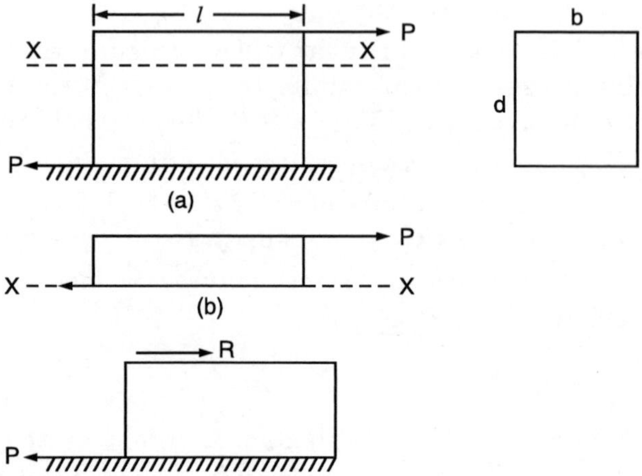

Fig. 2.8.

(*d*) **Bending Stress :** When a beam is subjected to external loading, bending moment is developed in the beam. In order to counteract the bending moment, internal resistance or stresses will developed in the beam. The stresses produced in the beam to resist the bending is called *bending stress.*

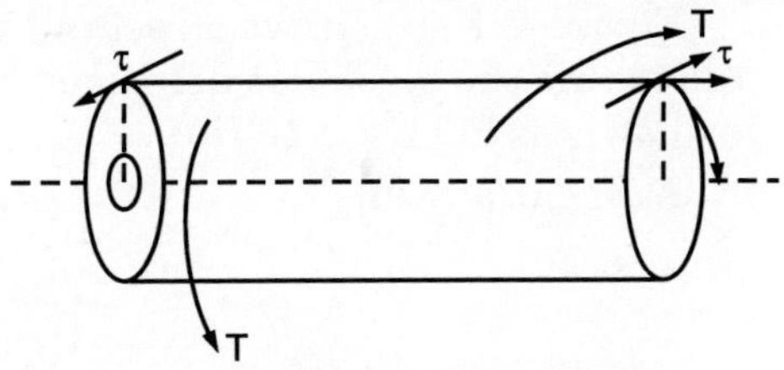

Fig. 2.9. Torsional stress.

(*e*) **Torsional Stress :** If a member is subjected to two equal and opposite torques, shear stresses will be set up in the member. These stresses are known as torsional stress.

(*f*) **Thermal Stress :** When the temperature of a rigidly fixed member is changed and the change in length is prevented due to fixidity, then stress will be induced in the member. This stress is known as **temperature stress** or **thermal stress.**

2.6 STRAIN

Strain is a measure of deformation produced in the member by the load.

It is the ratio of change in dimension of the member to the original dimension.

$$\therefore \quad \text{Strain,} \qquad e = \frac{\text{Change in dimension}}{\text{Original dimension}}$$

Unit of Strain : Strain has no unit as it is a ratio.

2.7 TYPES OF STRAINS

> **DO YOU KNOW?**
>
> The total strain energy stored in a strained member is termed as **resilience**.

Strains are of following types:

(*i*) Tensile strain
(*ii*) Compressive strain
(*iii*) Shear strain
(*iv*) Volumetric strain

(*i*) Tensile strain : It is the ratio of increase in length to the original length of a body when an axial pull acts on it. Refer figure 2.6.

$$\text{Tensile strain,} \quad e_t = \frac{\text{Increase in length}}{\text{Original length}} = \frac{\delta l}{l}.$$

(ii) Compressive strain : It is the ratio of decrease in length to the original length of a body when an axial push acts on it. Refer figure 2.7.

$$\text{Compressive strain, } e_c = \frac{\text{Decrease in length}}{\text{Original length}} = \frac{\delta l}{l}.$$

(*iii*) Shear Strain : Shear strain is the measure of the distortion produced in a body when shear force is acting on it. In other words, it is a measure of the angle through which a body is distorted when shear forces are acting on it. In figure 2.10, angle ϕ is the shear strain. Shear strain is measured in radian.

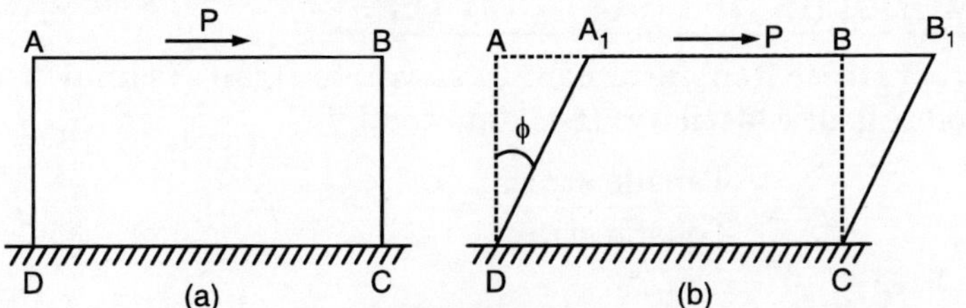

Fig. 2.10.

Figure 2.10(*a*) shows a rectangular block *ABCD* which is acted upon by a shear force *P*. Due to this the block has deformed to position A, B, C, D as shown in Figure 2.10 (*b*).

> **DO YOU KNOW?**
> Volumetric strain is the algebraic sum of the three principal strains.

∴ Shear strain,

$$e_s = \frac{AA_1}{AD} = \tan\phi$$

$$= \phi \quad (\tan\phi = \phi \text{ as } \phi \text{ is very small})$$

***(iv)* Volumetric Strain :** The ratio of change in volume to the original volume is known as volumetric strain. It is also known as bulk strain.

$$\text{Volumetric strain,} \quad e_v = \frac{\text{Change in volume}}{\text{Original volume}} = \frac{\delta V}{V}.$$

2.8 ELASTICITY

A material is said to be elastic if it undergoes deformation on application of a load and the deformation disappears on the removal of the load. Elasticity is the property of a material by virtue to which a material regain all its deformation after removal of the external loading.

2.9 ELASTIC LIMIT

Elastic limit is the value of stress beyond which the material will not return to its original position when unloaded but will retain a permanent deformation called "permanent set."

2.10 HOOKE'S LAW

Hooke's law states that within elastic limit stress is proportional to strain.

i.e. Stress ∝ strain

or Stress = A constant × strain

$$\therefore \quad \text{Constant} = \frac{\text{Stress}}{\text{Strain}}$$

> **DO YOU KNOW?**
> Hooke's law has been established by a scientist named Robert Hooke.

This constant is known as Modulus of elasticity.

Modulus of elasticity depends upon the followings:

(*i*) the nature of the material of the body; the metal which is very stiff has very high value of modulus of elasticity.

(*ii*) the manner in which the body is deformed.

2.11 PRINCIPLE OF SUPERPOSITION

Principle of superposition states that when an elastic material is acted upon number of forces, then the resultant strain will be the sum of the individual strains caused by each load acting separately.

2.12 YOUNG'S MODULUS OF ELASTICITY (*E*)

The ratio of the direct stress (tensile or compressive), to strain, is known as Young's modulus of Elasticity or Modulus of Elasticity. It is denoted by *E*.

$$E = \frac{\text{Tensile stress}}{\text{Tensile strain}} = \frac{\sigma_t}{e_t}$$

$$E = \frac{\text{Compressive stress}}{\text{Compressive strain}} = \frac{\sigma_c}{e_c}$$

Unit of E is N/mm^2.

2.13 MODULUS OF RIGIDITY (*G*)

The ratio of shear stress to shear strain is known as shear modulus or Modulus of Rigidity. It is denoted by G.

∴ Modulus of Rigidity,

$$G = \frac{\text{Shear stress}}{\text{Shear strain}} = \frac{\tau}{\phi}$$

Unit of G is N/mm^2.

2.14 BULK MODULUS (*K*)

The ratio of normal stress to the volumetric strain is known as Bulk Modulus. It is also known as volume modulus of elasticity. It is denoted by K.

Bulk modulus,

$$K = \frac{\text{Normal stress}}{\text{Volumetric strain}} = \frac{\sigma}{e_v}$$

Unit of K is N/mm^2.

TABLE 2.1. Values of Young's Modulus of Some Standard Materials

S. No.	*Material*	*Young's Modulus (E) in N/mm^2*
1.	Steel	2.1×10^5
2.	Cast Iron	1.3×10^5
3.	Wrought Iron	1.9×10^5
4.	Copper	1.0×10^5
5.	Brass	0.8×10^5
6.	Aluminium	0.7×10^5
7.	Timber	0.1×10^5

2.15 LONGITUDINAL AND LATERAL STRAIN

When a body is axially loaded, there is an axial deformation in the body and at the same time there is a change in the other dimensions of the body perpendicular to the line of action of the applied load. Longitudinal strain is the strain produced along the applied load and lateral strain is the strain produced at right angles to the applied load. Refer figure 2.11, in which a body is subjected to axial tensile load 'P'. Due to this the length of the body is increased from l to $(l + \delta l)$, at the same time the breadth and depth of the body decreases from b to $(b - \delta b)$ and d to $(d - \delta d)$ respectively.

DO YOU KNOW?

The material must be both homogeneous and isotropic in order for the lateral strains in a bar in tension to be the same at every point

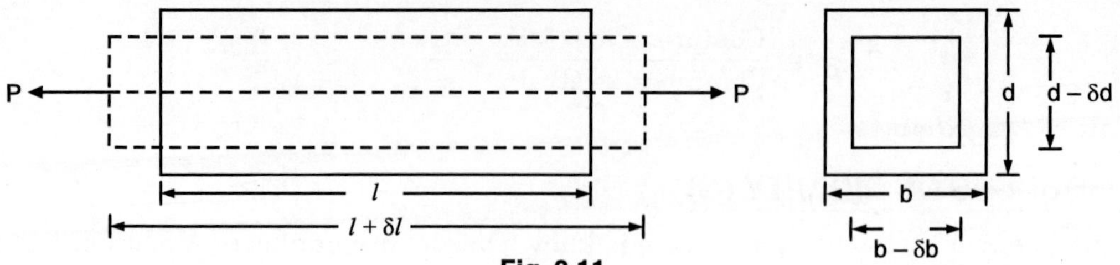

Fig. 2.11.

Then, longitudinal strain $= \dfrac{\delta l}{l}$

and lateral strain $= \dfrac{\delta b}{b}$ or $\dfrac{\delta d}{d}$.

2.16 POISSON'S RATIO (μ)

The ratio of lateral strain to the longitudinal strain, produced by a single stress, is known as Poisson's ratio. It is denoted by $\dfrac{1}{m}$ or μ. It has no unit.

Poisson's ratio, $$\frac{1}{m} = \frac{\text{Lateral strain}}{\text{Longitudinal strain}}$$

The value of m lies between 3 and 4.

DO YOU KNOW?

Material with an extremely low value of poisson's ratio, include cork for which 'μ' is practically zero.

2.17 RELATIONS BETWEEN *E, G* AND *K*

2.17.1 Relation between Modulus of Elasticity (*E*) and Modulus of Rigidity (*G*)

Consider a square block *ABCD* of unit thickness and sides x is subjected to shear stress 'τ' as shown in figure 2.12 (*a*). Due to this the shape of the body changed from *ABCD* to $A_1B_1C_1D$ as shown in Figure 2.12 (*b*). As a result the length of diagonal *AC* is increased and that of *BD* is decreased. Consider diagonal *AC*. Draw AA_2 perpendicular to A_1C.

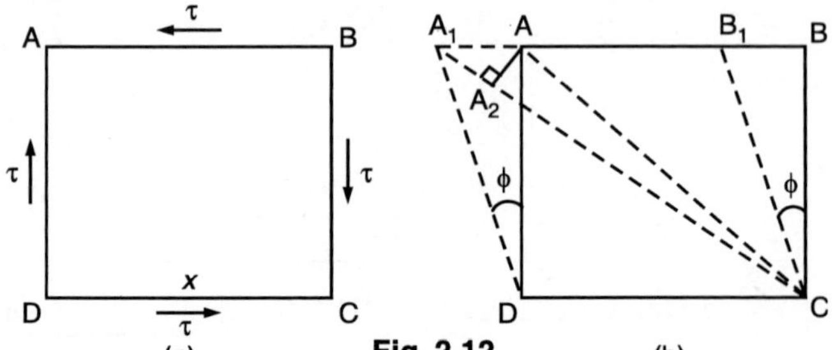

(a) Fig. 2.12 (b)

Increase in length of the diagonal $AC = A_1C - AC$

Since Angle ACA_2 is very small

$\therefore \quad AC = A_2C$

So increase in length of $AC = A_1C - A_2C$

$(\because AC = A_2C)$

$= A_1A_2 = A_1A \cos AA_1A_2$

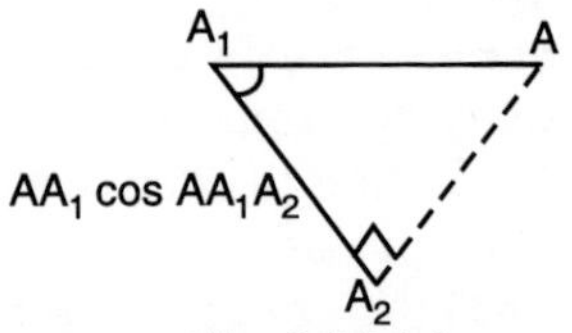

Fig. 2.12 (c).

But angle AA_1A_2 is nearly equal to $\angle BAC = 45°$

($\because$ ΔABC is isosceles and $\angle B = 90°$, $\therefore \angle BAC = 45°$)

$$\therefore \quad A_1A_2 = A_1A \cos 45° = A_1A \times \frac{1}{\sqrt{2}} = \frac{A_1A}{\sqrt{2}} \quad \ldots(i)$$

But shear strain, $\phi = \frac{A_1A}{AD} = \frac{A_1A}{x}$

$\therefore \quad A_1A = \phi x$

Putting the value of A_1A in (1), we get

$$\text{Increase in length of diagonal} = A_1A_2 = \frac{\phi x}{\sqrt{2}}$$

$$\text{Length of diagonal } AC = x\sqrt{2}$$

$$\therefore \quad \text{Strain in diagonal } AC = \frac{\text{Increase in length}}{\text{Original length}} = \frac{\phi x/\sqrt{2}}{x\sqrt{2}} = \frac{\phi x}{\sqrt{2}} \times \frac{1}{x\sqrt{2}} = \frac{\phi}{2} \quad ...(ii)$$

Increase in length of diagonal AC indicates that it is subjected to tensile stress and diagonal BD is subjected to compressive stress.

$$\text{Strain in diagonal } AC = \frac{\tau}{E}$$

and Strain in AC due to compressive stress along $BD = \frac{\tau}{Em}$

$$\left[\because \frac{1}{m} = \frac{\text{lateral strain}}{\text{longitudinal strain}}\right]$$

$$\therefore \quad \text{Total strain in } AC = \frac{\tau}{E} + \frac{\tau}{Em} = \frac{\tau}{E}\left(1 + \frac{1}{m}\right) \quad ...(iii)$$

From (*ii*) and (*iii*), we get

$$\frac{\phi}{2} = \frac{\tau}{E}\left(1 + \frac{1}{m}\right)$$

or

$$E = 2 \times \frac{\tau}{\phi}\left(1 + \frac{1}{m}\right)$$

$$\therefore \quad E = 2G\left(1 + \frac{1}{m}\right) \quad \left[\because G = \frac{\tau}{\phi}\right]$$

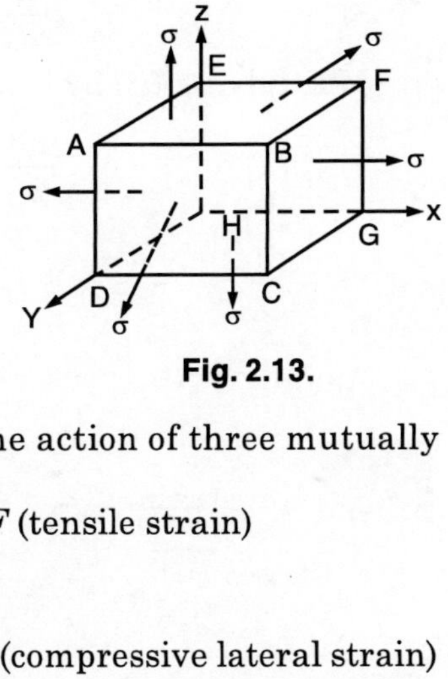

Fig. 2.13.

2.17.2 Relation between *E* and *K*

Consider a unit cube *ABCDEFGH* subjected to tensile stress 'σ' on all the faces.

Let us consider the strain of one side *AB* of the cube under the action of three mutually perpendicular stresses.

(*i*) Strain on *AB* due to stresses on the faces *ADHE* and *BCGF* (tensile strain)

$$= \frac{\sigma}{E}.$$

(*ii*) Strain on *AB* due to stresses on the faces *ABFE* and *DCGH* (compressive lateral strain)

$$= -\frac{\sigma}{mE}.$$

(*iii*) Strain on *AB* due to stresses on the faces *ABCD* and *EFGH* (compressive lateral strain)

$$= -\frac{\sigma}{mE} \quad \left[\because \frac{1}{m} = \frac{\text{Lateral strain}}{\text{Longitudinal strain}}\right]$$

$\therefore$ Net strain on $AB = \frac{\sigma}{E} - \frac{\sigma}{mE} - \frac{\sigma}{mE} = \frac{\sigma}{E}\left(1 - \frac{2}{m}\right)$

Volumetric strain in the unit cube,

$$e_v = 3 \times \text{strain in } AB$$

$$e_v = 3 \times \frac{\sigma}{E}\left(1 - \frac{2}{m}\right)$$

We know,

Bulk modulus, $K = \frac{\text{Stress}}{\text{Volumetric strain}} = \frac{\sigma}{e_v}$

$$K = \frac{\sigma \times E}{3\sigma\left(1 - \frac{2}{m}\right)}$$

or $$3K = \frac{E}{\left(1 - \frac{2}{m}\right)}$$

$\therefore$ $$E = 3K\left(1 - \frac{2}{m}\right)$$

DO YOU KNOW?

Stiffness of an axially loaded bar is defined as the force required to produce a unit deflection.

$$K = \frac{EA}{L}$$

DO YOU KNOW?

Reciprocal of stiffness is known as flexibility $= \frac{1}{K}$

Relation between *E*, *G* and *K*:

We know,

$$E = 2G\left(1 + \frac{1}{m}\right) \quad \text{...}(i)$$

or $$\frac{E}{2G} = 1 + \frac{1}{m} \quad \text{...}(ii)$$

Multiplying (*ii*) by 2 on both sides, we get

$$\frac{2E}{2G} = 2 + \frac{2}{m}$$

or $$\frac{E}{G} = 2 + \frac{2}{m} \quad \text{...}(iii)$$

Again, $$E = 3K\left(1 - \frac{2}{m}\right)$$

Note: Net strain, $e = \frac{\sigma}{E}\left(1 - \frac{2}{m}\right)$

$$\therefore Ee = \sigma\left(1 - \frac{2}{m}\right)$$

In the above expression, *Ee* and σ are positive members. Hence $\left(1 - \frac{2}{m}\right)$ must also be positive. This limits the value of $\frac{1}{m}$ to a maximum of 0.5.

$$\frac{E}{3K} = 1 - \frac{2}{m} \quad ...(iv)$$

Adding (*iii*) and (*iv*), we get

$$\frac{E}{G} + \frac{E}{3K} = 2 + \frac{2}{m} + 1 - \frac{2}{m}$$

or $$\frac{E}{G} + \frac{E}{3K} = 3$$

or $$\frac{3KE + GE}{3KG} = 3$$

or $$\frac{E(3K + G)}{3KG} = 3$$

or $$E(3K + G) = 9\,KG$$

$\therefore$ $$E = \frac{9\,KG}{3\,K + G}.$$

DO YOU KNOW?

The working stress of a material is not greater than its elastic limit.

2.18 STRESS-STRAIN CURVE

Stress-strain curve is a graph obtained from tensile test of steel. Figure 2.14 shows stress-strain curve for a mild steel.

From origin O to point A, the stress-strain diagram is a straight line *i.e.* stress is proportional to strain. Point A is known as *proportional limit.* Hooke's law holds good upto this limit. The stress is no longer proportional to the strain beyond this point (point A). Thus *proportional limit is the maximum stress at which stress is proportional to strain.*

Point B shows the elastic limit. From A to B, Hooke's law is not obeyed, although the material is elastic.

Thus, elastic limit is the stress beyond which the material will not return to its original shape when unloaded but will retain a permanent deformation.

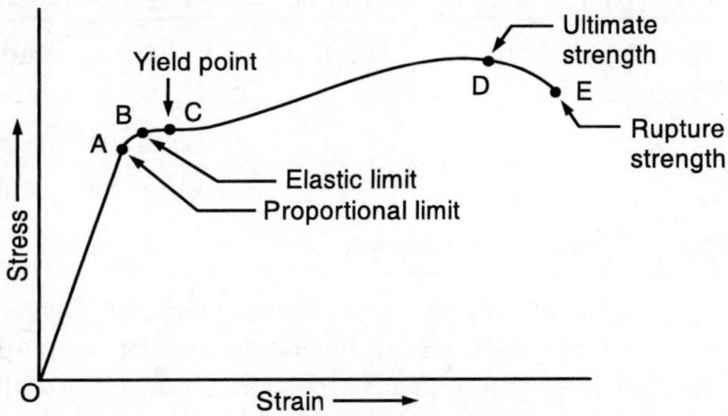

Fig. 2.14.

The next important occurrence is the *yield point C*. Between B to C, the metal shows an appreciable strain even without further increase in load (*i.e.*, stress). Thus, *yield point is the point at which there is an appreciable elongation i.e., yielding of the material without further increase in load* (*i.e.*, stress). Non-ferrous metals do not show a definite yield point. For such material, yield strength is determined by the offset method.

Beyond the point C, further straining can only be achieved by increasing the load, the stress-strain curve continuing to rise upto point D. This point D is the point of ultimate stress or ultimate strength or maximum stress. Thus, *ultimate strength is the highest ordinate of the stress-strain curve.* It is calculated by dividing the load at D by the original cross-sectional area of the bar. At this stage the bar begins to form a local neck.

Due to reduction in area at neck, the load falling off from the maximum until fracture at E. This point E is the breaking point. The stress corresponding to breaking point E is known as breaking stress or rupture strength. Thus *rupture strength is the stress at failure.*

2.19 MAXIMUM OR ULTIMATE STRESS

The maximum axial load that can be applied in a member without causing rupture is called *ultimate load.* Ultimate stress is the ratio of ultimate load to the original cross-sectional area.

$$\text{Ultimate stress} = \frac{\text{Ultimate load}}{\text{Original cross-sectional area}}$$

2.20 WORKING STRESS

The working stress is the maximum safe stress a material can carry. In design, the working stress should be limited to values not exceeding the proportional limit. This is calculated by dividing the ultimate stress by the factor of safety.

$$\text{Working stress} = \frac{\text{Ultimate stress}}{\text{Factor of safety}}$$

DO YOU KNOW?

Working stress is also called allowable stress.

2.21 FACTOR OF SAFETY

Factor of safety is defined as the ratio between ultimate stress and working stress.

$$\text{Factor of safety, } n = \frac{\text{Ultimate stress}}{\text{Working stress}}$$

DO YOU KNOW?

Margin of safety is defined as the factor of safety minus one.

Margin of safety $= n - 1$

The values of factor of safety in engineering design varies from 3 (for accurately known dead loads) and 12 (for shock loads of indefinite magnitude).

2.22 BREAKING STRESS OR RUPTURE STRENGTH

Breaking stress is defined as the ratio of load at the time of breaking or rupture to the original area of cross-section. Breaking stress is also known as rupture strength.

$$\text{Breaking stress} = \frac{\text{Load at breaking point}}{\text{Original area of cross-section}}$$

2.23 PROOF STRESS

The stress-strain curve for most of the non-ferrous materials do not have definite yield point or limit of proportionality. In such materials, proof stress is used instead of yield point.

Proof stress is the stress at which the strain has departed by 0.1% of the gauge length from the line of proportionality.*

This is obtained by drawing a line PQ parallel to line of proportionality OT, such that $OP = 0.1\%$, then the stress at Q is the proof stress.

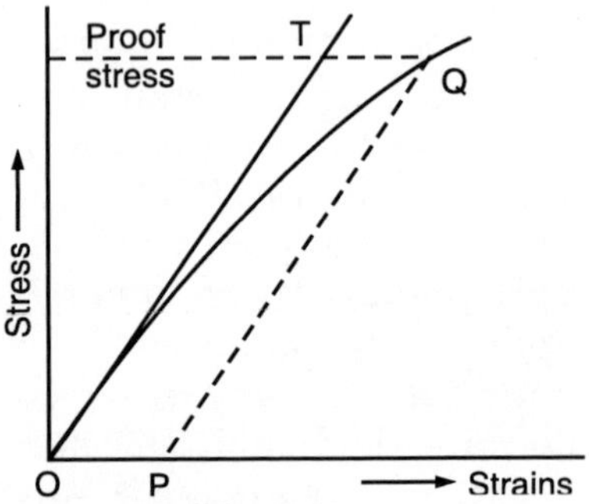

Fig. 2.15.

2.24 DEFORMATION OF A BODY SUBJECTED TO AXIAL LOAD

Consider a body which is subjected to axial load or force.

* Gauge length is the measured length of a specimen in tensile test for which strain is recorded by an 'extensometer'.

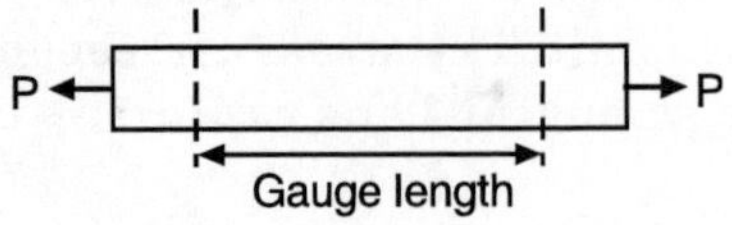

Let,
P = Axial load or force acting on it
A = Original area of cross-section
l = Original length of the body
δl = change in length of the body

Stress, $\sigma = \dfrac{P}{A}$

Strain, $e = \dfrac{\delta l}{l}$

We know $E = \dfrac{\sigma}{e}$

or $e = \dfrac{\sigma}{E}$

or $\dfrac{\delta l}{l} = \dfrac{P/A}{E} = \dfrac{P}{AE}$

$\therefore$ $\boxed{\delta l = \dfrac{Pl}{AE}}$.

DO YOU KNOW?

Optimum structure is a hypothetical structure that meets certain criterion, such as minimum volume or minimum weight.

EXAMPLE 2.1: A prismatic bar with rectangular cross-section (25 × 45 mm) and length 3.0 m is subjected to an axial tensile force of 70 KN. The measured elongation of the bar is 1.2 mm. Calculate the tensile stress and strain in the bar.

Solution: Given : Axial pull, P = 70 KN

Cross-sectional area, A = 25 × 45 = 1125 mm^2

Length of the bar, l = 3.0 m = 3000 mm

Elongation of the bar, δl = 1.2 mm

REMEMBER

Stress = $\dfrac{\text{Force}}{\text{Area}}$

Strain = $\dfrac{\text{Change in length}}{\text{Orignal length}}$

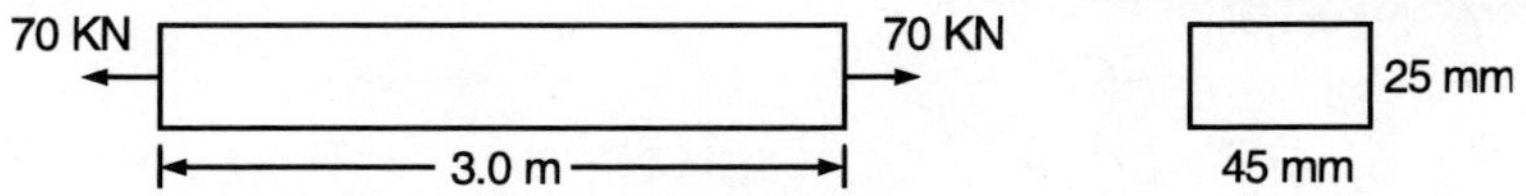

Fig. 2.16.

Stress, $\sigma = \dfrac{P}{A} = \dfrac{70 \times 1000}{1125} = 62.22$ N/mm^2 **Ans.**

Strain, $e = \dfrac{\delta l}{l} = \dfrac{1.2}{3000} = 0.0004$ **Ans.**

EXAMPLE 2.2. A rod 2 m long and of diameter 30 mm is subjected to an axial pull of 30 KN. If the young's modulus of the material of the rod is 2 × 10^5 N/mm^2, determine

(*i*) Stress

(*ii*) Strain

(*iii*) Elongation of the rod.

Soluton:

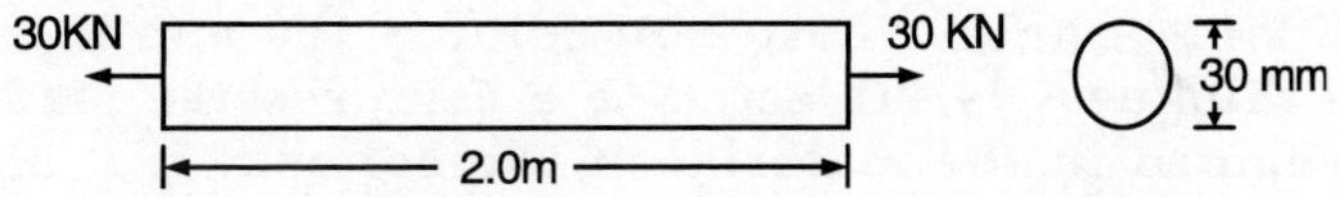

Fig. 2.17.

Given : Length of rod, $l = 2 \text{ m} = 2000 \text{ mm}$

Diameter of rod, $d = 30 \text{ mm}$

Axial pull, $P = 30 \text{ KN} = 30000 \text{ N}$

Young's modulus of elasticity,

$$E = 2 \times 10^5 \text{ N/mm}^2$$

Area of cross-section,

$$A = \frac{\pi}{4} d^2 = \frac{\pi}{4} \times (30)^2 = 706.86 \text{ mm}^2$$

Stress, $$\sigma = \frac{P}{A} = \frac{30000}{706.86} = 42.44 \text{ N/mm}^2 \quad \textbf{Ans.}$$

Young's modulus, $$E = \frac{\text{Stress}}{\text{Strain}} = \frac{\sigma}{e}$$

REMEMBER

$$E = \frac{\text{Stress}}{\text{Strain}}$$

$$2 \times 10^5 = \frac{42.44}{e}$$

∴ $$e = \frac{42.44}{2 \times 10^5} = 21.22 \times 10^{-5} \text{ mm}$$

Strain, $$e = \frac{\text{Change in length}}{\text{Original length}} = \frac{dl}{l}$$

$$21.22 \times 10^{-5} = \frac{dl}{2000}$$

∴ $$dl = 21.22 \times 10^{-5} \times 2000 = 0.424 \text{ mm} \quad \textbf{Ans.}$$

EXAMPLE 2.3: Find the minimum diameter of a steel wire, which is used to raise a load of 4000 N if the stress in wire is not to exceed 95 N/mm².

Solution: Given : Load, $W = 4000$ N

Stress, $\sigma = 95 \text{ N/mm}^2$

Stress, $$\sigma = \frac{W}{A}$$

∴ $$A = \frac{W}{\sigma} = \frac{4000}{95} = 42.10 \text{ mm}^2$$

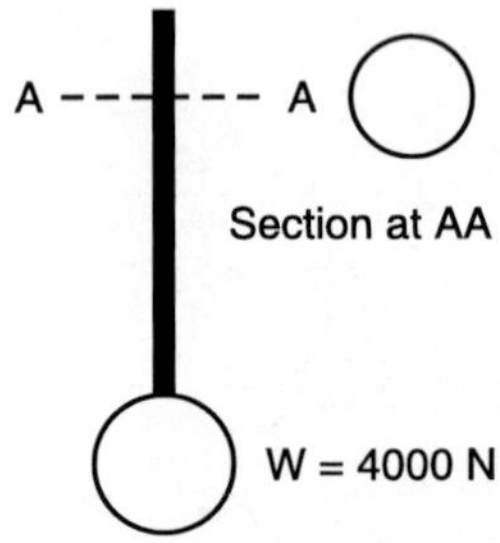

Fig. 2.18.

$$A = \frac{\pi}{4} d^2$$

$$42.10 = \frac{\pi}{4} \times d^2$$

or $$d^2 = \frac{42.10 \times 4}{\pi} = 53.60$$

∴ $$d = 7.32 \text{ mm.}$$

The minimum diameter of steel wire required is 7.32 mm. **Ans.**

EXAMPLE 2.4: A hollow cost iron column 100 mm in external diameter and 10 mm metal thickness is subjected to a compressive load of 40 KN. Calculate the stress developed in the material of the column. If it shorten's by 1.44 mm in a length of 8 m, calculate the Young's modulus.

Solution:

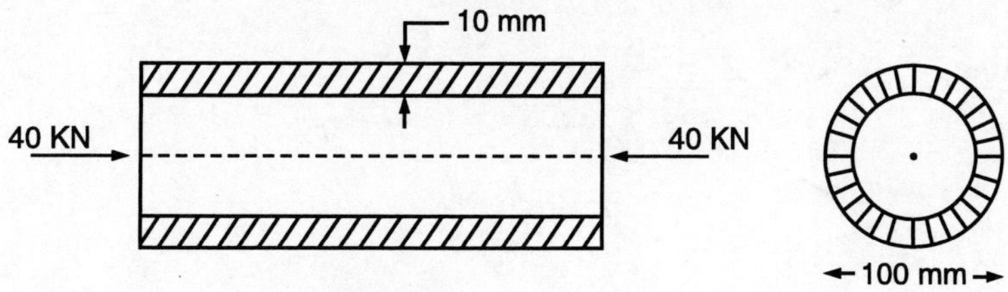

Fig. 2.19.

Given : Load, $P = 40 \text{ KN} = 40000 \text{ N}$

Outer diameter, $d_0 = 100 \text{ mm}$

Thickness, $t = 10 \text{ mm}$

Inner diameter, $d_i = 100 - 2 \times 10 = 80 \text{ mm}$

Original length, $l = 8 \text{ m} = 8000 \text{ mm}$

Change in length, $\delta l = 1.44 \text{ mm}$

Area of cross-section of column,

$$A = \frac{\pi}{4}(d_0^2 - d_i^2) = \frac{\pi}{4}(100^2 - 80^2) = 2827.43 \text{ mm}^3$$

Stress, $$\sigma = \frac{P}{A} = \frac{40000}{2827.43} = 14.15 \text{ N/mm}^2$$

Strain, $$e = \frac{\delta l}{l} = \frac{1.44}{8000} = 1.8 \times 10^{-4}$$

Young's modulus, $$E = \frac{\sigma}{e} = \frac{14.15}{1.8 \times 10^{-4}} = 7.86 \times 10^4 \text{ N/mm}^2$$ **Ans.**

EXAMPLE 2.5: A steel rod, 28 mm in diameter, 300 mm long is subjected to axial forces alternating between a maximum compression of 16 *KN* and a maximum tension of 7 KN. Find the difference between the greatest and least lengths of the rod. Take $E = 210$ *GN*/m^2.

Solution. Given : Diameter of rod,

$d = 28 \text{ mm}$

Length of rod, $l = 300 \text{ mm}$

Axial push, $P_C = 16 \, KN = 16000 \text{ N}$

Axial pull, $P_t = 7 \, KN = 7000 \text{ N}$

$$E = 210 \times 10^9 \text{ N/m}^2 = \frac{210 \times 10^9}{10^6} \text{ N/mm}^2 \qquad [\because 1 \, GN = 10^9 \text{ N}]$$

$$= 210 \times 10^3 \text{ N/mm}^2$$

Area of cross-section,

$$A = \frac{\pi}{4} d^2 = \frac{\pi}{4} \times (28)^2 = 615.75 \text{ mm}^2.$$

(*i*) When the rod is subjected to axial pull (tension)

Stress, $$\sigma = \frac{P_t}{A}$$

Strain, $$e = \frac{\delta l_1}{l}$$

We know, $E = \dfrac{\sigma}{e}$

or $E = \dfrac{\dfrac{P_t}{A}}{\dfrac{\delta l}{l}} = \dfrac{P_t \times l}{A \times \delta l}$

$\therefore$ δl_1 = Increase in length $= \dfrac{P_t \times l}{A \times E}$

$= \dfrac{7000 \times 300}{615.75 \times 210 \times 10^3}$

= 0.016 mm (+)

Greatest length after elongation = 300 + 0.016 = 300.016 mm.

(*ii*) When the rod is subjected to axial pull (compression)

δl_2 = Decrease in length

$= \dfrac{P_C \times l}{A \times E}$

$= \dfrac{16000 \times 300}{615.75 \times 210 \times 10^3}$

= 0.037 mm (–)

REMEMBER
Change in length, $\delta l = \dfrac{P \times L}{A \times E}$

Least length after compression

= 300 – 0.037

= 299.963 mm

Difference in length = 300.016 – 299.963

= 0.053 mm **Ans.**

EXAMPLE 2.6: The ultimate stress, for a hollow steel column which carries an axial load of 250 ***KN*** **is 600 N/mm². If the external diameter of the column is 200 mm, determine the internal diameter. Take factor of safety as 3.**

Solution: Given : Axial load,

P = 250 *KN* = 250000 N

Ultimate stress, σ_u = 600 N/mm²

External diameter, d_0 = 200 mm

Factor of safety, F = 3

Let the internal diameter be d_i

REMEMBER
Working stress = $\dfrac{\text{Ultimate stress}}{\text{F.O.S}}$

Working stress, $\sigma_w = \dfrac{\text{Ultimate stress } (\sigma_u)}{\text{Factor of safety } (F)} = \dfrac{600}{3} = 200 \text{ N/mm}^2$

Stress, $\sigma_w = \dfrac{P}{A}$

or $A = \dfrac{P}{\sigma_w} = \dfrac{250000}{200}$

$= 1250 \text{ mm}^2$

Note: For design purpose, working stress is considered which is less than the ultimate stress.

Now, $$A = \frac{\pi}{4}(d_0^2 - d_i^2)$$

$$1250 = \frac{\pi}{4}(200^2 - d_i^2)$$

$$\frac{1250 \times 4}{\pi} = 40000 - d_i^2$$

$$1591.55 = 40000 - d_i^2$$

$$d_i^2 = 40000 - 1591.55$$

$$d_i^2 = 38408.45$$

∴ $$d_i = 195.98 \simeq 196 \text{ mm} \quad \textbf{Ans.}$$

EXAMPLE 2.7: Show that volumetric strain for a cuboid is the algebraic sum of the linear strains of all the three sides.

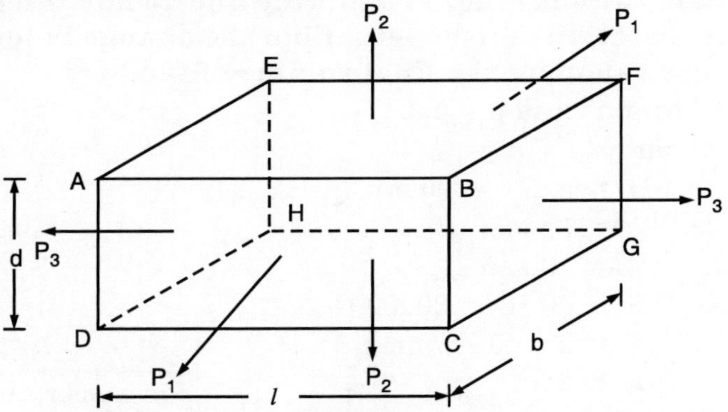

Fig. 2.20.

Solution: Consider a cuboid *ABCDEFGH* subjected to axial pull P_1, P_2 and P_3 as shown in the figure 2.20.

Let, Length of cuboid $= l$

Breadth of cuboid $= b$

Height of cuboid $= d$

Volume of cuboid, $v = lbd$

Linear strain along length $= e_l$

Linear strain along breadth $= e_b$

Linear strain along height $= e_d$

$$e_l = \frac{\text{Increase in length}}{\text{Original length}} = \frac{\delta l}{l}$$

∴ $$\delta l = l\, e_l$$

Increased length of the bar

$$= l + \delta l = l + l e_l$$

$$= l\,(1 + e_l)$$

Similarly,

Increased width of the bar $= b\,(1 + e_b)$

Increased height of the bar = $d\,(1 + e_d)$

$$\begin{aligned}\text{Increased volume} &= l\,(1 + e_l) \times b\,(1 + e_b) \times d\,(1 + e_d)\\ &= lbd\,(1 + e_l)\,(1 + e_b)\,(1 + e_d)\\ &= lbd\,(1 + e_l + e_b + e_d)\end{aligned}$$

(Neglecting other quantities being very small)

$$\begin{aligned}\text{Change in volume} &= \text{Increased volume} - \text{Original volume}\\ \delta V &= lbd\,(1 + e_l + e_b + e_d) - lbd\\ &= lbd\,(e_l + e_b + e_d)\end{aligned}$$

∴ Volumetric strain,

$$e_v = \frac{\delta_V}{V} = \frac{lbd\,(e_l + e_b + e_d)}{lbd} = e_l + e_b + e_d$$

i.e. volumetric strain for a cuboid is sum of all the linear strains.

EXAMPLE 2.8: A steel bar 4 m long, 35 mm wide and 20 mm thick is subjected to a pull of 30 *KN* in the direction of its length. Find the change in length, breadth and thickness of bar. Also calculate the final volume of the bar.

Take **$E = 2 \times 10^5$ N/mm^2 and $\mu = 0.3$.**

Solution: Length of bar,

l = 4 m = 4000 mm

Width of bar, b = 35 mm

Thickness of bar, d = 20 mm

Axial pull, P = 30 *KN* = 30000 N

$E = 2 \times 10^5$ N/mm^2

$\mu = 0.3$

REMEMBER
Poissons ratio $= \dfrac{\text{Lateral strain}}{\text{Longitudinal strain}}$

Area of cross-section,

$$A = b \times d = 35 \times 20 = 700 \text{ mm}^2$$

Stress, $\sigma = \dfrac{P}{A} = \dfrac{30000}{700} = 42.86$ N/mm^2

We know, $E = \dfrac{\sigma}{e}$

or strain (longitudinal), $e = \dfrac{\sigma}{E} = \dfrac{42.86}{2 \times 10^5} = 0.00021$

Now, longitudinal strain, $e = \dfrac{\delta l}{l}$

or $\delta l = e \times l = 0.00021 \times 4000 = 0.84$ mm (+)

∴ Increase in length, δl = 0.84 mm **Ans.**

Again,

Poisson's ratio, $\mu = \dfrac{\text{Lateral strain}}{\text{Longitudinal strain}}$

$$\begin{aligned}\text{or}\quad \text{Lateral strain} &= \text{Longitudinal strain} \times \mu\\ &= 0.00021 \times 0.3\\ &= 0.000063\end{aligned}$$

Lateral strain, $e_b = \dfrac{\delta b}{b}$

Change in width, $\delta b = e_b \times b = 0.000063 \times 35 = 0.0022$ mm (–) **Ans.**

Change in thickness,

$$\delta d = e_b \times d = 0.000063 \times 20 = 0.00126 \text{ mm (–)} \quad \textbf{Ans.}$$

Volumetric strain, $e_v = e_l - e_b - e_d$

$= 0.00021 - 0.000063 - 0.000063$ $\quad [\because e_b = e_d = 0.0000063]$

$= 0.000084$

Volume of bar, $v = lbd = 4000 \times 35 \times 20 = 2800000 \text{ mm}^3$

We know, $e_v = \dfrac{\delta v}{V}$

$\therefore$ $\delta v = e_v \times V = 0.000084 \times 2800000 = 235.2 \text{ mm}^3$ **Ans.**

2.25 BARS OF VARYING SECTIONS

Figure 2.21 shows a bar of varying sections subjected to axial load P. Let (A_1, l_1), (A_2, l_2) and (A_3, l_3) be the area of cross-section and length of portion AB, BC and CD of the bar.

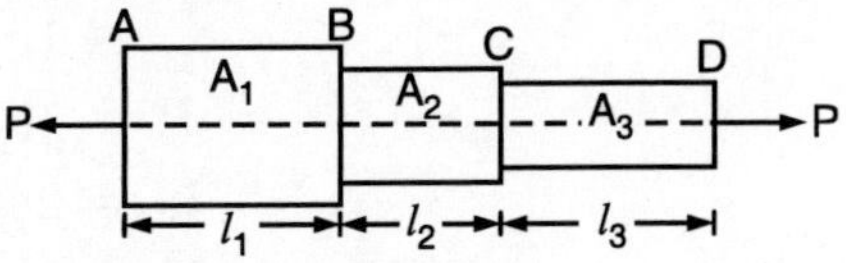

Fig. 2.21

Stress for portion AB, $\sigma_1 = \dfrac{P}{A_1}$

Stress for portion BC, $\sigma_2 = \dfrac{P}{A_2}$

Stress for portion CD, $\sigma_3 = \dfrac{P}{A_3}$

Strain in portion AB, $e_1 = \dfrac{\sigma_1}{E}$

Strain in portion BC, $e_2 = \dfrac{\sigma_2}{E}$

Strain in portion CD, $e_3 = \dfrac{\sigma_3}{E}$

where E is the young's modulus.

Change in length of portion AB,

$$\delta l_1 = e_1 l_1 = \frac{\sigma_1 l_1}{E} = \frac{Pl_1}{A_1 E}$$

Similarly, change in length of portion BC,

$$\delta l_2 = \frac{Pl_2}{A_2 E}$$

Change in length of portion CD,

$$\delta l_3 = \frac{Pl_3}{A_3 E}$$

Total change in length of the bar,

$$\delta l = \delta l_1 + \delta l_2 + \delta l_3 = \frac{Pl_1}{A_1 E} + \frac{Pl_2}{A_2 E} + \frac{Pl_3}{A_3 E}$$

$$\therefore \quad \delta l = \frac{P}{E}\left(\frac{l_1}{A_1}+\frac{l_2}{A_2}+\frac{l_3}{A_3}\right)$$

$$= \frac{1}{E}(\sigma_1 l_1 + \sigma_2 l_2 + \sigma_3 l_3).$$

EXAMPLE 2.9: A straight bar 500 mm long is 25 mm in diameter for the first 200 mm length and 15 mm diameter for the remaining length. If the bar is subjected to an axial pull of 15 *KN*, find the elongation of the bar. Take $E = 2 \times 10^5$ N/mm².

Solution:

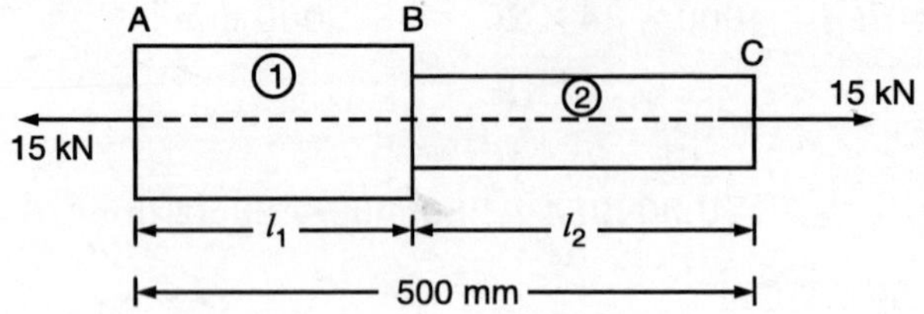

Fig. 2.22.

> **REMEMBER**
>
> In case of bars of varying section, total change in length of the bar = sum of change in length of different portions of the bar.

Given :

$$d_1 = 25 \text{ mm}$$

$$d_2 = 15 \text{ mm}$$

$$P = 15\ KN = 15000 \text{ N}$$

$$l_1 = 200 \text{ mm}$$

$$l_2 = 300 \text{ mm}$$

$$\therefore \quad A_1 = \frac{\pi}{4}d_1^2 = \frac{\pi}{4}\times(25)^2 = 490.87 \text{ mm}^2$$

$$A_2 = \frac{\pi}{4}d_2^2 = \frac{\pi}{4}\times(15)^2 = 176.71 \text{ mm}^2$$

Stress for portion *AB*,

$$\sigma_1 = \frac{P}{A_1} = \frac{15000}{490.87} = 30.56 \text{ N/mm}^2$$

Elongation of portion *AB*,

$$\delta l_1 = \frac{Pl_1}{A_1E} = \frac{15000\times 200}{490.87\times 2\times 10^5} = 0.0305 \text{ mm}$$

Elongation of portion *BC*,

$$\delta l_2 = \frac{Pl_2}{A_2E} = \frac{15000\times 300}{176.71\times 2\times 10^5} = 0.1273$$

$\therefore$ Total elongation of the bar,

$$\delta l = \delta l_1 + \delta l_2 = 0.0305 + 0.1273 = 0.1578 \text{ mm} \quad \textbf{Ans.}$$

EXAMPLE 2.10: Figure 2.23 shows a bar consisting of three lengths. Find the stresses in the three parts and the total extension of the bar for an axial pull of 50 *KN*. Take $E = 2 \times 10^5$ N/mm².

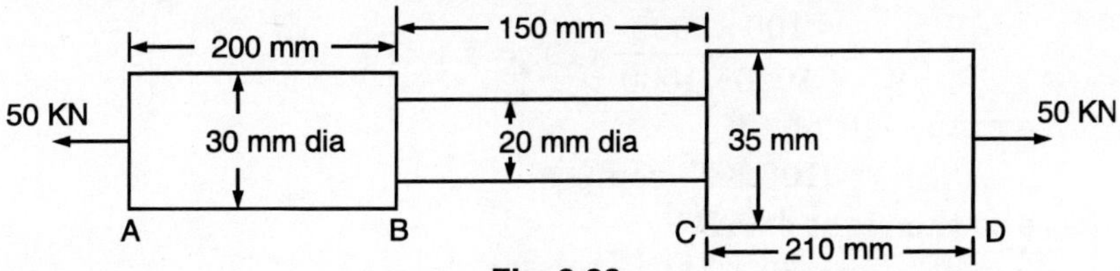

Fig. 2.23.

Solution: **Given :** $P = 50\ KN = 50000\ M;$ $E = 2 \times 10^5$ N/mm²

$d_1 = 30$ mm $\quad A_1 = \frac{\pi}{4} d_1^2 = \frac{\pi}{4} \times (30)^2 = 706.86 \text{ mm}^2$

$d_2 = 20$ mm $\quad A_2 = \frac{\pi}{4} d_2^2 = \frac{\pi}{4} \times (20)^2 = 314.16 \text{ mm}^2$

$d_3 = 35$ mm $\quad A_3 = \frac{\pi}{4} d_3^3 = \frac{\pi}{4} \times (35)^2 = 962.11 \text{ mm}^2$

$l_1 = 200$ mm

$l_2 = 150$ mm

$l_3 = 210$ mm

Elongation of portion AB,

$$\delta l_1 = \frac{Pl_1}{A_1 E} = \frac{50000 \times 200}{706.86 \times 2 \times 10^5} = 0.0707 \text{ mm}$$

Elongation of portion BC,

$$\delta l_2 = \frac{Pl_2}{A_2 E} = \frac{50000 \times 150}{314.16 \times 2 \times 10^5} = 0.1194 \text{ mm}$$

Elongation of portion CD,

$$\delta l_3 = \frac{Pl_3}{A_3 E} = \frac{50000 \times 210}{962.11 \times 2 \times 10^5} = 0.0546 \text{ mm}$$

Total elongation of the bar,

$$\delta l = 0.0707 + 0.1194 + 0.0546 = 0.2447 \text{ mm.} \quad \textbf{Ans.}$$

EXAMPLE 2.11: A member *ABC* is formed by connecting a steel bar of 20 mm ϕ to an aluminium bar of 30 mm ϕ and is subjected to forces as shown in the figure 2.24. Determine the change in length of the bar. Take $E_s = 205 \times 10^5$ KN/m² and $E_a = 100 \times 10^5$ *KN*/m².

(Delhi, Jan, 09)

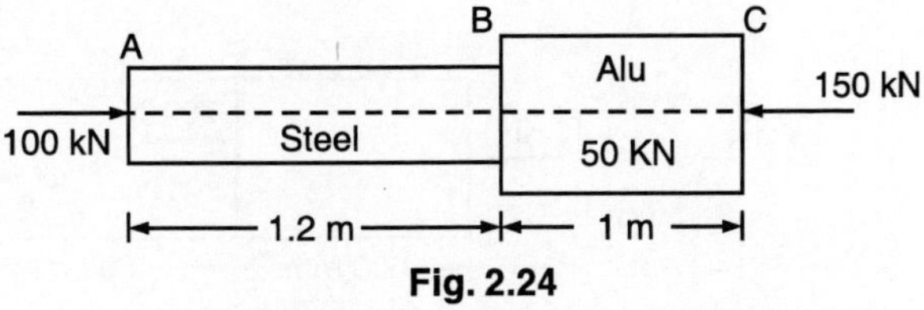

Fig. 2.24

Solution. **For Part *AB* :**

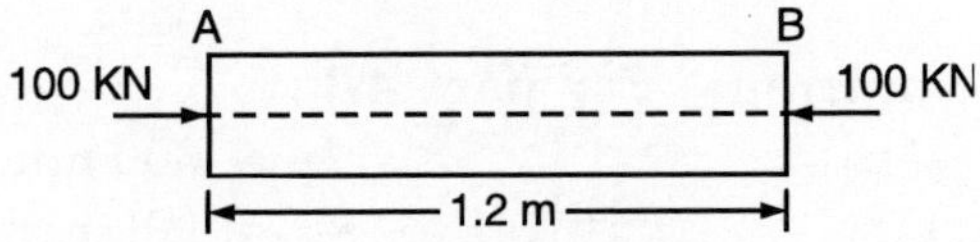

Fig. 2.25

$l_s = 1.2$ m = 1200 mm

$P_s = 100\ KN = 100000$ N

$A_s = \frac{\pi}{4} \times (20)^2 = 314.16 \text{ mm}^2$

$E_s = 205 \times 10^5 \text{ KN } m^2 = \frac{205 \times 1000}{1000 \times 1000} \times 10^5 \text{ N mm}^2$

$= 0.205 \times 10^5$ N/mm²

$$E_a = \frac{100 \times 1000}{1000 \times 1000} \times 10^5 = 0.1 \times 10^5 \text{ N/mm}^2$$

Resultant force to the left of AB

$= 100$ KN (comp.)

Resultant force to the right of AB

$= -50 + 150 = 100$ KN (comp.)

Change in length of the steel bar,

$$\delta_s = \frac{P_s l_s}{A_s E_s}$$

$$= \frac{100000 \times 1200}{314.16 \times 0.205 \times 10^5}$$

$= 18.63$ mm (comp.)

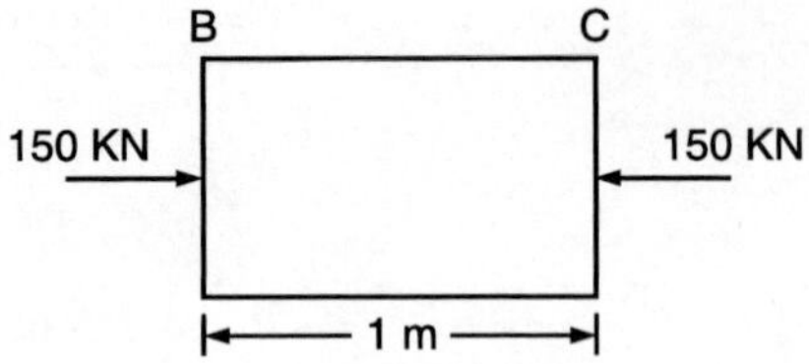

Fig. 2.26

For Part BC :

$P_a = 150 \text{ KN} = 150000 \text{ N}$

$l_a = 1 \text{ m} = 1000 \text{ m}$

$$A_a = \frac{\pi}{4} \times (30)^2 = 706.86 \text{ mm}^2$$

Resultant force to the left of BC

$= 100 + 50 = 150$ KN (comp.)

Resultant force to the right of BC

$= 150$ KN (comp.)

Change in length of the aluminium bar,

$$\delta_a = \frac{P_a l_a}{A_a E_a} = \frac{150000 \times 1000}{706.86 \times 0.1 \times 10^5} = 21.22 \text{ mm (comp.)}$$

Total decrease in length,

$\delta l = \delta_s + \delta_a = 18.63 + 21.22 = 39.85$ mm **Ans.**

EXAMPLE 2.12. A brass bar, having cross-sectional area of 900 mm^2 is subjected to axial forces as shown in Figure 2.27. Find the total change in length of the bar. Take $E = 1 \times 10^5$ N/mm^2.

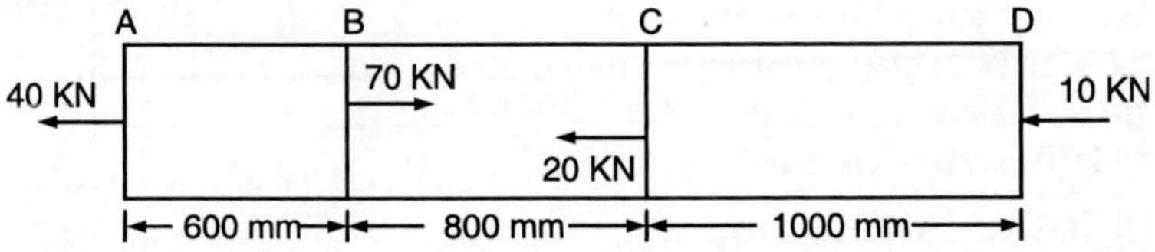

Fig. 2.27.

Solution: For part AB :

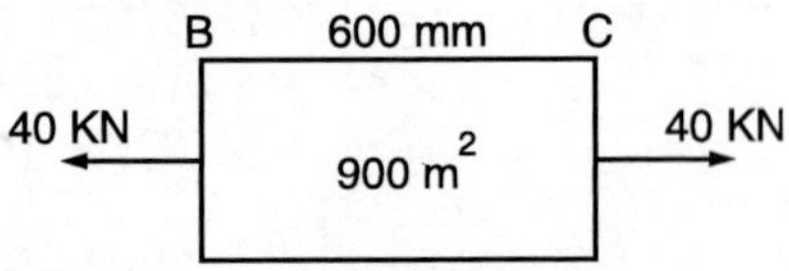

Fig. 2.27 (*i*)

$l_1 = 600$ mm

$A_1 = 900 \text{ mm}^2$

$E = 1 \times 10^5 \text{ N/mm}^2$

Resultant force to the left side of part $AB = 40$ KN (tensile)

Resultant force to the right side of part $AB = 70 - 20 - 10 = 40$ KN (tensile)

$$P_1 = 40 \text{ KN} = 40000 \text{ N}$$

Change in length of part *AB*,

$$\delta l_1 = \frac{P_1 l_1}{A_1 E} = \frac{40000 \times 600}{900 \times 1 \times 10^5} = 0.267 \text{ mm } (+)$$

For part *BC* :

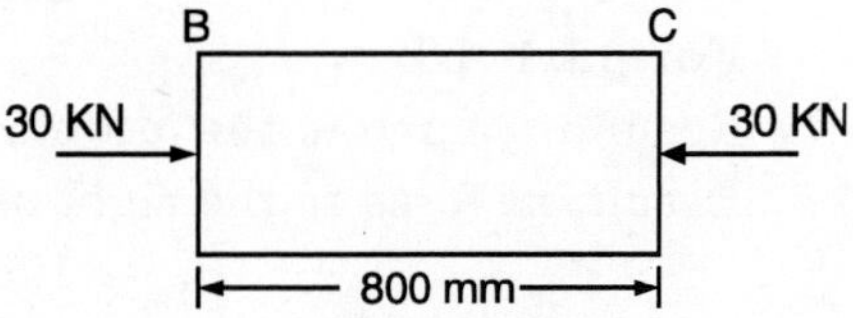

Fig. 2.27 (*ii*)

$$l_2 = 800 \text{ mm}$$
$$A_2 = 900 \text{ mm}^2$$
$$E = 1 \times 10^5 \text{ N/mm}^2$$

Resultant force to the left of part of *BC*

$$= 70 - 40 = 30 \text{ (compressive)}$$

Resultant force to the right of *BC*

$$= 20 + 10 = 30 \text{ (compressive)}$$
$$P_2 = 30 \text{ KN} = 30000 \text{ N}$$

Change in length of part *BC*,

$$\delta l_2 = \frac{P_2 l_2}{A_2 E} = \frac{30000 \times 800}{900 \times 1 \times 10^5} = 0.267 \text{ mm } (-)$$

For part *CD* :

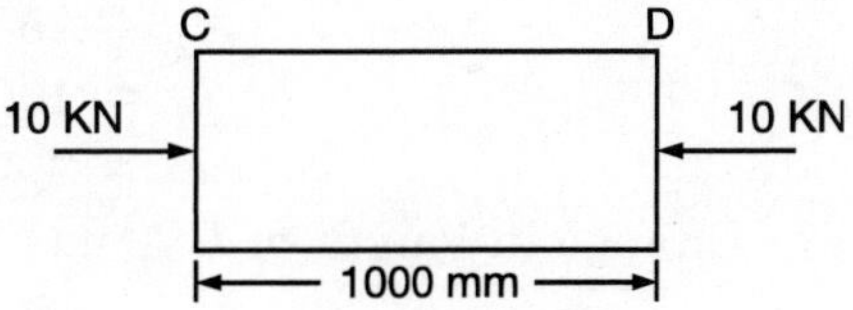

Fig. 2.27 (*iii*)

$$l_3 = 1000 \text{ mm}$$
$$A_3 = 900 \text{ mm}^2$$
$$E = 1 \times 10^5 \text{ mm}^2$$

Resultant force to the left of *CD*

$$= 40 - 70 + 20 = -10$$
$$= 10 \text{ KN (compressive)}$$

Resultant force to the right of *CD*

$$= 10 \text{ KN (compressive)}$$
$$P_3 = 10 \text{ KN} = 10000 \text{ N}$$

Change in length of part *CD*,

$$\delta l_3 = \frac{P_3 l_3}{A_3 E} = \frac{10000 \times 1000}{900 \times 1 \times 10^5} = 0.111 \text{ mm } (-)$$

Total change in length,

$$\delta l = \delta l_1 - \delta l_2 - \delta l_3 = 0.267 - 0.267 - 0.111$$
$$= -0.111 = 0.111 \text{ (decrease)}$$ **Ans.**

EXAMPLE 2.13. A steel bar *ABCD* is subjected to point loads as shown in the figure 2.28. Calculate the force *P* necessary for equilibrium. Also calculate the change in length of the bar. Take $E = 2 \times 10^5$ N/mm^2.

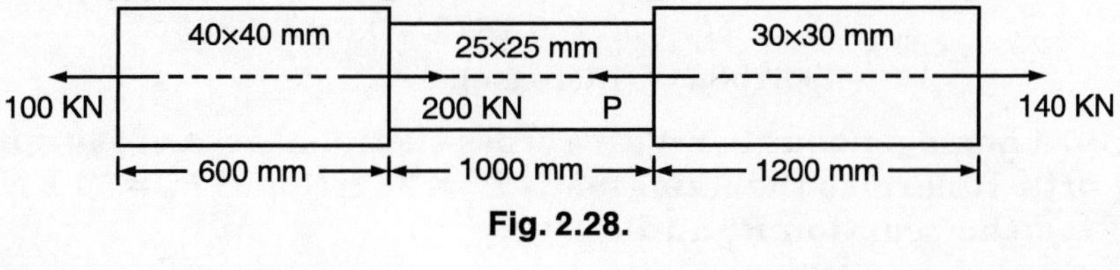

Fig. 2.28.

Solution: For equilibrium,

$$\Sigma F_x = 0$$

$$200 + 140 - 100 - P = 0$$

or $$240 - P = 0$$

$\therefore$ $P = 240$ KN **Ans.**

For part *AB*:

Resultant force to the left of $AB = 100$ (tensile)

Resultant force to the right of $AB = 200 - 240 + 140 = 100$ (tensile)

$$P_1 = 100\ KN = 100000\ \text{N}$$

$$A_1 = 40 \times 40 = 1600\ \text{mm}^2$$

$$l_1 = 600\ \text{mm}$$

$$E = 2 \times 10^5\ \text{N/mm}^2$$

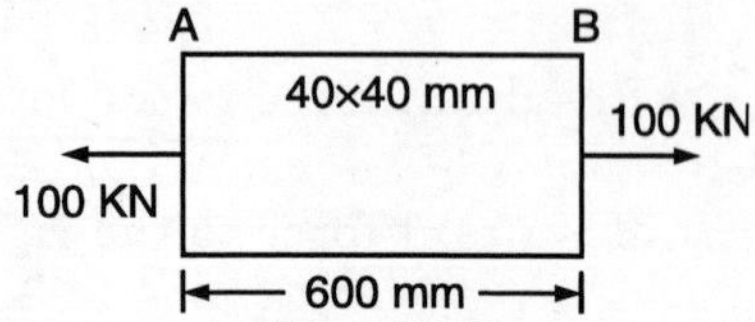

Fig. 2.28 (*i*)

Change in length of *AB*,

$$\delta l_1 = \frac{P_1 l_1}{A_1 E} = \frac{100000 \times 600}{1600 \times 2 \times 10^5} = 0.1875\ \text{mm}\ (+)$$

For part *BC*:

Resultant force to the left of $BC = 200 - 100 = 100$ (compressive)

Resultant force to the right of $BC = 240 - 140 = 100$ (compressive)

$$P_2 = 100\ \text{KN} = 100000\ \text{N}$$

$$A_2 = 25 \times 25 = 625\ \text{mm}^2$$

$$l_2 = 1000\ \text{mm}$$

$$E = 2 \times 10^5\ \text{N/mm}^2$$

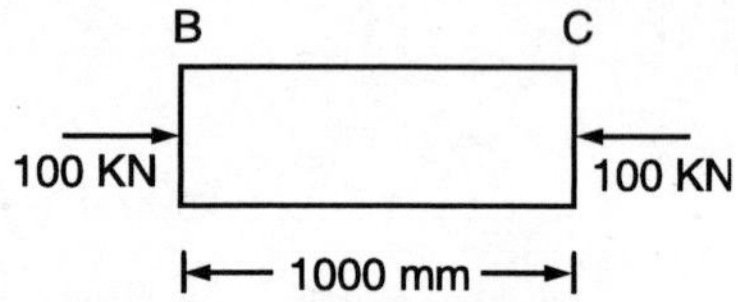

Fig. 2.28 (*ii*)

Change in length of *BC*,

$$\delta l_2 = \frac{P_2 l_2}{A_2 E} = \frac{100000 \times 1000}{625 \times 2 \times 10^5} = 0.8\ \text{mm}\ (-)$$

For part *CD*:

Resultant force to the left of

$$CD = 240 - 200 + 100 = 140\ \text{KN (tensile)}$$

Resultant force to the right of $CD = 140$ KN (tensile)

$$P_3 = 140\ \text{KN} = 140000\ \text{N}$$

$$l_3 = 1200\ \text{mm}$$

$$A_3 = 30 \times 30 = 900\ \text{mm}^2$$

$$E = 2 \times 10^5\ \text{N/mm}^2$$

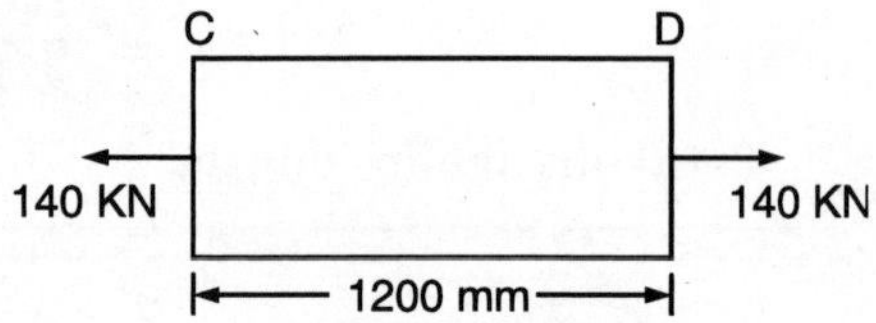

Fig. 2.28 (*iii*)

Change in length of *CD*,

$$\delta l_3 = \frac{P_3 l_3}{A_3 E} = \frac{140000 \times 1200}{900 \times 2 \times 10^5} = 0.9333\ \text{mm}\ (+)$$

Total change in length,

$$\delta l = \delta l_1 - \delta l_2 + \delta l_3 = 0.1875 - 0.8 + 0.9333$$

$$= 0.3208\ (+)\ \text{(Increase)}.$$

EXAMPLE 2.14. A homogeneous bar with a cross-sectional area of 500 mm^2 is attached to rigid supports. It carries the axial loads $P_1 = 25$ KN and $P_2 = 50$ KN as shown in figure 2.29. Find the reaction R_1 and R_2.

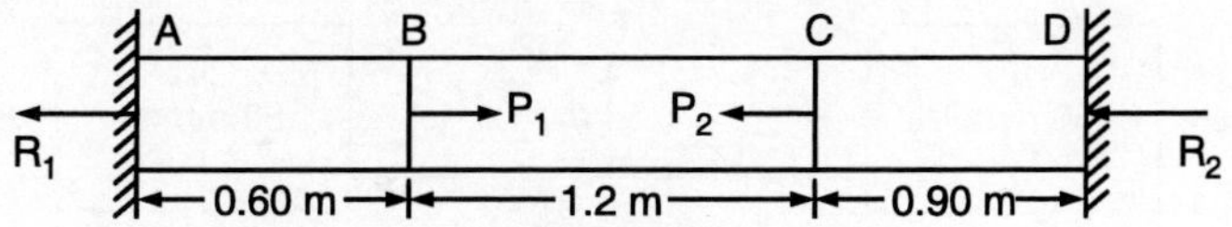

Fig. 2.29.

Solution: Given : $P_1 = 25 \text{ KN} = 25000 \text{ N}$

$P_2 = 50 \text{ KN} = 50000 \text{ N}$

$l_1 = 0.60 \text{ m} = 600 \text{ mm}$

$l_2 = 1.2 \text{ m} = 1200 \text{ mm}$

$l_3 = 0.90 = 900 \text{ mm}$

For equilibrium, $\Sigma F_x = 0$

$$P_1 - P_2 - R_1 - R_2 = 0$$

or $R_1 + R_2 = P_1 - P_2 = 25 - 50 = -25 \text{ KN}$

For part *AB*:

Elongation in length of *AB*,

$$\delta l_1 = \frac{R_1 l_1}{A_1 E} = \frac{R_1 \times 600}{500\, E} = \frac{6R_1}{5E} \ (+)$$

For part *BC*:

Contraction in length of *BC*,

$$\delta l_2 = \frac{(P_1 - R_1)\, l_2}{A_2 E} = \frac{(25 - R_1) \times 1200}{500E} = \frac{12\,(25 - R_1)}{5E} \ (-)$$

For part *CD*:

Contraction in length of *CD*, [$\because$ $(P_1 - R_1)$ is the resultant force in *BC*]

$$\delta l_3 = \frac{R_2 \times l_3}{A_3 E} = \frac{R_2 \times 900}{500E} = \frac{9R_2}{5E} \ (-)$$

Net change in length,

$$\delta l = \delta l_1 - \delta l_2 - \delta l_3$$

$$= \frac{6R_1}{5E} - \frac{12(25 - R_1)}{5E} - \frac{9R_2}{5E}$$

$$= \frac{6R_1}{5E} - \frac{12(25 - R_1)}{5E} - \frac{9(-25 - R_1)}{5E}$$

$$= \frac{6R_1 - 300 + 12R_1 + 175 + 9R_1}{5E} = 0$$

$$27R_1 - 125 = 0$$

$$R_1 = \frac{125}{27} = 4.63 \text{ KN} \quad \textbf{Ans.}$$

$$R_2 = -25 - R_1 = -25 - 4.63 = -29.63 \text{ KN} \quad \textbf{Ans.}$$

EXAMPLE 2.15: Figure 2.30 shows a bar subjected to tensile load of 150 KN. Find the diameter of the middle portion if the stress there is limited to 130 N/mm^2.

Find also the length of the middle portion if the total elongation of the bar is to be 0.20 mm. Take $E = 2 \times 10^5$ N/mm^2.

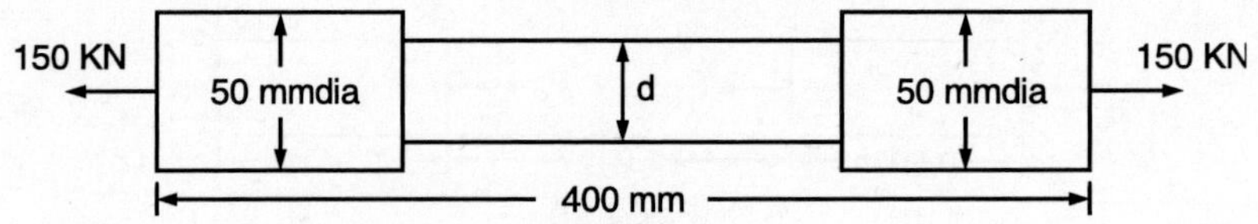

Fig. 2.30.

Solution: Given : $P = 150 \text{ KN} = 150000 \text{ N};$

$\sigma_m = 130 \text{ N/mm}^2$; σ_m = stress in middle portion

$l = 400 \text{ mm}$

$d_1 = d_2 = 50 \text{ mm}$

Let d and x be the diameter and length of middle portion of the bar.

Area of cross-section,

$$A_m = \frac{P}{\sigma_m} = \frac{150000}{130} = 1153.85 \text{ mm}^2$$

$$A_m = \frac{\pi}{4} d^2$$

$$1153.85 = \frac{\pi}{4} d^2$$

or $$d^2 = \frac{1153.85 \times 4}{\pi} = 1469.13$$

$\therefore$ $$d = 38.33 \text{ mm}$$

Area of cross-section of end portions,

$$A_e = \frac{\pi}{4} \times (50)^2 = \times 1963.49 \text{ mm}^2$$

Stress in the end portions,

$$\sigma_e = \frac{P}{A_e} = \frac{150000}{1963.49} = 76.39 \text{ N/mm}^2$$

Elongation of end portions

$$= \frac{Pl_e}{A_e E} = \frac{\sigma_e l_e}{E} = \frac{76.39 \times (400 - x)}{E} \qquad [\text{where } l_e = 400 - x]$$

Elongation of middle portion

$$= \frac{\sigma_m l_m}{E} = \frac{130 \times x}{E} \qquad [\text{where } l_m = x]$$

Total elongation of the bar = 0.20 mm

$$\frac{76.39\,(400 - x)}{E} + \frac{130x}{E} = 0.20$$

$$76.39\,(400 - x) + 130x = 0.20 \times E = 0.20 \times 2 \times 10^5$$

$$30556 - 76.39x + 130x = 40000$$

$$53.61x = 9444$$

$$x = \frac{9444}{53.61} = 176.16 \text{ mm.}$$

2.26 EXTENSION OF A TAPERING ROD

Consider a rod of length 'l' whose diameter changes uniformly from d_1 to d_2 as shown in the Figure 2.31. The rod is subjected to an axial pull 'P'.

Consider an elemental length 'dx' at a distance 'x' from the larger end.

Let the diameter of the bar be d_m at a distance 'x' from A.

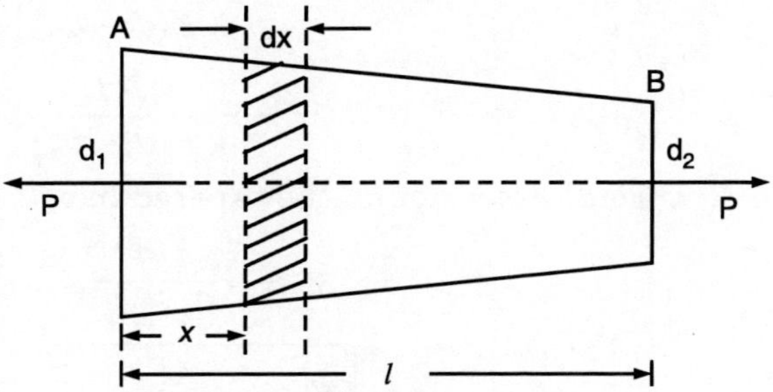

Fig. 2.31.

$$d_m = d_1 - \left(\frac{d_1 - d_2}{l}\right) x$$

$$= d_1 - Kx \qquad \text{where } K = \frac{d_1 - d_2}{l}$$

$$\frac{m}{x} = \frac{d_1 - d_2}{l}$$

$$\Rightarrow \quad m = \left(\frac{d_1 - d_2}{l}\right) x$$

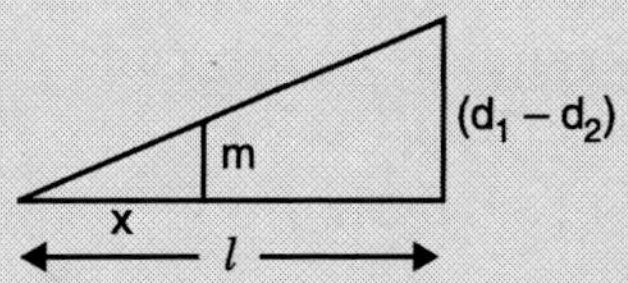

$$d_m = d_1 - m$$

$$= d_1 - \left(\frac{d_1 - d_2}{l}\right) x$$

Area of cross-section of the bar at x from A,

$$A_m = \frac{\pi}{4} d_m^{\,2} = \frac{\pi}{4} (d_1 - Kx)^2$$

Stress on the section,

$$\sigma_m = \frac{P}{A_m} = \frac{P}{\frac{\pi}{4}(d_1 - Kx)^2} = \frac{4P}{\pi (d_1 - Kx)^2}$$

Strain on the section,

$$e_m = \frac{\sigma_m}{E} = \frac{4P}{\pi E (d_1 - Kx)^2}$$

$\therefore$ Extension of the elemental portion 'dx' $= e_m \, dx = \dfrac{4P\, dx}{\pi E (d_1 - Kx)^2}$

Total extension of the bar,

$$\delta l = \int_0^l \frac{4P}{\pi E (d_1 - Kx)^2} dx = \frac{4P}{\pi EK}\left[\frac{1}{d_1 - Kx}\right]_0^l = \frac{4P}{\pi EK}\left[\frac{1}{d_1 - Kl} - \frac{1}{d_1}\right]$$

$$= \frac{4P}{\pi E} \frac{l}{d_1 - d_2}\left[\frac{1}{d_1 - \left(\frac{d_1 - d_2}{l}\right) l} - \frac{1}{d_1}\right]$$

$$\left[\text{By putting the value of } K = \frac{d_1 - d_2}{l}\right]$$

$$= \frac{4Pl}{\pi E(d_1 - d_2)} \left[\frac{1}{d_1 - d_1 + d_2} - \frac{1}{d_1} \right]$$

$$= \frac{4Pl}{\pi E(d_1 - d_2)} \left[\frac{1}{d_2} - \frac{1}{d_1} \right] = \frac{4Pl}{\pi E(d_1 - d_2)} \left[\frac{d_1 - d_2}{d_1 d_2} \right] = \frac{4Pl}{\pi E d_1 d_2}$$

∴ Total extension of the tapered bar,

$$\delta l = \frac{4Pl}{\pi E d_1 d_2}$$

∴

$$\boxed{\delta l = \frac{4Pl}{\pi E d_1 d_2}} \,.$$

Note: The formula obtained above is found to be satisfactory when the angle between the extreme faces of the bar does not exceed 20°.

EXAMPLE 2.16. A rod tapers uniformly from 40 mm diameter to 20 mm diameter in a length of 400 mm. If the rod is subjected to an axial load of 10 KN, find the extension of the rod. Take $E = 2 \times 10^5$ N/mm².

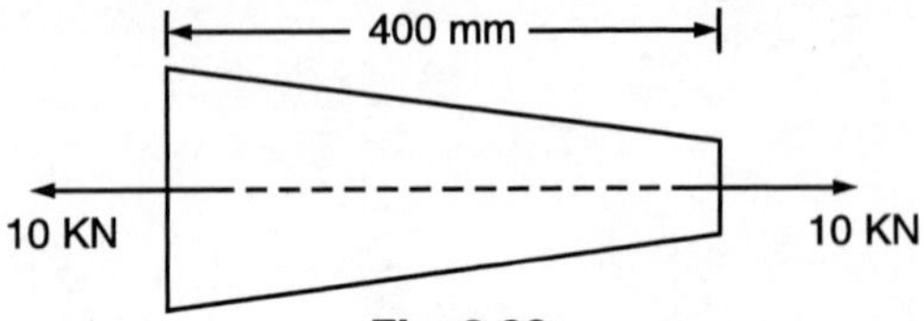

Fig. 2.32.

Solution: Given :

P = 10 KN = 10000 N

d_1 = 40 mm

d_2 = 20 mm

$E = 2 \times 10^5$ N/mm²

l = 400 mm

Extension of the tapered rod,

$$\delta l = \frac{4Pl}{\pi E d_1 d_2} = \frac{4 \times 10000 \times 400}{\pi \times 2 \times 10^5 \times 40 \times 20} = 0.0318 \text{ mm} \quad \textbf{Ans.}$$

EXAMPLE 2.17. Find the elongation of the bars shown in the Figure 2.33 under a tensile load of 30 KN. Take $E = 2 \times 10^5$ N/mm².

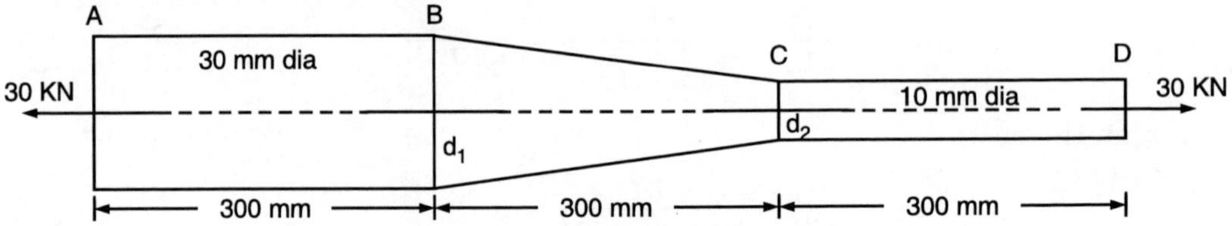

Fig. 2.33

Solution: Given :

P = 30 KN = 30000 N

d_1 = 30 mm

d_2 = 10 mm

l_1 = 300 mm

$l_2 = l_3$ = 300 mm

$$A_1 = \frac{\pi}{4} d_1^2 = \frac{\pi}{4} \times (30)^2 = 706.86 \text{ mm}^2$$

$$A_2 = \frac{\pi}{4} d_2^2 = \frac{\pi}{4} \times (10)^2 = 78.54 \text{ mm}^2$$

Elongation of portion AB,

$$\delta l_1 = \frac{Pl_1}{A_1 E} = \frac{30000 \times 300}{706.86 \times 2 \times 10^5} = 0.0637 \text{ mm}$$

Elongation of portion BC,

$$\delta l_2 = \frac{4\,Pl_2}{\pi E\,d_1 d_2} = \frac{4 \times 30000 \times 300}{\pi \times 2 \times 10^5 \times 30 \times 10} = 0.191 \text{ mm}$$

Elongation of portion CD,

$$\delta l_3 = \frac{Pl_3}{A_2 E} = \frac{30000 \times 300}{78.54 \times 2 \times 10^5} = 0.5729 \text{ mm}$$

Total elongation of the bar,

$$\delta l = \delta l_1 + \delta l_2 + \delta l_3 = 0.06371 + 0.191 + 0.5729$$
$$= 0.8276 \text{ mm} \quad \textbf{Ans.}$$

EXAMPLE 2.18. A rod circular in section tapers from 20 mm diameter at one end to 10 mm diameter at the other end and is 200 mm long. On applying an axial pull of 6000 N it was found to extend by 0.068 mm. Find the Young's modulus of the material of the rod.

Solution: Given : $P = 6000$ N
$d_1 = 20$ mm
$d_2 = 10$ mm
$l = 200$ mm
$\delta l = 0.068$ mm
$E = ?$

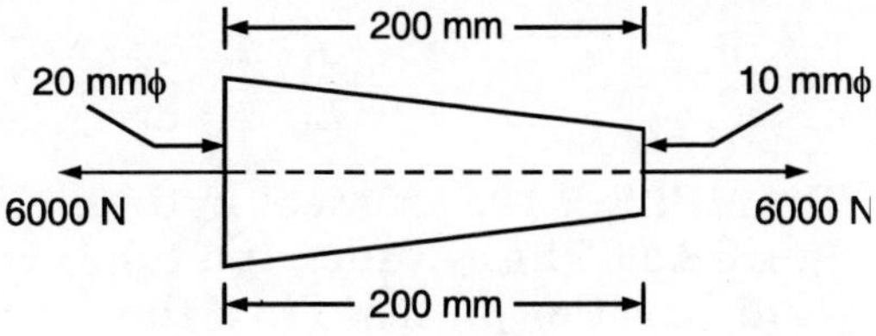

Fig. 2.34.

Elongation of a tapered rod,

$$\delta l = \frac{4Pl}{\pi E\,d_1 d_2}$$

$$0.068 = \frac{4 \times 6000 \times 200}{\pi \times E \times 20 \times 10}$$

or

$$E = \frac{4 \times 6000 \times 200}{\pi \times 0.068 \times 20 \times 10}$$
$$= 112344.67 \text{ N/mm}^2$$
$$= 1.12 \times 10^5 \text{ N/mm}^2 \quad \textbf{Ans.}$$

2.27 BARS OF COMPOSITE SECTION

A composite section is made up of two or more materials, joined together in such a way that the system is elongated or compressed as a single unit.

The following two points are important for a composite section :

(*i*) Elongation or contraction of individual materials of a composite section is equal. Hence strain produced in each material will be equal.

(*ii*) The total external load on composite section is equal to sum of the loads shared by individual materials.

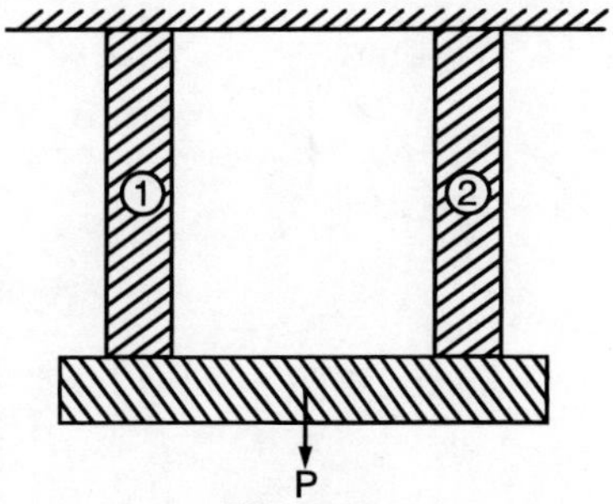

Fig. 2.35.

Figure 2.35 shows a composite section made up of two materials (1) and (2). Let P be the load acting on the section.

Let, A_1 = Area of cross-section of bar (1)
A_2 = Area of cross-section of bar (2)

$$E_1 = \text{Young's modulus of bar (1)}$$
$$E_2 = \text{Young's modulus of bar (2)}$$
$$P_1 = \text{Load carried by bar (1)}$$
$$P_2 = \text{Load carried by bar (2)}$$
$$\text{Total load} = \text{Load in bar (1) + Load in bar (2)}$$
$$P = P_1 + P_2 = \sigma_1 A_1 + \sigma_2 A_2$$

Again,

$$\text{Strain in bar (1)}, e_1 = \frac{\sigma_1}{E_1}$$

$$\text{Strain in bar (2)}, e_2 = \frac{\sigma_2}{E_2}$$

We know,

$$e_1 = e_2$$
$$\frac{\sigma_1}{E_1} = \frac{\sigma_2}{E_2}$$

DO YOU KNOW?

The ratio $\frac{E_1}{E_2}$ is called modular ratio.

EXAMPLE 2.19. A steel cylinder is enclosed in a copper tube as shown in the figure 2.36. The cylinder and tube are compressed between rigid parallel plates. Find the stresses in the steel and copper and also the compressive strain. Take E_s = 210 KN/mm², E_c = 110 KN/mm².

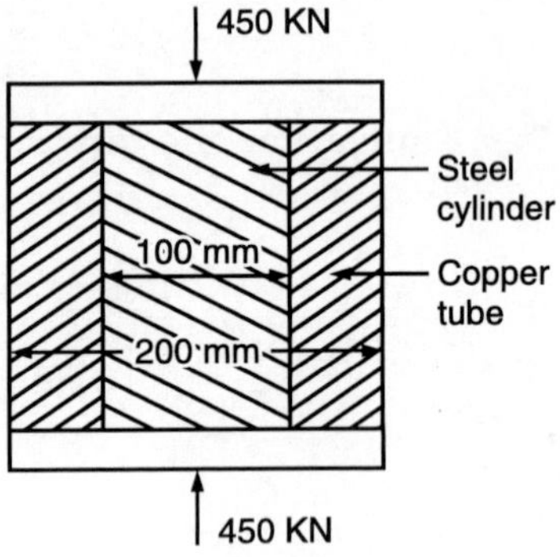

Fig. 2.36.

Solution: Given :

$$d_s = 100 \text{ mm}$$
$$(d_0)_c = 200 \text{ mm}$$
$$(d_i)_c = 100 \text{ mm}$$
$$P = 450 \text{ KN} = 450000 \text{ N}$$
$$E_s = 210 \text{ KN/mm}^2 = 210 \times 10^3 \text{ N/mm}^2$$
$$E_c = 110 \text{ KN/mm}^2 = 110 \times 10^3 \text{ N/mm}^2$$

Area of cross-section of steel cylinder,

$$A_s = \frac{\pi}{4} \times (100)^2 = 7853.98 \text{ mm}^2$$

Area of cross-section of copper tube,

$$A_c = \frac{\pi}{4} \{(200)^2 - (100)^2\} = 23561.94 \text{ mm}^2$$

We know

$$e_s = e_c$$
$$\frac{\sigma_s}{E_s} = \frac{\sigma_c}{E_c}$$
$$\sigma_s = \frac{E_s}{E_c}\sigma_c = \frac{210 \times 10^3}{110 \times 10^3}\sigma_c = 1.91\,\sigma_c$$

Again, Total load = load taken by steel cylinder + load taken by copper tube

i.e.

$$P = P_s + P_c$$
$$450000 = \sigma_s A_s + \sigma_c A_c$$
$$450000 = 1.91\,\sigma_c \times 7853.98 + \sigma_c \times 23561.94$$
$$450000 = \sigma_c \times (1.91 \times 7853.98 + 23561.94)$$
$$450000 = \sigma_c \times 38563.04$$

$$\therefore \qquad \sigma_c = \frac{450000}{38563.04} = 11.67 \text{ N/mm}^2 \quad \textbf{Ans.}$$

$$\sigma_s = 1.91\,\sigma_c = 1.91 \times 11.67 = 22.29 \text{ N/mm}^2 \textbf{ Ans.}$$

$$\text{Compressive strain, } e_c = \frac{\sigma_c}{E_c} = \frac{11.67}{110 \times 10^3} = 0.000106 \quad \textbf{Ans.}$$

EXAMPLE 2.20. A rigid bar is suspended from two wires of equal lengths 1.25 m. One wire is 1 mm in diameter and made of steel, other wire of 2 mm diameter is made of brass. A load of 98 N is placed on the rigid bar such that the bar remains horizontal. If the horizontal distance between wires is 200 mm, determine :

(*i*) stresses developed in steel and brass wires.

(*ii*) distance of the load from the steel wire.

Take $E_s = 2E_b = 210 \times 10^3$ N/mm².

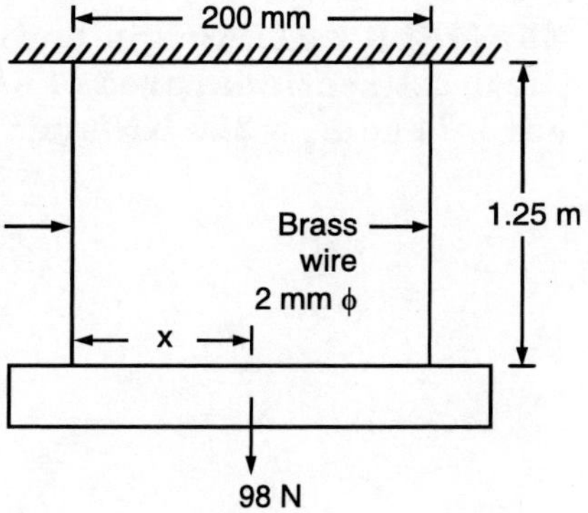

Fig. 2.37.

Solution: Given : $d_s = 1$ mm; $P = 98$ N

$d_b = 2$ mm; $E_s = 210 \times 10^3$ N/mm²

$E_b = 105 \times 10^3$ N/mm²

Area of cross-section of steel wire,

$$A_s = \frac{\pi}{4} \times (1)^2 = 0.785 \text{ mm}^2$$

Area of cross-section of brass wire,

$$A_b = \frac{\pi}{4} \times (2)^2 = 3.141 \text{ mm}^2$$

We know

$$e_s = e_b$$

$$\frac{\sigma_s}{E_s} = \frac{\sigma_b}{E_b}$$

$$\sigma_s = \frac{E_s}{E_b}\sigma_b = \frac{210 \times 10^3}{105 \times 10^3}\sigma_b$$

$$\therefore \qquad \sigma_s = 2\sigma_b$$

Again,

$$P = P_s + P_b$$

$$98 = \sigma_s A_s + \sigma_b A_b$$

$$98 = 2\sigma_b A_s + \sigma_b A_b$$

$$98 = \sigma_b (2 \times 0.785 + 3.141)$$

$$98 = \sigma_b \times 4.711$$

$$\therefore \qquad \sigma_b = \frac{98}{4.711} = 20.80 \text{ N/mm}^2 \quad \textbf{Ans.}$$

$$\sigma_s = 2 \times 20.80 = 41.6 \text{ N/mm}^2 \quad \textbf{Ans.}$$

$$P_b = 20.80 \times 3.141 = 65.33 \text{ N}$$

Let the load 98 N be placed at a distance x from steel wire.

Taking moment about A, we get

$$P \times x = P_b \times 200$$

$$x = \frac{P_b \times 200}{P} = \frac{65.33 \times 200}{98} = 133.3 \text{ mm} \quad \textbf{Ans.}$$

EXAMPLE 2.21. A rigid body of 50 KN hangs from three wires of equal lengths. If the cross-sectional area of each wire is 300 mm², calculate the load shared by each wire. Take E_s = 210 KN/mm² and E_c = 120 KN/mm².

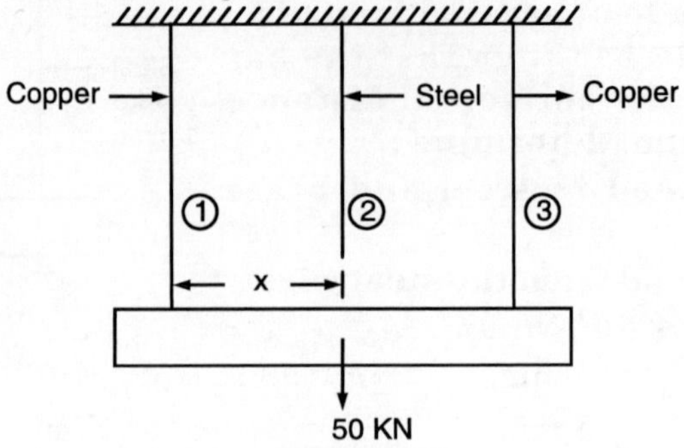

Fig. 2.38.

Solution: Given :

$$P = 50 \text{ KN} = 50000 \text{ N}$$

$$A_s = A_c = 300 \text{ mm}^2$$

$$E_s = 210 \times 10^3 \text{ N/mm}^2$$

$$E_c = 120 \times 10^3 \text{ N/mm}^2$$

We know $e_s = e_c$

$$\frac{\sigma_s}{E_s} = \frac{\sigma_c}{E_c}$$

$$\frac{P_s}{A_s E_s} = \frac{P_c}{A_c E_c}$$

$$P_s = \frac{E_s}{E_c} P_c = \frac{210 \times 10^3}{120 \times 10^3} P_c \qquad [\because A_s = A_c]$$

$$\therefore \quad P_s = 1.75\, P_c$$

$$P = P_s + 2P_c$$

[Total load = load taken by steel + 2 × load taken by copper wire]

$$50 = 1.75\, P_c + 2\, P_c$$

$$50 = 3.75\, P_c$$

$$\therefore \quad P_c = \frac{50}{3.75} = 13.33 \text{ KN} \quad \textbf{Ans.}$$

and $P_s = 1.75 \times 13.33 = 23.33$ KN **Ans.**

EXAMPLE 2.22. A steel rod of 30 mm diameter is enclosed centrally in a hollow copper tube of external diameter 50 mm and internal diameter of 40 mm. The composite member is then subjected to an axial pull of 45000 N. If length of each (rod and tube) is equal to 150 mm, determine :

(*a*) **the stress in the rod and tube.**

(*b*) **load carried by rod and tube.**

Take $E_s = 2.1 \times 10^5$ N/mm² and $E_c = 1.1 \times 10^5$ N/mm².

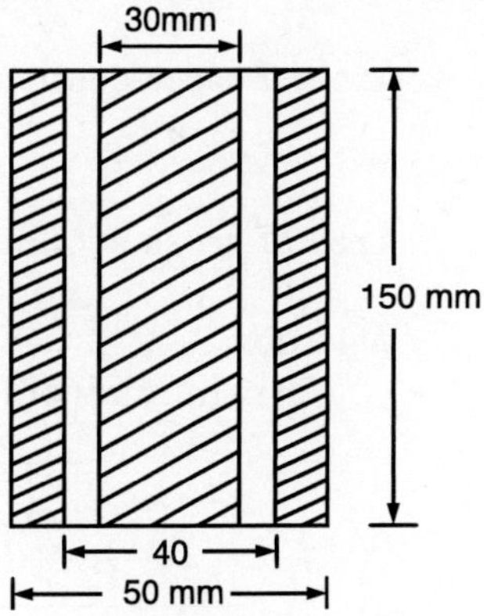

Fig. 2.39.

Solution: Given :

$$d_s = 30 \text{ mm}$$
$$l = 150 \text{ mm}$$
$$(d_i)_c = 40 \text{ mm}$$
$$(d_0)_c = 50 \text{ mm}$$
$$P = 45000 \text{ N}$$

Area of cross-section of steel rod,

$$A_s = \frac{\pi}{4} \times 30^2$$

$$\therefore \quad A_s = 706.86 \text{ mm}^2$$

Area of cross-section of copper tube,

$$A_c = \frac{\pi}{4}(d_0^2 - d_i^2) = \frac{\pi}{4}(50^2 - 40^2) = \frac{\pi}{4} \times 900 = 706.86 \text{ mm}^2$$

We know,

$$\text{Strain in tube} = \text{Strain in rod}$$

i.e.
$$e_c = e_s$$

$$\frac{\sigma_c}{E_c} = \frac{\sigma_s}{E_s}$$

$$\frac{\sigma_c}{1.1 \times 10^5} = \frac{\sigma_s}{2.1 \times 10^5}$$

$$\sigma_c = \frac{1.1}{2.1}\sigma_s = 0.524\,\sigma_s$$

We know,
$$P = P_s + P_c$$

$$45000 = \sigma_s A_s + \sigma_c A_c = \sigma_s \times 706.86 + 0.524\,\sigma_s \times 706.86$$

$$45000 = 706.86 \times 1.524\,\sigma_s$$

$$\frac{45000}{706.86 \times 1.524} = \sigma_s$$

$$\therefore \quad \sigma_s = 41.77 \text{ N/mm}^2 \quad \textbf{Ans.}$$

$$\sigma_c = 0.524\,\sigma_s = 0.524 \times 41.77 = 21.89 \text{ N/mm}^2 \quad \textbf{Ans.}$$

Load carried by rod,

$$P_s = \sigma_s A_s = 41.77 \times 706.86 = 29525.54 \text{ N} \quad \textbf{Ans.}$$

Load carried by tube,

$$P_c = \sigma_c A_c = 21.89 \times 706.86 = 15473.16 \text{ N} \quad \textbf{Ans.}$$

EXAMPLE 2.23. An R.C.C. circular column has a diameter of 300 mm and it is reinforced with 8 steel bars. The total area of steel bars is 2513.27 mm². The column carries a load of 250 KN. If the modulus of elasticity for steel is 18 times that of concrete, find the stresses in concrete and steel.

Solution: Given :

$$d_c = 300 \text{ mm}$$
$$A_s = 2513.27 \text{ mm}^2$$
$$P = 250 \text{ KN} = 250000 \text{ N}$$

$$E_s = 18\ E_c$$

Area of cross-section of column,

$$A_c = \frac{\pi}{4} \times (300)^2 = 70685.83\ \text{mm}^2$$

Area of cross-section of concrete,

$$A_c = A - A_s = 70685.83 - 2513.27 = 68172.56\ \text{mm}^2$$

We know,

$$\text{Strain in steel } (e_s) = \text{Strain in concrete } (e_c)$$

$$\frac{P_s}{A_s E_s} = \frac{P_c}{A_c E_c}$$

$$\frac{P_s}{2513.27 \times 18 E_c} = \frac{P_c}{68172.56 \times E_c}$$

$$\therefore \quad P_s = \frac{2513.27 \times 18}{68172.56} P_c = 0.66\ P_c$$

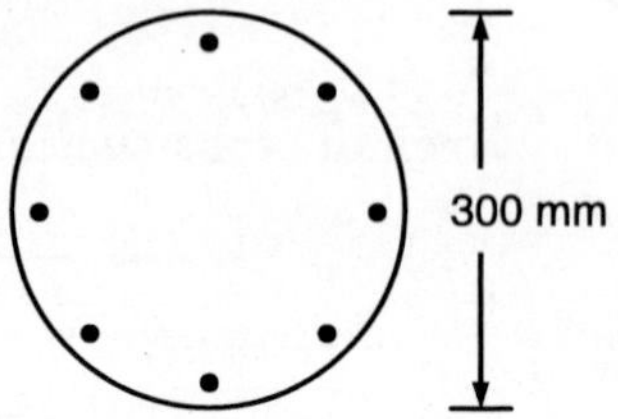

Fig. 2.40

Again,

$$P = P_s + P_c$$

$$250000 = 0.66\ P_c + P_c$$

$$250000 = 1.66\ P_c$$

$$\therefore \quad P_c = \frac{250000}{1.66} = 150602.41\ \text{N}$$

and

$$P_s = 0.66\ P_c = 0.66 \times 150602.41 = 99397.59\ \text{N}$$

Stress in concrete, $\sigma_c = \frac{P_c}{A_c} = \frac{150602.41}{68172.56} = 2.209\ \text{N/mm}^2$ **Ans.**

Stress in steel, $\sigma_s = \frac{P_s}{A_s} = \frac{99397.59}{2513.27} = 39.55\ \text{N/mm}^2$ **Ans.**

EXAMPLE 2.24. A member formed by connecting a steel bar to an aluminium bar is shown in the figure 2.41. Calculate the magnitude of the load P which will cause the total length of the member to decrease 0.25 mm. Take $E_s = 2.1 \times 10^5$ N/mm² and $E_a = 7 \times 10^4$ N/mm².

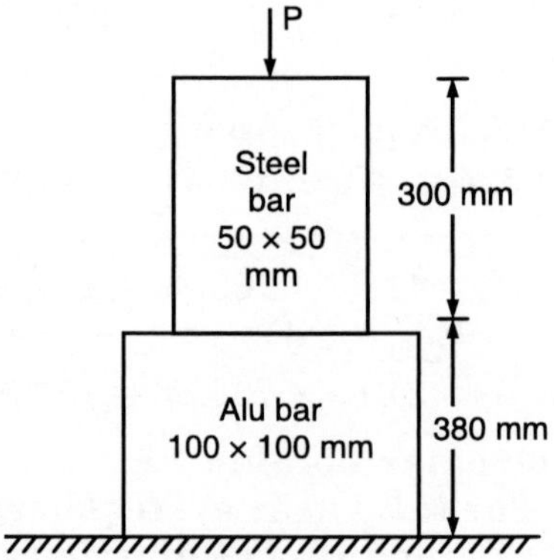

Fig. 2.41.

Solution: **Given :** Area of cross-section of steel bar,

$$A_s = 50 \times 50 = 2500\ \text{mm}^2$$

Area of cross-section of aluminium bar,

$$A_a = 100 \times 100 = 10000 \text{ mm}^2$$

Total decrease in length,

$$\delta l = 0.25 \text{ mm}$$
$$E_s = 2.1 \times 10^5 \text{ N/mm}^2$$
$$E_a = 7 \times 10^4 \text{ N/mm}^2$$
$$\delta l = \delta_s + \delta_a$$
$$0.25 = \frac{P_s l_s}{A_s E_s} + \frac{P_a \times l_a}{A_a E_a}$$
$$= \frac{P \times 300}{2500 \times 2.1 \times 10^5} + \frac{P \times 380}{10000 \times 7 \times 10^4} \qquad [\because P_s = P_a = P]$$
$$= (5.71 \times 10^{-7} + 5.42 \times 10^{-7})\, P$$
$$0.25 = 11.13 \times 10^{-7}\, P$$
$$\frac{0.25}{11.13 \times 10^{-7}} = P$$
$$\therefore \quad P = 224618.15 \text{ N} = 224.62 \text{ KN} \quad \textbf{Ans.}$$

2.28 THERMAL STRESS AND STRAIN

Thermal stress is the stress developed in a body due to change in temperature.

A change in temperature of a body tends to produce a change in its dimensions as shown in figure 2.42. The material undergoes a uniform thermal strain, e_T.

Thermal strain, $e_T = \alpha(\Delta T)$

where α = coefficient of thermal expansion

ΔT = change in temperature.

If the body is not restrained from expansion or contraction, no thermal stresses will be set up in the body.

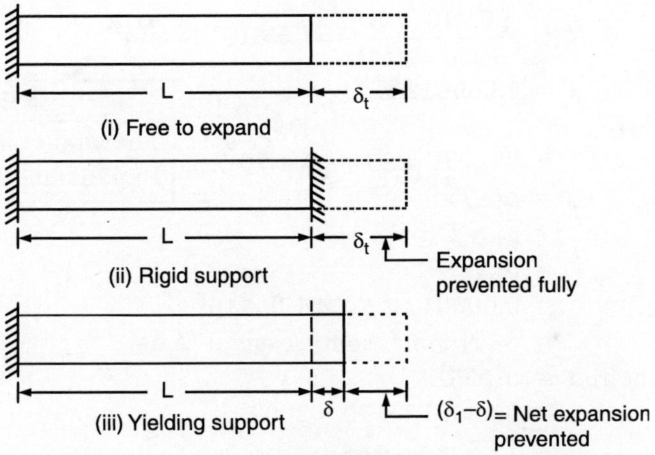

Fig. 2.42.

If a bar of length 'L' is free to expand due to rise in temperature, then extension of bar,

$$\delta_t = e_T \times L = \alpha(\Delta T)L$$

(*i*) **When supports are rigid**

If the ends of the bars are fixed to rigid support so that the entire expansion is prevented then thermal (compressive) stresses and strains will be set up.

$\therefore$ Thermal stress, σ_T = Thermal strain × E

$$\therefore \quad \sigma_T = \alpha\,(\Delta T)\,E \quad \text{(Compressive)}$$

Similarly, when the temperature of the bar is decreased and the bar is prevented from contracting, then tensile stresses will be induced in the bar.

(*ii*) **When supports yield**

In some cases, supports are unable to prevent expansion or contraction of the bar completely. Such supports are called yielding supports.

Refer figure 2.42 (*iii*).

$$\text{Expansion prevented} = \delta_t - \delta = \alpha(\Delta T)\,L - \delta$$

where δ is the yield of support

Thermal strain,

$$e_T = \frac{\alpha(\Delta T)L - \delta}{L} = \alpha(\Delta T) - \frac{\delta}{L}$$

Thermal stress,

$$\sigma_T = e_T \times E = \left\{\alpha(\Delta T) - \frac{\delta}{L}\right\} E.$$

EXAMPLE 2.25: A rod is 2 m long at a temperature of 10 K. Find the expansion of the rod, when the temperature is raised to 80 K. If this expansion is prevented, find the stress induced in the material of the rod.

Take $E = 1.0 \times 10^5$ MN/m² and $\alpha = 0.000012$/K.

Solution: Given : $L = 2\text{ m} = 2000\text{ mm}$

Initial temp., $T_1 = 10$ K

Final temp., $T_2 = 80$ K

$$E = \frac{1.0\times10^5\times10^6}{1000\times1000} = 1.0\times10^5\ \text{N/mm}^2$$

$$\alpha = 0.000012/\text{K}$$

Rise in temperature,

$$\Delta T = T_2 - T_1 = 80 - 10 = 70\ \text{K}$$

REMEMBER
Thermal stress = α (ΔT)E
Expansion of rod = α (ΔT) L

Thermal strain, $e_T = \alpha(\Delta T)$

Thermal stress, $\sigma_T = e_T \times E$

$$= \alpha(\Delta T)E$$

$$= 0.000012 \times 70 \times 1.0 \times 10^5$$

$$= 84\ \text{N/mm}^2\ \text{(compressive)}$$ **Ans.**

Expansion of the rod $= \alpha(\Delta T)L$

$$= 0.000012 \times 70 \times 2000$$

$$= 1.68\ \text{mm}$$ **Ans.**

EXAMPLE 2.26: A steel rod 30 mm diameter and 5 m long is connected to two grips and the rod is maintained at a temperature of 95 K. Determine the stress and pull exerted when the temperature falls to 30 K, if :

(*i*) the ends do not yield, and

(ii) the ends yield by 1.2 mm

Take $E = 2 \times 10^5$ MN/m² and $\alpha = 12 \times 10^{-6}$ K.

Solution: Given :

$d = 30$ mm

$L = 5$ m $= 5000$ m

Initial temp., $T_1 = 95$ K

Final temp., $T_2 = 30$ K

Fall in temp., $\Delta T = 95 - 30 = 65$ K

$$E = 2 \times 10^5 \text{ MN/m}^2 = \frac{2\times10^5\times10^6}{1000\times1000} \text{ N/mm}^2 = 2 \times 10^5 \text{ N/mm}^2$$

Area of cross-section, $A = \frac{\pi}{4} \times (30)^2 = 706.86$ mm²

Case I : When the ends do not yield.

Thermal stress, $\sigma_T = \alpha\,(\Delta T)\,E$

$= 12 \times 10^{-6} \times 65 \times 2 \times 10^5$

$= 156$ N/mm² (Tensile) **Ans.**

Pull exerted $= \sigma_T \times A = 156 \times 706.86 = 110270.16$ N **Ans.**

Case II : When the ends yield.

$\delta = 1.2$ mm

Total contraction, $\delta_t = \alpha\,(\Delta T)L = 12 \times 10^{-6} \times 65 \times 5000 = 3.9$ mm

Net contraction prevented,

$\delta_n = \delta_t - \delta = 3.9 - 1.2 = 2.7$ mm

Thermal strain, $e_T = \frac{2.7}{5000}$

Thermal stress, $\sigma_T = e_T \times E = \frac{2.7}{5000} \times 2 \times 10^5 = 108$ N/mm² **Ans.**

Pull exerted $= 108 \times 706.86 = 76340.88$ N **Ans.**

EXAMPLE 2.27. A steel bar of length 1 m and cross-sectional area 1600 mm² is heated from 20°C to 80°C and at the same time subjected to a pull *P*. If the total extension of the rod is 0.85 mm, what should be the magnitude of *P* ?

Take α for steel = 12 × 10⁻⁶/°C and *E* = 200 GPa.

Solution: Given :

$L = 1$ m $= 1000$ mm

$A = 1600$ mm²

Rise in temp., $\Delta T = 80 - 20 = 60$°C

$\delta = 0.85$ mm

$\alpha = 12 \times 10^{-6}$/°C

$E = 200$ GPa $= 200 \times 10^9$ N/m² [∵ 1 Pa = 1 N/m²]

$$= \frac{200\times10^9}{1000\times100} = 200 \times 10^3 \text{ N/mm}^2$$

Extension due to rise in temp.

$= \alpha\,(\Delta T)\,L = 12 \times 10^{-6} \times 60 \times 1000 = 0.72$ mm

Total extension is the sum of the extension due to rise in temperature and due to load.

$\therefore$ Total extension = Extension due to rise in temp. + Extension due to load

$$0.85 = 0.72 + \delta_l$$

$$\therefore \quad \delta_l = 0.85 - 0.72 = 0.13 \text{ mm}$$

$$\text{Tensile strain due to } P = \frac{\delta_l}{L} = \frac{0.13}{1000} = 1.3 \times 10^{-4}$$

$$\text{Tensile stress,} \quad \sigma = \text{E} \times \text{Tensile strain} = 200 \times 10^3 \times 1.3 \times 10^{-4} = 26 \text{ N/mm}^2$$

$$\sigma = \frac{P}{A}$$

$$\therefore \quad P = \sigma \times A = 26 \times 1600 = 41600 \text{ N}$$ **Ans.**

SUMMARY

1. Stress is defined as the internal force of resistance to deformation per unit cross-sectional area of the body.

$$\sigma = \frac{P}{A} \text{ N/mm}^2 \text{ or N/m}^2$$

2. Strain is defined as the ratio of change in dimension of a member to the original dimension.

$$e = \frac{\text{Change in dimension}}{\text{Original dimension}}$$

3. Strain has no unit.
4. Tensile strain is the ratio of increase in length to the original length of a body when an axial pull acts on it.
5. Compressive strain is the ratio of decrease in length to the original length of a body when an axial push acts on it.
6. Shear strain is the measure of the distortion produced in a body when shear force is acting on it.
7. Volumetric strain is the ratio of change in volume to the original volume.
8. Elasticity is the property of a material by virtue of which a material regain all its deformation after removal of the external loading.
9. Hooke's law states that within elastic limit stress is proportional to strain.

$$\frac{\text{Stress}}{\text{Strain}} = \text{Constant}$$

10. Young's modulus of elasticity (E) is the ratio of direct stress to strain.

$$E = \frac{\sigma}{e}.$$

11. Modulus of rigidity (G) is the ratio of shear stress to shear strain.
12. Bulk modulus (K) is the ratio of normal stress to the volumetric strain.
13. Poisson's ratio is the ratio of lateral strain to longitudinal strain.
14. Relation between E, G and K

(*i*) $E = 2G\left(1 + \frac{1}{m}\right)$ (*ii*) $E = 2K\left(1 - \frac{2}{m}\right)$

(*iii*) $E = \frac{9\,KG}{3K + G}$.

15. Ultimate stress is the maximum axial load that can be applied in a member without causing rupture.
16. Working stress is the maximum safe stress a material can carry.
17. Factor of safety is the ratio of ultimate stress and working stress.
18. Proof stress is the stress at which the strain has departed by 0.1% of the gauge length from the line of proportionality.
19. Thermal stress is the stress developed in a body due to change in temperature.

$$\text{Thermal stress} = \sigma_T = \alpha(\Delta T)E$$
$$\text{Thermal strain, } e_T = \alpha(\Delta T).$$

OBJECTIVE TYPE QUESTIONS

1. Fill in the Blanks :

(i) The load which is distributed over the length of a beam is known as ____________ .
(ii) ____________ is the internal force of resistance per unit area.
(iii) ____________ is a measure of deformation produced in the member by the load.
(iv) The property by virtue of which certain materials return back to their original position after the removal of the external forces is known as ____________ .
(v) According to Hooke's law, ____________ is proportional to ____________ .
(vi) Bulk modulus is the ratio of normal stress to ____________ .
(vii) The ratio of lateral strain to the longitudinal strain is ____________ .
(viii) $E = 2G\ (1 + –)$.
(ix) $E = \dfrac{9\ KG}{- + G}$.
(x) When a section is composed of more than one material, it is called a ____________ section.
(xi) Factor of safety = ____________
(xii) Young's modulus is indicative of the property called ____________ .

Answers: (i) u.d.l, (ii) stress, (iii) strain, (iv) elasticity, (v) stress, strain, (vi) volumetric strain, (vii) Poisson's ratio, (viii) $\dfrac{1}{m}$, (ix) 3K, (x) composite, (xi) $\dfrac{\text{ultimate stress}}{\text{working stress}}$, (xii) stiffness.

THEORETICAL QUESTIONS

(i) Define load. What are the different types of loads ?
(ii) Define the followings :
(a) Stress, (b) Strain, (c) Elasticity, (d) Elastic limit.
(iii) State Hooke's law.
(iv) Define the followings :
(a) Young's modulus of elasticity
(b) Bulk modulus
(c) Modulus of rigidity.
(v) Define the term 'factor of safety'.
(vi) Derive the relation $E = \dfrac{9\ KG}{3K + G}$.
(vii) What is Poisson's ratio ?
(viii) Define (a) temperature stress, (b) temperature strain.

NUMERICALS

1. A mild steel tie rod 3 m long having a cross-sectional area 4 sq. cm is subjected to an axial pull of 1000 kg. If E for steel = 2×10^6 kg/cm^2. Find (*i*) the stress, (*ii*) the strain, (*iii*) the elongation of the rod. (250 kg/cm^2; 0.000125; 0.0375 cm)
2. Find the minimum diameter of a steel wire which is used to raise a load of 5000 N if the stress in the wire is not to exceed 95 MN/m^2. (8.19 mm)
3. A hollow cast iron cylinder 4 m long, 300 mm outer diameter and thickness of metal 50 mm is subjected to a central load on the top when standing straight. The stress produced is 75000 KN/m^2. Assume Young's modulus for cast iron as 1.5×10^8 KN/m^2 and find (*i*) magnitude of the load, (*ii*) longitudinal strain produced, (*iii*) total decrease in length. (2945.2 KN; 0.0005; 2 mm)
4. A wooden tie is 75 mm wide, 150 mm deep and 2.5 m long. It is subjected to an axial pull of 40 KN. The extension of the member is found to be 0.80 mm. Find the Young's modulus for the tie material. (11.125 KN/mm^2)
5. The ultimate stress for a hollow steel column which carries an axial load of 2.2 MN is 500 N/mm^2. If the external diameter of the column is 250 mm, determine the internal diameter. Take the factor of safety as 4. (200.23 mm)
6. A brass bar having a cross-sectional area of 1000 mm^2 is subjected to axial forces as shown in below figure. Find the total change in length of the bar. Take $E = 1.05 \times 10^5$ N/mm^2. (0.1143 mm)

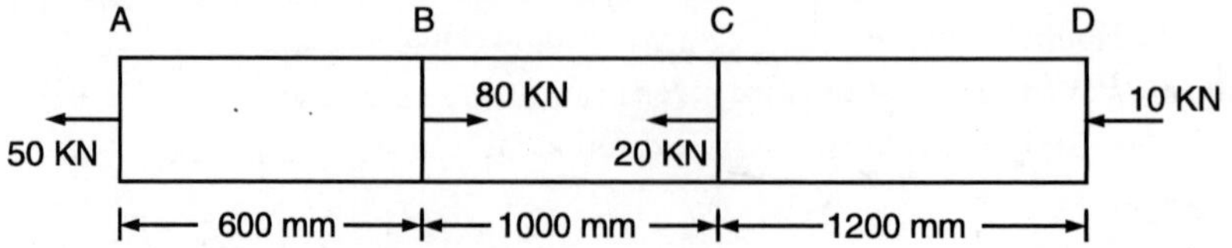

7. Figure below shows a steel bar consisting of three lengths. Find the stresses in the three parts and the total elongation of the bar for an axial pull of 40 KN. Take $E = 2.1 \times 10^5$ N/mm^2. (0.186 mm)

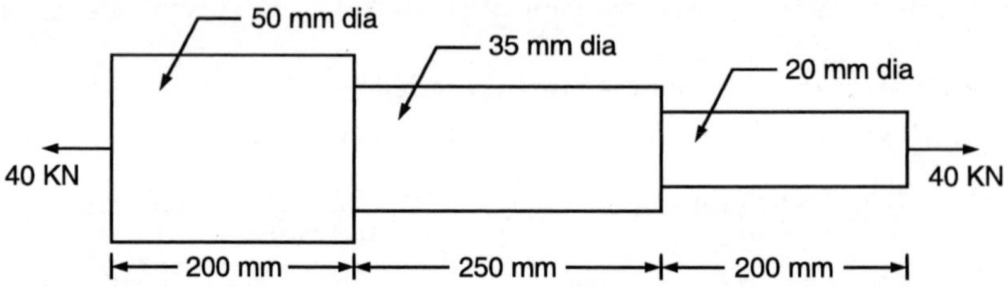

8. A member ABCD is subjected to points loads P_1, P_2, P_3 and P_4 as shown in below figure. Calculate the force P_3 necessary for equilibrium if P_1 = 120 KN, P_2 = 220 KN and P_4 = 160 KN. Determine also the net change in length of the member. Take $E = 2 \times 10^5$ N/mm^2. (0.55 mm)

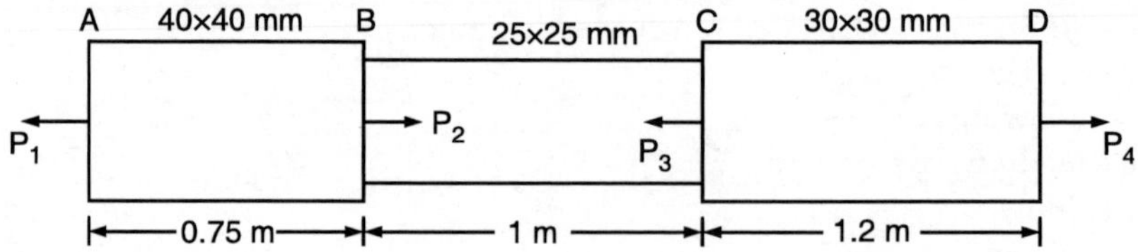

CHAPTER 3

Shear Force and Bending Moment

3.1 BEAM

A beam is a structural member that is designed to resist forces acting at right angles to its axis.

Types of beam : Beams are of following types depending upon the types of supports on which it rest :

(*i*) Simply supported beam
(*ii*) Cantilever beam
(*iii*) Fixed beam
(*iv*) Overhanging beam

(*i*) Simply Supported Beam : A beam is said to be simply supported beam if the ends of the beam are made to rest freely on supports. This is also known as simply beam.

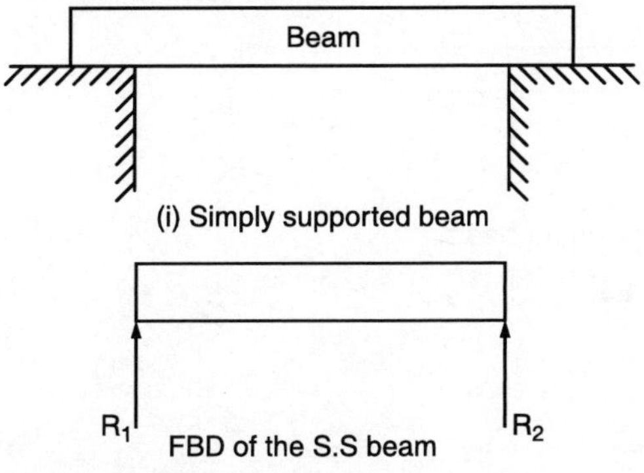

Fig. 3.1.

(*ii*) Cantilever Beam : A beam is said to be cantilever beam if the beam is fixed at one end and free at the other.

Chapter Outline

- Beam
- Types of Loading
- Span
- Shear Force (Abbreviated as S.F.)
- Bending Moment (Abbreviated as B.M.)
- Important Points for Drawing Shear Force and Bending Moment Diagrams
- B.M. and S.F. for Cantilever Beam
- B.M. and S.F. for Simply Supported Beam
- Shear Force and Bending Moment Diagrams for Overhanging Beans
- Load and Bending Diagrams From A Shear Force Diagram

DO YOU KNOW?

A beam which is provided with more than two supports is called a continuous beam.

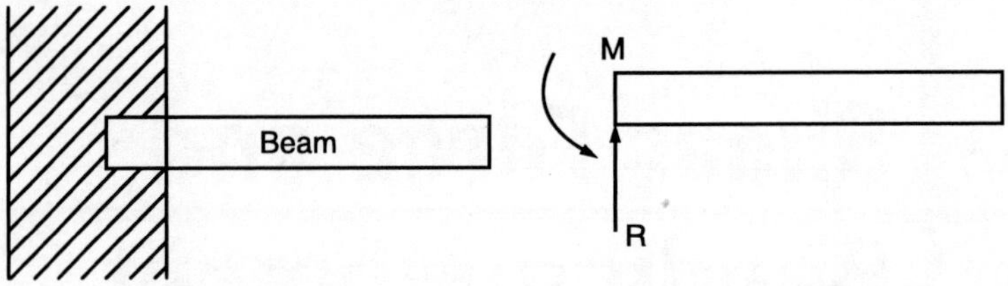

(ii) Cantilever beam FBD of cantilever beam

Fig. 3.2.

***(iii)* Fixed beam :** A beam is said to be fixed beam if both the ends of the beam are fixed or built in.

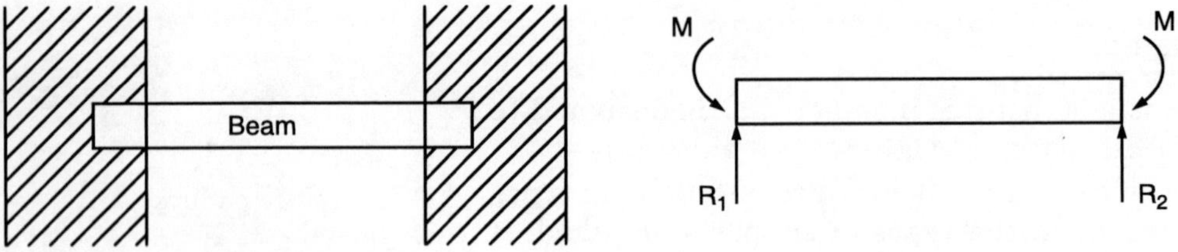

(iii) Fixed beam FBD of fixed beam

Fig. 3.3.

***(iv)* Overhanging Beam :** A beam is said to be overhanging beam if one end or both the ends of the beam projects beyond the supports.

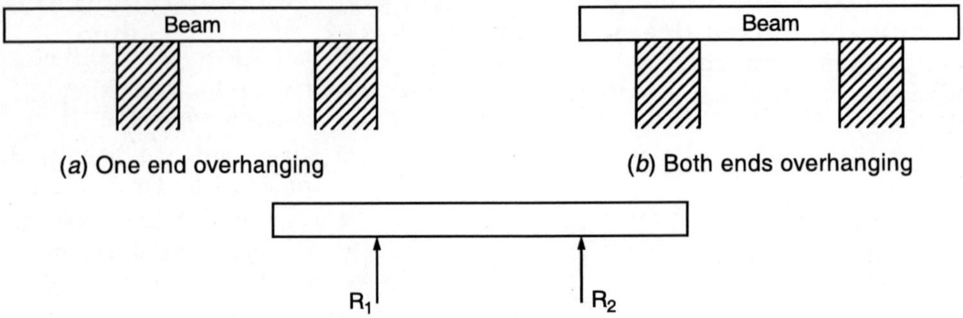

(*a*) One end overhanging (*b*) Both ends overhanging

(*c*) FBD of overhanging beam

Fig. 3.4.

3.2 TYPES OF LOADING

Usually loading on beams are of following types :

(*i*) Concentrated or Point load.

(*ii*) Uniformly distributed load (u.d.l)

(*iii*) Uniformly varying load

(*i*) Concentrated or Point Load

If a load is acting on a beam at a point, then the load is called a point or concentrated load. It is generally denoted by an arrow as shown in the figure 3.5.

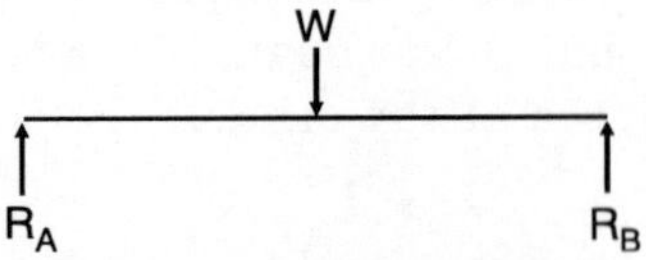

Fig. 3.5. Concentrated or point load.

(ii) Uniformaly Distributed Load

This is abbreviated as u.d.l. This type of load is uniformly distributed over a length of beam as shown in figure 3.6.

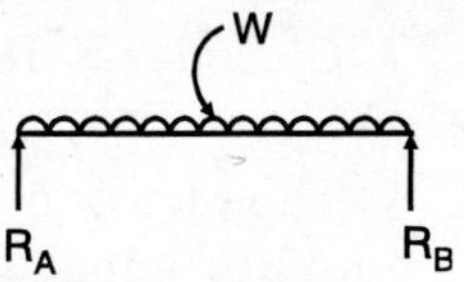

Fig. 3.6. Uniformaly distributed load.

(iii) Uniformly Varying Load

In this type of loading, the load varies at a uniform rate from one point to another. Triangular type or Trapezium type of loading falls under this category, as shown in figure 3.7 (*a*) and (*b*).

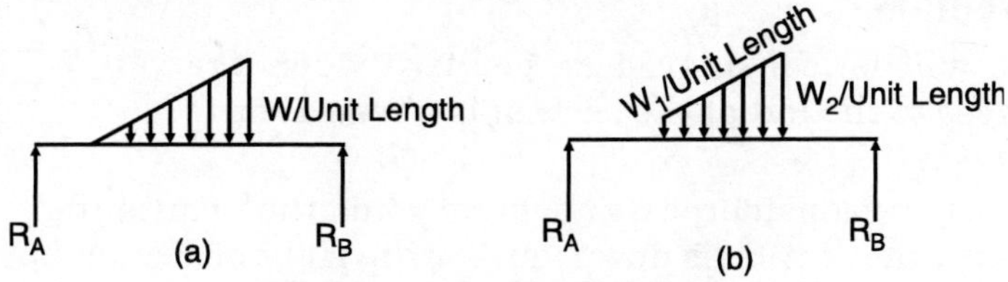

Fig. 3.7. Uniformly varying load.

3.3 SPAN

The horizontal distance between two consecutive supports of a beam is known as span of the beam.

3.3.1 Clear Span

The clear horizontal distance between two supports of a beam is known as clear span.

3.3.2 Effective Span

The horizontal distance between the centres of end bearings is known as effective span.

3.4 SHEAR FORCE (ABBREVIATED AS S.F.)

Shear force at any section of a beam is defined as the algebraic sum of the lateral components of all the forces acting an either side (left or right side) of the section. Refer figure 3.8

> **DO YOU KNOW?**
>
> If the intensity of the bearing reaction is not uniform the effective span is the horizontal distance between the lines of action of end reactions.

$$\text{Reaction at } A,\ R = 5000 \text{ N}$$

$$\text{Moment at } A,\ M_a = 5000 \times 3$$

$$= 15000 \text{ N-m}$$

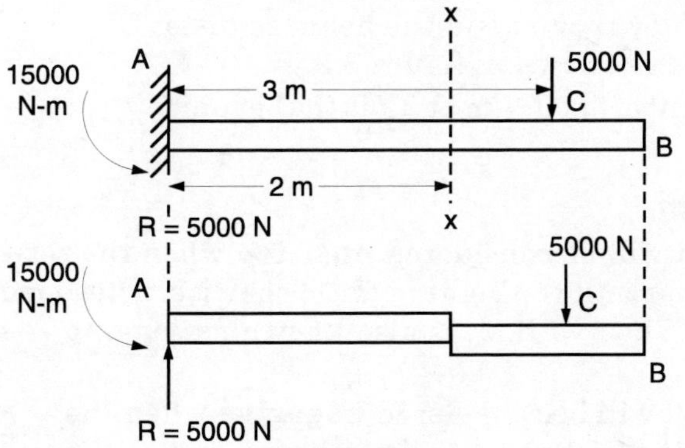

Fig. 3.8.

Consider a section *X-X* at a distance 2 m from fixed end. Consider LHS of the beam.

$$\text{Resultant force} = R = 5000 \text{ N (upward)}$$

Consider RHS of the beam

Resultant force = 5000 N (downward)

The resultant forces acting on left hand side and right hand side portion of the beam are equal and opposite in direction.

These equal and opposite forces tend to shear the beam at section *X-X*. This resultant force (*i.e.,* 5000 N) on either side of the section is known as shear force at that section.

3.4.1 Sign Convention

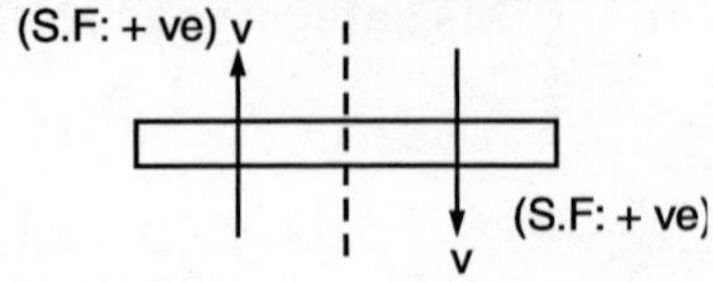

(*i*) Shear force will be considered as positive when the resultant of the forces to the left of the section is upwards or to the right is downwards.

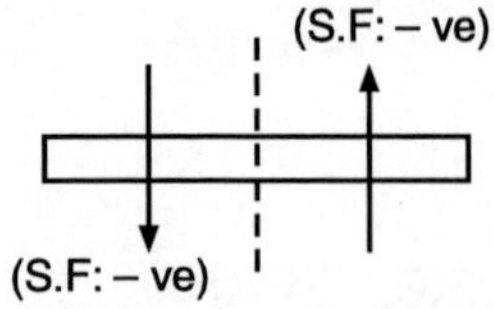

(*ii*) Shear force will be considered as negative when the resultant of the forces to the left of the section is downward or the right of section is upward.

3.4.2 Shear Force Diagram (S.F.D.)

A shear force diagram is the graphical representation of variation of shear force along the length of the beam.

3.5 BENDING MOMENT (ABBREVIATED AS B.M.)

Bending moment at any section of a beam is defined as the algebraic sum of the moments of all the forces acting on either side (left-hand or right-hand side) of the section.

Refer Figure 3.8.

Consider LHS portion of the beam

Resultant moment = 15000 – 10000

= 5000 N-m (Anticlockwise)

Consider RHS portion of the beam

Resultant moment = 5000 × 1

= 5000 N-m (Clockwise)

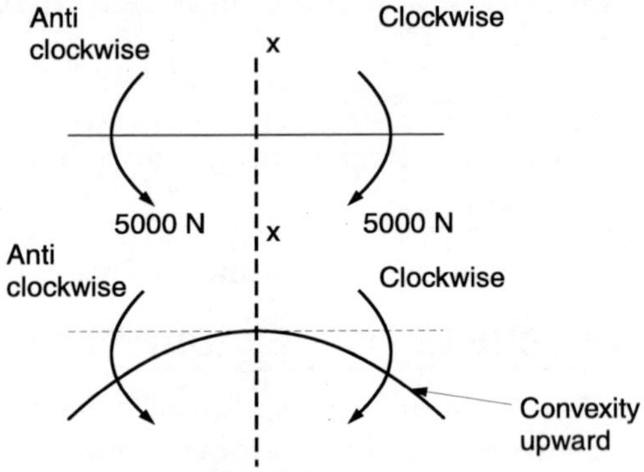

Fig. 3.9.

These resultant moments about *X-X* are equal and opposite in direction. They try to bend the beam so as to form convexity upwards as shown in figure 3.9.

This resultant moment of 5000 N-m at *XX* is the bending moment at *XX*.

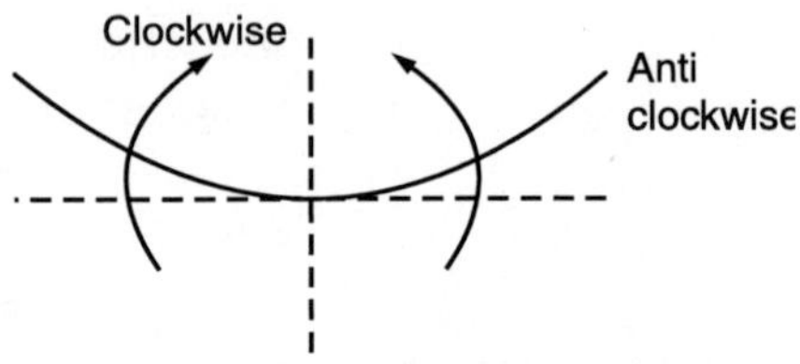

Fig. 3.10. Sagging or +ve B.M.

3.5.1 Sign Convention

(*i*) Bending moment will be considered **positive** when the resultant moment to the left of the section is clockwise and the right is anticlockwise. These moments tend to make the beam concave upward. The +ve B.M. is also known as sagging bending moment.

(*ii*) Bending moment will be considered **negative** when the moment on the left of the section is anti-clockwise and to the right of the section is clockwise. These moments tend to make the beam concave downward. The – ve *B.M* is also known as hogging B.M.

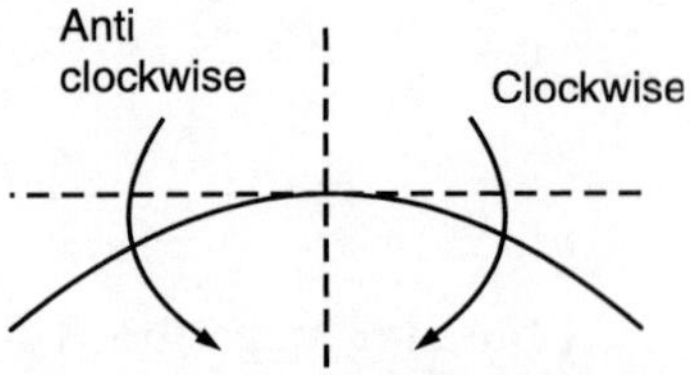

Fig. 3.11. Hogging or –ve B.M.

3.5.2 Bending Moment Diagram

A bending moment diagram is the graphical representation of variation of bending moment along the length of the beam.

3.6 IMPORTANT POINTS FOR DRAWING SHEAR FORCE AND BENDING MOMENT DIAGRAMS

(*i*) The positive values of shear force and bending moment are drawn above the base line and negative values below the base line.

(*ii*) The shear force diagram will change suddenly at a point load, the change is numerically equal to the load.

(*iii*) The shear force diagram between any two vertical load will be horizontal, as the shear force is constant. But bending moment varies at a constant rate and the bending moment diagram is an inclined straight line between two vertical loads.

(*iv*) For uniformly distributed load between two points, the shear force diagram will be an inclined straight line and bending moment diagram will be a parabolic curve.

(*v*) If there is uniformly varying load between two points, the shear force diagram and bending moment diagram will be parabolic and cubic respectively.

(*vi*) In case of simply supported beam, the bending moment at the two end supports will be zero. For cantilever, bending moment will be zero at the free end.

3.7. B.M. AND S.F FOR CANTILEVER BEAM

3.7.1 Cantilever Beam Carrying a Point Load at the Free End

Consider a cantilever beam *AB* of length '*l*' fixed at *A* and free at *B*. The beam is carrying a point load '*W*' at the free end as shown in the figure 3.12.

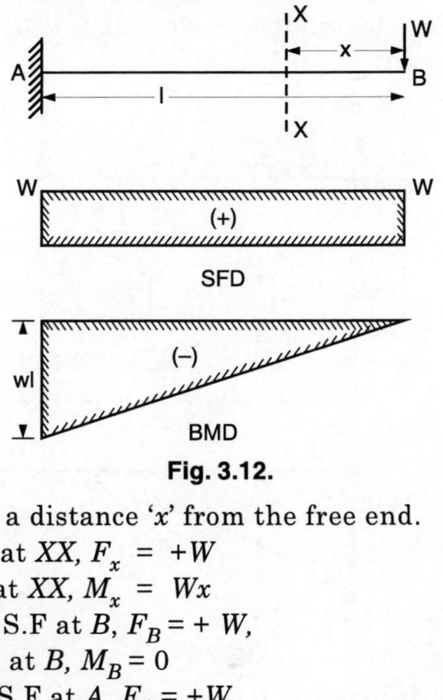

Fig. 3.12.

Consider a section *XX* at a distance '*x*' from the free end.

S.F. at XX, $F_x = +W$

B.M at XX, $M_x = Wx$

At point *A* $x = 0$, S.F at B, $F_B = +W$,

B.M at B, $M_B = 0$

At point *B* $x = l$, S.F at A, $F_A = +W$,

B.M at $A = -Wl$

Figure 3.12 shows the shear force and bending moment diagram.

3.7.2 Cantilever Beam Carrying UDL

Consider a cantilever beam AB of length 'l' carrying a UDL of w per unit length over the entire length.

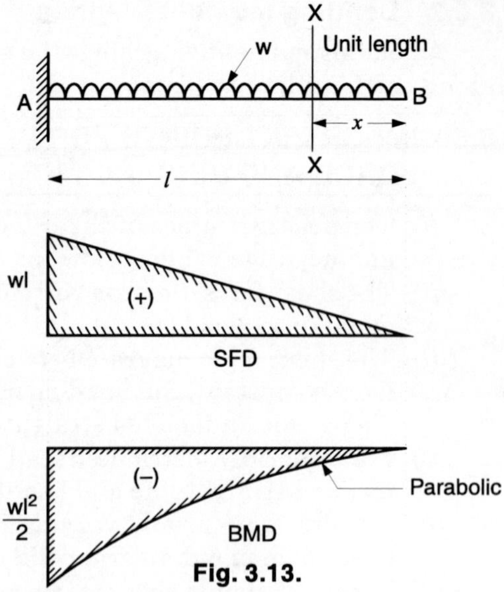

Fig. 3.13.

Consider a section XX at a distance 'x' from B.

$$\text{S.F. at } XX,\ F_x = +wx$$

$$\text{B.M. at } XX,\ M_x = -wx \times \frac{x}{2} = \frac{-wx^2}{2}$$

At point B

$$x = 0$$

$$\text{S.F. at } B,\ F_B = 0$$

$$\text{B.M. at } B,\ M_B = 0$$

At point A

$$x = l$$

$$\text{S.F. at } A,\ F_A = +wl$$

$$\text{B.M. at } B,\ M_A = -\frac{wl^2}{2}$$

Figure 3.13 shows the shear force and bending moment digram.

Note: For cantilever with u.d.l w/m length, maximum B.M will be at the fixed end and will be equal to $\frac{wl^2}{2}$.

3.7.3 Cantilever Beam Carrying UDL over Whole Length and Point Load at Free End

Consider a cantilever AB of length 'l' carrying UDL w per unit length over the entire length and point load 'W' at point B.

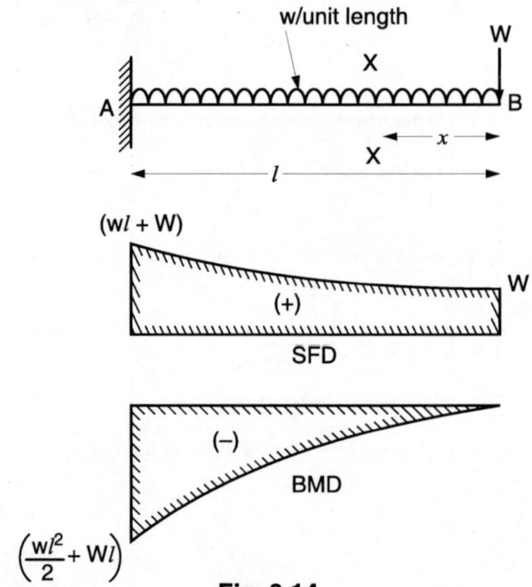

Fig. 3.14.

Consider a section X-X at a distance 'x' from point B.

$$\text{S.F. at } XX,\ F_x = +wx + W$$

$$\text{B.M. at } XX,\ M_x = -wx \times \frac{x}{2} - Wx = \frac{-wx^2}{2} - Wx$$

At point *B*

$$x = 0$$
$$\text{S.F. at } B, F_B = +W$$
$$\text{B.M. at } B, M_x = 0$$

At point *A*

$$x = l$$
$$\text{S.F. at } A, F_A = +wl + W$$
$$\text{B.M. at } A, M_A = -\frac{wl^2}{2} - Wl$$

3.7.4 Cantilever Beam Carrying a *UDL* over a Length '*a*' From Free End

Consider a cantilever *AB* of length '*l*' carrying *UDL* '*w*' per unit length over a length '*a*' of the beam from free end.

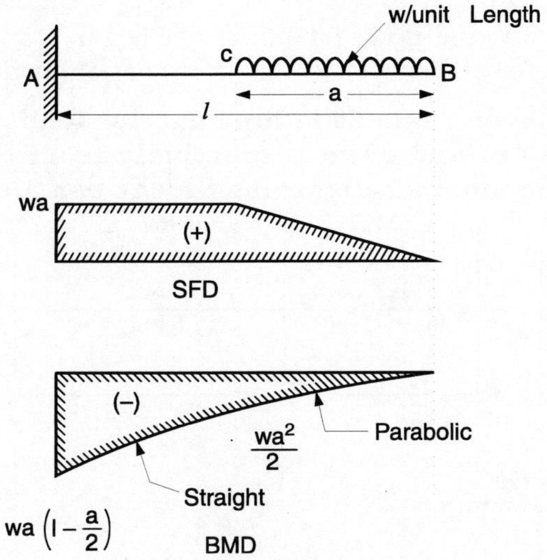

Fig. 3.15.

For part BC

Consider a section *X-X* at a distance *x* point *B*.

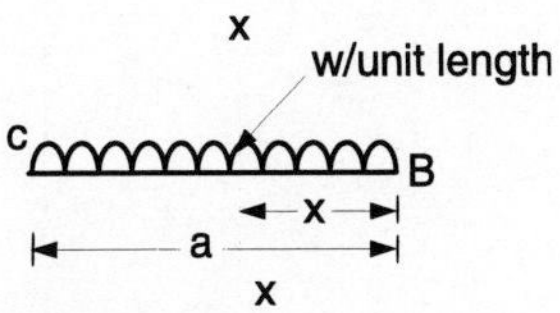

$$\text{S.F. at } XX, F_x = +wx$$
$$\text{B.M. at } XX, M_x = -wx.\frac{x}{2} = \frac{-wx^2}{2}$$

At *B*

$$x = 0$$
$$\text{S.F. at } B, F_B = 0$$
$$\text{B.M. at } B, M_B = 0$$

At *C*

$$x = a$$
$$\text{S.F. at } C, F_C = +wa$$
$$\text{B.M. at } C, M_C = \frac{-wa^2}{2}$$

For part *CA* : Consider a section *XX* at a distance *x* from *B*.

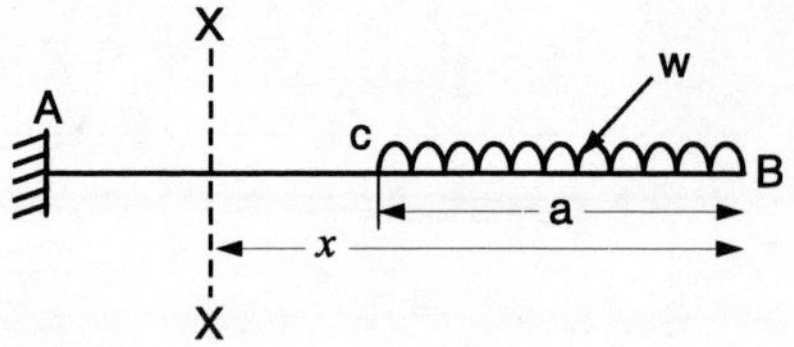

$$\text{SF at } XX,\ F_x = +wa$$

$$\text{BM at } XX,\ M_x = -wa\left(x - \frac{a}{2}\right)$$

At *C*
$$x = a$$

$$\text{S.F. at } C,\ F_C = +wa$$

$$\text{B.M. at } C,\ M_C = -wa\left(a - \frac{a}{2}\right) = \frac{-wa^2}{2}$$

At *A*
$$x = l$$

$$\text{S.F. at } A,\ F_A = +wa$$

$$\text{B.M. at } A,\ M_A = -wa\left(l - \frac{a}{2}\right)$$

Note: For calculation of S.F and B.M of a cantilever beam, we will proceed from free end towards fixed end for all cases, because the reaction at fixed end is unknown.

EXAMPLE 3.1. A cantilever beam 20 m long carries loads of 50 KN, 20 KN and 30 KN at distances 5 m, 15 m and 20 m respectively from the fixed end. Draw the shear force and bending moment diagrams for the cantilever.

Solution.

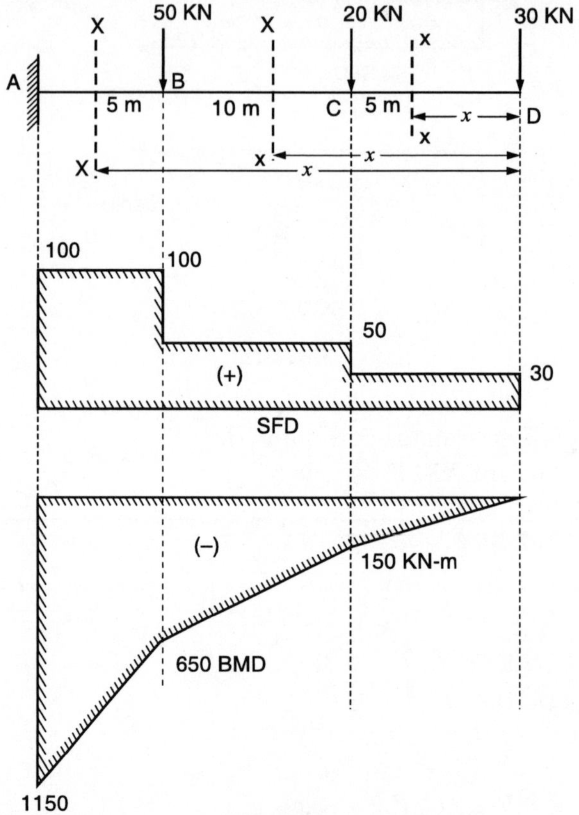

Fig. 3.16.

For part *CD* : Consider a section *XX* at a distance *x* from *D*.

$$\text{S.F. at } XX,\ F_x = +30 \text{ KN}$$
$$\text{B.M. at } XX,\ M_x = -30x \text{ KN-m}$$

At *D*

$$x = 0$$
$$\text{S.F. at } D,\ F_D = +30 \text{ KN}$$
$$\text{B.M. at } D,\ M_D = 0$$

At *C*

$$x = 5$$
$$\text{S.F. at } C,\ F_C = +30 \text{ KN}$$
$$\text{B.M. at } B,\ M_C = -30 \times 5 = -150 \text{ KN-m}$$

For part *BC* : Consider a section *XX* at a distance '*x*' from *D*.

$$\text{S.F. at } XX,\ F_x = +30 + 20 = +50 \text{ KN}$$
$$\begin{aligned}\text{B.M. at } XX,\ M_x &= -30 \times x - 20\,(x - 5)\\ &= -30x - 20x + 100\\ &= (-50x + 100) \text{ KN-m}\end{aligned}$$

At *C*

$$x = 5$$
$$\text{S.F. at } C,\ F_C = +30 \text{ KN}$$
$$\begin{aligned}\text{B.M. at } C,\ M_C &= -50 \times 5 + 100\\ &= -150 \text{ KN-m}\end{aligned}$$

At *B*

$$x = 15$$
$$\text{S.F. at } B,\ F_B = +50 \text{ KN}$$
$$\begin{aligned}\text{B.M. at } B,\ M_B &= (-50 \times 15 + 100)\\ &= -650 \text{ KN-m}\end{aligned}$$

For part *AB* : Consider a section *X-X* at a distance *x* from *D*.

$$\text{S.F. at } XX,\ F_x = +30 + 20 + 50 = +100 \text{ KN}$$
$$\begin{aligned}\text{B.M. at } XX,\ M_x &= -30x - 20\,(x - 5) - 50(x - 15)\\ &= -30x - 20x + 100 - 50x + 750\\ &= (-100x + 850) \text{ KN-m}\end{aligned}$$

At *B*

$$x = 15$$
$$\text{S.F. at } B,\ F_B = +100 \text{ KN}$$
$$\begin{aligned}\text{B.M. at } B,\ M_B &= -100 \times 15 + 850 = -1500 + 850\\ &= -650 \text{ KN-m}\end{aligned}$$

At *A*

$$x = 20$$
$$\text{S.F. at } B,\ F_A = +100 \text{ KN}$$
$$\begin{aligned}\text{B.M. at } A,\ M_A &= (-100 \times 20 + 850)\\ &= -1150 \text{ KN-m.}\end{aligned}$$

EXAMPLE 3.2. Cantilever beam AB of a span 4 m is fixed at *A* and free at *B*. It carries a UDL of 60 N/m for the entire span and two point loads of 200 N and 100 N at the centre and at the free end respectively.

Solution. Let *C* be the centre of the beam *AB* of length 4 m.

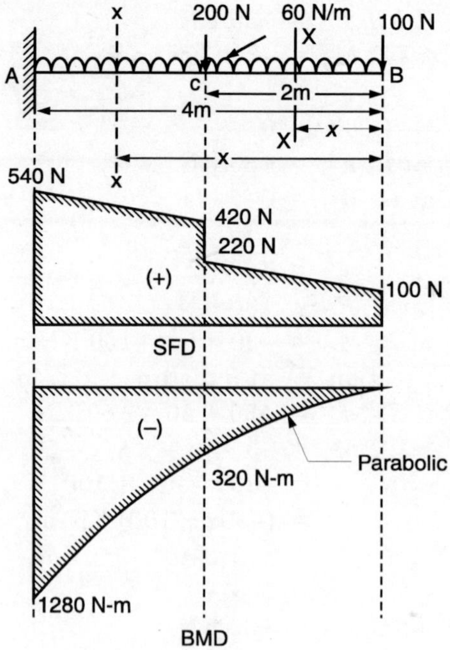

Fig. 3.17.

For part *BC* : Consider a section *XX* at a distance *x* from *B*.

$$\text{S.F. at } XX,\ F_x = +100 + 60 \times x$$
$$= (100 + 60x) \text{ N}$$

$$\text{B.M. at } XX,\ M_x = -100x - 60.x.\frac{x}{2}$$
$$= (-100x - 30x^2) \text{ N-m}$$

At *B*

$$x = 0$$
$$\text{S.F. at } B,\ F_B = (100 + 0)$$
$$= 100\text{N}$$
$$\text{B.M. at } B,\ M_B = (-100 \times 0 - 30 \times 0) = 0$$

At *C*

$$x = 2 \text{ m}$$
$$\text{S.F. at } C,\ F_C = (100 + 60 \times 2)$$
$$= 220 \text{ N}$$
$$\text{B.M. at } C,\ M_C = (-100 \times 2 - 30 \times 2^2)$$
$$= (-200 - 120) = -320 \text{ KN-m}$$

For part *AC* : Consider a section *XX* at a distance *x* from *B*.

$$\text{S.F. at } XX,\ F_x = +100 + 200 + 60x$$
$$= (300 + 60x) \text{ N}$$

$$\text{B.M. at } XX,\ M_x = -100x - 200(x - 2) - 60 \times x \times \frac{x}{2}$$
$$M_x = -100x - 200x + 400 - 30x^2$$
$$= 400 - 300x - 30x^2$$

At C

$$x = 2$$

$$\text{S.F. at } C, F_C = (300 + 60 \times 2) = 420 \text{ N}$$

$$\text{B.M. at } C, M_C = 400 - 300 \times 2 - 30 \times (2)^2$$

$$= 400 - 600 - 120 = -320 \text{ N-m}$$

At A

$$x = 4$$

$$\text{S.F. at } A, F_A = (300 + 60 \times 4) = 540 \text{ N}$$

$$\text{B.M. at } A, M_A = 400 - 300 \times 4 - 30 \times 4^2$$

$$= 400 - 1200 - 480 = -1280 \text{ N-m.}$$

EXAMPLE 3.3. A cantilever 1.5 m long is loaded with a UDL of 2 KN/m over a length of 1.25 m from the free end. It also carries a point load of 3 KN at a distance of 0.25 m from the free end. Draw the SF and BM diagrams of the cantilever.

Solution.

$$AB = 1.5 \text{ m}$$

$$BC = 1.25 \text{ m}$$

$$BD = 0.25 \text{ m}$$

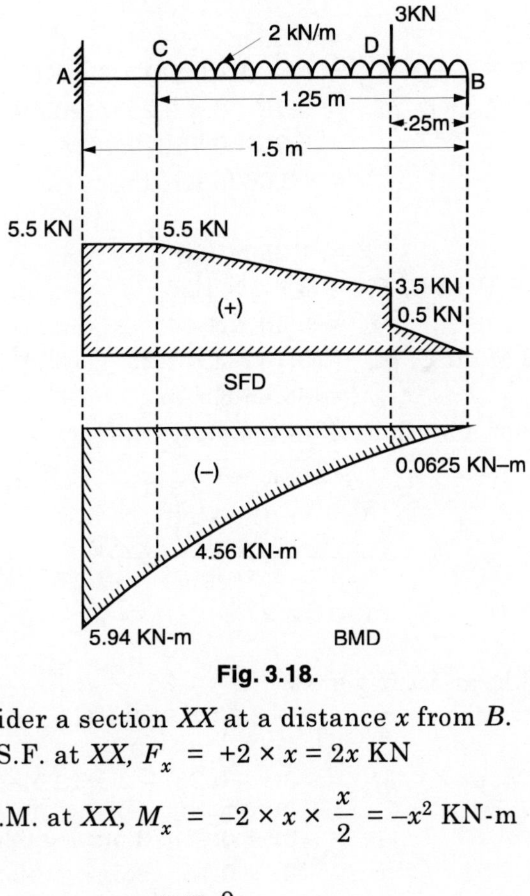

Fig. 3.18.

For part BD : Consider a section XX at a distance x from B.

$$\text{S.F. at } XX, F_x = +2 \times x = 2x \text{ KN}$$

$$\text{B.M. at } XX, M_x = -2 \times x \times \frac{x}{2} = -x^2 \text{ KN-m}$$

At B

$$x = 0$$

$$\text{S.F. at } B, F_B = 0$$

$$\text{B.M. at } B, M_B = 0$$

At *D*

$$x = 0.25 \text{ m}$$
$$\text{S.F. at } D,\ F_D = 2 \times 0.25 = 0.50 \text{ KN}$$
$$\text{B.M. at D},\ M_D = -(0.25)^2 = -0.0625 \text{ KN-m}$$

For part *DC* : Consider a section *XX* at a distance *x* from *B*.

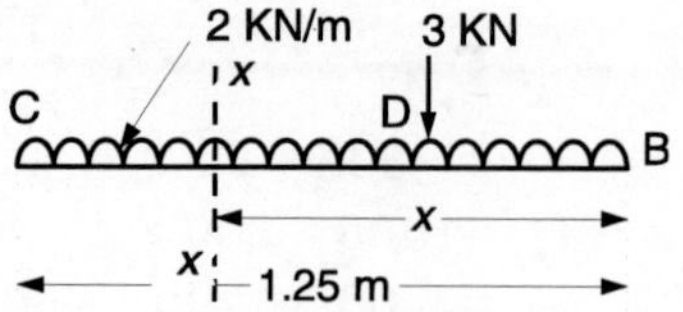

$$\text{S.F. at } XX,\ F_x = +3 + 2 \times x = (3 + 2x) \text{ KN}$$
$$\text{B.M. at } XX,\ M_x = -3 \times (x - 0.25) - 2 \times x \times \frac{x}{2}$$
$$= -3x + 0.75 - x^2$$
$$= 0.75 - 3x - x^2$$

At *D*

$$x = 0.25 \text{ m}$$
$$\text{S.F. at } D,\ F_D = (3 + 2 \times 0.25) = 3.5 \text{ KN}$$
$$\text{B.M. at } D,\ M_D = 0.75 - 3 \times 0.25 - (0.25)^2$$
$$= 0.75 - 0.75 - 0.0625$$
$$= -0.0625 \text{ KN-m}$$

At *C*

$$x = 1.25 \text{ m}$$
$$\text{S.F. at } C,\ F_C = 3 + 2 \times 1.25$$
$$= 5.50 \text{ KN}$$
$$\text{B.M. at } C,\ M_C = 0.75 - 3 \times 1.25 - (1.25)^2$$
$$= -4.56 \text{ KN-m}$$

For part *CA* : Consider a section *XX* at a distance *x* from *B*.

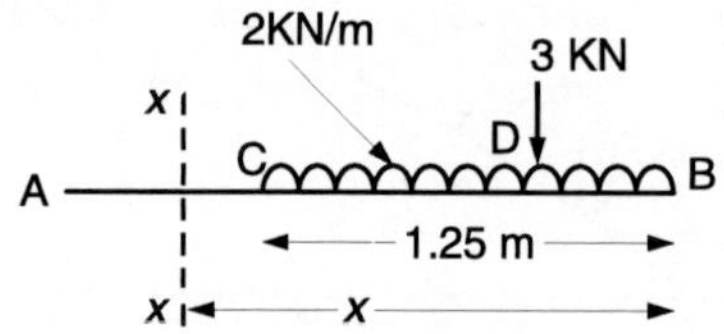

$$\text{S.F. at } XX,\ F_x = +3 + 2 \times 1.25$$
$$= 3 + 2.50 = 5.50 \text{ KN}$$
$$\text{B.M. at } XX,\ M_x = -3(x - 0.25) - 2 \times 1.25 \times \left(x - \frac{1.25}{2}\right)$$
$$= -3x + 0.75 - 2.5(x - 0.625)$$
$$= -3x + 0.75 - 2.5x + 1.56$$
$$= (-5.5x + 2.31) \text{ KN-m}$$

At C

$$x = 1.25 \text{ m}$$
$$\text{S.F. at } C, F_C = 5.50 \text{ KN}$$
$$\text{B.M. at } C, M_C = (-5.5 \times 1.25 + 2.31)$$
$$= -4.56 \text{ KN-m}$$

At A

$$x = 1.5 \text{ m}$$
$$\text{S.F. at } A, F_A = 5.50 \text{ KN}$$
$$\text{B.M. at } A, M_A = (-5.5 \times 1.5 + 2.31)$$
$$= -5.94 \text{ KN-m.}$$

EXAMPLE 3.4. A cantilever of length 6 m carries two point loads of 2 KN and 3 KN at a distance of 1 m and 6 m from the fixed end respectively. In addition to this the beam also carries a UDL of 1 KN/m over a length of 2 m at a distance of 3 m from the fixed end. Draw the S.F. and B.M. diagrams.

Solution.

$$AB = 6 \text{ m}$$
$$AC = 1 \text{ m}$$
$$CD = 2 \text{ m}$$
$$DE = 2 \text{ m}$$
$$EB = 1 \text{ m}$$

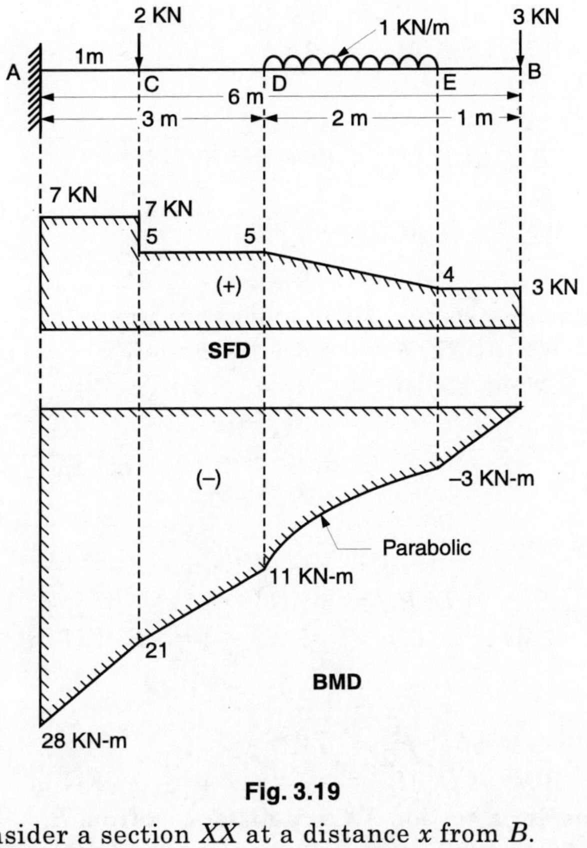

Fig. 3.19

For part BE : Consider a section XX at a distance x from B.

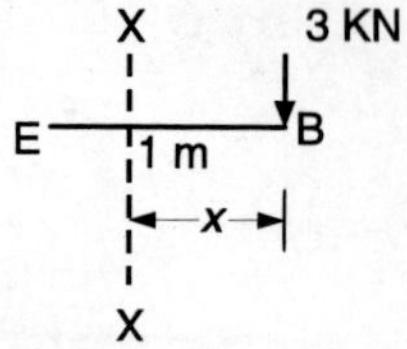

S.F. at XX, $F_x = +3$ KN

B.M. at XX, $M_x = -3x$ KN-m

At *B*

$x = 0$

S.F. at B, $F_B = +3$ KN

B.M. at B, $M_B = 0$

At *E*

$x = 1$

S.F. at E, $F_E = 3$ KN

B.M. at E, $M_E = -3 \times 1 = -3$ KN-m

For part *ED* : Consider a section XX at a distance x from B.

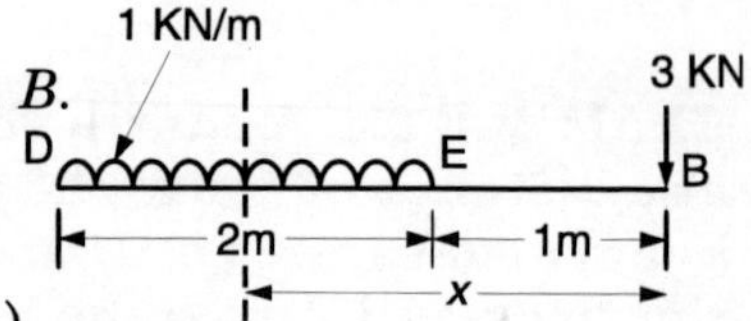

S.F. at XX, $F_x = +3 + 1\,(x - 1)$

$= (2 + x)$ KN

$$\text{B.M. at } XX,\ M_x = -3x - 1.(x-1).\frac{(x-1)}{2} \text{ KN-m}$$

$$= -3x - \frac{(x-1)^2}{2}$$

At *E*

$x = 1$ m

S.F. at E, $F_E = (2 + 1) = 3$ KN

B.M. at E, $M_E = -3 \times 1 - 0 = -3.0$ KN-m

At *D*

$x = 3$ m

S.F. at D, $F_D = (2 + 3) = 5$KN

$$\text{B.M. at } D,\ M_D = -3 \times 3 - \frac{(3-1)^2}{2}$$

$$= -9 - 2 = -11 \text{ KN-m}$$

For part *DC* : Consider a section XX at a distance x from B.

S.F. at XX, $F_x = +3 + 1 \times 2 = 5$ KN

B.M. at XX, $M_x = -3x - 1 \times 2 \times (x - 2)$

$= -3x - 2(x - 2)$

$= -3x - 2x + 4$

$= -5x + 4$

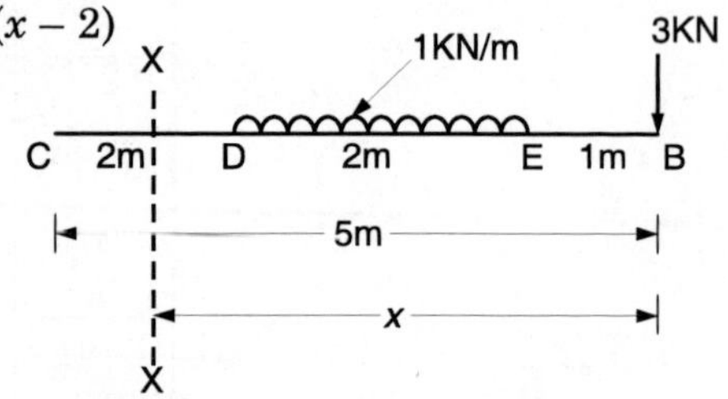

At *D*

$x = 3$ m

S.F. at D, $F_D = 5$ KN

B.M. at D, $M_D = -5 \times 3 + 4 = -11$ KN-m

At *C*

$x = 5$ m

S.F. at C, $F_C = 5$ KN

B.M. at C, $M_C = -5 \times 5 + 4 = -21$ KN-m

For part *CA* : Consider a section XX at a distance x from B.

S.F. at XX, $F_x = +3 + 1 \times 2 + 2 = 7$ KN

$$\begin{aligned} \text{B.M. at } XX,\ M_x &= -3x - 1 \times 2 \times (x-2) - 2 \times (x-5) \\ &= -3x - 2x + 4 - 2x + 10 \\ &= -7x + 14 \end{aligned}$$

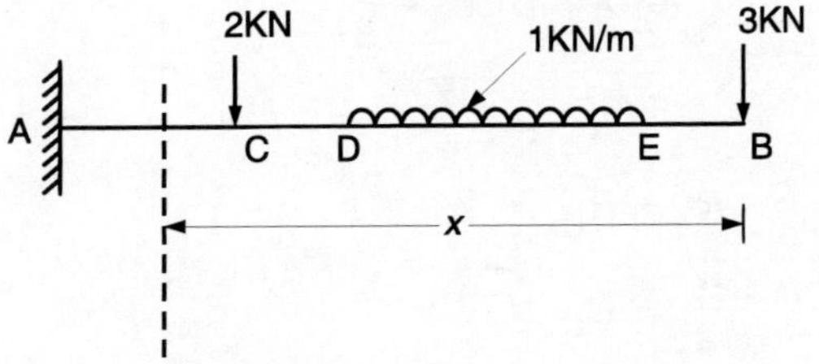

At C $x = 5$

$$\text{S.F. at } C,\ F_C = 7 \text{ KN}$$

$$\text{B.M. at } C,\ M_C = -7 \times 5 + 14 = -21 \text{ KN-m}$$

At A $x = 6$

$$\text{S.F. at } A,\ F_A = +7 \text{ KN}$$

$$\text{B.M. at } A,\ M_A = -7 \times 6 + 14 = -28 \text{ KN-m}$$

3.8 B.M. AND S.F. FOR SIMPLY SUPPORTED BEAM

3.8.1 Simply Supported Beam Carrying a Point Load at Mid Span

Consider a simply supported Beam *AC* of sapn '*l*' carrying a point load '*W*' at mid span, *B*, as shown in the figure 3.20.

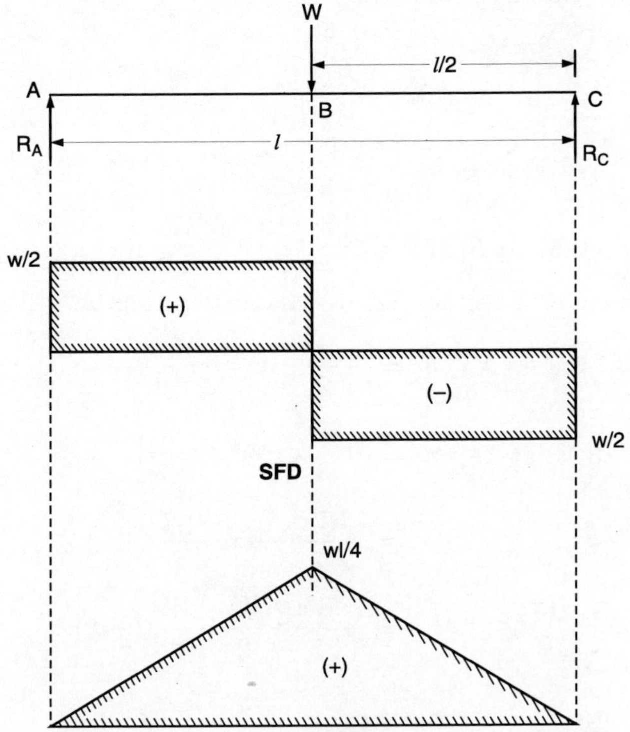

Fig. 3.20.

Let R_A and R_C be the reactions at A and C respectively.
Taking moment at C, we get

$$R_A \times l = W \times \frac{l}{2}$$

$$\therefore \quad R_A = \frac{W}{2}$$

$$R_A + R_C = W$$

$$\frac{W}{2} + R_C = W$$

$$\therefore \quad R_C = \frac{W}{2}$$

For part *AB* : Consider a section XX at a distance x from A.

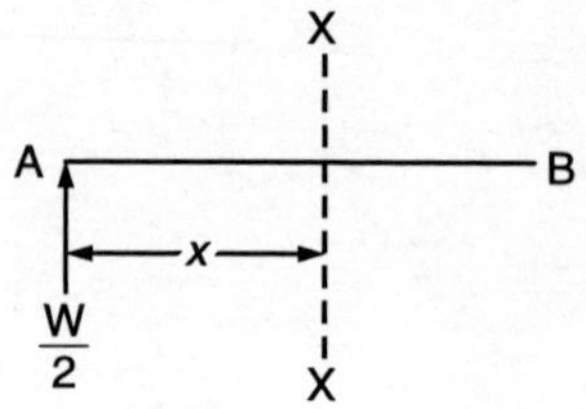

$$\text{S.F. at } XX,\ F_x = +\frac{W}{2}$$

$$\text{B.M. at } XX,\ M_x = -\frac{Wx}{2}$$

At *A*

$$x = 0$$

$$\text{S.F. at } A,\ F_A = +\frac{W}{2}$$

$$\text{B.M. at } A,\ M_A = 0$$

At *B*

$$x = \frac{l}{2}$$

$$\text{S.F. at } B,\ F_B = +\frac{W}{2}$$

$$\text{B.M. at } B,\ M_B = +\frac{W}{2}\left(\frac{l}{2}\right) = +\frac{Wl}{4}$$

For part *BC* : Consider a section XX at a distance x from A.

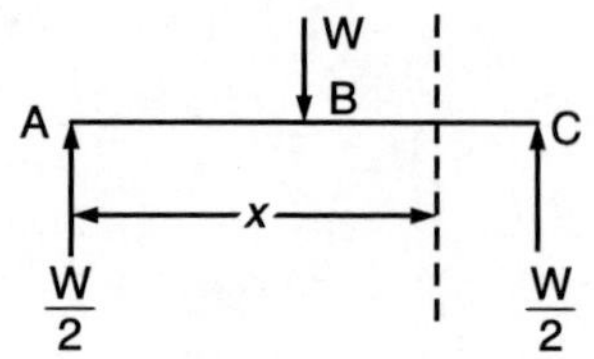

$$\text{S.F. at } XX,\ F_x = +\frac{W}{2} - W = -\frac{W}{2}$$

$$\text{B.M. at } XX,\ M_x = +\frac{W}{2}x - W\left(x - \frac{l}{2}\right)$$

$$= +\frac{Wx}{2} - Wx + \frac{Wl}{2}$$

$$= -\frac{Wx}{2} + \frac{Wl}{2} = \frac{W}{2}(l - x)$$

At *B*

$$x = \frac{l}{2}$$

$$\text{S.F. at } B, F_B = -\frac{W}{2}$$

$$\text{B.M. at } B, M_B = \frac{W}{2}\left(l - \frac{l}{2}\right) = +\frac{Wl}{4}$$

At *C*

$$x = l$$

$$\text{S.F. at } C, F_C = -\frac{W}{2}$$

$$\text{B.M. at } C, M_C = \frac{W}{2}(l - l) = 0$$

It is clear from the shear force and bending moment diagram that bending moment is maximum at the point where shear force changes its sign.

Note: For simply supported beam carrying a point load '*W*' at the mid span, maximum B.M. $= \frac{Wl}{4}$ and will be a the mid span.

3.8.2 Simply Supported Beam Carrying an Ecentric Point Load

Consider a simply supported beam *AC* carrying a point load '*W*' at a distance '*a*' from *A*. Let R_A and R_C be the reaction at *A* and *C* respectively.

$$\therefore \quad R_A + R_C = W \quad \text{... (i)}$$

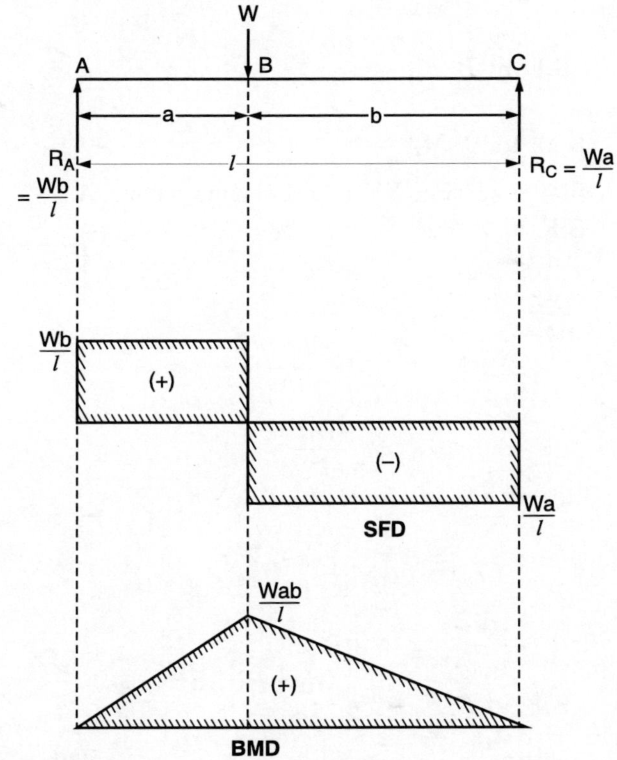

Fig. 3.21.

Taking moment about *C*, we get

$$R_A + l = W \times b$$

$$\therefore \qquad R_A = \frac{Wb}{l}$$

Putting the value of R_A is (i), we get

$$\frac{Wb}{l} + R_C = W$$

$$\therefore \qquad R_C = W - \frac{Wb}{l} = W\frac{(l-b)}{l} = \frac{Wa}{l} \qquad [\because l = a + b]$$

For part *AB* : Consider a section *AB* at a distance '*x*' from *A*.

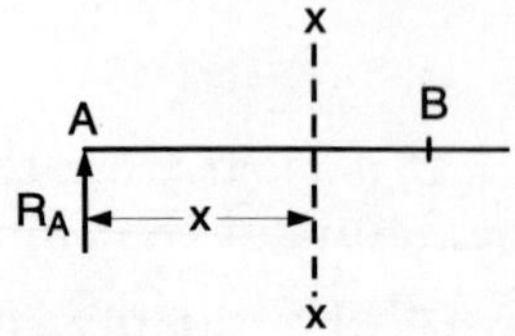

$$\text{S.F. at } XX,\ F_x = \frac{Wb}{l}$$

$$\text{B.M. at } XX,\ M_x = R_A x = \frac{Wb}{l}x$$

At *A*

$$x = 0$$

$$\text{S.F. at } A,\ F_A = \frac{Wb}{l}$$

$$\text{B.M. at } A,\ M_A = 0$$

At *B*

$$x = a$$

$$\text{S.F. at } B,\ F_B = \frac{Wb}{l}$$

$$\text{B.M. at } B,\ M_B = \frac{Wb}{l} \times a = \frac{Wab}{l}$$

For part *BC* : Consider a section *XX* at a distance *x* from *A*.

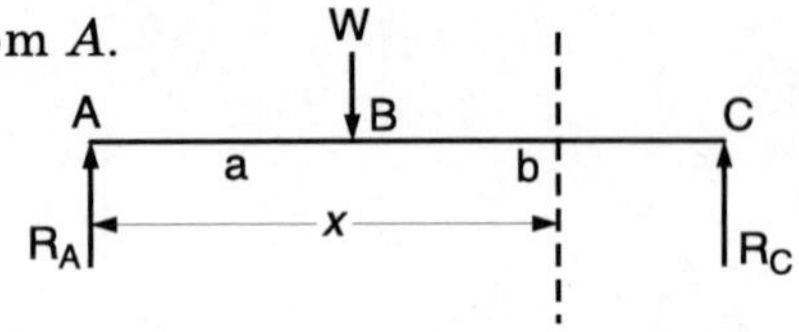

$$\text{S.F. at } XX,\ F_x = R_A - W$$

$$= \frac{Wb}{l} - W$$

$$= W\left(\frac{b-l}{l}\right)$$

$$= \frac{-Wa}{l} \qquad [\because l = a + b]$$

$$\text{B.M. at } XX,\ M_x = R_A x - W(x-a)$$

$$= \frac{Wb}{l}x - W(x-a)$$

At *B*

$$x = a$$

$$\text{S.F. at } B,\ F_B = -\frac{Wa}{l}$$

$$\text{B.M. at } B,\ M_B = \frac{Wb}{l}a \times - W(a-a) = \frac{Wab}{l}$$

At *C*

$$x = l$$

$$\text{S.F. at } C,\ F_C = -\frac{Wa}{l}$$

$$\text{B.M. at } B,\ M_C = \frac{Wb}{l} \times l - W(l-a)$$

$$= Wb - Wb \qquad [\because l = a + b]$$

$$= 0.$$

3.8.3 Simply Supported Beam Carrying a Uniformly Distributed Load

Consider a simply supported beam AB of length 'l' carrying a UDL of w/unit length over the entire length.

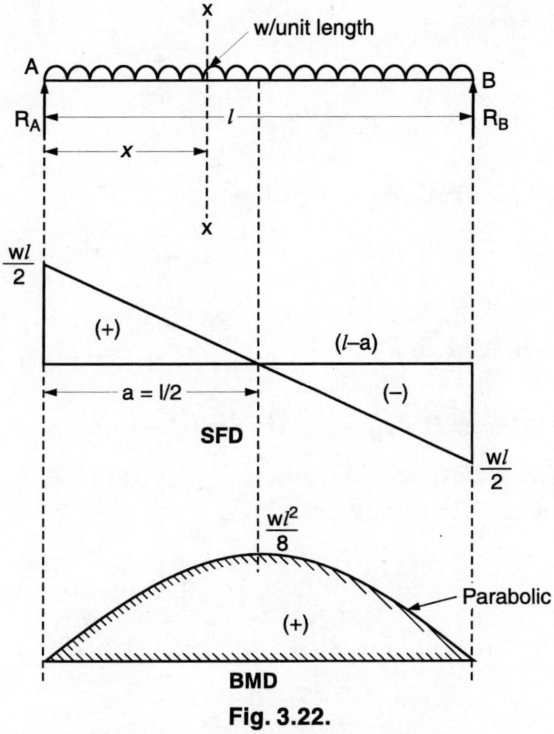

Fig. 3.22.

Let R_A and R_B be the reactions at A and B respectively.

$$R_A + R_B = wl \qquad \ldots(i)$$

Taking moment about B

$$R_A \times l = w \times l \times \frac{l}{2}$$

$$R_A = \frac{wl}{2}$$

Putting the value of R_A in (i), we get

$$\frac{wl}{2} + R_B = wl$$

$$\therefore \qquad R_B = wl - \frac{wl}{2} = \frac{wl}{2}$$

Consider a section XX at a distance x from A.

$$\text{S.F. at } XX,\ F_x = R_A - wx$$

$$= \frac{wl}{2} - wl = w\left(\frac{l}{2} - x\right)$$

$$\text{B.M. at } XX,\ M_x = R_A x - wx.\frac{x}{2}$$

$$= \frac{wl}{2}x - \frac{wx^2}{2}$$

$$= \frac{w}{2}(lx - x^2)$$

At *A*

$$x = 0$$

$$\text{S.F. at } A,\ F_A = w\left(\frac{l}{2} - 0\right) = \frac{wl}{2}$$

$$\text{B.M. at } A,\ M_A = \frac{w}{2}(0) = 0$$

At *B*

$$x = l$$

$$\text{S.F. at } B,\ F_B = w\left(\frac{l}{2} - l\right) = \frac{-wl}{2}$$

$$\text{B.M. at } B,\ M_B = \frac{w}{2}(l \times l - l^2) = 0$$

Let the S.F. be zero at a distance '*a*' from *A*.
Applying principle of similar triangles

$$\frac{a}{l-a} = \frac{\frac{wl}{2}}{\frac{wl}{2}} = 1$$

or $$a = l - a$$

or $$2a = l$$

$$\therefore \quad a = \frac{l}{2}$$

The S.F. is zero at the mid span of the beam.
The point where S.F. is zero, the B.M. will be maximum.

$$\therefore \quad \text{Maximum B.M.} = R_A \times \frac{l}{2} - w.\frac{l}{2}.\frac{l}{4}$$

$$= \frac{wl}{2}.\frac{l}{2} - \frac{wl^2}{8}$$

$$= \frac{wl}{4} - \frac{wl^2}{8} = \frac{2wl^2 - wl^2}{8} = \frac{wl^2}{8}$$

Note: For simply supported beam, we can proceed from either end, *i.e.*, LHS or RHS of the beam for calculation of S.F and B.M, because reactions at both the ends are known.

3.8.4 Simply Supported Beam Carrying a Load whose Intensity Varies Uniformly from Zero at One End to *w* per Unit Run at the other End

Consider a simply supported beam *AB* of length '*l*' carrying a uniformly varying load of intensity *w*/unit length over the entire length of the beam as shown in figure 3.23.

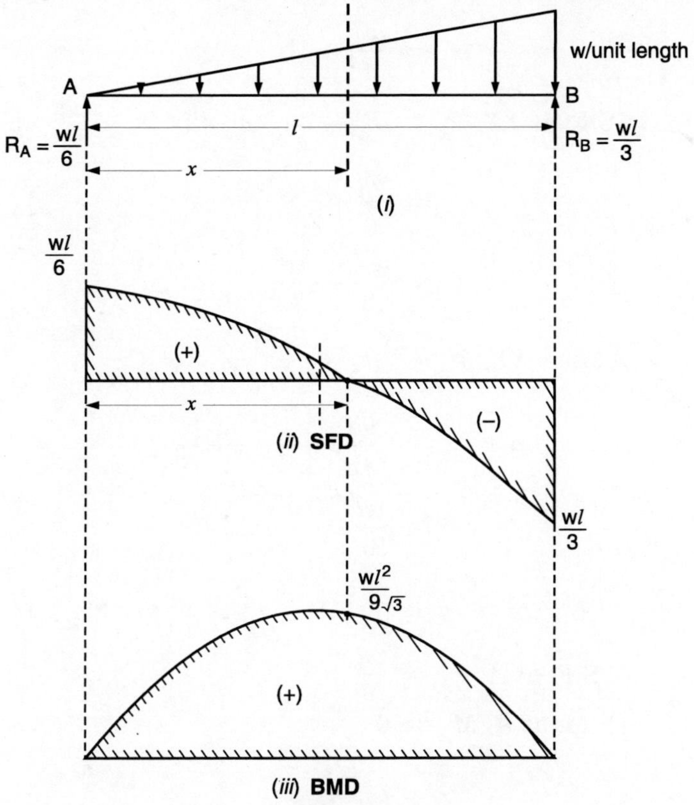

Fig. 3.23.

Let R_A and R_B be the reactions at A and B respectively.

$$R_A + R_B = \frac{1}{2} \times w \times l \qquad ...(i)$$

Taking moment about B, we get

$$R_A \times l = \frac{1}{2} \times w \times l \times \frac{l}{3}$$

$$\therefore \qquad R_A = \frac{\omega l^2}{6 \times l} = \frac{\omega l}{6}$$

Putting the value of R_A in (*i*), we get

$$\frac{wl}{6} + R_B = \frac{wl}{2}$$

$$\therefore \qquad R_B = \frac{wl}{2} - \frac{wl}{6}$$

$$= \frac{6wl - wl}{6} = \frac{wl}{3}$$

Consider a section XX at a distance 'x' from A.

Let the *Intensity of loading at XX* be w_x. (Refer figure 3.23 (*i*))

$$\frac{w_x}{w} = \frac{x}{l} \quad \text{[Applying laws of similar triangles]}$$

$$\therefore \quad w_x = \frac{wx}{l}$$

$$\text{S.F. at } XX,\ F_x = R_A - \frac{1}{2} \times w_x \times x$$

$$= \frac{wl}{6} - \frac{1}{2} \times \frac{wx}{l} \times x$$

$$= \frac{wl}{6} - \frac{wx^2}{2l}$$

$$\text{B.M. at } XX,\ M_x = R_A \times x - \frac{1}{2} w_x \times x \times \frac{x}{3}$$

$$= \frac{wl}{6}x - \frac{1}{2} \times \frac{wx}{l} . \frac{x^2}{3}$$

$$= \frac{wl}{6}x - \frac{wx^3}{6l}$$

At A

$$x = 0$$

$$\text{S.F. at } A,\ F_A = \frac{wl}{6}$$

$$\text{B.M. at } A,\ M_A = 0$$

At B

$$x = l$$

$$\text{S.F. at } B,\ F_B = \frac{wl}{6} - \frac{w \times l^2}{2l}$$

$$= \frac{wl}{6} - \frac{wl}{2}$$

$$= \frac{wl - 3wl}{6} = \frac{-2wl}{6} = \frac{-wl}{3}$$

$$\text{B.M. at } B,\ M_B = \frac{wl}{6} \times l - \frac{wl^3}{6l}$$

$$= \frac{wl^2}{6} - \frac{wl^2}{6} = 0$$

Let the S.F. be the zero at a distance 'x' from A.

Equating the S.F. at XX to zero

$$\frac{wl}{6} - \frac{wx^2}{2l} = 0$$

or $$\frac{wl}{6} = \frac{wx^2}{2l}$$

or $$\frac{wl^2}{3} = wx^2$$

or $$\frac{l^2}{3} = x^2$$

$\therefore$ $$x = \frac{l}{\sqrt{3}} = 0.577l$$

S.F. is zero at a distance of $\frac{l}{\sqrt{3}}$ from A. The B.M. will be maximum at this point.

$$\text{Maximum bending moment} = \frac{wl}{6}\left(\frac{l}{\sqrt{3}}\right) - \frac{w}{6l}\left(\frac{l}{\sqrt{3}}\right)^3$$

$$= \frac{wl^2}{6\sqrt{3}} - \frac{wl^2}{18\sqrt{3}} = \frac{3wl^2 - wl^2}{18\sqrt{3}}$$

$$= \frac{2wl^2}{18\sqrt{3}} = +\frac{wl^2}{9\sqrt{3}}$$

Note: For simply supported beam carrying a load whose intensity varies uniformly from zero at one end to some intensity at other end, we will proceed for calculation of S.F and B.M from the end at which intensity of loading is zero.

EXAMPLE 3.5. A simply supported beam of 16 m effective span carries concentrated loads of 400 N, 500 N and 300 N at distances 3 m, 7 m and 11 m respectively from left support. Calculate the maximum S.F. and B.M. Draw SF and BM diagrams also.

Solution. Let R_A and R_B be the reactions at A and B respectively.

$$R_A + R_B = 400 + 500 + 300$$
$$= 1200 \text{ N} \qquad \ldots(i)$$

Taking moment about B

$$R_A \times 16 = 300 \times 5 + 500 \times 9 + 400 \times 13$$
$$R_A \times 16 = 1500 + 4500 + 5200$$
$$R_A = \frac{11200}{16} = 700 \text{ N}$$

Putting the value of R_A in (*i*), we get

$$700 + R_B = 1200$$

$\therefore$ $$R_B = 500 \text{ N}$$

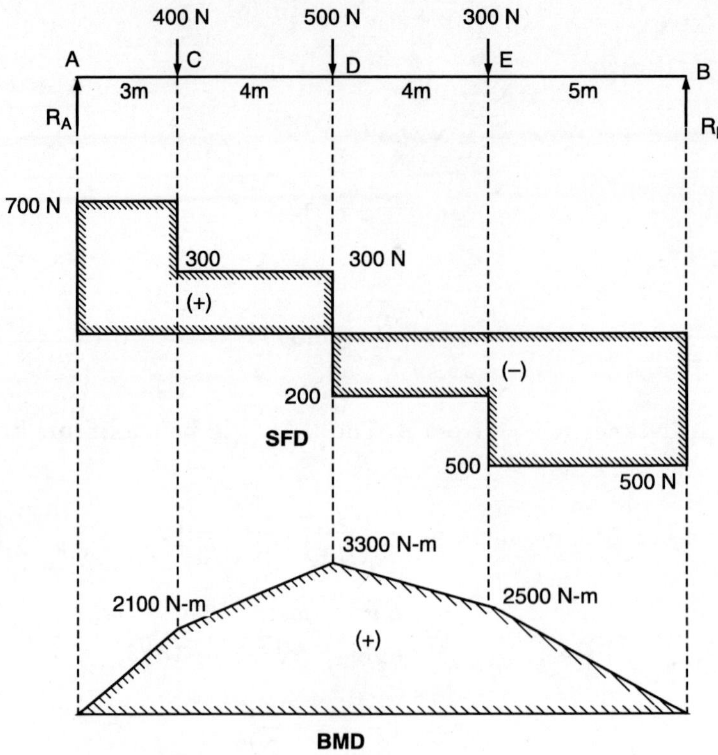

Fig. 3.24.

For part *AC* : Consider a section *XX* at a distance *x* from *A*.

$$\text{S.F. at } XX,\ F_x = R_A = 700 \text{ N}$$
$$\text{B.M. at } XX,\ M_x = R_A x = 700x$$

At *A*

$$x = 0$$
$$\text{S.F. at } A,\ F_A = 700 \text{ N}$$
$$\text{B.M. at } A,\ M_A = 0$$

At *C*

$$x = 3$$
$$\text{S.F. at } C,\ F_C = 700 \text{ N}$$
$$\text{B.M. at } C,\ M_C = 700 \times 3 = 2100 \text{ N-m}$$

For part *CD* : Consider a section *XX* at a distance *x* from *A*.

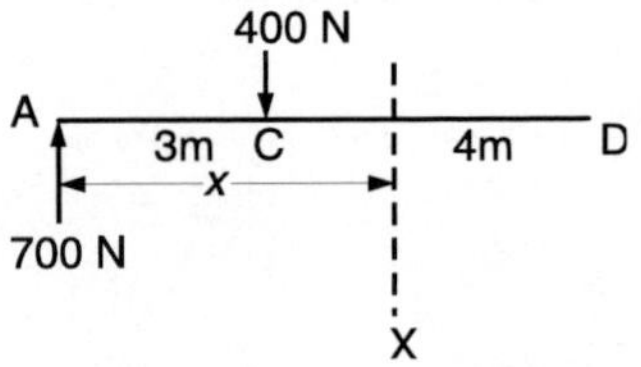

$$\text{S.F. at } XX,\ F_x = 700 - 400 = 300 \text{ N}$$
$$\begin{aligned}\text{B.M. at } XX,\ M_x &= 700x - 400(x - 3)\\ &= 700x - 400x + 1200\\ &= 300x + 1200\end{aligned}$$

At *C*

$$x = 3$$
$$\text{S.F. at } C,\ F_C = 300 \text{ N}$$
$$\text{B.M. at } C,\ M_C = 300 \times 3 + 1200 = 2100 \text{ N-m}$$

At *D*

$$x = 7$$
$$\text{S.F. at } D, F_D = 300 \text{ N}$$
$$\text{B.M. at } D, M_D = 300 \times 7 + 1200 = 3300 \text{ N-m}$$

For part *DE* : Consider a section *XX* at a distance x from A.

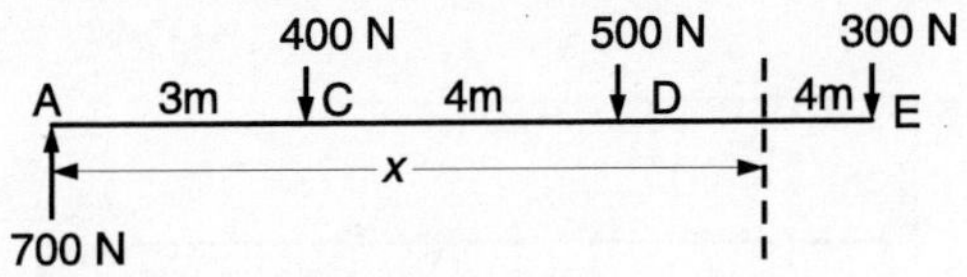

$$\text{S.F. at } XX, F_x = 700 - 400 - 500$$
$$= -200 \text{ N}$$
$$\text{B.M. at } XX, M_x = 700x - 400(x-3) - 500(x-7)$$
$$= 700x - 400x + 1200 - 500x + 3500$$
$$= -200x + 4700$$

At *D*

$$x = 7$$
$$\text{S.F. at } D, F_D = -200 \text{ N}$$
$$\text{B.M. at } D, M_D = -200 \times 7 + 4700$$
$$= -1400 + 4700$$
$$= 3300 \text{ N-m}$$

At *E*

$$x = 11$$
$$\text{S.F. at } E, F_E = -200 \text{ N}$$
$$\text{B.M. at } E, M_E = -200 \times 11 + 4700$$
$$= 2500 \text{ N-m}$$

For part *EB* : Consider a section *XX* at a distance x from A.

$$\text{S.F. at } XX, F_x = 700 - 400 - 500 = -500 \text{ N}$$
$$\text{B.M. at } XX, M_x = 700x - 400(x-3) - 500(x-7) - 300(x-11)$$
$$= 700x - 400x + 1200 - 500x + 3500 - 300x + 3300$$
$$= -500x + 8000$$

At *E*

$$x = 11$$
$$\text{S.F. at } E, F_E = -500 \text{ N}$$
$$\text{B.M. at } E, M_E = -500 \times 11 + 8000$$
$$= -5500 + 8000 = 2500 \text{ N-m}$$

At *B*

$$x = 16$$
$$\text{S.F. at } B, F_B = -500 \text{ N}$$
$$\text{B.M. at } B, M_B = -500 \times 16 + 8000 = 0.$$

EXAMPLE 3.6. A simply supported beam is carrying a UDL of 2 KN/m over a length of 3 m from the right end. The length of the beam is 6 m. Draw the SF and BM diagrams for the beam and also calculate the maximum BM at the section.

Solution. Let R_A and R_B be the reactions at A and B respectively.

$$R_A + R_B = 2 \times 3 = 6 \qquad ...(i)$$

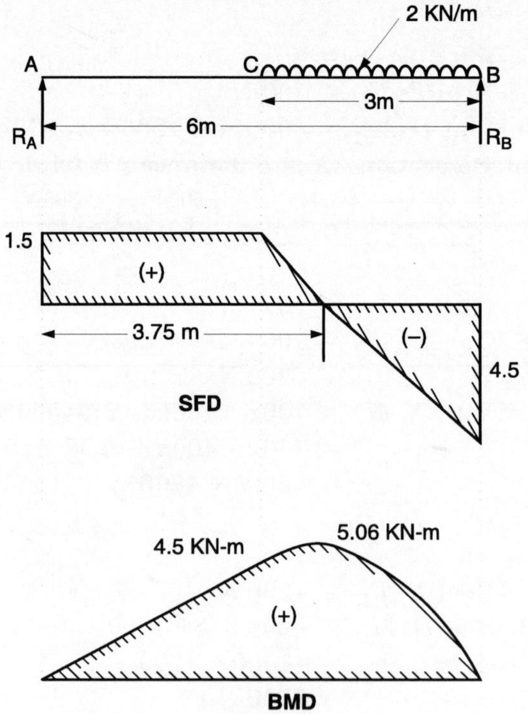

Fig. 3.25.

Taking moment about B, we get

$$R_A \times 6 = 2 \times 3 \times 1.5$$

$$R_A = \frac{9}{6} = 1.5 \text{ KN}$$

Putting the value of R_A in (*i*), we get

$$1.5 + R_B = 6$$

$$\therefore \quad R_B = 4.5 \text{ KN}$$

For part *AC*

$$\text{S.F. at } A, F_A = 1.5 \text{ KN}$$

$$\text{B.M. at } A, M_A = 0$$

$$\text{S.F. at } C, F_C = 1.5 \text{ KN}$$

$$\text{B.M. at } C, M_C = 1.5 \times 3 = 4.5 \text{ KN-m}$$

At part *CB*

$$\text{S.F. at } C, F_C = 1.5 \text{ KN}$$

$$\text{B.M. at } C, M_C = 4.5 \text{ KN-m}$$

$$\text{S.F. at } B, F_B = 1.5 - 2 \times 3 = -4.5 \text{ KN}$$

$$\text{B.M. at } B, M_B = 1.5 \times 6 - 2 \times 3 \times 1.5$$

$$= 9 - 9 = 0$$

Let the SF be zero at a distance x from A.

$$R_A - 2 \times (x - 3) = 0$$

$$1.5 - 2x + 6 = 0$$

$$2x = 7.5$$

$$x = \frac{7.5}{2} = 3.75 \text{ m}$$

$$\text{Maximum B.M.} = 1.5 \times 3.75 - 2 \times 0.75 \times \frac{0.75}{2}$$

$$= 5.625 - 0.56$$

$$= 5.065 \text{ KN-m.}$$

EXAMPLE 3.7. Draw the SF and BM diagrams for the beam loaded as shown in the figure 3.26. Find the maximum bending moment and its position.

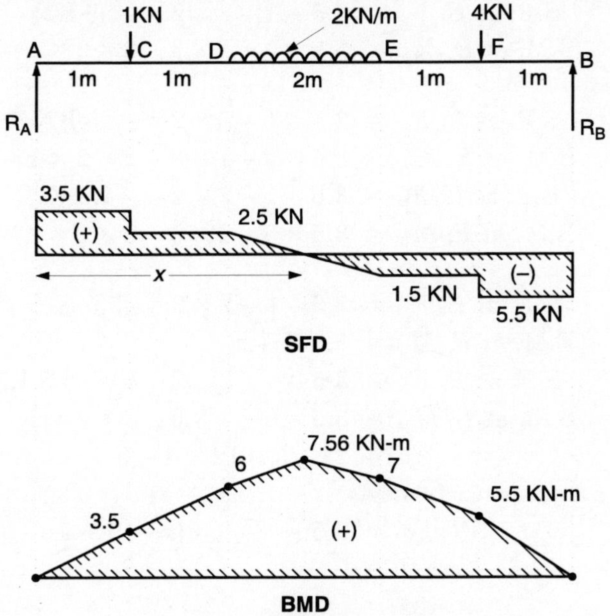

Fig. 3.26.

Solution. Let R_A and R_B be the reactions A and B respectively

$$R_A + R_B = 1 + 2 \times 2 + 4$$

$$R_A + R_B = 9 \text{ KN} \quad \ldots(i)$$

Taking moment about B

$$R_A \times 6 = 1 \times 5 + 2 \times 2 \times 3 + 4 \times 1$$

$$R_A \times 6 = 5 + 12 + 4 = 21$$

$$\therefore \quad R_A = \frac{21}{6} = 3.5 \text{ KN}$$

Putting the value of R_A in (*i*), we get

$$3.5 + R_B = 9$$

$$\therefore \quad R_B = 9 - 3.5 = 5.5 \text{ KN}$$

For part *AC*

$$\text{S.F. at } A,\ F_A = 3.5 \text{ KN}$$

$$\text{B.M. at } A,\ M_A = 0$$

$$\text{S.F. at } C,\ F_C = 3.5 \text{ KN}$$

$$\text{B.M. at } C,\ M_C = 3.5 \text{ KN-m}$$

For part *CD*

$$\text{S.F. at } C, F_C = 3.5 - 1 = 2.5 \text{ KN}$$
$$\text{B.M. at } C, M_C = 3.5 \times 1 = 3.5 \text{ KN-m}$$
$$\text{S.F. at } D, F_D = 2.5 \text{ KN}$$
$$\text{B.M. at } D, M_D = 3.5 \times 2 - 1 \times 1$$
$$= 7 - 1 = 6 \text{ KN-m}$$

For part *DE*

$$\text{S.F. at } D, F_D = 3.5 - 1 = 2.5 \text{ KN}$$
$$\text{B.M. at } D, M_D = 6 \text{ KN-m}$$
$$\text{S.F. at } E, F_E = 3.5 - 1 - 2 \times 2 = -1.5 \text{ KN}$$
$$\text{B.M. at } E, M_E = 3.5 \times 4 - 1 \times 3 - 2 \times 2 \times 1 = 7 \text{ KN-m}$$

For part *EF*

$$\text{S.F. at } E, F_E = 3.5 - 1 - 2 \times 2 = -1.5 \text{ KN}$$
$$\text{B.M. at } E, M_E = 3.5 \times 9 - 1 \times 3 - 2 \times 2 \times 1 = 7 \text{ KN-m}$$
$$\text{S.F. at } F, F_F = 3.5 - 1 - 2 \times 2 = -1.5 \text{ KN}$$
$$\text{B.M. at } F, M_F = 3.5 \times 5 - 1 \times 4 - 2 \times 2 \times 2 = 5.5 \text{ KN-m}$$

For part *FB*

$$\text{S.F. at } E, F_F = 3.5 - 1 - 2 \times 2 = -5.5 \text{ KN}$$
$$\text{B.M. at } F, M_F = 5.5 \text{ KN-m}$$
$$\text{S.F. at } B, F_B = 3.5 - 1 - 2 \times 2 - 4 = -5.5 \text{ KN}$$
$$\text{B.M. at } B, M_B = 3.5 \times 6 - 1 \times 5 - 2 \times 2 \times 3 - 4 \times 1$$
$$= 21 - 5 - 12 - 4 = 0$$

Let SF be zero at a distance x from A.

$$\text{S.F. at } x = 3.5 - 1 - 2 \times (x - 2) = 0$$
$$2.5 - 2x + 4 = 0$$
$$-2x = -6.5$$
$$\therefore \quad x = 3.25 \text{ m}$$

B.M. will be maximum at the point where S.F. is zero, *i.e.*, at 3.25 m from point A.

$$\text{Maximum BM} = 3.5 \times 3.25 - 1(3.25 - 1) - 2(3.25 - 2) \times \frac{(3.25 - 2)}{2}$$
$$= 11.375 - 2.25 - 1.562$$
$$= 7.563 \text{ KN-m.}$$

3.9 SHEAR FORCE AND BENDING MOMENT DIAGRAMS FOR OVERHANGING BEAMS

A beam is said to be overhanging beam if the end of the beam extend beyond the support. The bending moment is positive between the two supports, whereas the bending moment is negative for the overhanging portion. The bending moment is zero at some point while changing from +ve to –ve or –ve to +ve. The point where bending moment is zero, is known as **point of contraflexure**. Point of contraflexure is always in overhanging beam.

Point of contraflexure or *point of inflexion* is the point on a beam where the bending moment is zero after changing its sign from +ve to –ve or vice versa.

EXAMPLE 3.8. A cantilever of length 6 m carries a gradually varying load, zero at the free end to 2 KN/m at the fixed end. Draw the SF and BM diagrams for the cantilever.

Solution. Consider a section *XX* at a distance '*x*' from *B*.

Intensity of loading at *XX* $w_x = \frac{x}{6} \times 2 = \frac{x}{3}$ KN/m

$$\text{S.F. at } XX,\ F_x = \frac{1}{2} \times \left(\frac{x}{3}\right) \times x$$

$$= \frac{x^2}{6} \text{ KN}$$

$$\text{B.M. at } XX,\ M_x = \frac{1}{2}\left(\frac{x}{3}\right) \times x \times \frac{x}{3}$$

$$= \frac{x^3}{18} \text{ KN-m}$$

At *B*

$$x = 0$$

$$\text{S.F. at } B = 0$$

$$\text{B.M. at } B = 0$$

At *C*

$$x = 6$$

$$\text{S.F. at } C,\ F_C = \frac{6^2}{6} = 6 \text{ KN}$$

$$\text{B.M. at } C,\ M_C = \frac{6^3}{18} = 12 \text{ KN-m.}$$

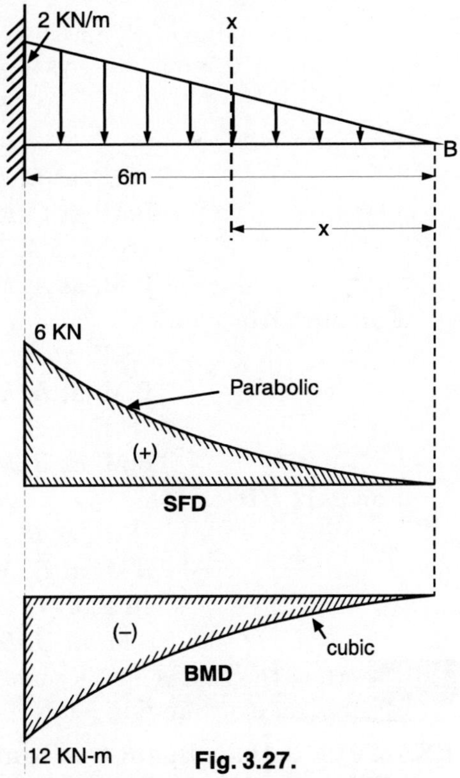

Fig. 3.27.

EXAMPLE 3.9. A beam of 10 m length is supported at two ends 6 m apart and has overhangs of 2 m on each end and carries point loads of 15 N at each free end. Draw SF and BM diagrams for the beam.

Solution. Let R_A and R_B be the reactions at *A* and *B* respectively.

$$R_A + R_B = 30 \text{ N} \quad \ldots(i)$$

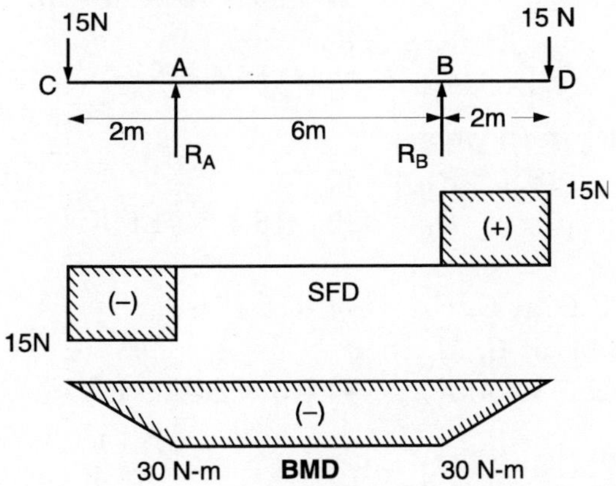

Fig. 3.28.

Taking moment about B,

$$R_A \times 6 = 15 \times 8 - 15 \times 2 = 90$$

$$\therefore \quad R_A = \frac{90}{6} = 15 \text{ N}$$

Putting the value of R_A in (*i*) we get

$$15 + R_B = 30$$

$$\therefore \quad R_B = 15 \text{ N}$$

For part *CA*

$$\text{S.F. at } C,\ F_C = -15 \text{ N}$$
$$\text{B.M. at } C,\ M_C = 0$$
$$\text{S.F. at } A,\ F_A = -15 \text{ N}$$
$$\text{B.M. at } A,\ M_A = -15 \times 2 = -30 \text{ N-m}$$

For part *AB*

$$\text{S.F. at } A,\ F_A = -15 + 15 = 0$$
$$\text{B.M. at } A,\ M_A = -30 \text{ N-m}$$
$$\text{S.F. at } B,\ F_B = -15 + 15 = 0$$
$$\text{B.M. at } B,\ M_B = -15 \times 8 + 15 \times 6 = -30 \text{ N-m}$$

For part *BD*

$$\text{S.F. at } B,\ F_B = -15 + 15 + 15 = +15 \text{ N}$$
$$\text{B.M. at } B,\ M_B = -15 \times 8 + 15 \times 6 = -30 \text{ N-m}$$
$$\text{S.F. at } D,\ F_D = -15 + 15 + 15 = +15$$
$$\text{B.M. at } D,\ M_D = -15 \times 10 + 15 \times 8 + 15 \times 2 = 0.$$

Note: Here the B.M remains constant between A and B. The B.M.D is entirely – ve though it is an overhanging beam.

EXAMPLE 3.10. A beam 8 m long is loaded as shown in the figure 3.29. Draw SF and BM diagrams for the beam and find the points of contraflexure.

Solution. Let R_A and R_B be the reactions at A and B respectively.

$$R_A + R_B = 4 + 6 + 5 \times 8 = 50 \text{ KN} \qquad \ldots(i)$$

Taking moment about B, we get

$$R_A \times 5 = 4 \times 6 + 5 \times 6 \times 3 - 6 \times 2 - 5 \times 2 \times 1$$
$$= 24 + 90 - 12 - 10 = 92$$

$$R_A = \frac{92}{5} = 18.4 \text{ KN}$$

Putting the value of R_A in (*i*) we get

$$18.4 \times R_B = 50$$

$$\therefore \quad R_B = 50 - 18.4 = 31.6 \text{ KN}$$

For part *CA*

$$\text{S.F. at } C,\ F_C = -4 \text{ KN}$$
$$\text{B.M. at } C,\ M_C = 0$$
$$\text{S.F. at } A,\ F_A = -4 - 5 \times 1 = -9 \text{ KN}$$
$$\text{B.M. at } A,\ M_A = -4 \times 1 - 5 \times 1 \times \frac{1}{2}$$
$$= -4 - 2.5 = -6.5 \text{ KN-m}$$

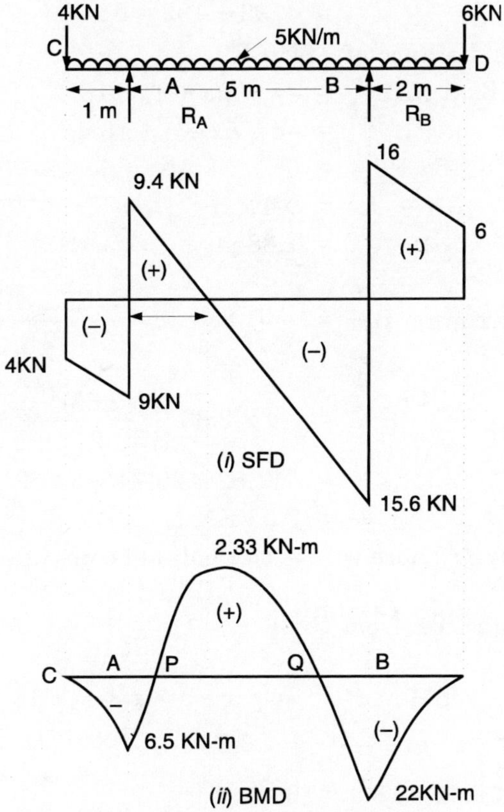

Fig. 3.29.

For part *AB*

$$\text{S.F. at } A, F_A = -4 - 5 \times 1 + 18.4$$
$$= 9.4 \text{ KN}$$
$$\text{B.M. at } A, M_A = -6.5 \text{ KN-m}$$
$$\text{S.F. at } B, F_B = -4 + 18.4 - 5 \times 6$$
$$= -34 + 18.4 = -15.6 \text{ KN}$$
$$\text{B.M. at } B, M_B = -4 \times 6 + 18.4 \times 5 - 5 \times 6 \times 3$$
$$= -22 \text{ KN-m}$$

For part *BD*

$$\text{S.F. at } B, F_B = -4 - 5 \times 6 + 18.4 + 31.6 = 16$$
$$\text{B.M. at } B, M_B = -22 \text{ KN-m}$$
$$\text{S.F. at } D, F_D = -4 - 5 \times 8 + 18.4 + 31.6$$
$$= -44 + 50 = 6 \text{ KN}$$
$$\text{B.M. at } D, M_D = -4 \times 8 - 5 \times 8 \times 4 + 18.4 \times 7 + 31.6 \times 2$$

$$= -32 - 160 + 128.8 + 63.2$$
$$= -192 + 192 = 0$$

Let the SF be zero at a distance of x from A.

$$\text{S.F. at } XX,\ F_x = -4 - 5(x + 1) + 18.4$$
$$\Rightarrow \quad = -4 - 5x - 5 + 18.4$$
$$\Rightarrow \quad -5x + 9.4 = 0$$
$$5x = 9.4$$
$$x = 1.88 \text{ m}$$

$$\text{Maximum BM} = -4(1.88 + 1) - \frac{5(1.88 + 1)^2}{2} + 18.4 \times 1.88$$
$$= -4 \times 2.88 - \frac{5(2.88)^2}{2} + 34.59$$
$$= -11.52 - 20.74 + 34.59$$
$$= 2.33 \text{ KN-m}$$

Point of contraflexure : There will be two point of contraflexure P and Q where BM is zero (Refer figure 3.26 (*ii*)).

Let point P be at a distance x from C.

$$\text{BM at } P = -4 \times x - 5 \times x \times \frac{x}{2} + 18.4 \times (x - 1) = 0$$

$$-4x - \frac{5x^2}{2} + 18.4x - 18.4 = 0$$
$$-8x - 5x^2 + 36.8x - 36.8 = 0$$
$$-5x^2 + 28.8x - 36.8 = 0$$
$$5x^2 - 28.8x + 36.8 = 0$$

$$\therefore \quad x = \frac{+28.8 \pm \sqrt{(28.8)^2 - 4 \times 5 \times 36.8}}{2 \times 5}$$
$$= \frac{28.8 \pm \sqrt{93.44}}{10}$$
$$= \frac{28.8 \pm 9.7}{10} = \frac{28.8 + 9.7}{10} \text{ or } \frac{28.8 - 9.7}{10}$$
$$= 3.85 \text{ m or } 1.91 \text{ m}$$

$\therefore$ Points of contraflexure are at 1.91 m and 3.85 m from C.

Note: Point of contraflexure occurs for only overhanging becams on one side or both the sides.

EXAMPLE 3.11. A beam 10 m long carries load as shown in the figure 3.30. Draw SF and BM diagrams for the beam and determine the points of contraflexure, if any.

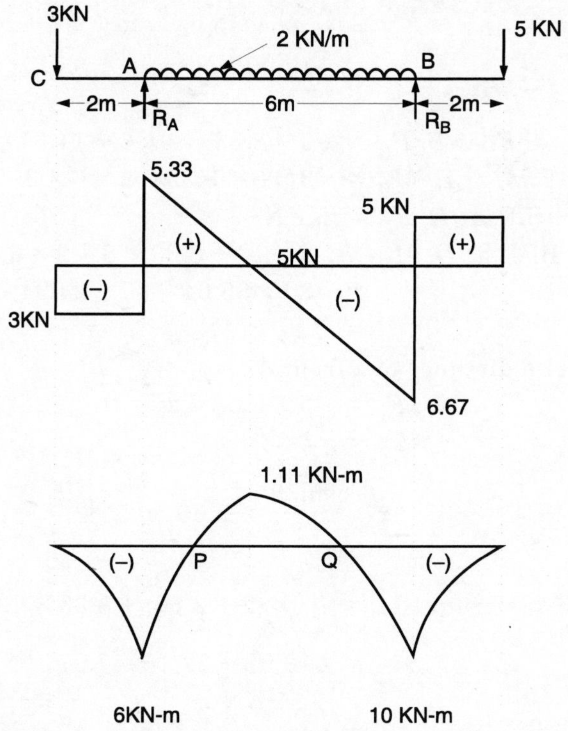

Fig. 3.30.

Solution. Let R_A and R_B be the reactions at A and B respectively.

$$R_A + R_B = 3 + 5 + 2 \times 6 = 20 \quad \ldots(i)$$

Taking moments about B, we get

$$R_A \times 6 = 3 \times 8 + 2 \times 6 \times 3 - 5 \times 2$$
$$= 24 + 36 - 10 = 50$$
$$R_A = \frac{50}{6} = 8.33 \text{ KN}$$

Putting the value of R_A in (*i*), we get

$$8.33 + R_B = 20$$
$$\therefore \quad R_B = 20 - 8.33 = 11.67 \text{ KN}$$

For part *CA*

$$\text{S.F. at } C, F_C = -3 \text{ KN}$$
$$\text{B.M. at } C, M_C = 0$$
$$\text{S.F. at } A, F_A = -3 \text{ KN}$$
$$\text{B.M. at } A, M_A = -3 \times 2 = -6 \text{ KN-m}$$

For part *AB*

$$\text{S.F. at } A, F_A = -3 + 8.33 = 5.33 \text{ KN}$$
$$\text{B.M. at } A, M_A = -6 \text{ KN} - \text{m}$$
$$\text{S.F. at } B, F_B = -3 + 8.33 - 2 \times 6 = -6.67 \text{ KN}$$

$$\text{B.M. at } A,\ M_B = -3 \times 8 + 8.33 \times 6 - 2 \times 6 \times 3$$
$$= -24 + 49.98 - 36$$
$$= -10.00 \text{ KN-m}$$

For part *BD*

$$\text{S.F. at } B,\ F_B = -3 + 8.33 - 2 \times 6 + 11.67 = 5 \text{ KN}$$
$$\text{B.M. at } B,\ M_B = -3 \times 8 + 8.33 - 2 \times 6 \times 3 = -10.00 \text{ KN-m}$$
$$\text{S.F. at } D,\ F_D = 5 \text{ KN}$$
$$\text{B.M. at } D,\ M_D = -3 \times 10 + 8.33 \times 8 - 2 \times 6 \times 5 + 11.67 \times 2$$
$$= -30 + 66.64 - 60 + 23.34$$
$$= 0$$

Let the SF be zero at a distance of x from A.

$$F_x = -3 + 3.88 - 2x = 0$$
$$-2x = -5.33$$
$$x = \frac{5.33}{2} = 2.66 \text{ m}$$

$$\text{Maximum BM} = -3 \times (2 + 2.66) + 8.33 \times 2.66 - 2 \times 2.66 \times \frac{2.66}{2}$$
$$= -13.98 + 22.16 - 7.07$$
$$= 1.11 \text{ KN-m}$$

Point of contraflexure : There will be two point of contraflexure P and Q where B.M. is zero.

Let the point of contraflexure be at a distance 'x' from C.

$$\text{B.M. at } P = -3x + 8.33(x - 2) - 2(x - 2)\frac{(x - 2)}{2} = 0$$

$$-3x + 8.33x - 16.66 - (x^2 - 4x + 4) = 0$$
$$5.33x - 16.66 - x^2 + 4x - 4 = 0$$
$$-x^2 + 9.33x - 20.66 = 0$$
$$x^2 - 9.33x + 20.66 = 0$$

$$\therefore \quad x = \frac{+9.33 \pm \sqrt{(9.33)^2 - 4 \times 1 \times 20.66}}{2} = \frac{9.33 \pm 2.1}{2}$$

or
$$x = \frac{9.33 + 2.1}{2} \text{ or } \frac{9.33 - 2.1}{2}$$
$$= 5.71 \text{ m or } 3.61 \text{ m}$$

$\therefore$ The two point of contraflexure will be at 3.61 m and 5.71 m from point C.

EXAMPLE 3.12. Draw the shear force and bending moment diagrams for an overhanging beam with loads as shown in figure 3.31. Also locate point of contraflexure of any.

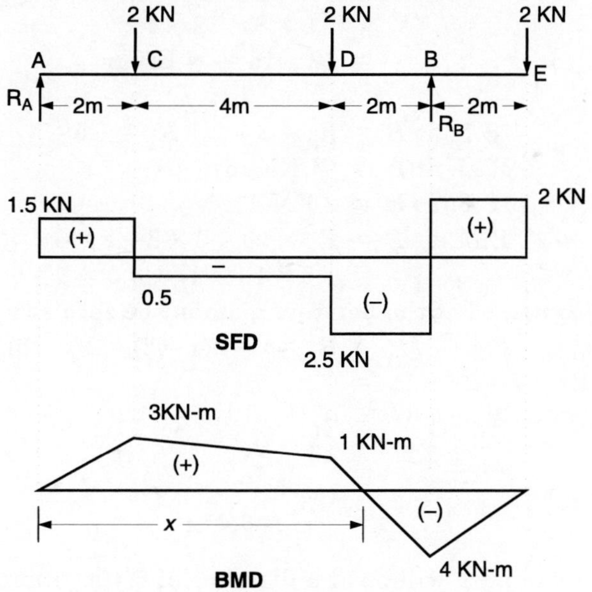

Fig. 3.31.

Solution. Let R_A and R_B be the reactions at A and B respectively.

$$R_A + R_B = 2 + 2 + 2 = 6 \text{ KN} \quad \ldots(i)$$

Taking moment about B, we get

$$R_A \times 8 = 2 \times 6 + 2 \times 2 - 2 \times 2$$

$$R_A \times 8 = 12 + 4 - 4 = 12$$

$$\therefore \quad R_A = \frac{12}{8} = 1.5 \text{ KN}$$

Putting the value of R_A in (*i*), we get

$$1.5 + R_B = 6$$

$$\therefore \quad R_B = 4.5 \text{ KN}$$

For part *CA*

$$\text{S.F. at } A = 1.5 \text{ KN}$$

$$\text{B.M. at } A = 0$$

$$\text{S.F. at } C = 1.5 \text{ KN}$$

$$\text{B.M. at } C = R_A \times 2 = 1.5 \times 2 = 3 \text{ KN-m}$$

For part *CD*

$$\text{S.F. at } C = R_A - 2 = 1.5 - 2 = -0.5 \text{ KN}$$

$$\text{B.M. at } C = R_A \times 2 = 3 \text{ KN-m}$$

$$\text{S.F. at } D = R_A - 2 = 1.5 - 2 = -0.5 \text{ KN}$$

$$\text{B.M. at } D = R_A \times 6 - 2 \times 4$$

$$= 1.5 \times 6 - 8 = 1 \text{ KN-m}$$

For part *DB*

$$\text{S.F. at } D = R_A - 2 - 2 = 1.5 - 4 = -2.5 \text{ KN}$$

$$\text{B.M. at } D = 1 \text{ KN-m}$$

$$\text{S.F. at } B = R_A - 2.2 = 1.5 - 4 = -2.5 \text{ KN}$$

$$\text{B.M. at } B = R_A \times 8 - 2 \times 6 - 2 \times 2$$

$$= 1.5 \times 8 - 12 - 4$$
$$= 12 - 16 = -4 \text{ KN-m}$$

For part *BE*

$$\text{S.F. at } B = R_A - 2 - 2 + R_B = 1.5 - 4 + 4.5 = 2 \text{ KN}$$
$$\text{B.M. at } B = -4 \text{ KN-m}$$
$$\text{S.F. at } E = 2 \text{ KN}$$
$$\text{B.M. at } E = R_A \times 10 - 2 \times 8 - 2 \times 4 + R_B \times 2$$
$$= 1.5 \times 10 - 24 + 4.5 \times 2 = 0$$

Point of contraflexure : Let the bending moment be zero at a distance x from A.

$$M_x = R_A \times x - 2(x - 2) - 2(x - 6) = 0$$
$$1.5x - 2x + 4 - 2x + 12 = 0$$
$$-2.5x + 16 = 0$$
$$2.5x = 16$$
$$\therefore \quad x = \frac{16}{2.5} = 6.4 \text{ m}$$

$\therefore$ The point of contraflexure will be at a distance of 6.4 m from A.

EXAMPLE 3.13. Draw the SF and BM diagram shown in the figure 3.32. Show the points of maximum sagging and hagging BM. Also find point of contraflexure.

[*Delhi, Dec. 2009*]

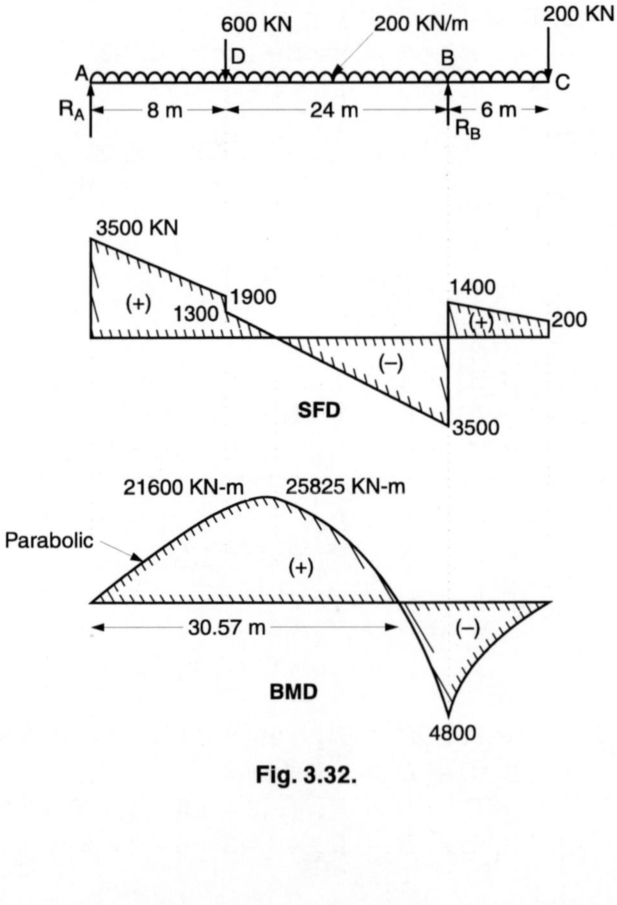

Fig. 3.32.

Solution. Let R_A and R_B be the reactions at A and B respectively.

$$R_A + R_B = 600 + 200 + 200 \times 38$$
$$= 600 + 200 + 7600$$
$$= 8400 \text{ KN} \quad ...(i)$$

Taking moments about B, we get

$$R_A \times 32 = 600 \times 24 + 200 \times 32 \times 16 - 200 \times 6 - 200 \times 6 \times 3$$
$$= 14400 + 102400 - 1200 - 3600$$
$$= 112000$$

$$\therefore \quad R_A = \frac{112000}{32} = 3500 \text{ KN}$$

Putting the value of R_A in (i), we get

$$3500 + R_B = 8400$$
$$\therefore \quad R_B = 4900 \text{ KN}$$

For part *AD*

$$\text{S.F. at } A = R_A = 3500.00 \text{ KN}$$
$$\text{B.M. at } A = 0$$
$$\text{S.F. at } D = R_A - 200 \times 8$$
$$= 3500 - 1600 = 1900 \text{ KN}$$
$$\text{B.M. at } D = R_A \times 8 - 200 \times 8 \times 4$$
$$= 3500 \times 8 - 200 \times 32$$
$$= 21600 \text{ KN-m}$$

For part *DB*

$$\text{S.F. at } D = R_A - 200 \times 8 - 600$$
$$= 3500 - 2200 = 1300 \text{ KN}$$
$$\text{B.M. at } D = 21600 \text{ KN-m}$$
$$\text{S.F. at } B = R_A - 600 - 200 \times 32$$
$$= 3500 - 7000 = -3500 \text{ KN}$$
$$\text{B.M. at } B = R_A \times 32 - 600 \times 24 - 200 \times 32 \times 16$$
$$= 3500 \times 32 - 14400 - 102400$$
$$= -4800 \text{ KN-m}$$

For part *BC*

$$\text{S.F. at } B = R_A - 600 + R_B - 200 \times 32$$
$$= 3500 - 600 + 4900 - 6400$$
$$= 1400 \text{ KN}$$
$$\text{B.M. at } B = 4800 \text{ KN-m}$$
$$\text{S.F. at } C = 3500 + 4900 - 200 \times 38 - 600$$
$$= 8400 - 8200 = 200 \text{ KN}$$
$$\text{B.M. at } C = 3500 \times 38 + 4900 \times 6 - 600 \times 30 - 200 \times 38 \times 19$$
$$= 133000 + 29400 - 18000 - 144400$$
$$= 0$$

Let the S.F. be zero at a distance 'x' from A.

$$F_x = 3500 - 600 - 200x = 0$$
$$2900 - 200x = 0$$

$$\therefore \qquad x = \frac{2900}{200} = 14.5 \text{ m from } A$$

$$\text{Maximum BM} = 3500 \times 14.5 - 600(14.5 - 8) - 200 \times 14.5 \times \frac{14.5}{2}$$

$$= 50750 - 3900 - 21025$$

$$= 25825 \text{ KN-m}$$

Point of contraflexure : Let the B.M. be zero at a distance x from A.

$$M_x = 3500 \times x - 600(x - 8) - 200 \times x \times \frac{x}{2} = 0$$

$$3500x - 600x + 4800 - 100x^2 = 0$$

$$2900x + 4800 - 100x^2 = 0$$

$$100x^2 - 2900x - 4800 = 0$$

$$x^2 - 29x - 48 = 0$$

$$x = \frac{+29 \pm \sqrt{(29)^2 + 4 \times 1 \times 48}}{2 \times 1} = \frac{29 \pm 32.14}{2}$$

$$= \frac{29 + 32.14}{2} = 30.57 \text{ m}$$

or
$$x = \frac{29 - 32.14}{2} = \text{Not possible being negative}$$

$\therefore$ The point of contraflexure will be at 30.57 m from A as shown in figure 3.32.

EXAMPLE 3.14. A simply supported beam is loaded as shown in the figure 3.33. Draw the shear force and bending moment diagrams.

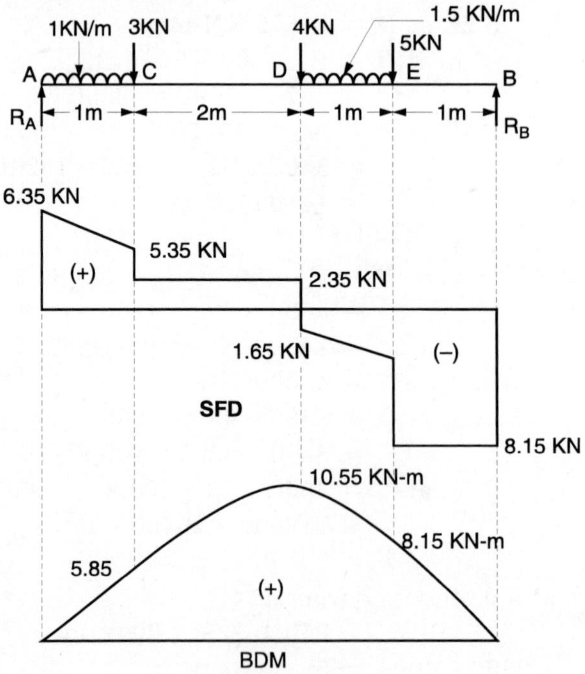

Fig. 3.33.

Solution. Let R_A and R_B be the reactions at A and B respectively.

$$R_A + R_B = 1 \times 1 + 3 + 4 + 1.5 \times 1 + 5$$
$$= 1 + 3 + 4 + 1.5 + 5$$
$$R_A + R_B = 14.5 \text{ KN} \quad ...(i)$$

Taking moments about B, we get,

$$R_A \times 5 = 1 \times 1 \times 4.5 + 3 \times 4 + 4 \times 2 + 1.5 \times 1 \times 1.5 + 5 \times 1$$
$$R_A \times 5 = 4.5 + 12 + 8 + 2.25 + 5$$

or $$R_A \times 5 = 31.75$$

$\therefore$ $$R_A = \frac{31.75}{5} = 6.35 \text{ KN}$$

Putting the value of R_A in (i), we get

$$6.35 + R_B = 14.5$$

$\therefore R_B = 14.5 - 6.35 = 8.35 = 8.15$ KN

For part *AC*

$$\text{SF at } A = 6.35 \text{ KN}$$
$$\text{BM at } A = 0$$
$$\text{SF at } C = 6.35 - 1 \times 1 = 5.35 \text{ KN}$$
$$\text{BM at } C = 6.35 \times 1 - 1 \times 1 \times 0.5 = 5.85 \text{ KN-m}$$

For part *CD*

$$\text{SF at } C = 6.35 - 1 \times 1 - 3 = 6.35 - 4 = 2.35 \text{ KN}$$
$$\text{SF at } D = 6.35 - 1 \times 1 - 3 = 2.35 \text{ KN}$$
$$\text{BM at } D = 6.35 \times 3 - 1 \times 1 \times 2.5 - 3 \times 2$$
$$= 19.05 - 2.5 - 6 = 10.55 \text{ KN} - \text{m}$$

For part *DE*

$$\text{SF at } D = 6.35 - 1 \times 1 - 3 - 4 = 6.35 - 8 = -1.65 \text{ KN}$$
$$\text{SF at } E = 6.35 - 1 \times 1 - 3 - 4 - 1.5 \times 1 = 6.35 - 9.5$$
$$= -2.65 \text{ KN-m}$$
$$\text{BM at } E = 6.35 \times 4 - 1 \times 1 \times 3.5 - 3 \times 3 - 4 \times 1 - 1.5 \times 1 \times 0.5$$
$$= 25.4 - 3.5 - 9 - 4 - 0.75 = 8.15 \text{ KN–m}$$

For part *EB*

$$\text{SF at } E = 6.35 - 1 \times 1 - 3 - 4 - 1.5 \times 1 - 5 = 6.35 - 14.5$$
$$= -8.15 \text{ KN}$$
$$\text{SF at } B = -8.15 \text{ KN}$$

BM at $B = 6.35 \times 5 - 1 \times 1 \times 4.5 - 3 \times 4 - 4 \times 2 - 1.5 \times 1 \times 1.5 - 5 \times 1$

$$= 31.75 - 4.5 - 12 - 8 - 2.25 - 5 = 31.75 - 31.75 = 0$$

EXAMPLE 3.15. Draw the SF and BM diagrams indicating principal values for overhang beam shown in the figure 3.34. ***(Delhi, Jan. 2009)***

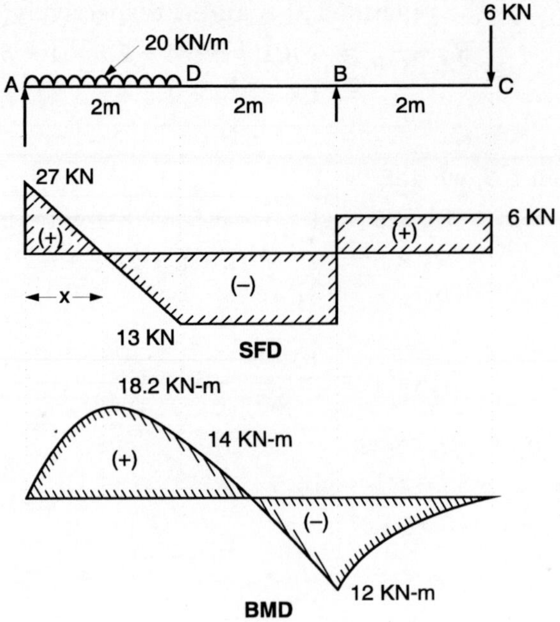

Fig. 3.34.

Solution: Let R_A and R_B be the reactions at A and B respectively.

$$R_A + R_B = 20 \times 2 + 6 = 46 \text{ KN} \quad \ldots(i)$$

Taking moments about B

$$R_A \times 4 = 20 \times 2 \times 3 - 6 \times 2$$
$$R_A \times 4 = 120 - 12 = 108$$
$$\therefore \quad R_A = \frac{108}{4} = 27 \text{ KN}$$

Putting the value of R_A in (i), we get

$$27 + R_B = 46$$
$$\therefore \quad R_B = 46 - 27 = 19 \text{ KN}$$

For part *AD*

$$\text{S.F. at } A = R_A = 27 \text{ KN}$$
$$\text{B.M. at } A = 0$$
$$\text{S.F. at } D = R_A - 20 \times 2 = 27 - 40 = -13 \text{ KN}$$
$$\text{B.M. at } D = 27 \times 2 - 20 \times 2 \times 1$$
$$= 54 - 40 = 14 \text{ KN-m}$$

For part *DB*

$$\text{S.F. at } D = 27 - 20 \times 2 = -13 \text{ KN}$$
$$\text{S.F. at } B = -13 \text{ KN}$$
$$\text{B.M. at } B = 27 \times 4 - 20 \times 2 \times 3$$
$$= 108 - 120 = -12 \text{ KN-m}$$

For part *BC*

$$\text{S.F. at } B = R_A - 20 \times 2 + R_B$$
$$= 27 - 40 + 19 = 6 \text{ KN}$$

$$\text{S.F. at } C = 6 \text{ KN}$$
$$\text{B.M. at } C = 27 \times 6 - 20 \times 2 \times 5 + 19 \times 2$$
$$= 200 - 200 = 0$$

Let the S.F. be zero at a distance 'x' from A.

$$F_x = 27 - 20x = 0$$
$$20x = 27$$
$$\therefore \quad x = \frac{27}{20} = 1.35 \text{ m}$$
$$\text{Maximum BM} = R_A \times 1.35 - 20 \times 1.35 \times \frac{1.35}{2}$$
$$= 27 \times 1.35 - 18.225$$
$$= 18.2 \text{ KN-m.}$$

EXAMPLE 3.16. A beam 6 m long is supported on two points 4 m apart overhanging by 1 m on each end and carries *a* UDL of 3000 N/m throughout its length. Draw SF and BM diagrams and determine the points of contraflexure, if any.

Solution. Let R_A and R_B be the reactions at A and B respectively.

$$R_A + R_B = 3 \times 6 = 18 \text{ KN} \quad \ldots(i)$$

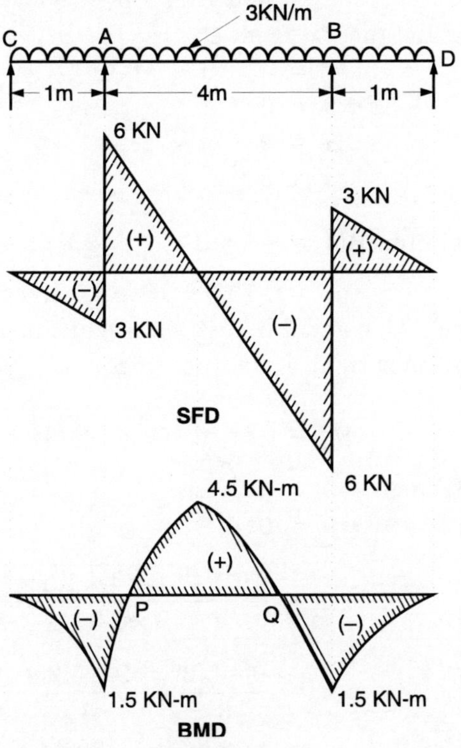

Fig. 3.35.

Taking moments about B

$$R_A \times 4 = 3 \times 5 \times 2.5 - 3 \times 1 \times 0.5$$
$$= 37.5 - 1.5 = 36$$
$$\therefore \quad R_A = \frac{36}{4} = 9 \text{ KN}$$

Putting the value of R_A in (*i*), we get

$$9 + R_B = 18$$

$$\therefore \quad R_B = 9 \text{ KN}$$

For part *CA*

$$\text{S.F. at } C = 0$$

$$\text{B.M. at } C = 0$$

$$\text{S.F. at } A = -3 \times 1 = -3 \text{ KN}$$

$$\text{B.M. at } A = -3 \times 1 \times 0.5 = -1.5 \text{ KN-m}$$

For part *AB*

$$\text{S.F. at } A = -3 \times 1 + 9 = 6 \text{ KN}$$

$$\text{S.F. at } B = -3 \times 5 + 9 = -6 \text{ KN}$$

$$\text{B.M. at } B = -3 \times 5 \times 2.5 + 9 \times 4$$

$$= -37.5 + 36 = -1.5 \text{ KN-m}$$

Fors part *BD*

$$\text{S.F. at } B = -3 \times 5 + 9 + 9 = 3 \text{ KN}$$

$$\text{S.F. at } D = -3 \times 6 + 9 + 9 = 0$$

$$\text{B.M. at } D = -3 \times 6 \times 3 + 9 \times 5 + 9 \times 1$$

$$= -54 + 54 = 0$$

Let the S.F. be zero at a distance '*x*' from *C*.

$$F_x = -3 \times x + 9 = 0$$

$$-3x + 9 = 0$$

$$3x = 9$$

$$\therefore \quad x = 3$$

$$\text{Maximum BM} = -3 \times 3 \times \frac{3}{2} + 9 \times (3 - 1)$$

$$= -13.5 + 18 = 4.5 \text{ KN-m}$$

Point of contraflexure : There will be two points of contraflexure *P* and *Q*. Let the bending moment be zero at a distance *x* from *C*.

$$M_x = -3x.\frac{x}{2} + 9 \times (x - 1) = 0$$

$$-1.5x^2 + 9x - 9 = 0$$

$$1.5x^2 - 9x + 9 = 0$$

$$\therefore \quad x = \frac{+9 \pm \sqrt{81 - 4 \times 1.5 \times 9}}{2 \times 1.5} = \frac{9 \pm 5.2}{3}$$

$$\text{or} \quad x = \frac{9 + 5.2}{3} \text{ or } \frac{9 - 5.2}{3}$$

$$= 4.73 \text{ m or } 1.27 \text{ m}$$

$\therefore$ The points of contraflexure will be at 1.27 m and 4.73 m from point *C*.

EXAMPLE 3.17. Draw shear force and bending moment diagrams for the loaded beam shown in figure 3.36. Locate the point of contraflexure, if any.

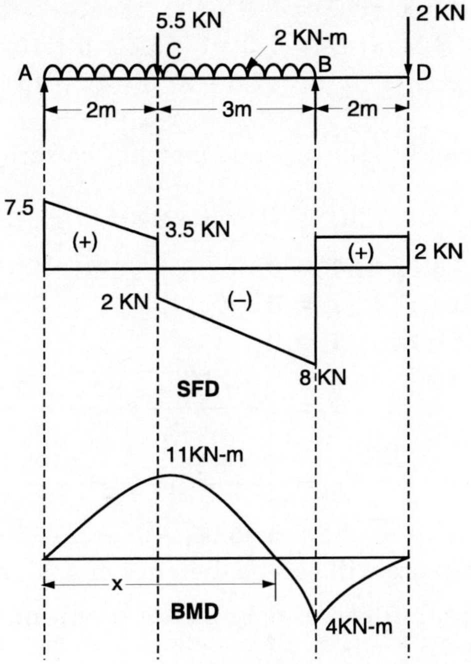

Fig. 3.36.

Solution. Let R_A and R_B be the reactions at A and B respectively.

$$R_A + R_B = 5.5 + 2 + 2 \times 5 = 17.5 \text{ KN}$$

Taking moment about B, we get

$$R_A \times 5 = 5.5 \times 3 + 2 \times 5 \times 2.5 - 2 \times 2$$

$$R_A \times 5 = 41.5 - 4$$

$$\therefore \quad R_A = \frac{37.5}{5} = 7.5 \text{ KN}$$

Putting the value of R_A in (*i*), we get

$$7.5 + R_B = 17.5$$

$$\therefore \quad R_B = 17.5 - 7.5 = 10 \text{ KN}$$

For part *AC*

$$\text{S.F. at } A = R_A = 7.5 \text{ KN}$$

$$\text{B.M. at } A = 0$$

$$\text{S.F. at } C = R_A - 2 \times 2 = 7.5 - 4 = 3.5 \text{ KN}$$

$$\text{B.M. at } C = 7.5 \times 2 - 2 \times 2 \times 1 = 15 - 4 = 11 \text{ KN-m}$$

For part *CB*

$$\text{S.F. at } C = R_A - 2 \times 2 - 5.5 = 7.5 - 4 - 5.5 = -2.0 \text{ KN}$$

$$\text{S.F. at } B = 7.5 - 5.5 - 2 \times 5 = 7.5 - 15.5 = -8.0 \text{ KN}$$

$$\begin{aligned} \text{B.M. at } B &= R_A \times 5 - 2 \times 5 \times 2.5 - 5.5 \times 3 \\ &= 7.5 \times 5 - 25 - 16.5 \\ &= 37.5 - 25 - 16.5 = -4 \text{ KN-m} \end{aligned}$$

For part *BD*

$$\begin{aligned} \text{S.F. at } B &= 7.5 - 5.5 + 10 - 2 \times 5 \\ &= 17.5 - 15.5 = 2 \text{ KN} \end{aligned}$$

$$\text{S.F. at } D = 8 \text{ KN}$$

$$\begin{aligned}\text{B.M. at } D &= 7.5 \times 7 - 5.5 \times 5 + 10 \times 2 - 2 \times 5 \times 4.5 \\ &= 52.5 - 27.5 + 20 - 45 \\ &= 72.5 - 72.5 = 0\end{aligned}$$

Point of contraflexure : Let the bending moment be zero at a distance x from A.

$$M_x = R_A \times x - 2 \times x \times \frac{x}{2} - 5.5(x - 2) = 0$$

$$\begin{aligned}7.5x - x^2 - 5.5x + 11 &= 0 \\ 2x - x^2 + 11 &= 0 \\ x^2 - 2x - 11 &= 0\end{aligned}$$

$$\therefore \quad x = \frac{+2 \pm \sqrt{4 + 4 \times 11}}{2} = \frac{2 \pm 6.7}{2}$$

or

$$x = \frac{2 + 6.7}{2} \text{ or } x = \frac{2 - 6.7}{2} = -2.35 \text{ (Not possible)}$$

$$= 4.35 \text{ m}$$

$\therefore$ The points of contraflexure will be at a distance of 4.35 m from point A.

EXAMPLE 3.18. Draw shear force and bending moment diagrams for the loaded beam shown in figure 3.37.

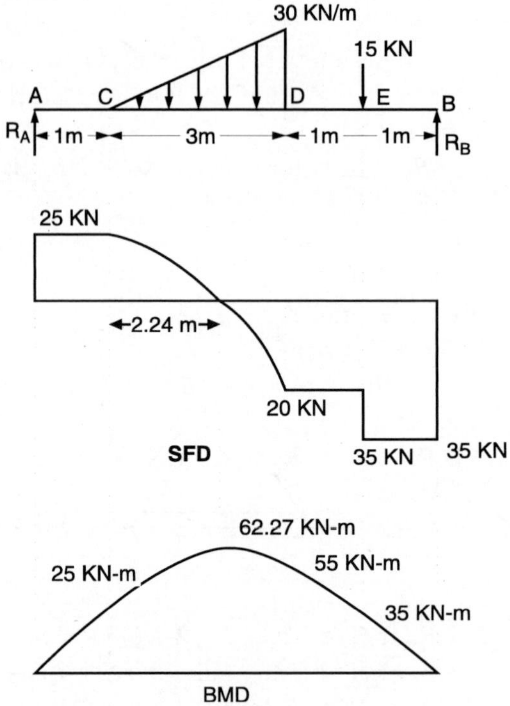

Fig. 3.37.

Sol. Let R_A and R_B be the reactions at A and B respectively.

$$R_A + R_B = \frac{1}{2} \times 30 \times 3 + 15$$

$$R_A + R_B = 60 \text{ KN} \qquad \ldots(i)$$

Taking moment about B, we get

$$R_A \times 6 = \frac{1}{2} \times 30 \times 3 \times 3 + 15 \times 1$$

$$R_A \times 6 = 150$$

$$\therefore \quad R_A = \frac{150}{6} = 25 \text{ KN}$$

Putting the value of R_A in (i), we get

$$25 + R_B = 60$$

$$\therefore \quad R_B = 35 \text{ KN}$$

For part *AC*

$$\text{S.F. at } A = 25 \text{ KN}$$

$$\text{B.M. at } A = 0$$

$$\text{S.F. at } C = 25 \text{ KN}$$

$$\text{B.M. at } C = 25 \times 1 = 25 \text{ KN-m}$$

For part *CD*

$$\text{S.F. at } C = 25 \text{ KN}$$

$$\text{S.F. at } D = 25 - \frac{1}{2} \times 30 \times 3 = -20 \text{ KN}$$

$$\text{B.M. at } D = 25 \times 4 - \frac{1}{2} \times 30 \times 3 \times 1$$

$$= 100 - 45 = 55 \text{ KN-m}$$

For part *DE*

$$\text{S.F. at } D = -20 \text{ KN}$$

$$\text{S.F. at } E = -20 \text{ KN}$$

$$\text{B.M. at } E = 25 \times 5 - \frac{1}{2} \times 30 \times 3 \times 2 = 35 \text{ KNm}$$

For part *EB*

$$\text{S.F. at } E = 25 - \frac{1}{2} \times 30 \times 3 - 15$$

$$= 25 - 60 = -35 \text{ KN}$$

$$\text{S.F. at } B = -35 \text{ KN}$$

$$\text{B.M. at } B = 25 \times 6 - \frac{1}{2} \times 30 \times 3 \times 3 - 15 \times 1$$

$$= 150 - 135 - 15 = 0$$

Let the SF be zero at a distance x from C

$$F_x = 25 - \frac{1}{2} \times 10x \times x = 0$$

$$25 - 5x^2 = 0$$

$$5x^2 = 25$$

$$x^2 = 5$$

$$\therefore \quad x = 2.24 \text{ m}$$

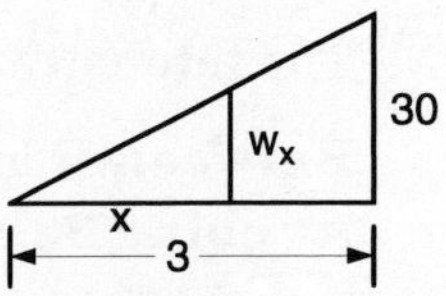

$$\frac{w_x}{30} = \frac{x}{3}$$

$$w_x = \frac{30x}{3} = 10x$$

$$\text{Maximum BM} = 25 \times 3.24 - \frac{1}{2} \times (10 \times 2.24) \times 2.24 \times \frac{2.24}{3}$$

$$= 81 - 18.73 = 62.27 \text{ KN-m}$$

3.10 LOAD AND BENDING MOMENT DIAGRAMS FROM A SHEAR FORCE DIAGRAM

In some questions, SF diagram is given and we have to find out the loading and have to draw the BM diagram. For these, first load diagram is drawn from the given SF diagram and then from the load diagram bending moment diagram is drawn. The following points should be kept in mind while drawing the load diagram from bending moment diagram:

(*i*) If there is a sudden increase or decrease in SF at a point, if indicates that there is a reaction or point load at the point.

(*ii*) If the SFD shows a horizontal line between the two points, this indicates that there is no loading between the two points.

(*iii*) If the SFD shows an inclined line between the two points. This indicates that there is UDL between the points.

(*iv*) Presence of parabolic curve between two points indicates the presence of uniformly varying load between the two points.

EXAMPLE 3.19. The shear force diagram of a beam 4 m in length is shown in figure 3.38. Determine the loading and hence draw the bending moment diagrams.

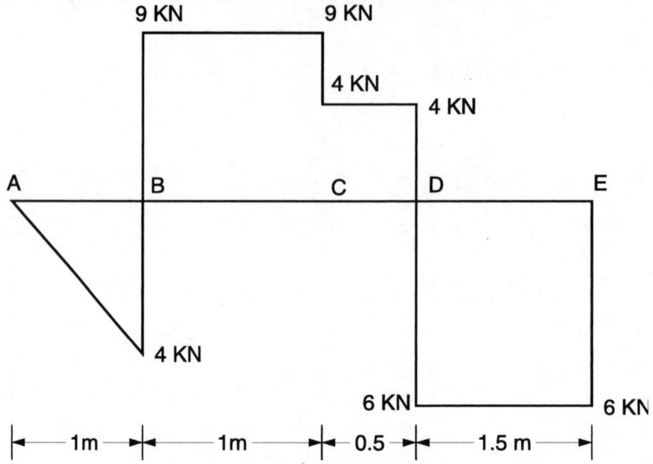

Fig. 3.38. SFD

Solution. At A : SF is zero *i.e.,* there is no load at *A*.

Between *A* and *B* : The SFD is an inclined straight line between *A* and *B*. Therefore, the beam is carrying UDL between *A* and *B*.

The intensity of UDL between *A* and $B = \frac{4}{1} = 4$ KN/m

Between *B* and *C* :

At point *B* : SF suddenly changes from –4 KN to 9 KN. Therefore, there is a reaction at *B*.

$$\therefore \qquad R_B = 9 + 4 = 13 \text{ KN}$$

Between *B* and *C*, SFD is horizontal line *i.e.,* there is no loading between *B* and *C*.

At *C* : SF changes from +9 KN to +4 KN; *i.e.,* there is a point load at $C = 9 - 4 = 5$ KN.

Between *C* and *D* : SFD is horizontal *i.e.*, there is no loading between *C* and *D*.

At point *D* : SF changes suddenly from +4 KN to –6 KN *i.e.*, there is a point load at $D = 4 + 6 = 10$ KN.

Between *D* and *E* : SFD is horizontal *i.e.*, there is no loading at *E*.

At *E* : SF changes from 6 KN to 0 *i.e.*, there is a reaction at $E = 16$ KN

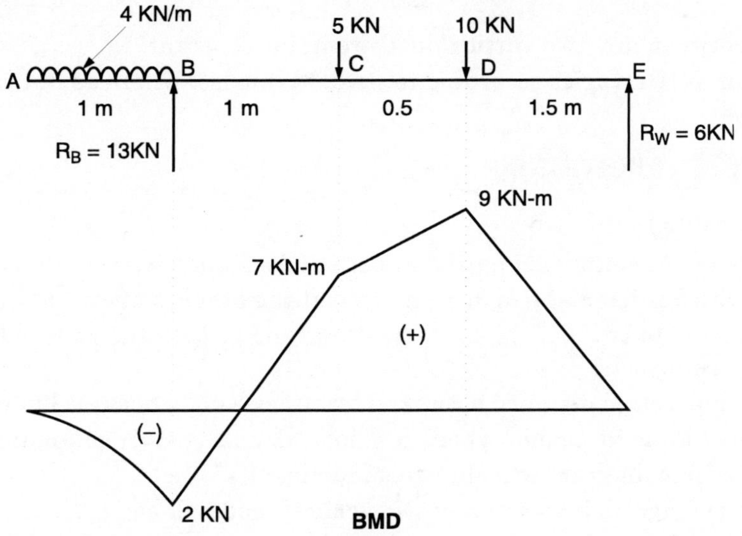

Fig. 3.39.

$$\begin{aligned}
\text{B.M. at } A &= 0 \\
\text{B.M. at } B &= -4 \times 1 \times 0.5 \\
&= -2 \text{ KN-m} \\
\text{B.M. at } C &= -4 \times 1 \times 1.5 + 13 \times 1 \\
&= -6 + 13 = 7 \text{ KN-m} \\
\text{B.M. at } D &= -4 \times 1 \times 2 + 13 \times 1.5 - 5 \times 0.5 \\
&= -8 + 19.5 - 2.5 = 9 \text{ KN-m} \\
\text{B.M. at } E &= -4 \times 1 \times 3.5 + 13 \times 3 - 5 \times 2 - 10 \times 1.5 \\
&= -14 + 39 - 10 - 15 \\
&= 0
\end{aligned}$$

SUMMARY

1. Shear force at any section of a beam is defined as the algebraic sum of the lateral components of all the forces acting on either side of the section.
2. Shear force will be considered as positive when the resultant of the forces to the left of the section is upward or to the right downward.
3. Shear Force Diagram (SFD) is the graphical representation of variation of shear force along the length of the beam.
4. Bending moment at any section of a beam is the algebraic sum of the moments of all the forces acting on either side of the section.
5. Bending Moment Diagram (BMD) is the graphical representation of variation of bending moment along the length of the beam.
6. The positive bending moment is known as sagging B.M.

7. The negative bending moment is known as hogging BM.
8. In case of cantilever the maximum B.M. will be at the fixed end.
9. Bending moment is maximum at a point where SF is zero.
10. In case of simply supported beam, BM is maximum at the point of load.
11. Point of contraflexure is the point on a beam where bending moment is zero after changing its sign.
12. Shear force between any two vertical loads remains constant.
13. Shear force for a UDL varies according to straight line law where as BM varies according to parabolic law.

OBJECTIVE TYPE QUESTIONS

1. Fill in the Blanks :

(*i*) A beam which is simply supported at both ends is known as .... beam.
(*ii*) A beam which is fixed at one end and free at the other is known as .... beam.
(*iii*) A beam is said to be .... beam if one end or both the ends of the beam projects beyond the supports.
(*iv*) The clear horizontal distance between two supports of a beam is known as ....
(*v*) B.M. is maximum at a point where S.F. is .... after changing of sign.
(*vi*) The shear force diagram will change suddenly at a ....
(*vii*) Shear force diagram between any two vertical load will be ....
(*viii*) For UDL the SFD will be an .... line.
(*ix*) For UDL the BMD will be a ....
(*x*) For cantilever, the BM will be zero at the ....
(*xi*) The –ve BM is also known as ....
(*xii*) Maximum BM for a simply supported beam with point load , (*W*) at mid span is equal to ....
(*xiii*) The point of contraflexure is a point on the beam where .... is zero.
(*xiv*) The rate of change of BM is equal to .... at the section.
(*xv*) The rate of change of .... is equal to rate of loading.

Answers

(*i*) Simply supported, (*ii*) cantilever, (*iii*) overhanging, (*iv*) clear span,
(*v*) zero, (*vi*) point load, (*vii*) horizontal, (*viii*) inclined,
(*ix*) parabolic curve, (*x*) free end, (*xi*) hogging BM, (*xii*) $\frac{WL}{4}$,
(*xiii*) BM, (*xiv*) shear force, (*xv*) shear force.

THEORETICAL QUESTIONS

1. Define the followings :
   (*a*) shear force, (*b*) Bending moment,
   (*c*) Point of contraflexure.
2. What are shear force diagram and bending moment diagram ?
3. Draw the shear force and bending moment diagrams for a simply supported beam carrying a point load *W* at mid span.

NUMERICALS

1. A cantilever 2 m long carries point loads of 4 KN, 5 KN and 3 KN at distances 1 m, 1.5 m and 2 m respectively from the fixed end side. Draw the shear force and bending moment diagrams for the beam. [Max. SF = 12 KN, Max. BM = –17.5 KN-m]
2. A cantilever of length 5 m carries a UDL of 2 KN/m over the whole length and a point load of 4 KNm at the free end. Draw the shear force and bending moment diagrams for the cantilever. [Max. SF = 14 KN; Max. BM = –45 KN-m]
3. A cantilever 2.5 m long carries a UDL of 60 KN/m at a length of 1.8 m from the free end. Draw the shear force and bending moment diagram. [Max. S.F. = 108 KN; Max. BM = –172.8 KN-m]
4. Draw the S.F. and B.M. diagrams for cantilever loaded as shown in figure 3.40. [Max. S.F. = 6 KN; Max. BM = –18 KN-m]

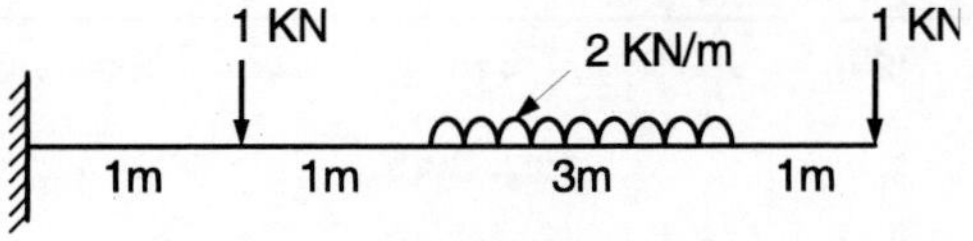

Fig. 3.40.

5. Draw the SF and BM diagrams for simply supported beam loaded as shown in figure 3.41. [Max. S.F. = 4 KN; Max. BM = 6 KN-m]

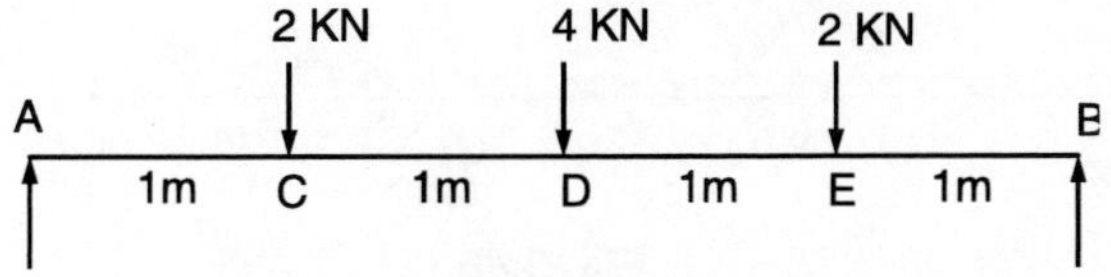

Fig. 3.41.

6. A simply supported beam 6 m long is carrying a UDL of 2 KN/m over a length of 3 m from the left end. Draw the S.F. and B.M. diagrams for the beam. [Max. S.F. = 4.5 KN; Max. B.M. = 5.06 KN-m]
7. Draw the shear force and bending moment diagrams for the beam shown in figure 3.42. [Max. S.F. = 5.5 KN; Max. B.M. = 7.56 KN-m]

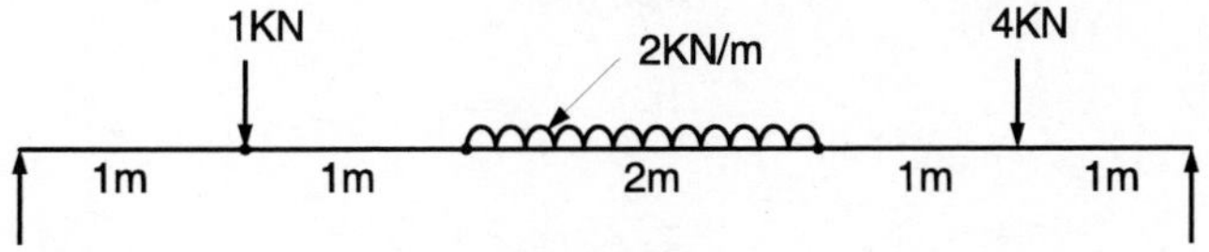

Fig. 3.42.

8. A beam *AB* 10 m long has supports at its ends *A* and *B*. It carries a point load 5 KN at 3 m from *A* and a point load of 5 KN at 7 m from *A* and a UDL of 1 KN/m between the point loads. Draw the SF and BM diagrams. [Max. S.F. = 7 KN; Max. BM = 23 KN-m]
9. Draw the SF and BM diagrams for the beam shown in the figure 3.43.

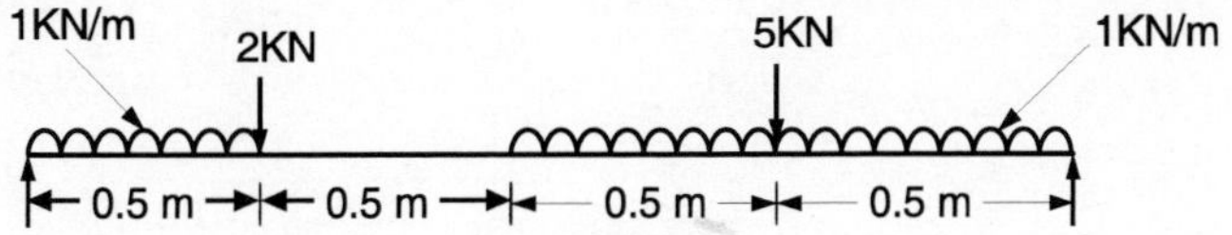

Fig. 3.43.

10. Draw SF and BM diagrams for a simply supported beam, loaded as shown in figure 3.44. Find the position and value of the maximum bending moment, that will occur in the beam. [Max. BM = 3.47 KN-m at a distance 2.3 m from *A*]

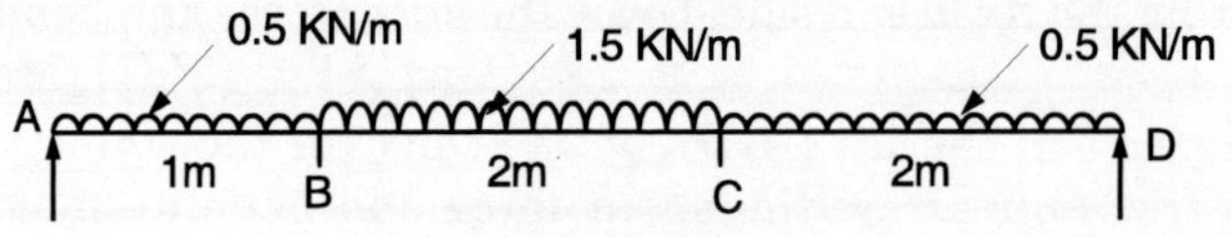

Fig. 3.44.

11. An overhanging beam of length 8 m is simply supported over 4 m apart and overhang of 2 m at each end. The beam carries two concentrated loads 20 KN at each end. Draw the shear force and bending moment diagrams. [Max. SF = 20 KN; Max. BM = –40 KN-m]

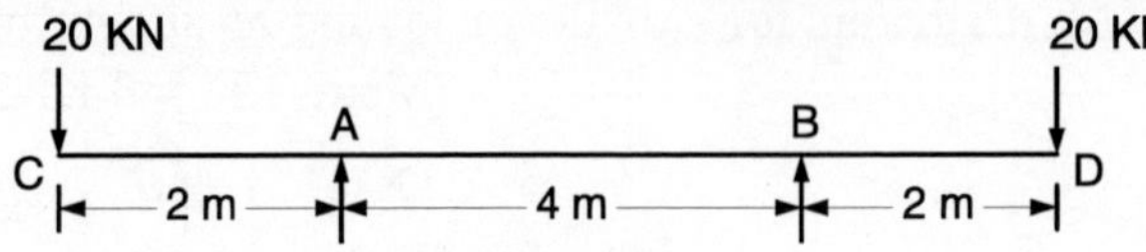

Fig. 3.45.

12. A beam *ABC* is subjected to loads as shown in figure 3.46. Draw SF and BM diagrams. Also find position and magnitude of maximum moment. [Max. BM = 3125 Nm at a distance 2.5 m from *A*]

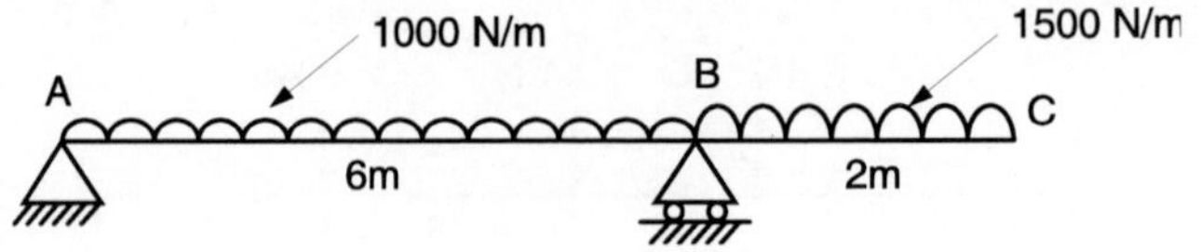

Fig. 3.46.

13. Draw the shear force and bending moment diagrams for the beam shown in figure 3.47.

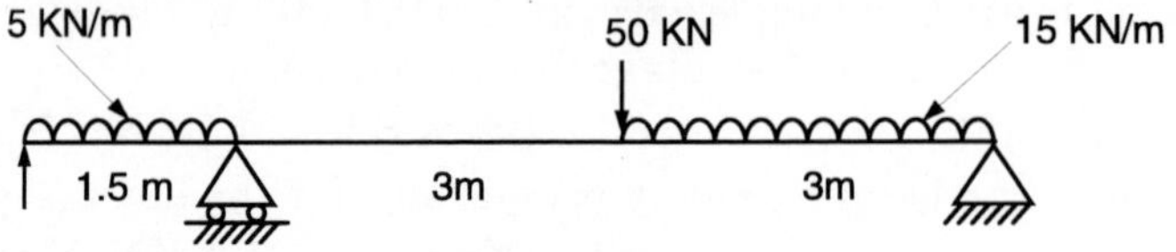

Fig. 3.47.

CHAPTER 4

Centre of Gravity

4.1 CENTRE OF GRAVITY (C.G.)

Chapter Outline

- Centre of Gravity (C.G.)
- Centroid or Centre of Area
- Methods of determining Centre of Gravity
- C.G. by Geometrical Method
- Centre of Gravity by Graphical Method
- Centre of Gravity by Moment Method
- Axis of Reference
- Steps to find out the Centre of Gravity of any Section by Moment

It is the point through which the resultant of parallel force system formed by the weights of different particles of a body acts.

In other words, it is the point through which weight of the body acts. Consider a body of weight 'W' which is made up of number of small particles having weights W_1, W_2, W_3, W_4 etc. We know that the weights of a body or particle will always act vertically downward. Hence the weights of the particles shown in the figure 4.1 form a parallel force system, whose resultant is W acting at G.

Then $W = W_1 + W_2 + W_3 + W_4 + ...$

and point G is called the centre of gravity of the body.

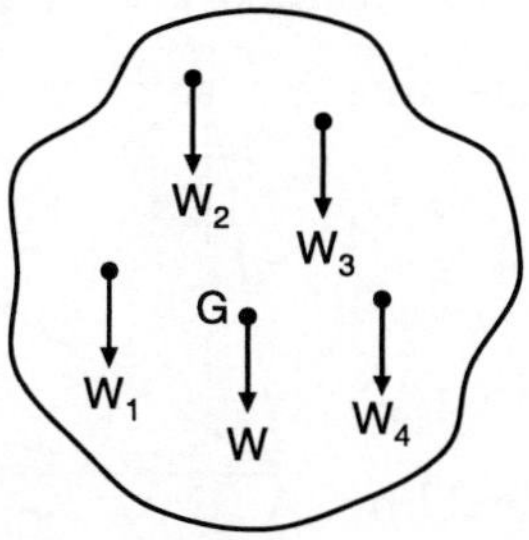

Fig. 4.1.

DO YOU KNOW?

The concept of C.G was first introduced by the ancient Greek mathematician, physicist and engineer Archimedes of syracuse.

Some important points regarding C.G.

(a) A body is always having only one C.G.

(b) The C.G. of a body is independent of the position of the body, *i.e.*, the C.G. of a body does not change if its position is changed.

(c) The C.G. of a body depends upon the shape of the body.

(*d*) The C.G. of a body is an imaginary point which may or may not lie within the body.

4.2 CENTROID OR CENTRE OF AREA

The centroid or centre of area is the point through which whole area of the lamina is supposed to act.

The term centre of gravity is used for bodies which have mass, but centroid is used for bodies which does not have mass *i.e.*, lamina figures.

DO YOU KNOW?

The barycentre is the point between two objects where they balance each other. In other words, the C.G where two or more celestial bodies orbit each other.

4.3 METHODS OF DETERMINING CENTRE OF GRAVITY

The centre of gravity (or centroid) may be found out by any one of the following methods :

(*a*) Geometrical method
(*b*) Graphical method
(*c*) Method of moments

DO YOU KNOW?

A body's centre of gravity may coincide with its geometrical centre, especially if the body is symmetric and composed of homogeneous material

4.4 C.G. BY GEOMETRICAL METHOD

The C.G. of simple figures can be found out from the geometry of the figure. In general, if a figure is symmetrical about a point or line then C.G. will lie on that point or line.

The C.G. (or centroid) of some simple figures and solids are given below.

(a) Rectangle

The C.G. of a rectangle lies at point where the two diagonals intersect each other, *i.e.*, point *G*. It is also at the middle point of length and breadth of the rectangle.

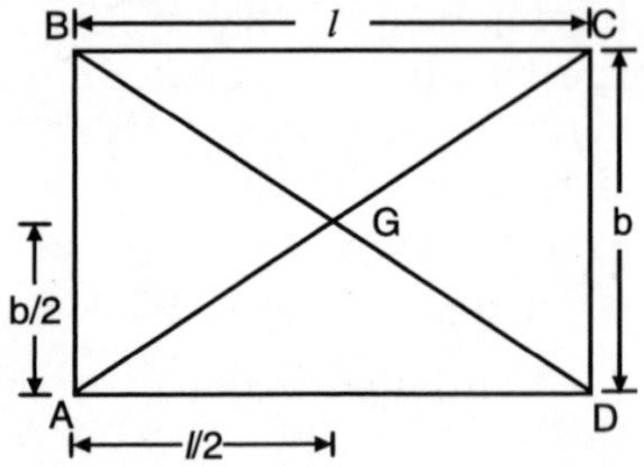

Fig. 4.2.

$\overline{X}$ = Distance of C.G. from point A along the horizontal line = l/2

$\overline{Y}$ = Distance of C.G. from point A along the vertical line = b/2

(*b*) Parallelogram

Area $= l \times h$

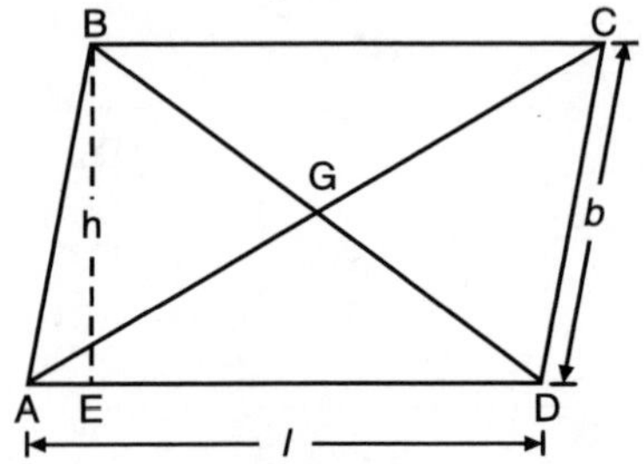

Fig. 4.3.

$\overline{X}$ = Distance of C.G. (*G*) from point A along *AD* = l/2

$\overline{Y}$ = Distance of C.G. (*G*) from point A along vertical line = *h*/2

(c) Triangle

C.G. lies at point where the three medians of the triangle meet (A median is the line joining the vertex and middle point of the opposite side).

AD, BE and CF are the medians.

$$\text{Area} = \frac{1}{2} \times b \times h$$

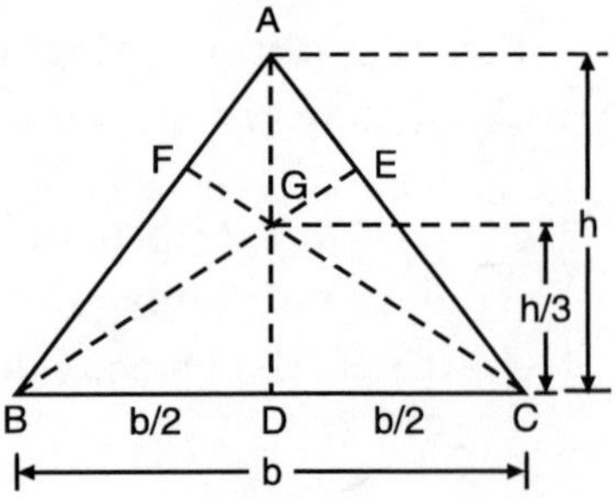

Fig. 4.4.

$$\overline{X} = \frac{b}{2}$$

and $$\overline{Y} = \frac{h}{3}$$

Similarly,

$$\overline{X} = \frac{l + a}{3} \text{ from o.}$$

$$\overline{Y} = \frac{h}{3} \text{ from ox.}$$

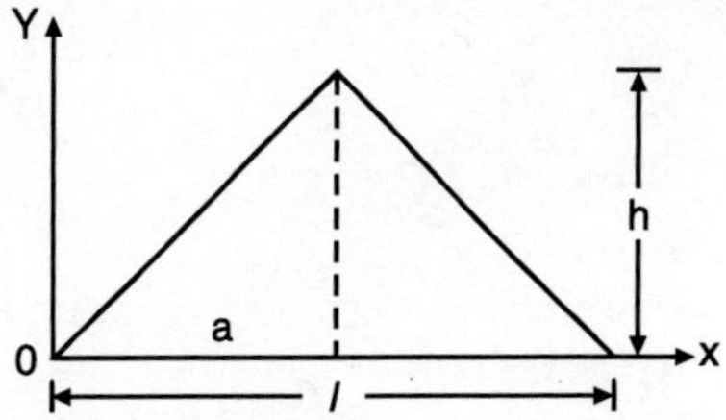

Fig. 4.5(i).

$$\text{Area} = \frac{1}{2} bh$$

$$\overline{X} = b/3,$$

$$\overline{Y} = h/3$$

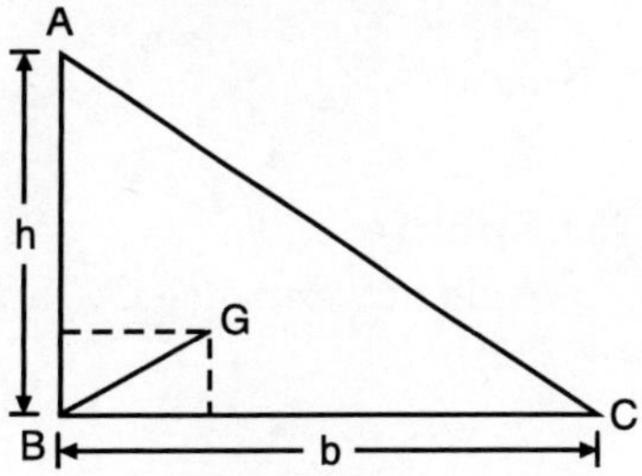

Fig. 4.5(ii).

(d) Circle

C.G. lies at the centre.

$$\text{Area} = \frac{\pi}{4} d^2$$

$$\overline{X} = d/2$$

and $$\overline{Y} = d/2$$

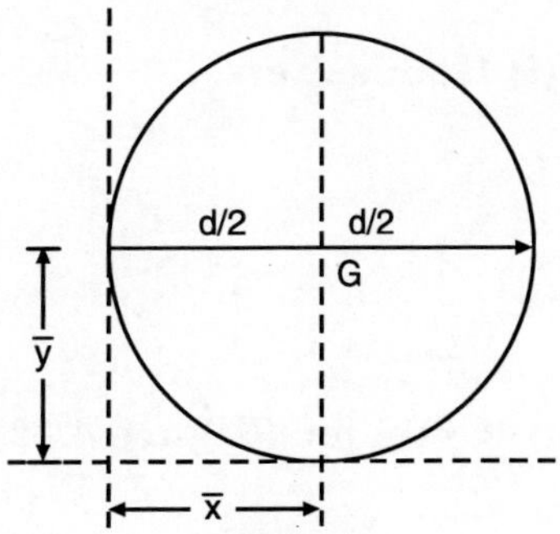

Fig. 4.6.

(e) Semicircle

$$OB = BA = BC = r$$

$$\text{Area} = \frac{\pi d^2}{8}$$

$$\overline{X} = d/2$$

and $$\overline{Y} = \frac{4r}{3\pi}$$

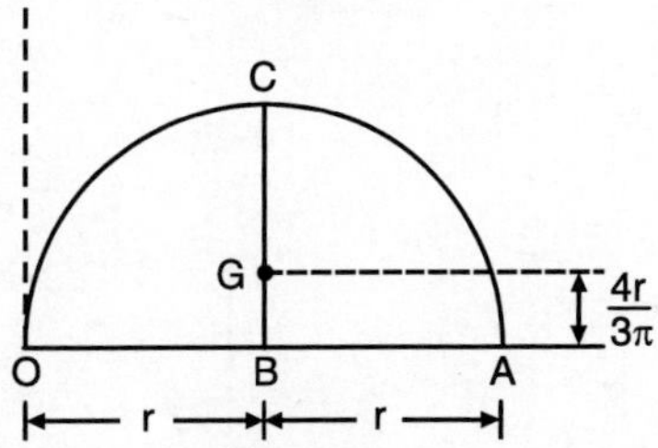

Fig. 4.7.

(f) Quadrant

$$\text{Area} = \frac{1}{16} \pi d^2$$

$$\overline{X} = \frac{4r}{3\pi}$$

$$\overline{Y} = \frac{4r}{3\pi}$$

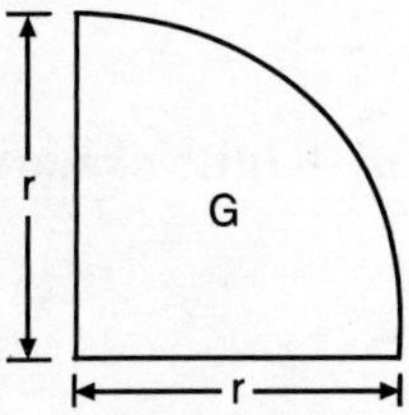

Fig. 4.8.

(g) Trapezium

$$\text{Area} = \frac{1}{2}(a + b) \times h$$

$$\overline{X} = \frac{a^2 + b^2 + ab}{3(a + b)}$$

$$\overline{Y} = \left(\frac{b+2a}{a+b}\right) \times \frac{h}{3}$$

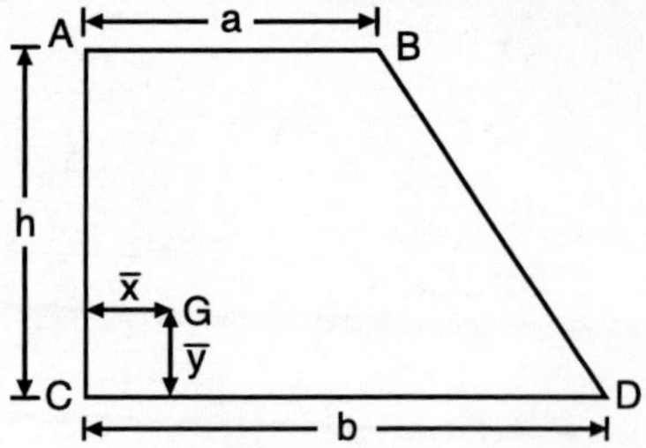

Fig. 4.9(i).

Similarly,

$$\text{Area} = \frac{1}{2}(a+b) \times h$$

$$\overline{X} = \left(\frac{a+2b}{a+b}\right) \times \frac{h}{3}$$

$$\overline{Y} = \frac{a^2+b^2+ab}{3(a+b)}$$

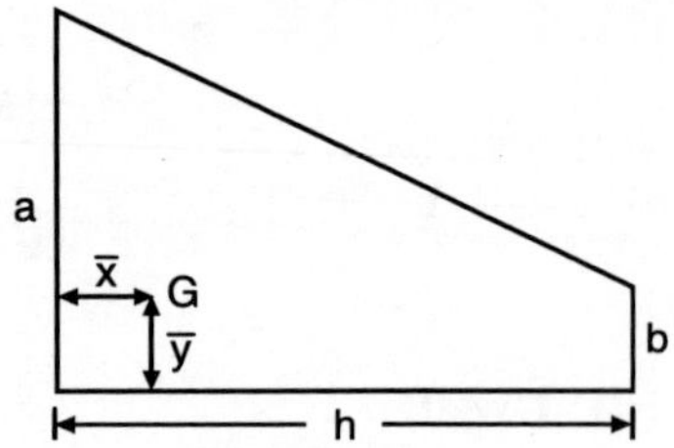

Fig. 4.9(ii).

(*h*) Sphere

Axis of reference OX and OY/C.G. lies at the centre

$$\text{Volume} = \frac{4}{3}\pi r^3$$

$$\overline{Y} = r$$

$$\overline{X} = 0$$

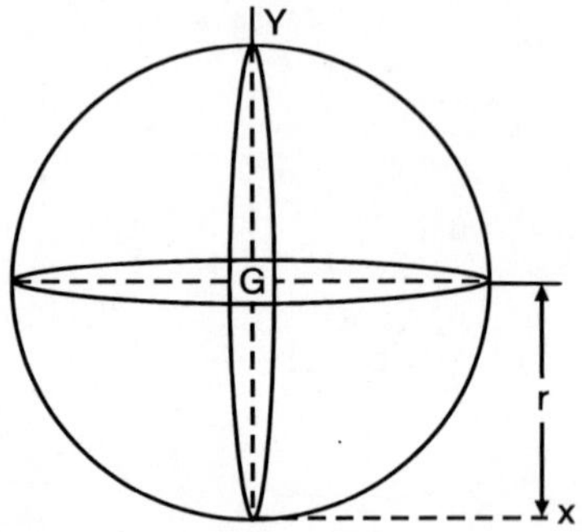

Fig. 4.10.

(*i*) Hemisphere

$$\text{Volume} = \frac{2}{3}\pi r^3$$

$$\overline{Y} = \frac{3r}{8} \text{ from } O.$$

$$\overline{X} = 0$$

(*j*) Cylinder (Figure 4.12)

$$\text{Volume} = \pi r^2 h$$

$$\overline{Y} = h/2$$

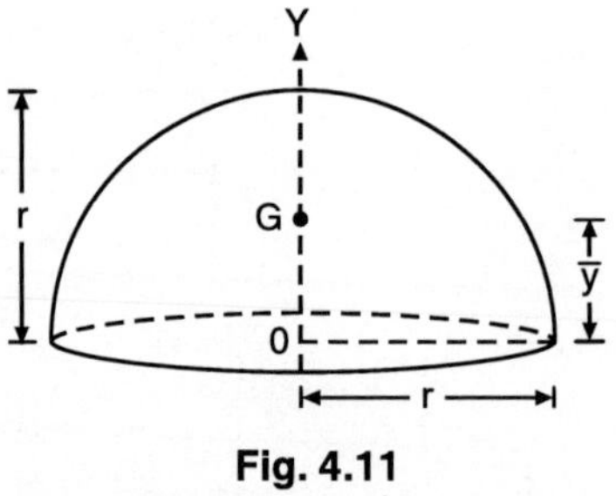

Fig. 4.11

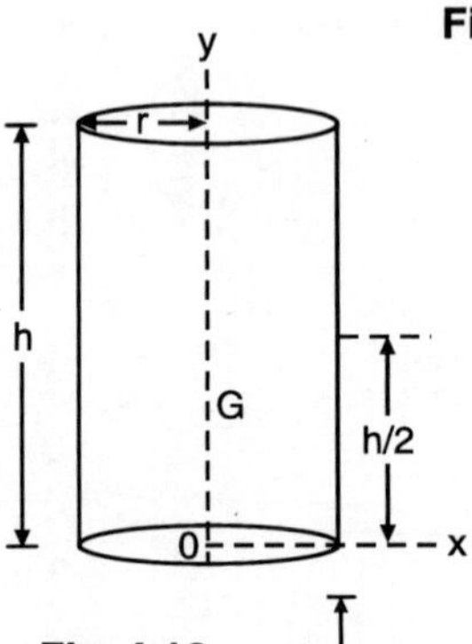

Fig. 4.12

(*k*) Right circular cone (Figure 4.13)

$$\text{1Volume} = \frac{1}{3}\pi r^2 h$$

$$\overline{Y} = h/4$$

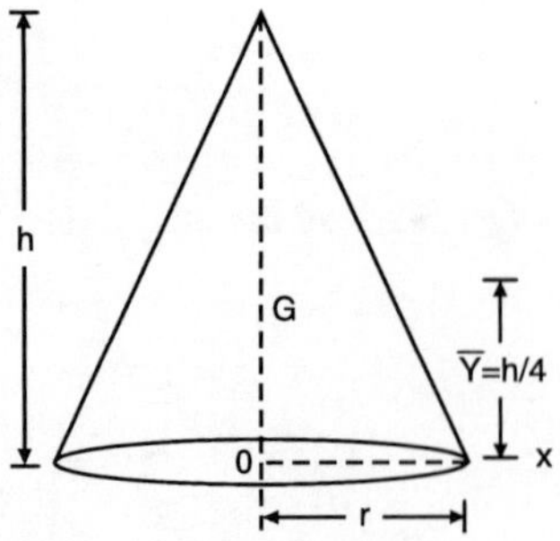

Fig. 4.13.

4.5 CENTRE OF GRAVITY BY GRAPHICAL METHOD STEPS

1. Consider a *Z*-section as shown in the figure 4.14 (*a*) Divide this *Z*-section in three rectangles A B IH , M IC J and JD E F having a_1, a_2 and a_3 respectively.
2. Treat the area of the rectangles a_1, a_2 and a_3 as three vertical forces acting through the centroids of the respective rectangle. Now draw vectors *ab*, *bc* and *cd* to represent a_1, a_2, a_3 to some suitable scale. Suppose a_1 = 20 cm², a_2 = 30 cm² and a_3 = 25 cm² then Choose scale 1 1 cm = 10 cm² and draw vectors *ab* = 2 cm, *bc* = 3 cm and *cd* = 2.5 cm.
3. Select suitable point o and join *ao*, *bo*, *co* and *do*.
4. Draw vertical lines through the centroids of the rectangles.
5. Select a suitable point A_1 on the vertical line of rectangles ABIH and draw a line A_1Z_1 parallel to *oa*. Similarly draw lines A_1 A_2, A_2A_3 and A_3Z_2 parallel to lines *ob*, *oc* and *od*.

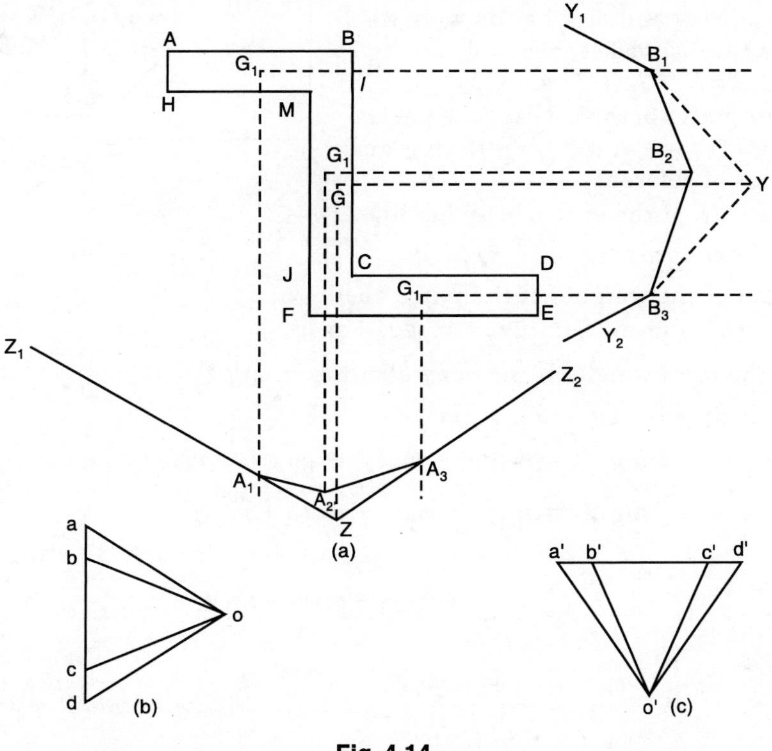

Fig. 4.14.

6. Now produce the lines A_1Z_1 and A_3Z_2 to meet at *Z*. Draw a vertical line through *Z*.
7. Now treat the area of the rectangles a_1, a_2 and a_3 as three parallel horizontal forces acting through the centroid of the respective rectangles. Draw horizontal vectors *a′b′*, *b′c′* and *c′d′* to represent a_1, a_2 and a_3 to some suitable scale.
8. Select any point *o′* and join *o′a′*, *o′b′*, *o′c′*, *o′d′*.
9. Draw horizontal lines form the centroids of the three rectangles.
10. Select any suitable point B_1 on the horizontal line of rectangle ABIH and draw a line B_1Y_1 parallel to *o′a′*. Draw a line B_1B_2 parallel to *o′b′* from point B_1. Similarly draw line B_2B_3 and B_3Y_2 parallel to *o′c′* and *o′d′* respectively.
11. Produce lines B_1Y_1 and B_3Y_2 to meet at *Y*. Now draw a horizontal line through *Y*.

12. The point at which the vertical line through Z and horizontal line through Y meet is the required C.G. *i.e.*, point G in this figure.

Note: This method can be used for any type of section for finding out its C.G. Suppose if a section is symmetrical about x axis then the C.G. will lie on that axis *i.e.*, $\overline{Y}$ is known. Hence to calculate the $\overline{X}$ only vertical part of the graph is to be drawn.

4.6 CENTRE OF GRAVITY BY MOMENT METHOD

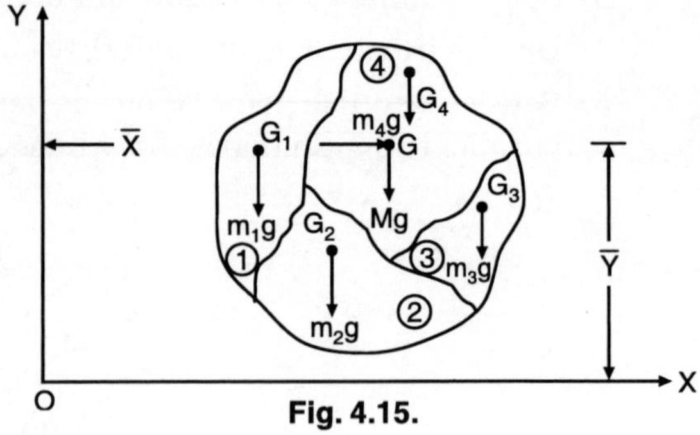

Fig. 4.15.

Consider a body of mass M whoe C.G. is to be found out. Divide the body into small parts as shown in the figure 4.15.

Let m_1, m_2, m_3 and m_4 be the masses of the four parts in which the body is divided.

The m_1g, m_2g, m_3g and m_4g be the weights of the four parts 1, 2, 3 and 4 respectively acting through their C.G's G_1, G_2, G_3 and G_4.

Let the co-ordinates of the C.G's of the parts be (x_1, y_1), (x_2, y_2) (x_3, y_3) and (x_4, y_4) taking ox and oy as the axis of reference.

Let G be the C.G. of the entire body having weight Mg and its co-ordinate is $(\bar{x}, \bar{y})$.

Now, moment of the weights of the parts about

$$OX = m_1gy_1 + m_2gy_2 + m_3gy_3 + m_4gy_4$$

Moment of the total weight of the body about $ox = Mg\,\overline{Y}$

From the principle of moments, we get

$$Mg\,\overline{Y} = m_1gy_1 + m_2gy_2 + m_3gy_3 + m_4gy_4$$

or

$$Mg\,\overline{Y} = m_1y_1 + m_2y_2 + m_3y_3 + m_4y_4$$

(By dividing both sides by g)

or

$$\overline{Y} = \frac{m_1y_1 + m_2y_2 + m_3y_3 + m_4y_4}{M}$$

or

$$\overline{Y} = \frac{\Sigma my}{M}$$

(where $M = m_1 + m_2 + m_3 + m_4$)

Similarly, Moment of the weights of the parts about OY

$$= m_1gx_1 + m_2gx_2 + m_3gx_3 + m_4gx_4$$

Moment of the weight of the body about OY

$$= Mg\,\overline{X}$$

From principle of moments, we get

$$Mg\,\overline{X} = m_1gx_1 + m_2gx_2 + m_3gx_3 + m_4gx_4$$

or

$$Mg\,\overline{X} = m_1x_1 + m_2x_2 + m_3x_3 + m_4x_4$$

or

$$\overline{X} = Mg\,\frac{m_1x_1 + m_2x_2 + m_3x_3 + m_4x_4}{M}$$

or $$\bar{X} = \frac{\Sigma mx}{M} \text{ where } M = m_1 + m_2 + m_3 + m_4$$

If we consider a body which does not have mass but have area only *i.e.*, lamina then the centroid of the lamina may be found out in the similar way. In this case, instead mass of the parts we consider area of the parts in which the body is divided.

Let a lamina of area A be divided into n number of small parts having areas $a_1, a_2, a_3 \ldots. a_n$.

Let the co-ordinates of the points at which these areas are acting, *i.e.*, centroid of the small parts be (x_1, y_1) (x_2, y_2), (x_3, y_3) ...(x_n, y_n) from a reference axes and $(\bar{x}, \bar{y})$ be the co-ordinate of the centroid of the entire lamina then

$$\bar{X} = \frac{a_1x_1 + a_2x_2 + a_3x_3 + \ldots\ldots\ldots a_nx_n}{A}$$

$$\bar{X} = \frac{\Sigma ax}{A} \text{ where } A = a_1 + a_2 + a_3 + \ldots a_n$$

Similarly, $$\bar{Y} = \frac{a_1y_1 + a_2y_2 + a_3y_3 + \ldots\ldots a_ny_n}{A}$$

or $$\bar{Y} = \frac{\Sigma ay}{A} \text{ where } A = a_1 + a_2 + a_3 + \ldots. a_n$$

4.7 AXIS OF REFERENCE

The C.G. of a body is always determined with respect to some assumed axis. The axis from which the distance of the C.G. is measured is called axis of reference. For different axis of reference, the distance of the C.G. will be different.

EXAMPLE : Consider an angle section {Figure 4.16 (*i*)} whose C.G is at point *G*. If the axis of reference is *ox* then, the distance of C.G. from *ox* is 3 cm.

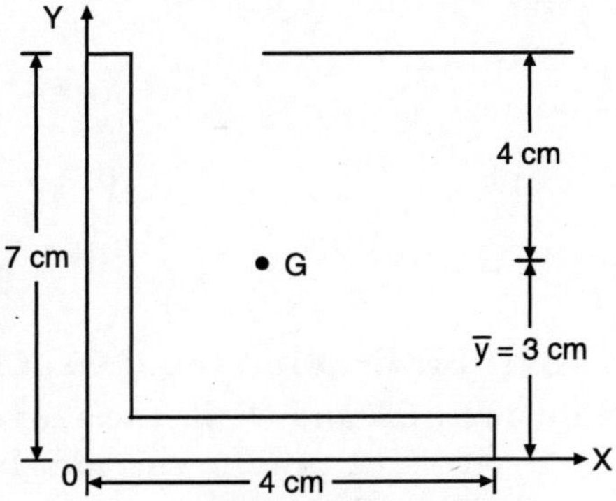

Fig. 4.16 (*i*).

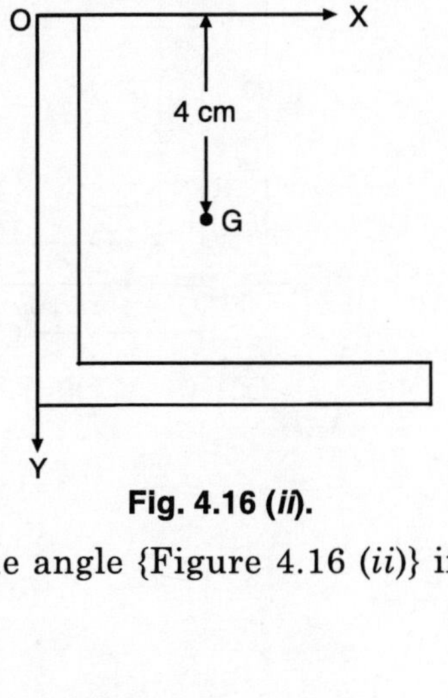

Fig. 4.16 (*ii*).

> **DO YOU KNOW?**
>
> The location of centre of gravity for a body remains fixed in relation to the body regardless of the orientation of the body.

For the same angle {Figure 4.16 (*ii*)} if the axis of reference is ox, then the distance of C.G. is 4 cm.

4.8 STEPS TO FIND OUT THE CENTRE OF GRAVITY OF ANY SECTION BY MOMENT METHOD

(*a*) Choose the axis of reference. The axes should be so chosen that it should touch or pass through the nearest line of the figure.

(*b*) Check whether the section is symmetrical about any axis. If it is symmetrical about any axis then the C.G. will lie on that axis.

(*c*) The complete figure should be divided into small regular figures such as rectangles, triangles, circles etc.

(*d*) Find the position of the C.G. by calculating $\bar{X}$ and $\bar{Y}$ using the formula.

(*e*) Suppose if any area is removed from the figure (as shown in figure 4.17) then the area of the figure removed is substracted from the total area and the moment of removed area is taken as negative.

A circular portion is removed from the square as shown in figure 4.17.

Let a_1 and a_2 be the area of the square and circle respectively having co-ordinates of the centroids (x_1, y_1) and (x_2, y_2).

Let $(\bar{X}, \bar{Y})$ be the co-ordinate of the centroid of the remainder having area A. Then $A = a_1 - a_2$

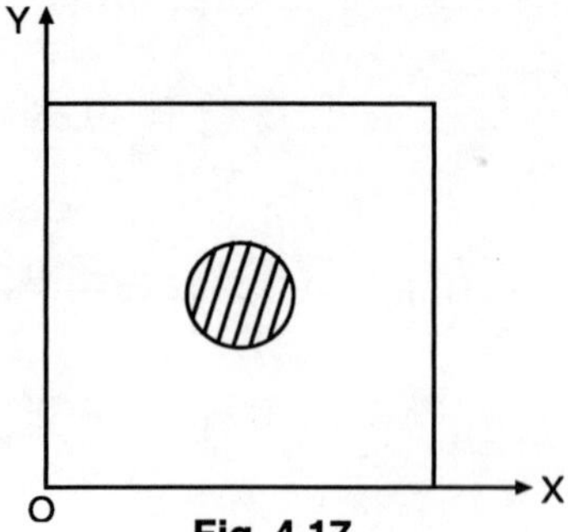

Fig. 4.17.

$$\bar{X} = \frac{a_1 x_1 - a_2 x_2}{A}$$

and

$$\bar{Y} = \frac{a_1 y_1 - a_2 y_2}{A}$$

SOLVED EXAMPLES

EXAMPLE 4.1. Find the C.G. of an angle section 100 mm × 60 mm × 20 mm.

Solution : OX and OY be the axes of reference.

Divide the angle section into two rectangles ① and ②

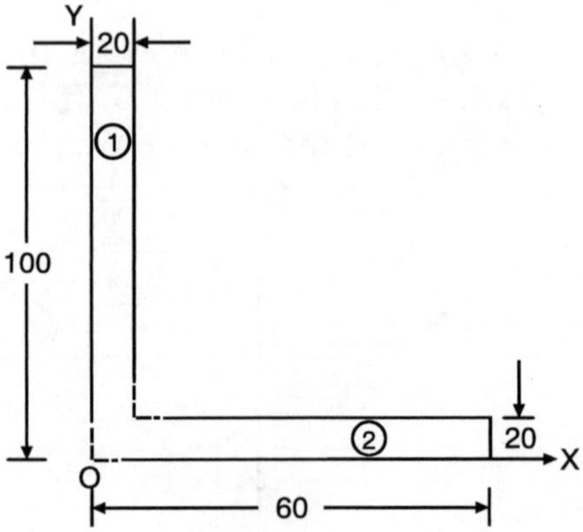

Fig. 4.18.

Area of rectangle ①

$$a_1 = 80 \times 20 = 1600 \text{ mm}^2$$

Area of rectangle ②

$$a_2 = 60 \times 20 = 1200 \text{ mm}^2$$

Let (x_1, y_1) and (x_2, y_2) be the co-ordinates of the centroids of rectangles ① and ② respectively.

$$\therefore \quad x_1 = \frac{20}{2} = 10 \text{ mm}$$

$$x_2 = \frac{60}{2} = 30 \text{ mm}$$

$$y_1 = \frac{80}{2} + 20 = 60 \text{ mm}$$

$$y_2 = \frac{20}{2} = 10 \text{ mm}$$

We know,

$$\bar{X} = \frac{a_1x_1 + a_2x_2}{a_1 + a_2} \quad ...(1)$$

Putting the values of a_1, x_1, a_2, x_2 in (1) we get.

$\bar{X}$ = Distance of the C.G. of the angle sections from OY axis

$$= \frac{1600 \times 10 + 1200 \times 30}{1600 + 1200} = \frac{52000}{2800} = 18.57 \text{ mm. } \textbf{Ans.}$$

Again $\bar{Y}$ = Distance of the C.G. of the angle section from OX axis.

$$= \frac{a_1y_1 + a_2y_2}{a_1 + a_2} = \frac{1600 \times 60 + 1200 \times 10}{1600 + 1200} = 38.57 \text{ mm. } \textbf{Ans.}$$

The same problem can be solved in the tabular from, shown below :

Components	**Area (a) mm²**	**Dist. of C.G. from OY (x) mm**	**Dist. of C.G. from OX (y) mm**	ax	ay
Rectangle ①	20 × 80 = 1600	10	60	1600 × 10 = 16000	96000
Rectangle ②	20 × 60 = 1200	30	10	1200 × 30 = 36000	12000
	Σa = 2800			Σax = 52000	Σay = 108000

$$\bar{X} = \frac{\Sigma ax}{\Sigma a} = \frac{52000}{2800} = 18.57 \text{ mm } \textbf{Ans.}$$

$$\bar{Y} = \frac{\Sigma ay}{\Sigma a} = \frac{108000}{2800} = 38.57 \text{ mm } \textbf{Ans.}$$

EXAMPLE 4.2. Find the C.G of the *z*-section shown in the Fig. 4.19.

(B.T.E., Delhi, Feb. 1997)

Solution: Divide the z-section in three rectangles (1), (2) and (3) as shown in the figure 4.19.

Let OX and OY be the axes of reference.

Let a_1, a_2 and a_3 be the area of rectangles ①, ② and ③ respectively having co-ordinates of the centroids of the rectangles (x_1, y_1), (x_2, y_2) and (x_3, y_3). wr.t to the reference axes.

Let $(\bar{X}, \bar{Y})$ be the co-ordinate of the centroid of the z-section. Then,

$$a_1 = 10 \times 2.5 = 25 \text{ cm}^2$$

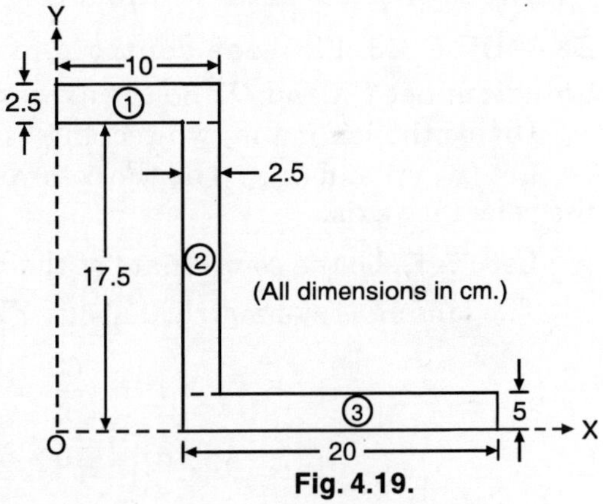

Fig. 4.19.

$$a_2 = 12.5 \times 2.5 = 31.25 \text{ cm}^2$$

$$a_3 = 20 \times 5 = 100 \text{ cm}^2$$

$$x_1 = \frac{10}{2} = 5 \text{ cm}$$

$$x_2 = (10 - 2.5) + \frac{2.5}{2} = 7.5 + 1.25 = 8.75 \text{ cm}$$

$$x_3 = (10 - 2.5) + \frac{20}{2} = 7.5 + 10 = 17.5 \text{ cm}$$

and $$y_1 = 17.5 + \frac{2.5}{2} = 18.75 \text{ cm}$$

$$y_2 = 5 + \frac{(17.5 - 5)}{2} = 5 + \frac{12.5}{2} = 5 + 6.25 = 11.25 \text{ cm}$$

$$y_3 = \frac{5}{2} = 2.5 \text{ cm.}$$

We know,

$$\bar{X} = \frac{a_1x_1 + a_2x_2 + a_3x_3}{a_1 + a_2 + a_3}$$

or $$\bar{X} = \frac{25 \times 5 + 31.25 \times 8.75 + 100 \times 17.5}{25 + 31.25 + 100}$$

$$\bar{X} = \frac{125 + 273.44 + 1750}{156.25} = \frac{2148.44}{156.25}$$

$\therefore$ $$\bar{X} = 13.75 \text{ cms}$$

and $$\bar{Y} = \frac{a_1y_1 + a_2y_2 + a_3y_3}{a_1 + a_2 + a_3}$$

or $$\bar{Y} = \frac{25 \times 18.75 + 31.25 \times 11.25 + 100 \times 2.5}{25 + 31.25 + 100}$$

or $$\bar{Y} = \frac{468.75 + 351.56 + 250}{156.25} = \frac{1070.31}{15.25}$$

$\therefore$ $$\bar{Y} = 6.85 \text{ cm.}$$

The co-ordinate of the centroid is (13.75, 6.85). **Ans.**

EXAMPLE 4.3. Find the centroid of the lamina shown in the figure 4.20.

Solution: Let *OX* and *OY* be the axes of reference.

Divide the lamina in two rectangles (1) and (2) having area a_1 and a_2 respectively.

Let (x_1, y_1) and (x_2, y_2) be the co-ordinates of the centroids of the rectangles 1 and 2 from the reference axis.

Let $(\bar{x}, \bar{y})$ be the co-ordinate of the centroid of the lamina from the reference axis.

The lamina is symmetrical about *YY* axis and hence centroid will lie on that axis.

$\therefore$ $$\bar{X} = \frac{60}{2} = 30 \text{ mm}$$

$$a_1 = 10 \times 50 = 500 \text{ mm}^2$$

$$a_2 = 60 \times 10 = 600 \text{ mm}^2$$

$$y_1 = 10 + \frac{50}{2} = 35 \text{ mm}$$

$$y_2 = \frac{10}{2} = 5 \text{ mm}$$

We know, $\bar{Y} = \dfrac{a_1 y_1 + a_2 y_2}{a_1 + a_2}$

or $\bar{Y} = \dfrac{500 \times 35 \times 600 \times 5}{500 + 600}$

$$= \frac{17500 + 3000}{1100}$$

$\therefore$ $\bar{Y} = 18.64$ mm

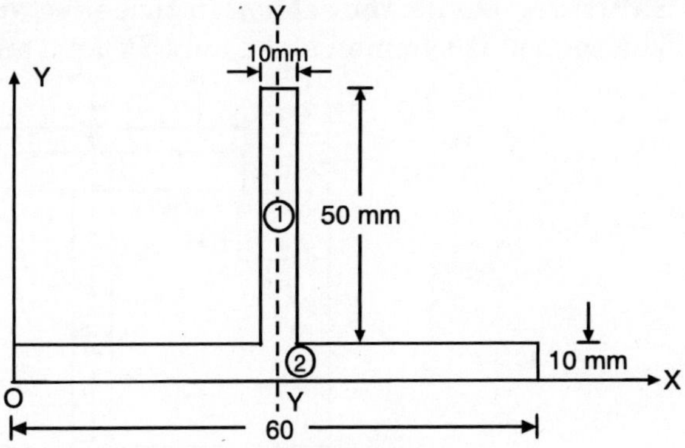

Fig. 4.20.

Hence centroid will lie at (30, 18.64) from reference axis. **Ans.**

EXAMPLE 4.4. Find the C.G. of a channel section shown in figure 4.21.

Solution: Let OX and OY be the axis of reference.

Te channel section is symmetrical about xx axis and hence the C.G. will lie on that axis.

$\therefore$ $\bar{Y} = \dfrac{400}{2} = 200$ mm.

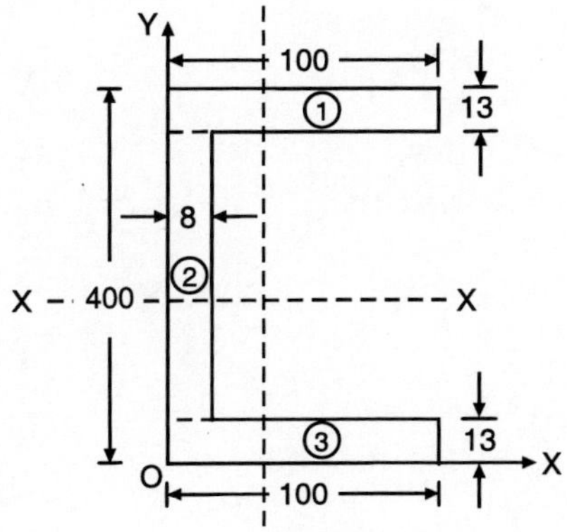

Fig. 4.21.

Divide the channel section into three rectangles ①, ② and ③ as shown in figure 4.21.

Let a_1, a_2, a_3 be the area of rectangles ①, ② and ③ and x_1, x_2, x_3 be distance of C.G of rectangles ①, ② and ③ from OY axis.

$$a_1 = 100 \times 13 = 1300 \text{ mm}^2$$

$$a_2 = 374 \times 8 = 2992 \text{ mm}^2 \ (374 = 400 - 13 - 13)$$

$$a_3 = 100 \times 13 = 1300 \text{ mm}^2$$

$$x_1 = \frac{100}{2} = 50 \text{ mm}$$

$$x_2 = \frac{8}{2} = 4\text{mm}$$

$$x_3 = \frac{100}{2} = 50 \text{ mm}$$

$$\bar{x} = \frac{a_1 x_1 + a_2 x_2 + a_3 x_3}{a_1 + a_2 + a_3}$$

$$= \frac{1300 \times 50 + 2992 \times 4 + 1300 \times 50}{(1300 + 2992 + 1300)} = 25.38 \text{ mm}$$

The C.G. of the channel section lies at (25.38, 200) from the reference axes. **Ans.**

EXAMPLE 4.5. The figure 4.22 shows an unsymmetrical I-section, the upper flange is 60 mm × 7.5 mm, the lower flange 120 × 10 mm, the overall depth is 160 mm and the thickness of the metal web is 5 mm. Find the C.G. of the section.

Solution: Divide the section in three rectangles (1), (2) and (3) as shown in the figure 4.22. The section is symmetrical about YY axis and the C.G. will lie on that axis.

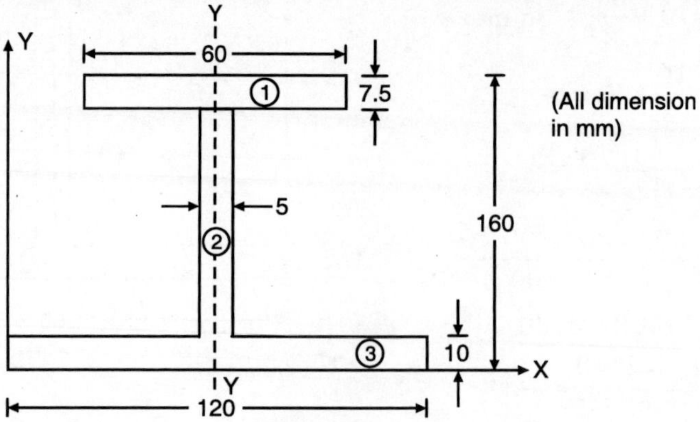

Fig. 4.22.

$\therefore \quad \bar{x}$ = Distance of C.G. from OY axis

$$X = \frac{120}{2} = 60 \text{ mm} \quad \textbf{Ans.}$$

Now,
$$a_1 = \text{Area of rectangle (1)} = 60 \times 7.5 = 450 \text{ mm}^2$$
$$a_2 = \text{Area of rectangle (2)} = 142.5 \times 5 = 712.5 \text{ mm}^2$$
$$a_3 = \text{Area of rectangle (3)} = 120 \times 10 = 1200 \text{ mm}^2.$$

$$y_1 = 10 + 142.5 + \frac{7.5}{2} = 156.25 \text{ mm}$$

$$y_2 = 10 + \frac{142.5}{2} = 81.25 \text{ mm}$$

$$y_3 = \frac{10}{2} = 5 \text{ mm}$$

We know
$$\bar{Y} = \frac{a_1 y_1 + a_2 y_2 + a_3 y_3}{a_1 + a_2 + a_3}$$

or
$$\bar{Y} = \frac{450 \times 156.25 + 712.5 \times 81.25 + 1200 \times 5}{450 + 712.5 + 1200}$$

or
$$\bar{Y} = \frac{70312.5 + 57890.6 + 6000}{2362.5} = 56.80 \text{ mm} \quad \textbf{Ans.}$$

EXAMPLE 4.6. Find the position of centroid of plane area as shown in figure 4.23.
(B.T.E., Delhi, June 2000)

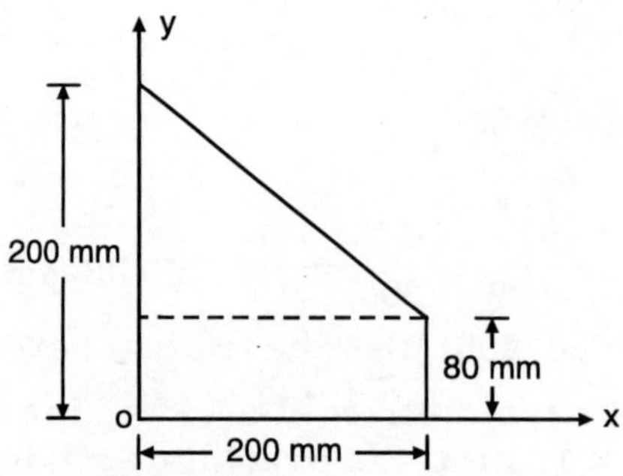

Fig. 4.23.

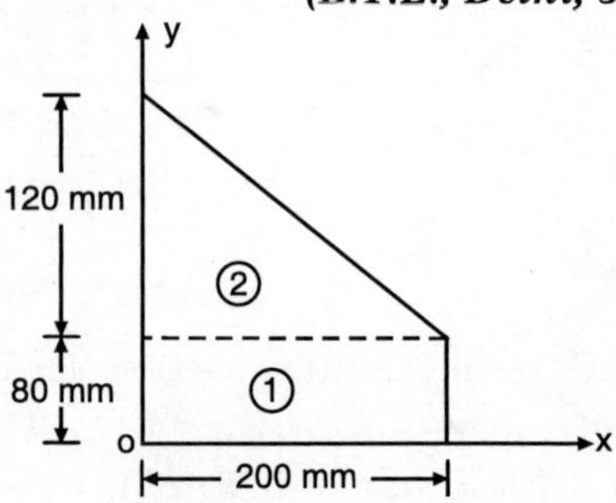

Fig. 4.24.

Solution: Let OX and OY be the axis of reference

Divide the plane area into two sections

① Rectangle

② Triangle

Let a_1 and a_2 be the area of rectangle ① and triangle ② respectively.

$$a_1 = 200 \times 80 = 16000 \text{ mm}^2$$

$$a_2 = \frac{1}{2} \times 200 \times 120 = 12000 \text{ mm}^2$$

Let (x_1, y_1) and (x_2, y_2) be the co-ordinates of the centroids of rectangle ② and triangle ② respectively.

Let $(\overline{x}, \overline{y})$ be the co-ordinate of the centroid of the entire plane area.

$$x_1 = \frac{200}{2} = 100 \text{ mm}$$

$$y_1 = \frac{80}{2} = 40 \text{ mm}$$

$$x_2 = \frac{200}{3} = 66.67 \text{ mm}$$

$$y_2 = 80 + \frac{120}{3} = 120 \text{ mm}$$

We know,

$$\overline{x} = \frac{a_1 x_1 + a_2 x_2}{a_1 + a_2}$$

$$\overline{x} = \frac{16000 \times 100 + 12000 \times 66.67}{16000 + 12000} = 85.71 \text{ mm}$$

and

$$\overline{y} = \frac{a_1 y_1 + a_2 y_2}{a_1 + a_2}$$

$$\overline{y} = \frac{16000 \times 40 + 12000 \times 120}{16000 + 12000} = 74.28 \text{ mm}$$

The centroid of the plane area with respect to the reference axis is at (85.71, 74.28). **Ans.**

EXAMPLE 4.7. Locate the centroid of the piece of sheet metal shown in the figure 4.25.

Solution: Divide the sheet metal into the three sections

1. Semicircle
2. Rectangle
3. Triangle

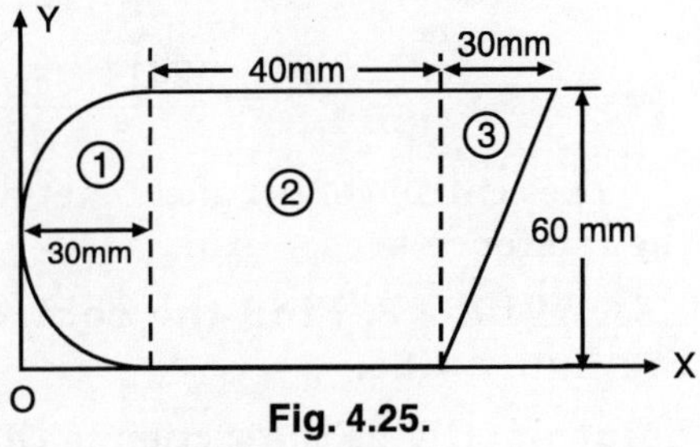

Fig. 4.25.

a_1 =Area of semicircle

$$= \frac{1}{2}\pi r^2 = \frac{1}{2} \times \pi \times (30)^2$$

$$= 1413.72 \text{ mm}^2$$

a_2 = Area of rectangle

$$= 40 \times 60 = 2400 \text{ mm}^2$$

$$a_3 = \text{Area of triangle} = \frac{1}{2} \times 30 \times 60 = 900 \text{ mm}^2$$

x_1 = Distance of C.G. of section (1) from OY

$$= 30 - \frac{4r}{3\pi} = 30 - \frac{4 \times 30}{3\pi}$$

($\because$ C.G. of semicircle is $\frac{4r}{3\pi}$ from base)

$$= 30 - 12.73 = 17.27 \text{ mm}$$

x_2 = Distance of C.G. of section (2) from OY

$$= 30 + \frac{40}{2} = -50 \text{ mm}$$

x_3 = Distance of C.G. of section (3) from OY

$$30 + 40 + \frac{30}{3} = 80 \text{ mm}$$

y_1 = Distance of C.G. of section (1) from OX

$$= \frac{60}{2} = 30 \text{ mm}$$

$$y_2 = \frac{60}{2} = 30 \text{ mm}$$

$$y_3 = 60 - \frac{60}{3} = 40 \text{ mm}$$

Now we know,

$$\bar{X} = \frac{a_1 x_1 + a_2 x_2 + a_3 x_3}{a_1 + a_2 + a_3}$$

or
$$\bar{X} = \frac{1413.72 \times 17.27 + 2400 \times 50 + 900 \times 80}{1413.72 + 2400 + 900}$$

$$= \frac{24414.94 + 120000 + 72000}{4713.72} = 45.91 \text{ mm}$$

and
$$\bar{y} = \frac{a_1 y_1 + a_2 y_2 + a_3 y_3}{a_1 + a_2 + a_3}$$

or
$$\bar{y} = \frac{1413.72 \times 30 + 2400 \times 30 + 900 \times 40}{1413.72 + 2400 + 900}$$

$$= \frac{42411.6 + 72000 + 36000}{4713.72} = 31.91 \text{ mm}$$

The centroid of the sheet metal will lie at (45.91, 31.91) from the reference axes. **Ans.**

EXAMPLE 4.8. Find the centroid of a section shown in the figure 4.26.

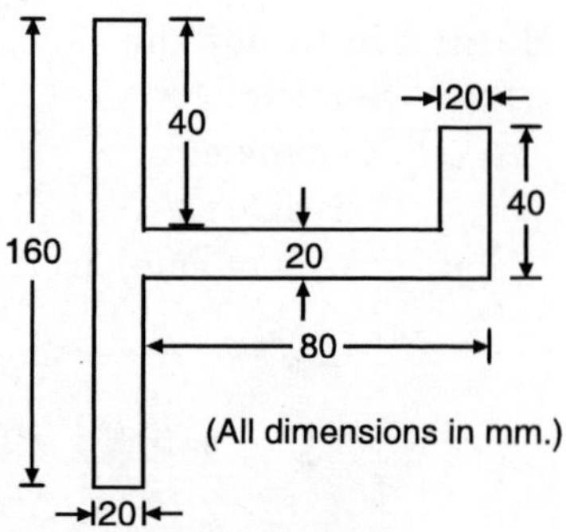

Fig. 4.26 (i)

Solution: Divide the section in three rectangles (1), (2) and (3) as shown in the figure 4.26 (i).

Let OX and OY be the axes of reference.

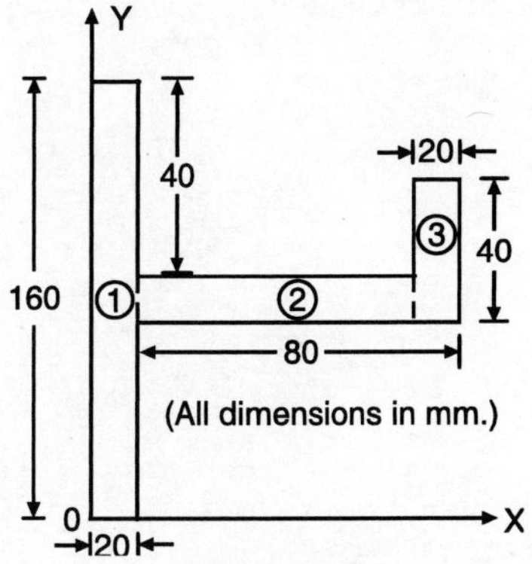

Fig. 4.26 (ii).

$$a_1 = 160 \times 20 = 3200 \text{ mm}^2$$

$$a_2 = 60 \times 20 = 1200 \text{ mm}^2$$

$$a_3 = 20 \times 40 = 800 \text{ mm}^2$$

$$x_1 = \frac{20}{2} = 10 \text{ mm}$$

$$x_2 = 20 + \frac{60}{2} = 50 \text{ mm}$$

$$x_3 = 20 + 60 + \frac{20}{2} = 90 \text{ mm}$$

$$y_1 = \frac{160}{2} = 80 \text{ mm}$$

$$y_2 = 100 + \frac{20}{2} = 110 \text{ mm}$$

$$y_3 = 100 + \frac{40}{2} = 120 \text{ mm}$$

Now, $$\bar{X} = \frac{a_1 x_1 + a_2 x_2 + a_3 x_3 y_3}{a_1 + a_2 + a_3}$$

$$= \frac{3200 \times 10 + 1200 \times 50 + 800 \times 90}{3200 + 1200 + 800}$$

or, $$\bar{X} = \frac{32000 + 60000 + 72000}{5200}$$

$\therefore$ $$\bar{X} = 31.54 \text{ mm} \quad \textbf{Ans.}$$

and $$\bar{y} = \frac{a_1 y_1 + a_2 y_2 + a_3 y_3}{a_1 + a_2 + a_3}$$

or, $$\bar{y} = \frac{3200 \times 80 + 1200 \times 110 + 800 \times 120}{3200 + 1200 + 800}$$

or, $$\bar{y} = \frac{256000 + 132000 + 96000}{5200} = 93.08 \text{ mm} \quad \textbf{Ans.}$$

EXAMPLE 4.9. Find the position of C.G. of a wall section in the Fig. 4.27.

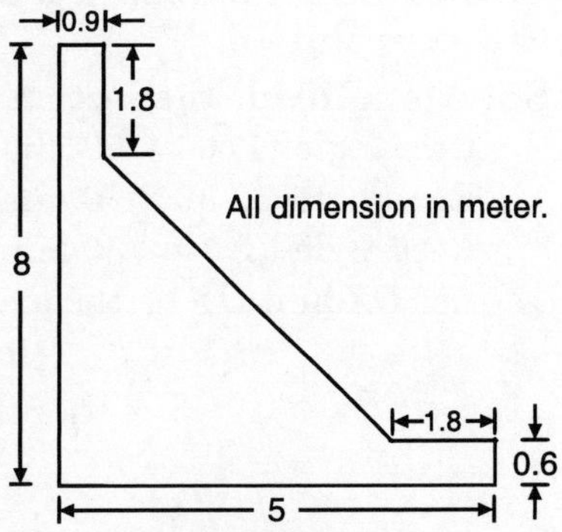

Fig. 4.27 (*i*).

Solution: Divide the figure in three sections as shown in the figure 4.27 (*i*).

Section (1) Rectangle

Section (2) Triangle

Section (3) Rectangle

$$a_1 = 8 \times 0.9 = 7.2 \text{ m}^2$$

$$a_2 = \frac{1}{2} \times 2.3 \times 5.6 = 6.44 \text{ m}^2$$

$$a_3 = 4.1 \times 0.6 = 2.46 \text{ m}^2$$

$$x_1 = \frac{0.9}{2} = 0.45 \text{ m}$$

$$x_2 = 0.9 + \frac{2.3}{3} = 1.67 \text{ m}$$

$$x_3 = 0.9 + \frac{4.1}{3} = 2.95 \text{ m}$$

$$y_1 = \frac{8}{2} = 4 \text{ m}$$

$$y_2 = 0.6 + \frac{5.6}{3} = 2.47 \text{ m}$$

$$y_3 = \frac{0.6}{2} = 0.3 \text{ m}$$

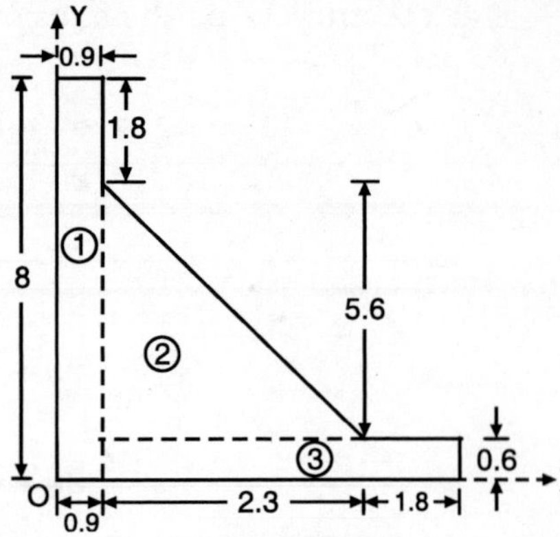

Fig. 4.27 (*ii*).

Now, $$\bar{X} = \frac{a_1 x_1 + a_2 x_2 + a_3 x_3}{a_1 + a_2 + a_3}$$

or $$\bar{X} = \frac{7.2 \times 0.45 + 6.44 \times 1.67 + 2.46 \times 2.95}{7.2 + 6.44 + 2.46}$$

or $$\bar{X} = \frac{3.24 + 10.75 + 7.26}{16.1}$$

$\therefore$ $\bar{X} = 1.32$ m from OY. **Ans.**

and $$\bar{y} = \frac{a_1 y_1 + a_2 y_2 + a_3 y_3}{a_1 + a_2 + a_3}$$

or $$\bar{y} = \frac{7.2 \times 4 + 6.44 \times 2.47 + 2.46 \times 0.3}{7.2 + 6.44 + 2.46}$$

or $$\bar{y} = \frac{28.8 + 15.91 + 0.74}{16.1}$$

$$= \frac{45.45}{16.1} = 2.82 \text{ from } OX.$$ **Ans.**

EXAMPLE 4.10. Locate the centroid of the section in the figure 4.28 with respect to the axes shown.

Solution: Divide the section in three rectangles (1), (2) and (3) as shown in the figure 4.28.

Rectangle (1) 60 × 50 mm
Rectangle (2) 30 × 50 mm
Rectangle (3) 10 × 30 mm
Let OX and OY be the axes of reference.

$$a_1 = 60 \times 50 = 3000 \text{ mm}^2,$$

$$x_1 = \frac{60}{2} = 30 \text{ mm}$$

$$a_2 = 30 \times 50 = 1500 \text{ mm}^2$$

$$x_2 = \frac{50}{2} = 25 \text{ mm}$$

$$a_3 = 10 \times 30 = 300 \text{ mm}^2$$

$$x_3 = \frac{10}{2} = 5 \text{ mm}$$

$$y_1 = 30 + \frac{50}{2} = 55 \text{ mm}$$

$$y_2 = 40 + \frac{30}{2} = 55 \text{ mm}$$

$$y_3 = \frac{30}{2} = 15 \text{ mm}$$

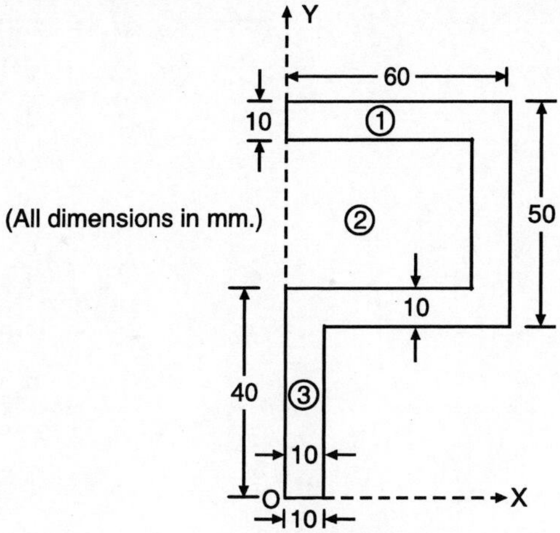

Fig. 4.28.

Let $(\bar{X}, \bar{Y})$ be the co-ordinate of C.G. of the section from the reference axis.

$$\bar{X} = \frac{a_1 x_1 - a_2 x_2 + a_3 x_3}{a_1 - a_2 + a_3}$$

$$= \frac{3000 \times 30 - 1500 \times 25 + 300 \times 5}{3000 - 1500 + 300} = \frac{54000}{1800}$$

∴ $\bar{X} = 30 \text{ mm}$ **Ans.**

and $\bar{y} = \frac{a_1 y_1 - a_2 y_2 + a_3 y_3}{a_1 - a_2 + a_3}$

$$= \frac{3000 \text{ Ÿ } 55 - 1500 \text{ Ÿ } 55 + 300 \text{ Ÿ } 15}{3000 - 1500 + 300} = \frac{87000}{1800}$$

∴ $\bar{y} = 48.33 \text{ mm}$ **Ans.**

EXAMPLE 4.11. From a rectangular lamina ABCD in which AB = 40 cms and BC = 60 cms, a triangular piece ABC is removed such that CO = BO = 25 cms. Calculate the C.G. of the remainder. ***(B.T.E., Delhi, 1997, 1998)***

Solution: Let *ABCD* be the rectangle from which Δ*OBC* is removed. We have to calculate the C.G. of the shaded portion.

Let rectangle be taken as section (1) and triangle as section (2),

In Δ *OBC*, *OE* is perpendicular to BC.

$$BE = EC = 20 \text{ cm}$$

∴ $OE = \sqrt{25^2 - 20^2}$

∴ $OE = \sqrt{625 - 400}$

∴ $OE = \sqrt{225} = 15 \text{ cm}$

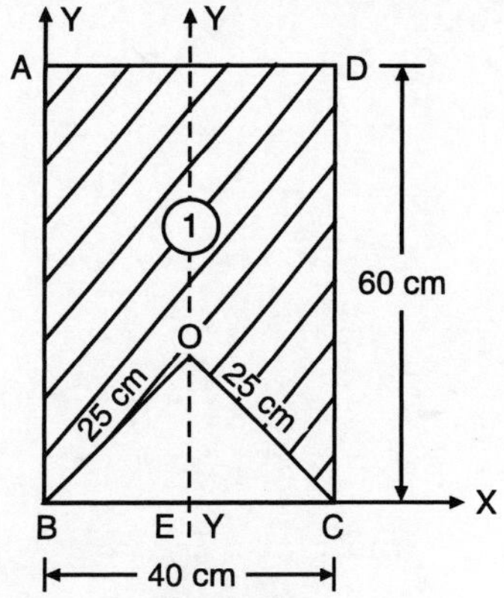

Fig. 4.29.

As the lamina is symmetrical about vertical axis *YY*, the C.G. will lie on this axis.

$$\bar{X} = \frac{40}{2} = 20 \text{ cms}$$

Now, $a_1 = 40 \times 60 = 2400 \text{ cm}^2$

$$a_2 = \frac{1}{2} \times 40 \times 15 = 300 \text{ cm}^2$$

$$y_1 = \frac{60}{2} = 30 \text{ cms}$$

$$y_2 = \frac{15}{3} = 5 \text{ cms}$$

$\bar{y}$ = Distance of C.G. of the shaded area from BX

$$\bar{y} = \frac{a_1 y_1 - a_2 y_2}{a_1 - a_2} = \frac{2400 \times 30 - 300 \times 5}{2400 - 300}$$

or, $$\bar{y} = \frac{72000 - 1500}{2100}$$

$\therefore$ $\bar{y} = 33.57$ cms

The C.G of the shaded portion will be at (20, 33.57) from the reference axis. **Ans.**

EXAMPLE 4.12. A semicircle of diameter 100 mm is drawn on a paper sheet. A triangle with diameter of the semicircle as base is cut out of the semicircle as shown in the figure 4.30. Find the centroid of the shaded area.

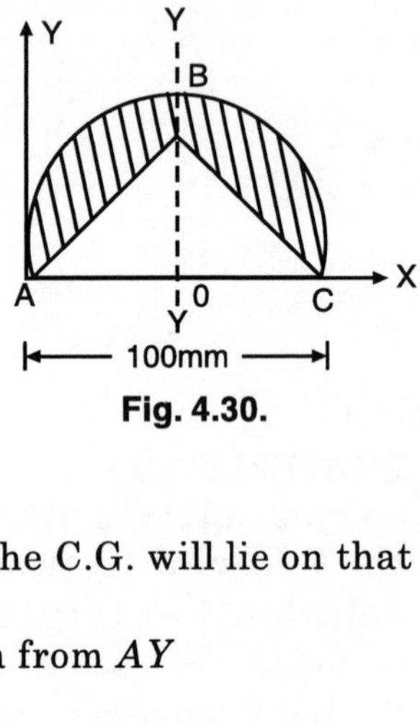

Fig. 4.30.

Solution: Let r be radius of the semicircle.

$\therefore$ $$r = \frac{100}{2} = 50 \text{ mm}$$

In the ΔABC, $OA = OC = 50$ mm and $OB = 50$ mm

Let semicircle be considered as section (1) and triangle as section (2)

$\therefore$ $$a_1 = \frac{1}{2} \times \pi r^2 = \frac{1}{2} \times \pi \times 50^2 = 3927 \text{ mm}^2$$

and $$a_2 = \frac{1}{2} \times 100 \times 50 = 2500 \text{ mm}^2$$

As the diagram is symmetrical about the vertical axis *YY* hence the C.G. will lie on that axis.

$\therefore$ $\bar{X}$ = Distance of C.G. of the shaded area from *AY*

$$= \frac{100}{2} = 50 \text{ mm}$$

$$y_1 = \frac{4r}{3\pi} = \frac{4 \times 50}{3\pi} = 21.22 \text{ mm}$$

$$y_2 = \frac{1}{3} \times 50 = 16.67 \text{ mm}$$

and $$\bar{y} = \frac{a_1 y_1 - a_2 y_2}{a_1 - a_2} = \frac{3927 \times 21.22 - 2500 \times 16.67}{3927 - 2500}$$

or, $$\bar{y} = \frac{83330.94 - 41675}{1427}$$

$\therefore$ $\bar{y} = 29.19$ mm

$$\bar{X} = 50 \text{ mm} \quad \textbf{Ans.}$$

$$\bar{y} = 29.19 \text{ mm} \quad \textbf{Ans.}$$

EXAMPLE 4.13. The corner of the plane lamina is folded along the parent plate as shown in figure 4.31. Find the centroid.

Solution: Divide the lamina in three parts, as shown in figure 4.31.

Square (1) — 120 mm × 120 mm

Triangle (2) — dotted portion

Triangle (3) — folded portion (shown in firm line)

Let OX and OY be the axes of reference.

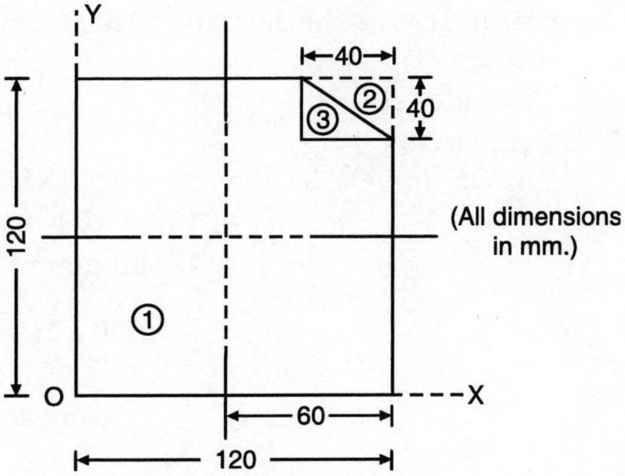

Fig. 4.31.

$$a_1 = 120 \times 120 = 14400 \text{ mm}^2$$

$$a_2 = \frac{1}{2} \times 40 \times 40 = 800 \text{ mm}^2$$

$$a_3 = \frac{1}{2} \times 40 \times 40 = 800 \text{ mm}^2$$

$$x_1 = \frac{120}{2} = 60 \text{ mm}$$

$$x_2 = 80 + \frac{2}{3} \times 40 = 106.67 \text{ mm}$$

$$x_3 = 80 + \frac{1}{3} \times 40 = 93.33 \text{ mm}$$

$$y_1 = \frac{120}{2} = 60 \text{ mm}$$

$$y_2 = 80 + \frac{2}{3} \times 40 = 106.67 \text{ mm}$$

$$y_3 = 80 + \frac{2}{3} \times 40 = 93.33 \text{ mm}$$

Let $(\bar{X}, \bar{Y})$ be the co-ordinate of C.G. from the reference axis,

$$\bar{X} = \frac{a_1 x_1 - a_2 x_2 + a_3 x_3}{a_1 - a_2 + a_3}$$

or

$$\bar{X} = \frac{14400 \times 60 - 800 \times 106.67 + 800 \times 93.33}{14400 - 800 + 800}$$

$$\bar{X} = \frac{853328}{14400}$$

$\therefore$

$$\bar{X} = 59.26 \text{ mm} \quad \textbf{Ans.}$$

$$\bar{y} = \frac{a_1 y_1 - a_2 y_2 + a_3 y_3}{a_1 - a_2 + a_3}$$

$$\bar{y} = \frac{14400 \times 60 - 800 \times 106.67 + 800 \times 93.33}{14400 - 800 + 800} = \frac{853328}{14400}$$

$\therefore \quad \bar{y} = 59.26$ mm **Ans.**

EXAMPLE 4.14. Find the C.G. of the section shown in figure 4.32.

Solution: Divide the section in three parts as shown in figure 4.32.

Square (1)
Triangle (2)
Semicircle (3)

Let OX and OY be the reference axis.

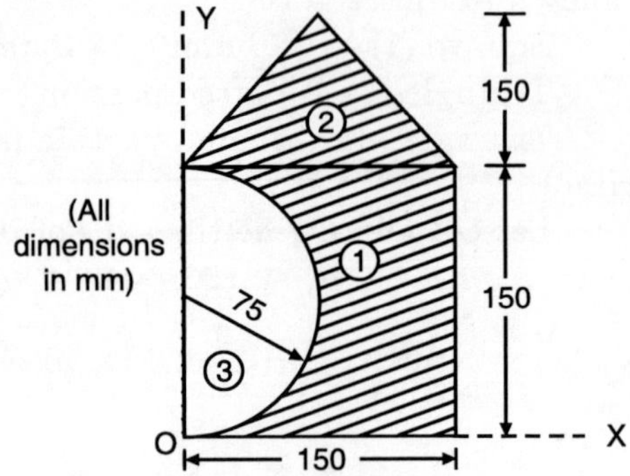

Fig. 4.32.

$$a_1 = 150 \times 150 = 22500 \text{ mm}^2$$

$$a_2 = \frac{1}{2} \times 150 \times 150 = 11250 \text{ mm}^2$$

$$a_3 = \frac{1}{2} \times \pi \times (75)^2 = 8835.73 \text{ mm}^2$$

$$y_1 = \frac{150}{2} = 75 \text{ mm}$$

$$x_1 = \frac{150}{2} = 75 \text{ mm}$$

$$y_2 = 150 + \frac{150}{3} = 200 \text{ mm}$$

$$x_2 = \frac{150}{2} = 75 \text{ mm}$$

$$y_3 = \frac{150}{2} = 75 \text{ mm}$$

$$x_3 = \frac{4 \times 75}{3 \times \pi} = 31.83 \text{ mm}$$

$$\bar{X} = \frac{a_1x_1 + a_2x_2 - a_3x_3}{a_1 + a_2 - a_3}$$

$$= \frac{22500 \times 75 + 11250 \times 75 - 8835.73 \times 31.83}{22500 + 11250 - 8835.73}$$

$$= \frac{2250008.7}{24914.27} = 90.31 \text{ mm} \quad \textbf{Ans.}$$

$$\bar{y} = \frac{a_1y_1 + a_2y_2 - a_3y_3}{a_1 + a_2 - a_3}$$

$$= \frac{22500 \times 75 + 11250 \times 200 - 8835.73 \times 75}{22500 + 11250 - 8835.73}$$

$$= \frac{3274820.3}{24914.27} = 131.44 \text{ mm} \quad \textbf{Ans.}$$

EXAMPLE 4.15. Find the centroid of the shaded area of the lamina shown in figure 4.33.

Solution : Divide the lamina into four sections.

Sections (1) — Triangle
Sections (2) — Rectangle
Sections (3) — Triangle
Section (4) — Circle

Let OX and OY be the axis of reference.

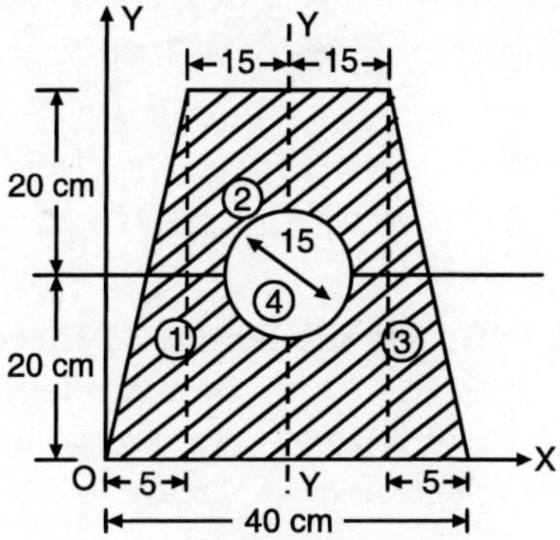

Fig. 4.33.

Then,

$$a_1 = \frac{1}{2} \times 5 \times 40 = 100 \text{ cm}^2$$

$$a_2 = 30 \times 40 = 1200 \text{ cm}^2$$

$$a_3 = \frac{1}{2} \times 5 \times 40 = 100 \text{ cm}^2$$

$$a_4 = \frac{\pi \times 15^2}{4} = 176.71 \text{ cm}^2$$

$$x_1 = 5 - \frac{5}{3} = 3.33 \text{ cm}$$

$$x_2 = 5 + \frac{30}{2} = 20 \text{ cm}$$

$$x_3 = 5 + 30 + \frac{5}{3} = 36.67 \text{ cm}$$

$$x_4 = 20 \text{ cm}$$

and
$$y_1 = \frac{40}{3} = 13.33 \text{ cm}$$

$$y_2 = \frac{40}{2} = 20 \text{ cm}$$

$$y_3 = \frac{40}{3} = 13.33 \text{ cm}$$

$$y_4 = 20 \text{ cm}$$

We know
$$\bar{X} = \frac{a_1 x_1 + a_2 x_2 + a_3 x_3 - a_4 x_4}{a_1 + a_2 + a_3 - a_4}$$

or
$$\bar{X} = \frac{100 \times 3.33 + 1200 \times 20 + 100 \times 36.67 - 176.71 \times 20}{100 + 1200 + 100 - 176.71}$$

or
$$\bar{X} = \frac{333 + 24000 + 3667 - 3534.2}{1223.29} = \frac{24465.8}{1223.29}$$

∴
$$\bar{X} = 20 \text{ cm}$$

$$\bar{y} = \frac{100 \times 13.33 + 1200 \times 20 + 100 \times 13.33 - 176.71 \times 20}{100 + 1200 + 100 - 176.71}$$

or
$$\bar{y} = \frac{1333 + 24000 + 1333 - 3534.2}{1223.29} = 18.91 \text{ cm}$$

The centroid will lie at (20, 18.91) from the reference axis. **Ans.**

EXAMPLE 4.16. Find the height of the centroid from the reference axis for the lamina shown in the figure 4.34.

Solution : As $\angle CAD = 90°$

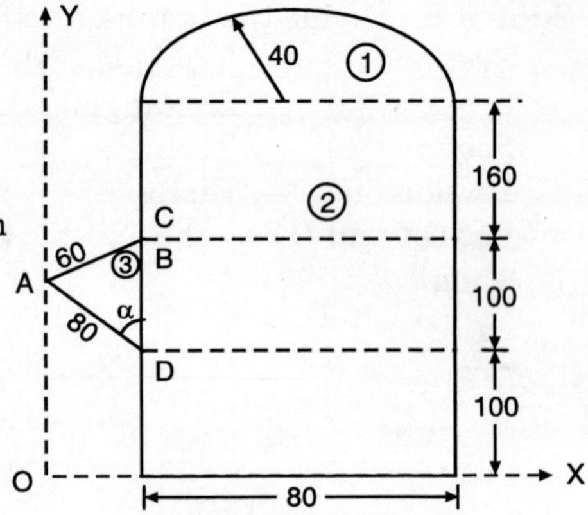

(All dimensions in mm.)

Fig. 4.34.

$$\therefore \quad \tan \alpha = \frac{60}{80} = \frac{3}{4}$$

$$\therefore \quad \alpha = 36.87°$$

$$AB = 80 \sin 36.87° = 48 \text{ mm}$$

Divide the lamina in three sections

Section 1 — Semicircle

Section 2 — Rectangle

Section 3 — Triangle

$$a_1 = \frac{1}{2} \times \pi \times 40^2 = 2513.27 \text{ mm}^2$$

$$a_2 = 360 \times 80 = 28800 \text{ mm}^2$$

$$a_3 = \frac{1}{2} \times 100 \times 48 = 2400 \text{ mm}^2$$

$$x_1 = 48 + 40 = 88 \text{ mm}$$

$$x_2 = 48 + 40 = 88 \text{ mm}$$

$$x_3 = 2 \times \frac{48}{3} = 32 \text{ mm}$$

$$y_1 = 100 + 100 + 160 + \frac{4 \times 40}{3\pi} = 376.98 \text{ mm}$$

$$y_2 = \frac{360}{2} = 180 \text{ mm}$$

$$y_3 = 100 + \frac{100 + 64}{3}$$

$$= 154.67 \text{ mm} \qquad [\because BD = 80 \cos 36.87° = 64]$$

$$\bar{X} = \frac{a_1 x_1 + a_2 x_2 + a_3 x_3}{a_1 + a_2 + a_3}$$

$$\bar{X} = \frac{2513.27 \times 88 + 28800 \times 88 + 2400 \times 32}{2513.27 + 28800 + 2400}$$

or $$\bar{X} = \frac{2832367.76}{33713.27}$$

$$\therefore \quad \bar{X} = 84.01 \text{ mm} \quad \textbf{Ans.}$$

and $$\bar{y} = \frac{a_1 y_1 + a_2 y_2 + a_3 y_3}{a_1 + a_2 + a_3}$$

or $$\bar{y} = \frac{2513.27 \times 376.98 + 28800 \times 180 + 2400 \times 154.67}{2513.27 + 28800 + 2400}$$

$$\therefore \quad \bar{y} = 192.88 \text{ mm.} \quad \textbf{Ans.}$$

EXAMPLE 4.17. Determine the centroid of the shaded area shown in the figure 4.35.

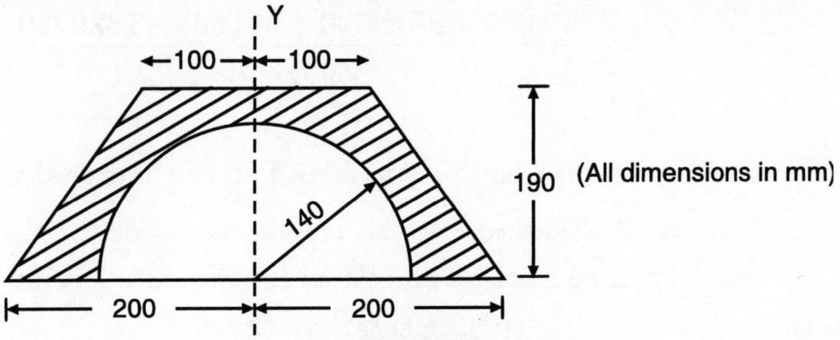

Fig. 4.35.

Solution: The given lamina is symmetrical about YY axis and so the centroid will lie on that axis.

$\therefore \quad \bar{X} = 200$ mm

Let OX and OY be the axes of reference.

Divide the lamina into four parts

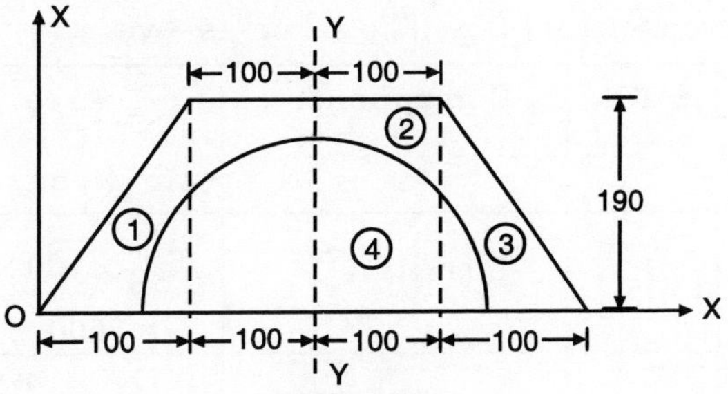

Fig. 4.36.

(*i*) Triangle ①, area = a_1

(*ii*) Rectangle ②, area = a_2

(*iii*) Triangle ③ area = a_3

(*iv*) Semicircle ④, area = a_4

and a_1, a_2, a_3, are + ve and a_4 is – ve.

Let (x_1, y_1), (x_2, y_2), (x_3, y_3) and (x_4, y_4) be the co-ordinates of the centroids of triangle ①, rectangle ②, triangle ③ and semicircle ④ w.r.t. the reference axis.

Let $\bar{Y}$ be the distance of the centroid of the entire lamina from the OX axis.

Now
$$a_1 = \frac{1}{2} \times 100 \times 190 = 9500 \text{ mm}^2$$

$$a_2 = 200 \times 190 = 38000 \text{ mm}^2$$

$$a_3 = \frac{1}{2} \times 100 \times 190 = 9500 \text{ mm}^2$$

$$a_4 = \frac{1}{2} \times \pi r^2 = \frac{1}{2} \times \pi \times (140)^2 = 30787.61 \text{ mm}^2$$

$$y_1 = \frac{190}{3} = 63.33 \text{ mm}$$

$$y_2 = \frac{190}{2} = 95 \text{ mm} ; \quad y_3 = \frac{190}{3} = 63.33 \text{ mm}$$

$$y_4 = \frac{4r}{3\pi} = \frac{4 \times 140}{3 \times \pi} = 59.42 \text{ mm}$$

$$\bar{y} = \frac{a_1 y_1 + a_2 y_2 + a_3 y_3 - a_4 y_4}{a_1 + a_2 + a_3 - a_4}$$

$$= \frac{9500 \times 63.33 + 38000 \times 95 + 9500 \times 63.33 - 30787.61 \times 59.42}{9500 + 38000 + 9500 - 30787.61}$$

$$= \frac{601635 + 3610000 + 601635 - 1829399.8}{26212.39} = 113.83$$

Tabular Method

The given lamina is symmetrical about *YY* axis and so the centroid will lie on that axis.

$\therefore \qquad \bar{X} = 200$ mm

Let *OX* and *OY* be the reference axes as shown in figure 4.36. Divide the lamina.

(*i*) Triangle ① (*ii*) Rectangle ②

(*iii*) Triangle ③ (*iv*) Semicircle ④

The area of triangle ①, rectangle ②, triangle ③ will be taken as + ve and that of semicircle ④ will be taken as – ve.

S. No.	Components	Area (*a*) (mm²)	Distance of C.G. from OX axis (*y*) (mm)	*ay*
1.	Triangle ①	$\frac{1}{2} \times 100 \times 190$ = 9500	$\frac{190}{3} = 63.33$	601635
2.	Rectangle ②	200 × 190 = 38000	$\frac{190}{2} = 95$	3610000
3.	Triangle ③	$\frac{1}{2} \times 100 \times 190$ = 9500	$\frac{190}{3} = 63.33$	601635
4.	Semicircle ④	$\frac{1}{2} \times \pi \times (140)^2$ = 30787.61	$\frac{4 \times r}{3\pi} = \frac{4 \times 140}{3 \times \pi}$	1829399.8(–)
		Σa = 26212.39		Σay= 2983870.2

$$\therefore \qquad \bar{y} = \frac{\Sigma ay}{\Sigma a} = \frac{2983870.2}{26212.39} = 113.83 \quad \textbf{Ans.}$$

EXAMPLE 4.18. Determine the centroid of shaded area shown in the figure 4.37.

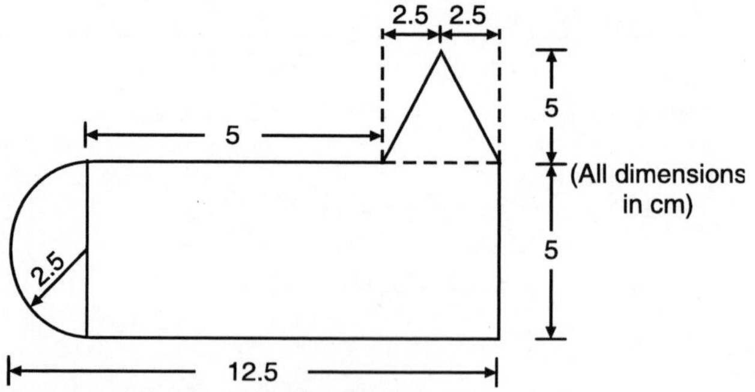

Fig. 4.37.

Solution: Let *OX* and *OY* be the axes of reference.

Divide the lamina into three parts :

(*i*) Semicircle ①

(*ii*) Rectangle ②

(*iii*) Triangle ③

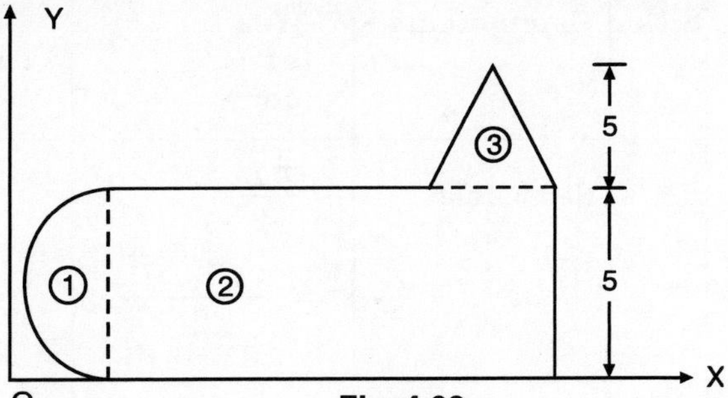

Fig. 4.38.

Now, a_1 = area of the semicircle $\frac{1}{2}\pi r^2$

$= \frac{1}{2} \times \pi \times (2.5)^2 = 9.82 \text{ cm}^2$

a_2 = area of the rectangle = $10 \times 5 = 50 \text{ cm}^2$

a_3 = area of triangle = $\frac{1}{2} \times 5 \times 5 = 12.5 \text{ cm}^2$

x_1 = Distance of C.G. of the semicircle from OY

$= 2.5 - \frac{4r}{3\pi} = 2.5 - \frac{4 \times 2.5}{3 \times \pi} = 1.44 \text{ cm}$

x_2 = Distance of C.G. of the rectangle from $OY = 2.5 + \frac{10}{2} = 7.5 \text{ cm}$

x_3 = Distance of C.G. of the triangle from $OY = 2.5 + 5 + 2.5 = 10 \text{ cm}$

y_1 = Distance of C.G. of the triangle from $OX = \frac{5}{2} = 2.5 \text{ cm}$

y_2 = Distance of C.G of the rectangle from $OX = \frac{5}{2} = 2.5 \text{ cm}$

y_3 = Distance of C.G of the rectangle from OX

$= 5 + \frac{5}{3} = 5 + 6.67 \text{ cm} = 6.67 \text{ cm}.$

Let $\bar{X}$ and $\bar{y}$ be the distance of the C.G of the lamina from OY and OX respectively.

$$\bar{X} = \frac{a_1 x_1 + a_2 x_2 + a_3 x_3}{a_1 + a_2 + a_3} = \frac{9.82 \times 1.44 + 50 \times 7.5 + 12.5 \times 10}{9.82 + 50 + 12.5}$$

$$= \frac{14.14 + 375 + 125}{72.32} = 7.11 \text{ cm} \quad \textbf{Ans.}$$

and

$$\bar{y} = \frac{a_1 y_1 + a_2 y_2 + a_3 y_3}{a_1 + a_2 + a_3} = \frac{9.82 \times 2.5 + 50 \times 2.5 + 12.5 \times 6.67}{9.82 + 50 + 12.5}$$

$$= \frac{24.55 + 125 + 83.37}{72.32} = 3.22 \quad \textbf{Ans.}$$

Tabular Method

Let OX and OY be the axes of reference

Divide the lamina into three parts :

(*i*) Semicircle ①

(*ii*) Rectangle ②

(*iii*) Triangle ③

S.No.	Components	Area (a) (cm^2)	Distance of C.G. from OY (x cm)	Distance of C.G. from OX (y cm)	ax	ay
1.	Semicircle	$\frac{\pi r^2}{2}$ $= \frac{\pi \times (2.5)^2}{2}$ $= 9.82$	$2.5 - \frac{4r}{3\pi}$ $= 2.5 - \frac{4 \times 2.5}{3 \times \pi}$ $= 1.44$	$\frac{5}{2} = 2.5$	14.14	24.55
2.	Rectangle	$10 \times 5 = 50$	$2.5 + \frac{10}{2} = 7.5$	$\frac{5}{2} = 2.5$	375	125
3.	Triangle	$\frac{1}{2} \times 5 \times 5$ $= 12.5$	$2.5 + 5 + 2.5$ $= 10$	$5 + \frac{5}{3}$ $= 6.67$	125	83.37
	$\Sigma a = 72.32$			$\Sigma ax = 514.14$		$\Sigma ay = 232.92$

$$\overline{X} = \frac{\Sigma ax}{\Sigma a} = \frac{514.14}{72.32} = 7.11 \text{ cm} \quad \textbf{Ans.}$$

and

$$\overline{Y} = \frac{\Sigma ay}{\Sigma a} = \frac{232.92}{72.32} = 3.22 \quad \textbf{Ans.}$$

EXAMPLE 4.19. Find the centroid of the shaded are w.r.t. the axis shown in the figure 4.39.

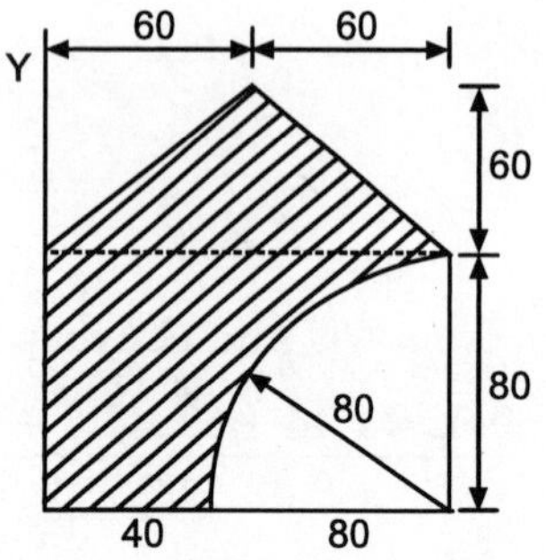

(All dimensions in mm)

Fig. 4.39.

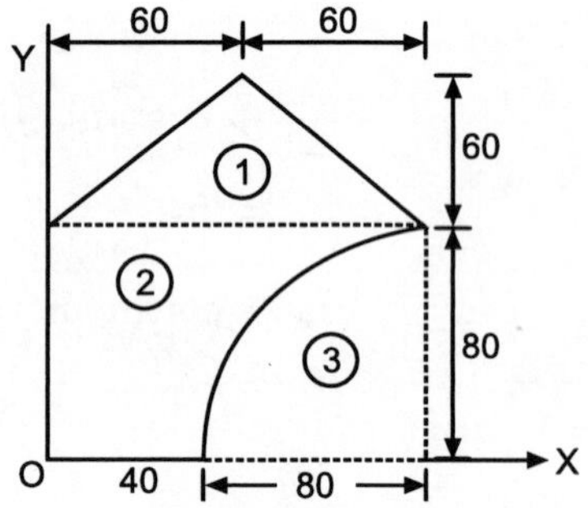

Fig. 4.40.

Solution : Let OX and OY be the axes of reference.

Divide the lamina into three parts

(*i*) Triangle (1)

(*ii*) Rectangle (2)

(*iii*) Quadrant (3)

$$\overline{X} = \frac{\Sigma ax}{\Sigma a} = \frac{359465.37}{8173.45} = 43.98 \text{ mm}$$

$$\overline{Y} = \frac{\Sigma ay}{\Sigma a} = \frac{573348.63}{8173.45} = 70.15 \text{ mm.}$$

S.No.	Component	Area (*a*)	Distance of C.G from OY axis (*x*)	Distance of C.G from OX-axis (*y*)	ax	ay
1.	Triangle	$\frac{1}{2}\times 120\times 60$ = 3600	60	$80+\frac{60}{3}=100$	216000	360000
2.	Rectangle	120×80 = 9600	$\frac{120}{2}=60$	$\frac{80}{2}=40$	576000	384000
3.	Quadrant	$\frac{1}{4}\times\pi\times 80^2$ = 5026.55 (–)	$120-\frac{4\times 80}{3\pi}$ = 86.05	$\frac{4\times 80}{3\pi}=33.95$	432534.63 (–)	170651.37 (–)
		Σa = 8173.45			Σax = 359465.37	Σay = 573348.63

EXAMPLE 4.20. Locate the centroid of the shaded area w.r.t the co-ordinate axes shown in the figure 4.41

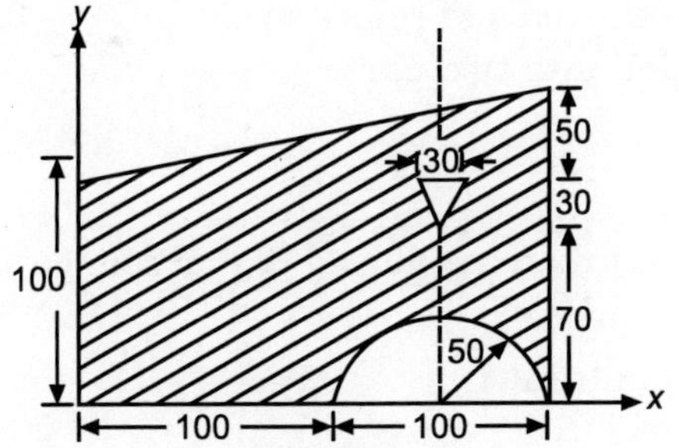

Fig. 4.41.

Solution: Let OX and OY be the reference axes. (Figure 4.41(*i*))

Divide the lamina into four parts :

(*i*) Triangle (1)

(*ii*) Rectangle (2)

(*iii*) Quadrant (3)

(*iv*) Semicircle (4)

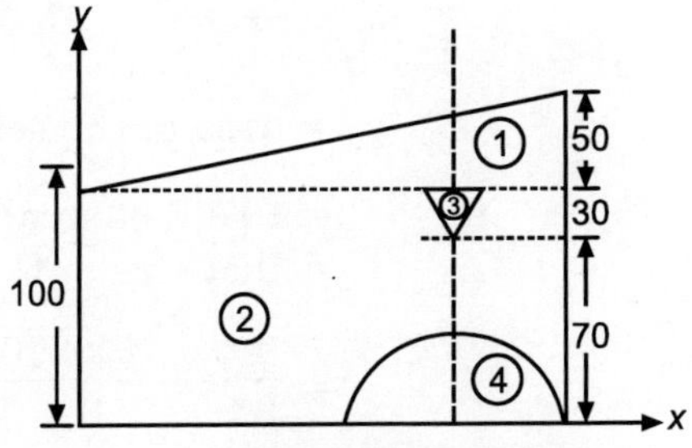

Fig. 4.41. (i)

(All dimensions in mm)

$$\overline{X}=\frac{\Sigma ax}{\Sigma a}=\frac{2010100}{20623}=97.47\text{ mm}$$

$$\overline{Y}=\frac{\Sigma ay}{\Sigma a}=\frac{1459519.1}{20623}=70.77\text{ mm.}$$

S.No.	Component	area (*a*)	Dist. of C.G from OY axis (*x*)	Dist. of C.G from OX-axis (*y*)	ax	ay
1.	Triangle	$\frac{1}{2} \times 50 \times 200$ = 5000	$\frac{2}{3} \times 200$ = 133.33	$100 + \frac{50}{3}$ = 116.67	666650	583350
2.	Rectangle	100 × 200 = 20000	$\frac{200}{2} = 100$	$\frac{100}{2} = 50$	2000000	1000000
3.	Quadrant	$\frac{1}{2} \times 30 \times 30$ = 450 (–)	100 + 150 = 150	$70 + \frac{2 \times 30}{3}$ = 90	67500 (–)	40500 (–)
4.	Semicircle	$\frac{1}{2} \times \pi \times 50^2$ = 3927 (–)	= 150	$\frac{4 \times 50}{3\pi}$ = 21.22	589050 (–)	83330.94(–)
		Σ*a* = 20623			Σ*ax* = 2010100	Σ*ay* =1459519.1

EXAMPLE 4.21. Find the co-ordinates of the centroid of the area obtained after removing a semicircle of radius 100 mm from a quadrant of a circle of radius 200 mm.

Solution: Let *OX* and *OY* be the axes of reference.

The given lamina is divided into two parts:

(i) Quadrant ①

(ii) Semicircle ②

The area of quadrant will be taken as +ve and that of semicircle will be taken as –ve.

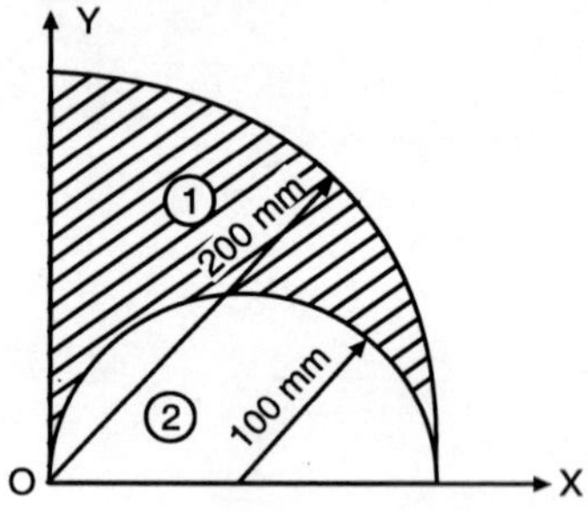

Fig. 4.42.

$$a_1 = \text{Area of quadrant } ①$$

$$= \frac{1}{4} \times \pi r^2 = \frac{1}{4} \times \pi \times (200)^2 = 31415.93 \text{ mm}^2 \ (+)$$

$$a_2 = \text{Area of the semicircle } ② = \frac{1}{2} \pi r_1^2 = \frac{1}{2} \times \pi \times (100)^2$$

$$= 15707.96 \text{ mm}^2 \ (–)$$

x_1 = Distance of C.G. of the quadrant from *OY* axis

$$= \frac{4r}{3\pi} = \frac{4 \times 200}{3\pi} = 84.88 \text{ mm}$$

x_2 = Distance of C.G. of the semicircle from *OX* axis = 100 mm

y_1 = Distance of C.G. of the quadrant from *OY* axis

$$= \frac{4r}{3\pi} = \frac{4 \times 200}{3\pi} = 84.88 \text{ mm}$$

y_2 = Distance of C.G of the semicircle from *OX* axis

$$= \frac{4r_1}{3\pi} = \frac{4 \times 100}{3\pi} = 42.44 \text{ mm}$$

$\bar{X}$ = Distance of C.G. of the entire lamina from OY axis

$$= \frac{a_1 x_1 - a_2 x_2}{a_1 - a_2} = \frac{31415.93 \times 84.88 - 15707.96 \times 100}{31415.93 - 15707.96}$$

= 69.76 mm **Ans.**

$\bar{y}$ = Distance of C.G. of the entire lamina from OX axis

$$= \frac{a_1\ y_1 - a_2\ y_2}{a_1 - a_2} = \frac{31415.93 \times 84.88 - 15707.96 \times 42.44}{31415.96 - 15707.96}$$

= 127.32 mm **Ans.**

Tabular Method

Divide the lamina into two parts :

(*i*) Quadrant ①

(*ii*) Semicircle ②

S.No.	Components	Area (a) (mm²)	Distance of C.G. from OY axis (x) (mm)	Distance of C.G. from OX axis (mm)	ax	ay
1.	Quadrant	$\frac{1}{4} \times \pi \times (200)^2$ = 31415.93	$= \frac{4 \times 200}{3\pi}$ = 84.88	= 84.88	2666584.1	2666584.1
2.	Semicircle	$\frac{1}{2} \times \pi \times (100)^2$ = 15707.96 (–)	100	$\frac{4 \times 100}{3\pi}$ = 42.44	1570796 (–)	666645.82 (–)
	Σa = 15707.97				Σax = 1095788.1	Σay = 1999938.3

$$\therefore \quad \bar{X} = \frac{\Sigma ax}{\Sigma a} = \frac{1095788.1}{15707.97} = 69.76 \text{ mm} \quad \textbf{Ans.}$$

and

$$\bar{y} = \frac{\Sigma ay}{\Sigma a} = \frac{1999938.3}{15707.97} = 127.32 \text{ mm} \quad \textbf{Ans.}$$

EXAMPLE 4.22. A built up section of structural steel consists of a flange plate 400 mm × 20 mm, a web plate 600 mm × 15 mm and two angles 150 mm × 150 mm × 10 mm assembled to form a section as shown in the figure 4.43. Determine the C.G. of the composite section by neglecting the weight of the section.

Solution: Let OX and OY be the axes of reference.

Divide the composite section into four parts as shown in the figure 4.43.

(*i*) Rectangle ①

(*ii*) Rectangle ②

(*iii*) Rectangle ③

(*iv*) Rectangle ④

a_1 = area of rectangle ①
= $400 \times 20 = 8000 \text{ mm}^2$

a_2 = area of rectangle ②
= $(150 + 15 + 150) \times 10$
= $315 \times 10 = 3150 \text{ mm}^2$

a_3 = area of rectangle ③
= $(10 + 15 + 10) \times 140$
= $35 \times 140 = 4900 \text{ mm}^2$

a_4 = area of rectangle ④
= $(600 - 150) \times 15$
= $450 \times 15 = 6750 \text{ mm}^2$

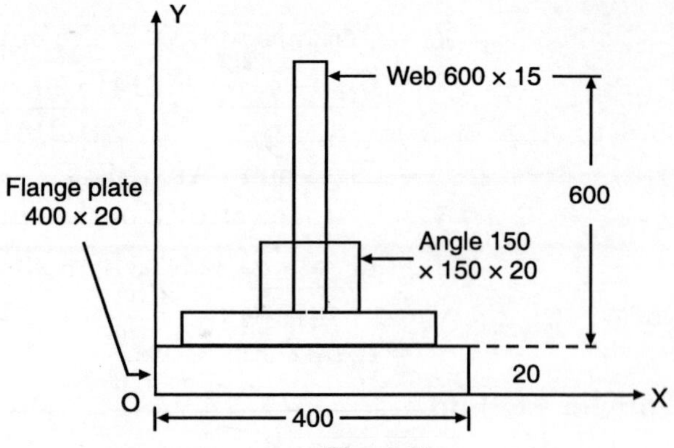

Fig. 4.43.

As the section is symmetrical about YY axis, the C.G will lie on YY-axis figure 4.44.

$$\therefore \quad \bar{X} = \frac{400}{2} = 200 \text{ mm} \quad \textbf{Ans.}$$

Let y_1, y_2, y_3 and y_4 be the the distance of C.G of rectangle ① ② ③ and ④ respectively from OX axis.

$$y_1 = \frac{20}{2} = 10 \text{ mm}$$

$$y_2 = 20 + \frac{10}{2} = 25 \text{ mm}$$

$$y_3 = 20 + 10 + \frac{140}{2}$$
$$= 100 \text{ mm}$$

$$y_4 = 20 + 10 + 140 + \frac{450}{2} = 395 \text{ mm}$$

$\bar{y}$ = Distance of C.G. of the section from OX axis.

$$= \frac{a_1 y_1 + a_2 y_2 + a_3 y_3 + a_4 y_4}{a_1 + a_2 + a_3 + a_4}$$

$$= \frac{8000 \times 10 + 3150 \times 25 + 4900 \times 100 + 6750 \times 395}{8000 + 3150 + 4900 + 6750}$$

$$\therefore \quad \bar{y} = \frac{80000 + 78750 + 490000 + 2666250}{22800} = 145.39 \text{ mm} \quad \textbf{Ans.}$$

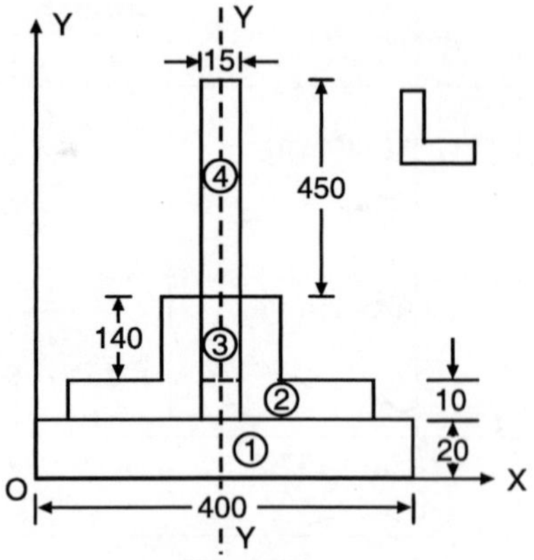

Fig. 4.44.

Tabular Method

S.No.	Components	Area (a) mm^2	Distance of C.G from OX axis (y) (mm)	ay
1.	Rectangle ①	400 × 20 = 80000	10	80000
2.	Rectangle ②	315 × 10 = 3150	25	78750

3.	Rectangle ③	140 × 35 = 4900	100	490000
4.	Rectangle ④	450 × 15 = 6750	395	2666250
		$\Sigma a = 22800$		$\Sigma ay = 3315000$

$$\therefore \quad \bar{Y} = \frac{\Sigma ay}{\Sigma a} = \frac{3315000}{22800} = 145.39 \text{ mm} \quad \textbf{Ans.}$$

EXAMPLE 4.23. A circular plate of uniform thickness and of diameter 500 mm as shown in the Fig. 4.45 has two circular holes of 40 mm diameter each. Where should a 80 mm diameter hole be drilled so that centre of gravity of the plate will be at the geometric centre.

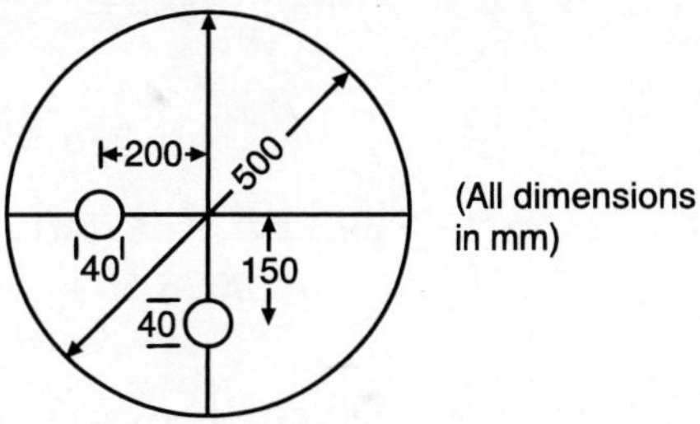

Fig. 4.45.

Solution: Let a hole of 80 mm diameter be drilled at a distance x and y from OY and OX axis as shown in the figure 4.46.

a_1 = area of circle of 500 mm dia

$$= \frac{\pi}{4} \times (500)^2 = 196349.54 \text{ mm}^2$$

$a_2 = a_3$ = area of circle of 40 mm dia

$$= \frac{\pi}{4} \times (40)^2 = 1256.64 \text{ mm}^2(-)$$

a_4 = area of circle of 80 mm dia

$$= \frac{\pi}{4} \times (80)^2 = 5026.55 \text{ mm}^2(-)$$

Let (x_1, y_1), (x_2, y_2), (x_3, y_3) and (x_4, y_4) be the co-ordinates of the C.G. of the circle ① ② ③ and ④ from the reference axes.

$$x_1 = \frac{500}{2} = 250 \text{ mm}$$

$x_2 = 250 - 200 = 50$ mm

$x_3 = 250$ mm

$x_4 = x$ mm

$$y_1 = \frac{500}{2} = 250 \text{ mm},$$

$y_2 = 250$ mm

$y_3 = 250 - 150 = 100$ mm

$y_4 = y$ mm

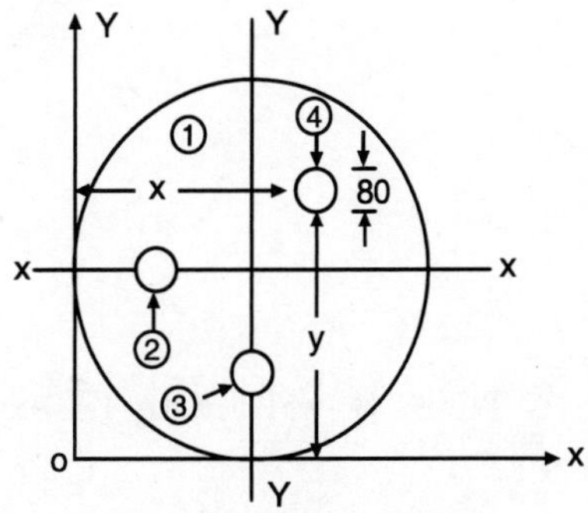

Fig. 4.46.

$\bar{X}$ = Distance of C.G. of the plate from OY axix

= 250 mm (given)

$\bar{y}$ = Distance of C.G. of the plate from OX-axis

= 250 mm (given)

$$\bar{X} = \frac{a_1 x_1 - a_2 x_2 - a_3 x_3 - a_4 y_4}{a_1 - a_2 - a_3 - a_4}$$

$$250 = \frac{196349.54 \times 250 - 1256.64 \times 50 - 1256.64 \times 250 - 5026.44 \times x}{196349.54 - 1256.64 - 1256.64 - 5026.44}$$

$$250 = \frac{49087385 - 62832 - 314160 - 5026.55x}{188809.71}$$

$$250 \times 188809.71 = 48710393 - 5026.55x$$

$$5026.55x = 1507965.5$$

$$\therefore \quad x = 300 \text{ mm from } OX \text{ axis} \quad \textbf{Ans.}$$

$$\bar{y} = \frac{a_1y_1 - a_2y_2 - a_3y_3 - a_4y_4}{a_1 - a_2 - a_3 - a_4}$$

$$250 = \frac{196349.54 \times 250 - 1256.64 \times 250 - 1256.64 \times 100 - 5026.55 \times y}{196349.54 - 1256.64 - 1256.64 - 5026.44}$$

$$250 = \frac{48647561 - 5026.55y}{188809.71}$$

$$250 \times 188809.71 = 48647561 - 5026.55\, y$$

$$5026.55\, y = 1445133.5$$

$$y = 287.50 \text{ mm from } OY \text{ axis} \quad \textbf{Ans.}$$

EXAMPLE 4.24. The cross-section of a gantry girder is as shown in the figure 4.44. It is made up of an I-section of depth 450 mm, flange width 200 mm and a channel of size 400 mm × 150 mm. Thickness of all members is 10 mm. Find the C.G. of the section.

Solution: The section is symmetrical about YY.

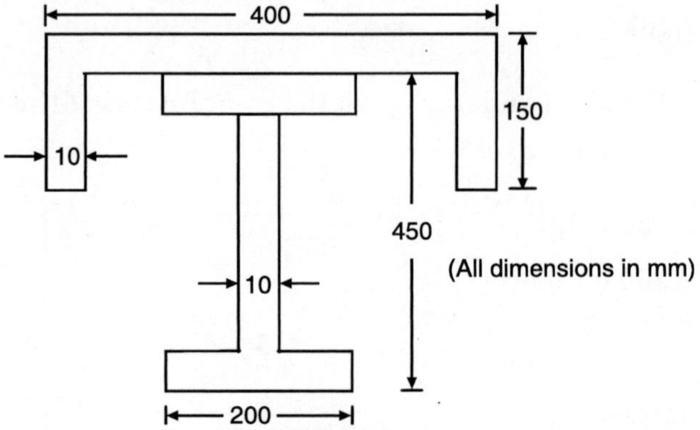

Fig. 4.47.

The C.G. will lie on *YY*-axis.

$$\bar{X} = \frac{400}{2} = 200 \text{ mm} \quad \textbf{Ans.}$$

Divide the section into six rectangles as shown in the figure 4.48.

a_1 = area of rectangle ① = 400 × 10 = 4000 mm^2

a_2 = area of rectangle ② = 140 × 10 = 1400 mm^2

a_3 = area of rectangle ③ = 140 × 10 = 1400 mm^2

a_4 = area of rectangle ④ = 200 × 10 = 2000 mm^2

$$a_5 = \text{area of rectangle } (5) = 430 \times 10 = 4300 \text{ mm}^2$$

$$a_6 = \text{area of rectangle } (6) = 200 \times 10 = 2000 \text{ mm}^2$$

Let $y_1, y_2, y_3, y_4, y_5, y_6$ be the the distance of C.G of rectangle (1), (2), (3), (4), (5) and (6) from OX axis respectively.

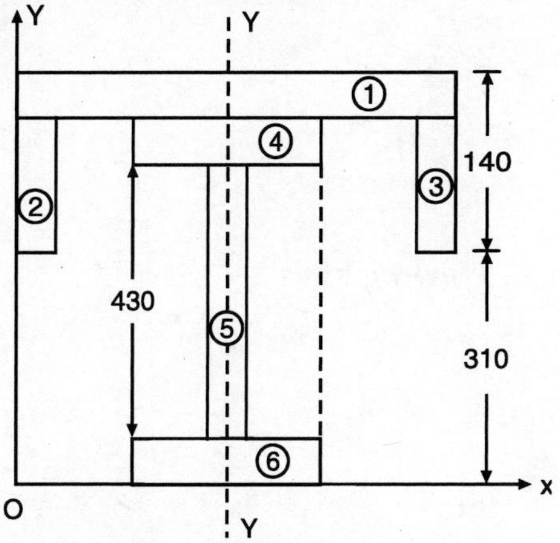

Fig. 4.48.

$$y_1 = 450 + \frac{10}{2} = 455 \text{ mm}$$

$$y_2 = y_3 = 310 + \frac{140}{2} = 380 \text{ mm}$$

$$y_4 = 440 + \frac{10}{2} = 445 \text{ mm}$$

$$y_5 = 10 + 215 \text{ mm} = 225 \text{ mm}$$

$$y_6 = \frac{10}{2} = 5 \text{ mm}$$

$\bar{y}$ = Distance of C.G. of the section from OX axis

$$= \frac{a_1 y_1 + a_2 y_2 + a_3 y_3 + a_4 y_4 + a_5 y_5 + a_6 y_6}{a_1 + a_2 + a_3 + a_4 + a_5 + a_6}$$

$$= \frac{4000 \times 455 + 1400 \times 380 + 1400 \times 380 + 2000 \times 445 + 4300 \times 225 + 2000 \times 5}{4000 + 1400 + 1400 + 2000 + 4300 + 2000}$$

$$= \frac{4751500}{15100} = 314.67 \text{ mm} \quad \textbf{Ans.}$$

EXAMPLE 4.25. A body consists of a solid hemisphere of radius 100 mm and a right circular solid cone of height 120 mm. The hemisphere and cone have a common base and are made of the same material. Find the position of the C.G. of the compound body.

Solution: Let ρ be the density of the material

Section (1) Cone

Section (2) hemisphere

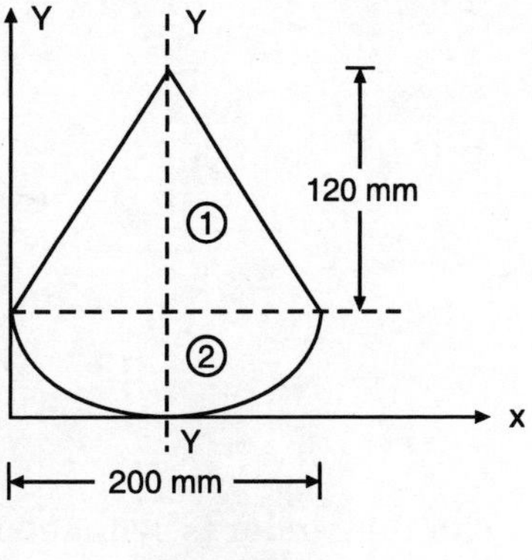

Fig. 4.49.

$$V_1 = \frac{1}{3} \pi r^2 h$$

$$= \frac{1}{3} \times \pi \times 100^2 \times 120$$

$$= 1256637.1 \text{ mm}^3$$

$$V_2 = \frac{1}{2} \times \frac{4}{3} \pi r^3 = \frac{2}{3} \pi r^3$$

$$V_2 = \frac{2}{3} \times \pi \times (100)^3 = 2094395.1 \text{ mm}^3$$

The body is symmetrical about YY axis and hence C.G. will lie on this axis.

$$\therefore \quad \bar{X} = \frac{200}{2} = 100 \text{ mm}$$

Again $y_1 = 100 + \frac{120}{4} = 130$ mm

($\because$ C.G. of the cone lies at $h/4$ from the base)

$y_2 = 100 - \frac{3}{8} \times 100$

($\because$ C.G. of the cone lies at $\frac{3}{8}\frac{3}{8}$ r from the base)

$= 62.5$ mm.

Now we know

$$\bar{y} = \frac{m_1 y_1 + m_2 y_2}{m_1 + m_2}$$

(where m_1 and m_2 be the mass of Cone and hemisphere)

or $$\bar{y} = \frac{V_1 \rho y_1 + V_2 \rho y_2}{(V_1 + V_2)\rho}$$

or $$\bar{y} = \frac{V_1 y_1 + V_2 y_2}{V_1 + V_2}$$

or $$\bar{y} = \frac{1256637.1 \times 130 + 2094395.1 \times 62.5}{1256637.1 + 2094395.1}$$

or $\bar{y} = 87.81$ mm **Ans.**

EXAMPLE 4.26. A solid steel cone is made lighter by removing a part of the material from it in the from of a cylinder as shown in the figure 4.50. Determine the position of the C.G. of the remaining metal.

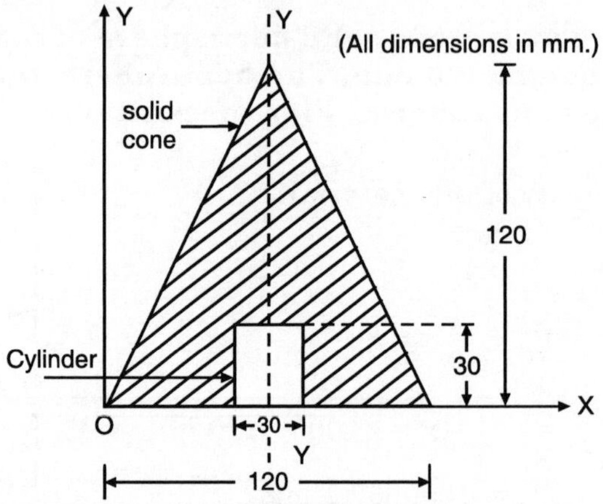

Fig. 4.50.

The remainder is symmetrical about YY axis and hence will lie on that axis.

$\therefore$ $$\bar{X} = \frac{120}{2} = 60 \text{ mm}$$

V_1 = Volume of the cone

$$= \frac{1}{3} \times \pi r^2 \times h = \frac{1}{3} \times \pi \times 60^2 \times 120 = 452389.34 \text{ mm}^2$$

$$V_2 = \text{Volume of the cylinder}$$
$$= \pi r^2 h = \pi \times (15)^2 \times 30 = 21205.75 \text{ mm}^3$$
$$y_1 = \frac{120}{4} = 30 \text{ mm}$$
$$y_2 = \frac{30}{2} = 15 \text{ mm}$$

Let ρ be the density of the material

$$\bar{y} = \frac{m_1 y_1 - m_2 y_2}{m_1 - m_2}$$

or
$$\bar{y} = \frac{V_1 \rho y_1 - V_2 \rho y_2}{(V_1 - V_2) \rho}$$

or
$$\bar{y} = \frac{V_1 y_1 - V_2 y_2}{V_1 - V_2} = \frac{452389.34 \times 30 - 21205.75 \times 15}{452389.34 - 21205.75}$$
$$\bar{y} = 30.74 \text{ mm. } \textbf{Ans.}$$

EXAMPLE 4.27. A solid right circular cone of height 12 cm and base 4 cm is cut by a plane perpendicular to its axis so that its upper half portion is removed. Find the position of the C.G. of the remaining portion from its base.

Solution: Let *AX* and *AY* be reference axis. From the figure 4.51.

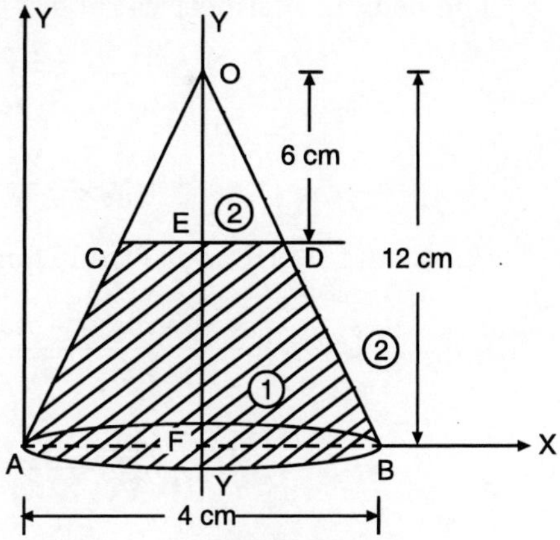

Fig. 4.51.

$$\frac{CD}{AB} = \frac{OE}{OF}$$

or
$$\frac{CD}{4} = \frac{6}{12}$$

or
$$CD = \frac{6}{12} \times 4$$

∴
$$CD = 2 \text{ cm}$$

i.e., the diameter of the removed portion of the cone *OCD* is 2 cm. As the remainder is symmetrical about the centroid axis *YY* the C.G. will lie on that axis.

∴
$$\bar{X} = \frac{4}{2} = 2 \text{ cm}$$

Again,
$$V_1 = \text{Volume of the cone OAB}$$
$$= \frac{1}{3} \times \pi r^2 h = \frac{1}{3} \times \pi \times 2^2 \times 12^2 = 603.18 \text{ cm}^3$$
$$V_2 = \text{Volume of the cone OCD}$$
$$= \frac{1}{3} \times \pi \times 1^2 \times 6^2 = 37.7 \text{ cm}^3$$
$$y_1 = \frac{12}{4} = 3 \text{ cm}$$
$$y_2 = 6 + \frac{6}{4} = 7.5 \text{ cm}$$

Now we know

$$\bar{y} = \frac{V_1 y_1 - V_2 y_2}{V_1 - V_2}$$

or
$$\bar{y} = \frac{603.18 \times 3 - 37.7 \times 7.5}{603.18 - 37.7} = 2.7 \text{ cm from AX. } \textbf{Ans.}$$

EXAMPLE 4.28. Locate the centre of gravity of an idealized bullet of 10 mm diameter with a cone in the front and a hemisphere cut from the back as shown in the figure. Assume the material to be homogeneous.

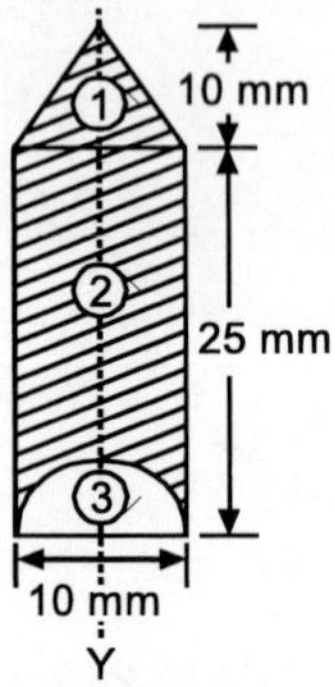

Fig. 4.52.

Solution: Let ρ be the density of the material.

Divide the figure into three parts :

(*i*) Cone (1)

(ii) Cylinder (2)

(*iii*) Hemisphere (3)

$$V_1 = \frac{1}{3}\pi r^2 h = \frac{1}{3}\pi \times 5^2 \times 10 = 261.80 \text{ mm}^3$$

$$V_2 = \pi r^2 h = \pi \times 5^2 \times 25 = 1963 \text{ mm}^3$$

$$V_3 = \frac{2}{3}\pi r^3 = \frac{2}{3}\pi \times 5^3 = 261.80 \text{ mm}^3$$

The body is symmetrical about *YY* axis

$$\therefore \quad \bar{X} = \frac{10}{2} = 5 \text{ mm} \qquad y_1 = 25 + \frac{10}{4} = 27.5 \text{ mm}$$

$$y_2 = \frac{25}{2} = 12.5 \text{ mm} \qquad y_3 = \frac{3 \times 5}{8} = 1.875 \text{ mm}$$

(y_1, y_2 and y_3 are measured from the base)

$$\bar{Y} = \frac{m_1 y_1 + m_2 y_2 - m_3 y_3}{m_1 + m_2 - m_3} = \frac{\rho V_1 y_1 + \rho V_2 y_2 - \rho V_3 y_3}{\rho V_1 + \rho V_2 - \rho V_3}$$

$$= \frac{V_1 y_1 + V_2 y_2 - V_3 y_3}{V_1 + V_2 - V_3} = \frac{261.80 \times 27.5 + 1963 \times 12.5 - 261.80 \times 1.875}{261.80 + 1963 - 261.80}$$

$$= \frac{31246.12}{1963} = 15.92 \text{ mm} \quad \textbf{Ans.}$$

EXAMPLE 4.29. A solid consists of a cylinder and a hemisphere of equal radius fixed base to base. Find the ratio of the radius to the height of the cylinder so that the C.G. of the entire solid is lying at the common face to hmisphere and cylinder.

(B.T.E. Delhi 2000)

Solution: Let *OX* and *OY* be the reference axis.

Let r be the radius of hemisphere and cylinder and $h(= nr)$ be the height of the cylinder.

$$V_1 = \text{Volume of cylinder}$$
$$= \pi r^2 h = \pi r^2 \times nr = n \pi r^3$$
$$V_2 = \text{Volume of hemisphere}$$
$$= \frac{2}{3}\pi r^3$$

$$y_1 = \text{distance of C.G. of cylinder from } OX \text{ axis} = \frac{nr}{2}$$

$$y_2 = \text{distance of C.G of hemisphere from } OX \text{ axis}$$

$$y_2 = nr + \frac{3r}{8} = r\left(n + \frac{3}{8}\right)$$

$$\bar{y} = \text{Distance of C.G. of compound solid from } OX \text{ axis}$$

$$\bar{y} = nr$$

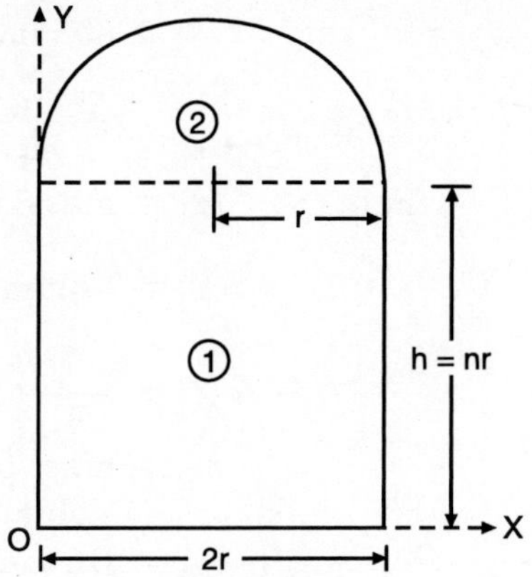

Fig. 4 .53.

We know $\bar{y} = \dfrac{V_1 y_1 + V_2 y_2}{V_1 + V_2}$

$$n r = \frac{n \pi r^3 \times \frac{n r}{2} + \frac{2}{3}\pi r^3 \times \left(n + \frac{3}{8}\right) r}{n \pi r^3 + \frac{2}{3}\pi r^3}$$

$$n r = \frac{\frac{n^2 r}{2} + \frac{2r}{3} \times \left(n + \frac{3}{8}\right)}{n + \frac{2}{3}}$$

$$n = \frac{\frac{n^2}{2} + \frac{2}{3}n + \frac{1}{4}}{n + \frac{2}{3}}$$

$$n^2 + \frac{2n}{3} = \frac{n^2}{2} + \frac{2n}{3} \times \frac{1}{4}$$

$$n^2 - \frac{n^2}{2} = \frac{1}{4}$$

$$n^2 = \frac{1}{2}$$

$$\therefore \qquad n = \frac{1}{\sqrt{2}}$$

The ratio of radius to height of cylinder

$$= \frac{r}{h} = \frac{r}{n r} = \frac{1}{n} = \sqrt{2} \quad \textbf{Ans.}$$

EXAMPLE 4.30. A right circular cylinder and a hemisphere having the same radius of 60 mm are joined together face to face. The density of the material of the hemisphere is twice that of the material of the cylinder. Find the greatest height of the cylinder so that the compound body is in stable equilibrium on its base.

Solution: Method I : Let the density of the material of the cylinder be ρ.

∴ The density of the material of hemisphere = 2 ρ.

Let h be the greatest height of the cylinder *X* and *OX* and *OY* be the reference axes.

The body will be stable on its base as long as its centre of gravity is in its hemispherical base *X* i.e.,

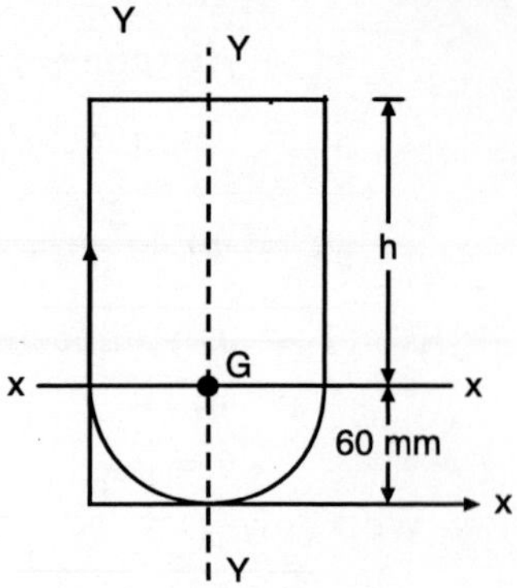

Fig. 4.54

$$\bar{y} = 60 \text{ mm}$$

$$\bar{y} = \frac{m_1 y_1 + m_2 y_2}{m_1 + m_2}$$

where m_1 = mass of the hemispherical portion

$$X_1 = \text{Volume} \times \text{density} = \frac{1}{2} \times \frac{4}{3} \pi \times (60)^3 \times 2\rho$$

$$X_1 = \frac{4P}{8} \pi (60)^3$$

m_2 = mass of the cylindrical portion

$= \text{Volume} \times \text{density} = \pi \times (60)^2 \, h \, \rho$

y_1 = Distance of the C.G. of the hemispherical portion from OX axis

$$y_1 = r - \frac{3r}{8} = \left(60 - \frac{3 \times 60}{8}\right) = \frac{300}{8}$$

[$\because$ Distance of the C.G. of the hemisphere from its base is $\frac{3r}{8}$]

y_2 = Distance of C.G. of the cylindrical portion from OX axis

$$= r + \frac{h}{2} = \left(60 + \frac{h}{2}\right)$$

$$\therefore \quad \bar{y} = 60 = \frac{\left[\frac{4\rho\pi}{3}(60)^3 \times \left(\frac{300}{8}\right) + \pi (60)^2 h \rho \times \left(60 + \frac{h}{2}\right)\right]}{\left[\frac{4}{3}\rho \pi (60)^3 + \pi (60)^2 h\right]}$$

or
$$60 = \frac{\pi \rho (60)^2 \left[\frac{4}{3} \times 60 \times \frac{300}{8} + h\left(60 + \frac{h}{2}\right)\right]}{\pi \rho (60)^2 \left[\frac{4}{3} \times 60 + h\right]}$$

or
$$60(80 + h) = 3000 + 60h + \frac{h^2}{2}$$

or
$$4800 + 60h = 3000 + 60h + \frac{h^2}{2}$$

or
$$\frac{h^2}{2} = 4800 - 3000 = 1800$$

or
$$h^2 = 3600 \Rightarrow h = \pm 60 \text{ mm}$$

$\therefore$ $h = 60$ mm **Ans.** (neglecting the – ve sign)

Method II

The problem can be solved by considering *XX* and *YY* axis as reference axes.

The body will be stable on its base as long as its centre of gravity is in hemispherical base *i.e.*, on *x-x* axis.

Since the C.G. is on the x-x axis, $\bar{y} = 0$

$$\therefore \ \bar{y} = 0 = \frac{\left[-\frac{1}{2}\times\frac{4}{3}\pi\times(60)^3\times 2\rho\times\left(\frac{3}{8}\times 60\right)+(\pi\times(60)^2\times h\times\rho)\times\frac{h}{2}\right]}{\left[-\frac{1}{2}\times\frac{4}{3}\pi\,(60)^3\times\ 2\,\rho+\pi\times(60)^2\times h\times\rho\right]}$$

[∵ The hemispherical portion is taken as –ve as it lies below the xx axis]

$$\text{or}\quad \pi\times(60)^2\,\rho\times\frac{h^2}{2} = \frac{1}{2}\times\pi\times(60)^4\,\rho$$

$$\text{or}\quad \frac{h^2}{2} = \frac{1}{2}\times(60)^2$$

$$\text{or}\quad h^2 = (60)^2$$

$$\therefore\quad h = \mathbf{60\ mm}\qquad \textbf{Ans.}$$

EXAMPLE 4.31. A rod of length 7.5 m is bent so as to form five sides of a regular hexagon. Find the distance of its C.G. from either of the rod.

Solution: Let ABCDEF be the bar in which $AB = BC = CD = DE = EF = 1.5$ m so that the total length = 7.5 m.

Let G be the C.G. of the rod when bent in the form of a hexagon.

Let W be the weight of each portion of the bar forming the hexagon.

∴ Total weight of the five bars = 5 W and it acts through point G.

Now introduce a bar AF of weight W, so as to complete the hexagon. Then the C.G. of the hexagon is shifted to point H.

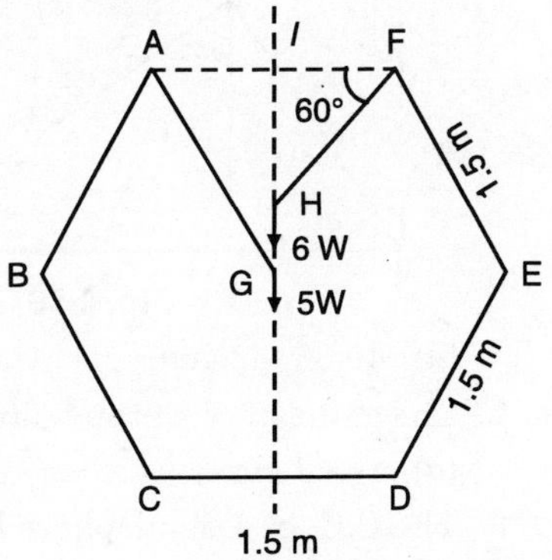

Fig. 4.55.

Taking moments of the weights about AF, we get

$$6\ W\times HI - W\times 0 = 5\ W\times GI$$

$$\text{or}\quad 6\ W\times HI = 5\ W\times GI \qquad ...(i)$$

From $\Delta\ HFI$

$$HI = \text{HF}\sin 60° = 1.5\times\frac{\sqrt{3}}{2} = 1.3\text{ m}$$

$$\text{and}\quad AI = \text{IF} = 0.75\text{ m}$$

Putting the value of HI in equation (1) we get

$$6\text{ W}\times 1.3 = 5\text{ W}\times\text{GI}$$

$$\text{or}\quad GI = \frac{6\,W\times 1.3}{5\,W}$$

$$\therefore\quad GI = 1.56\text{ m}$$

Now from Δ AIG $\quad AG^2 = AI^2 + IG^2$

$$\text{or}\quad AG = \sqrt{AI^2 + IG^2} = \sqrt{(0.75)^2+(1.56)^2}$$

$$\text{or}\quad AG = \sqrt{0.56+2.43} = \sqrt{2.99}$$

$$\therefore\quad AG = 1.73\text{ m}$$

The C.G. will lie at 1.73 m from either end. **Ans.**

SUMMARY

1. The point at which the entire weight of the body is supposed to be concentrated, is called **centre of gravity** of the body.
2. The **centroid** is the point through which whole area of the lamina is supposed to act. The term C.G. is used for bodies which have mass, but centroid is used for laminas.
3. (*a*) The C.G. of a rectangle lies at the point on intersection of the diagonals.
   (*b*) The C.G. of a parallelogram lies at the point of intersection of the diagonals.
   (*c*) C.G. of a triangle lies at the point where the three medians of the triangle meet.
   (*d*) C.G. of a circle lies at the centre.
   (*e*) C.G. of a semicircle lies at 0.42 r from the base.

OBJECTIVE TYPE QUESTIONS

I. Multiple Choice Questions

1. The co-ordinate of the C.G. from point B of the triangle shown in the figure 4.56 is

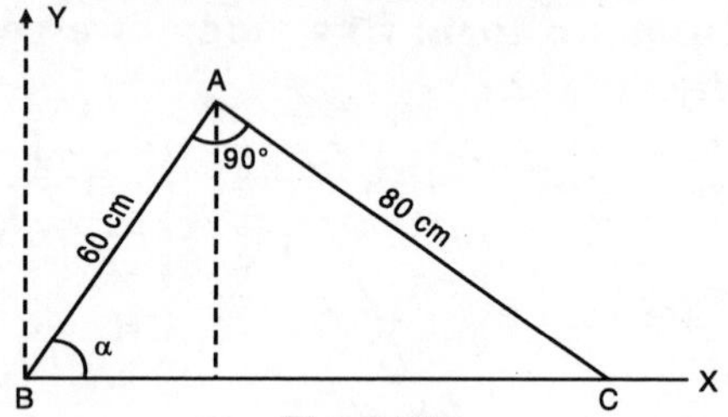

Fig. 4.56.

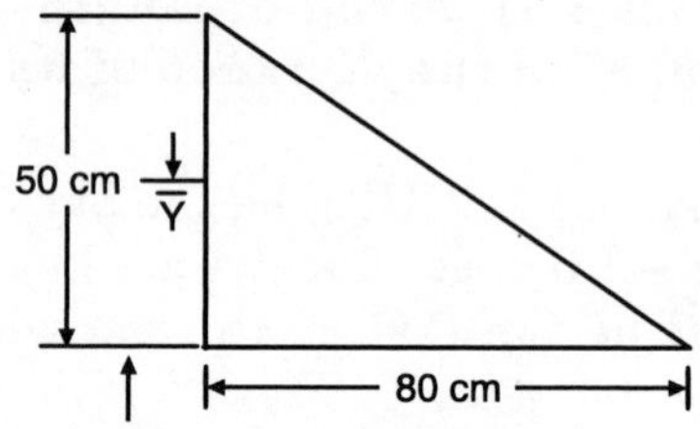

Fig. 4.57.

(*a*) 45.33, 48 cm (*b*) 33.33, 48 cm (*c*) 49.33, 48 cm (*d*) 40, 48 cm

2. The value of $\overline{y}$ of the triangle shown in figure 4.57. is
   (*a*) 26.67 cm (*b*) 16.67 cm (*c*) 33.33 cm (*d*) none of these
3. The C.G. of a hemisphere lies at the following distance form its base along vertical radius 'r'.
   (*a*) $\frac{3r}{8}$ (*b*) $\frac{8}{3r}$ (*c*) $\frac{8r}{3}$ (*d*) $\frac{3}{8r}$
4. The C.G. of a semicircle of radius 10 cm lies at following distance above its base.
   (*a*) 4.24 cm (*b*) 2.39 cm (*c*) 1.33 cm (*d*) 0.23 cm.
5. The C.G. of a right circular cone or radius 'r' and height 'h' lies at the following distance from the base measured along the vertical radius.
   (*a*) $\frac{h}{3}$ (*b*) $\frac{h}{2}$ (*c*) $\frac{h}{4}$ (*d*) $\frac{h}{6}$
6. Which of the following statement is correct.
   (*a*) The term centroid is used for lamina and C.G. for the solids.
   (*b*) The term C.G. is used for lamina and centroid for solids.
   (*c*) Both centroid and C.G. is used for solids.
   (*d*) Both centroid and C.G. is used for lamina.
7. Centre of gravity
   (*a*) must lie at the geometric centre of the body provided it is of uniform density.
   (*b*) located at the geometric centre of the body.
   (*c*) must lie some where inside the body.
   (*d*) none of the above.

8. The centre of gravity of a hemisphere or radius R from its base measured along the vertical radius is

(a) $\frac{3R}{2}$ (b) $\frac{4R}{3}$ (c) $\frac{4R}{7}$ (d) $\frac{3R}{8}$

9. Centre of gravity of a triangle lies at the point of intersection of

(a) medians (b) altitudes (c) diagonals (d) bisector of angles.

Answers

1. (a) 2. (b) 3. (a) 4. (a) 5. (c) 6. (a) 7. (a)
8. (d) 9. (a)

II. Fill in the Blanks

(a) Centre of gravity is the point at which the whole .... of a body is supposed to be concentrated.
(b) .... is the point at which the whole area of a lamina is supposed to be concentrated.
(c) The C.G. of a solid cone lies at a distance of .... its height from the base.
(d) Centroid is a term used for bodies having ....
(e) The axis about which a plane figure is symmetrical on both of its sides is called ....

Answers

(a) weight (b) centroid (c) $h/4$ (d) area (e) axis of symmetry.

THEORETICAL QUESTIONS

1. What do you mean by centre of gravity of a body ?
2. Differentiate between centroid and C.G. **(B.T.E., Delhi June 2000)**
3. Enumerate the steps involved in finding out the C.G of a body by graphical method.
4. Derive the formula for finding out the C.G by the method of moments.

EXERCISES

1. Find the centroid of T-shaped lamina having flange of 150 × 20 mm and web of 100 × 20 mm. **(Ans.** 34 mm)
2. Find the centroid of an angle section 100 mm × 80 mm × 20 mm. **(Ans.** $\bar{X}$ = 47.25mm, $\bar{y}$ = 35.2mm)
3. Find the C.G. of a channel section 60 mm web × 40 mm flange × 10 mm thickness. **(Ans.** 15 mm & 30 mm)
4. Find the C.G. of the I-section shown in the figure 4.42. **(Ans.** $\bar{y}$ = 55 mm from the lower flange)

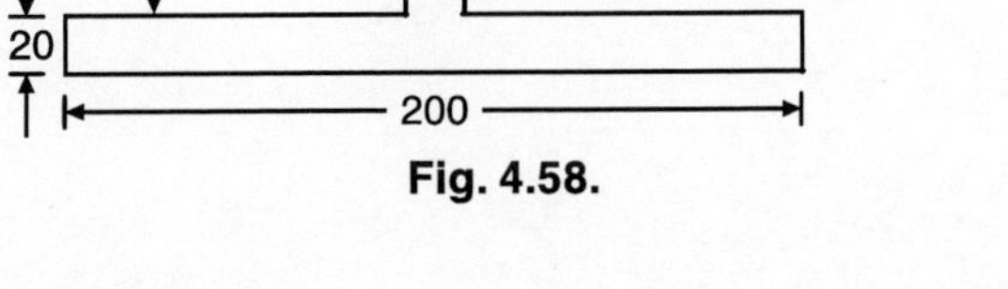

Fig. 4.58.

5. Find the C.G. of an I-section with top flange 100 mm × 20 mm, web 200 mm × 30 mm and bottom flange 300 mm × 40 mm. (**Ans.** 79 mm from the bottom flange)
6. In a rectangular lamina 100 mm × 120 mm a rectangle opening PQRS 30 mm × 40 mm is made. Find the centsroid of the lamina shown in figure 4.53. (**Ans.** $\bar{X}$ = 47.2 mm, $\bar{Y}$ = 62.2 mm)

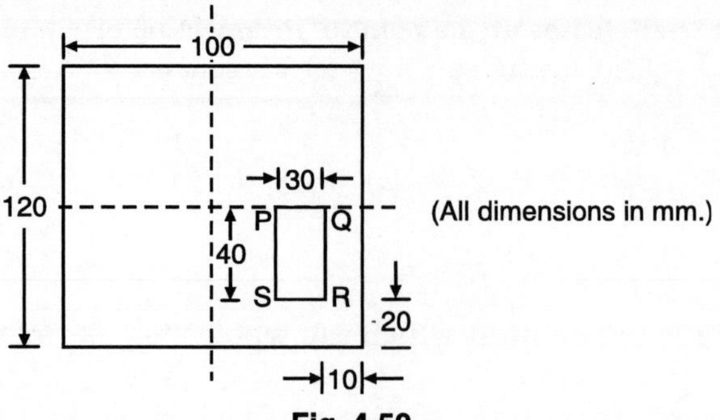

Fig. 4.59.

7. The corner of the plane lamina is folded along the parent plate as shown in the figure 4.60. Find the centroid of the lamina. **(Ans. $\bar{X}$ = 49.55 mm, $\bar{y}$ = 49.55 mm)**

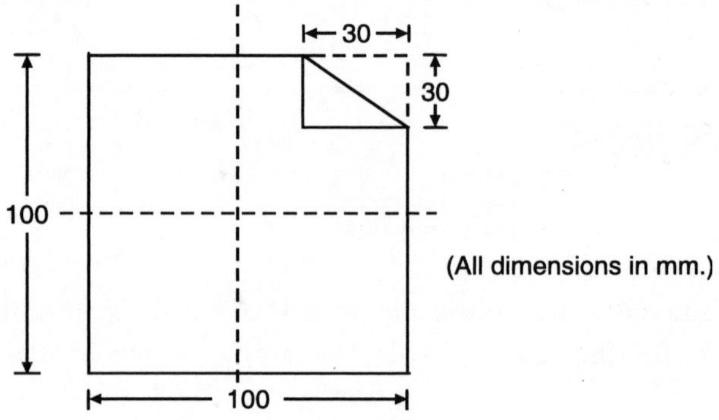

Fig. 4.60.

8. A solid right circular cone has its base scooped out, so that the hollow is right cone on the same base. Find the height up to which the cone should be scooped out so that the C.G. of the remainder may coincide with the vertex of the hollow. (**Ans.** h/3)
9. Find the centroid of the plane area shown in the figure 4.61.

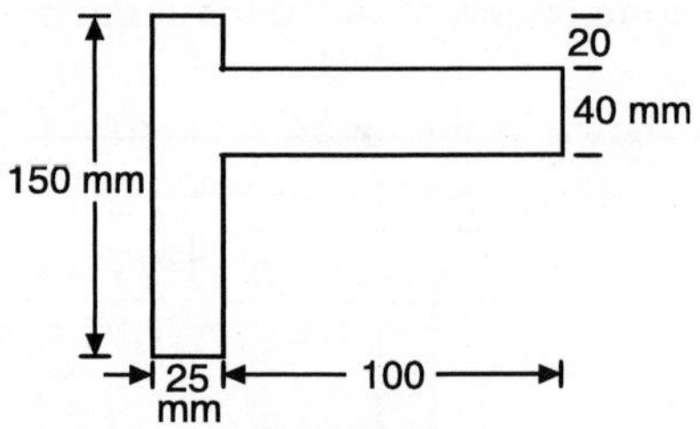

Fig. 4.61.

(**Ans.** $\bar{X}$ = 44.76 mm, $\bar{Y}$ = 56.94 mm)

❑❑❑

CHAPTER 5

Moment of Inertia

5.1 MOMENT OF INERTIA

We know that the product of force (F) and the perpendicular distance (x) between the point or line and line of action of the force is called *moment* (Fx) of the force. This moment (Fx) is also called *first moment of the force*. If this moment (Fx) is again multiplied by the distance (x) between the point and the line of action of the force then the product is called *moment of moment* (Fx^2) of the force. This quantity (Fx^2) is also called *second moment of the force or Moment of Inertia* (M.I.).

When, instead of force, the area or mass of a figure or body is taken into consideration, then the second moment is known as second moment of area or second moment of mass. But all such second moments are broadly termed as moment of inertia. It is generally denoted by I.

So, moment of inertia of a body is defined as the product of area or mass and the square of the perpendicular distance of its C.G. from the axis (about which moment of inertia is to be calculated).

Let us consider a lamina of area A with its C.G. at G. Let x and y be perpendicular distances of the C.G. from OY and OX axis. (as shown in the figure 5.1)

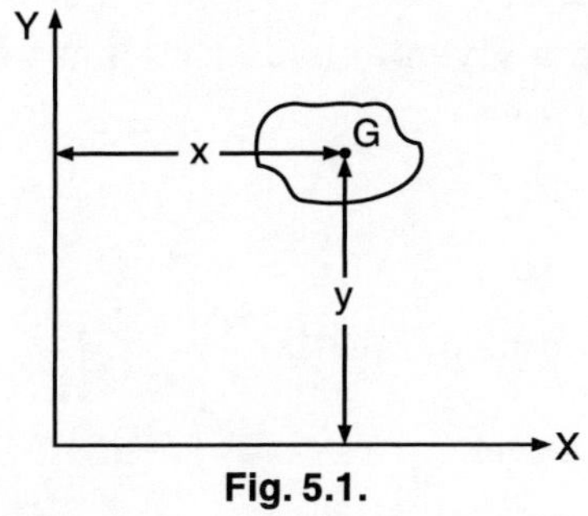

Fig. 5.1.

Chapter Outline

DO YOU KNOW?

Moment of inertia is a property of a shape that is used to predict its resistance to bending and deflection

∴ M.I. of the lamina about OX axis

$$= I_{OX} = A \cdot y^2$$

and M.I. of the lamina about OY axis

$$= I_{OY} = A \cdot x^2$$

5.2 MOMENT OF INERTIA OF A PLANE AREA

Consider a plane area (A) whose M.I. is to be found out. Spilt the area into number of small elements as shown in the figure 5.2.

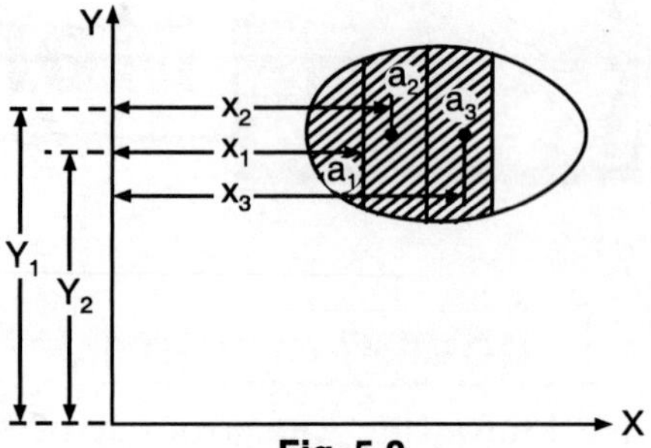

Fig. 5.2.

Let a_1, a_2, $\mathbf{a}_3$ .... be the area of the small elements so that

$$A = a_1 + a_2 + a_3 + \ldots.$$

Let (x_1, y_1), (x_2, y_2), (x_3, y_3) be the co-ordinates of the centroid from the reference axis ox and oy.

∴ Moment of inertia of the plane area about OY

$$= I_{OY} = a_1 x_1^{\,2} + a_2 x_2^{\,2} + a_3 x_3^{\,2} + \ldots$$

and $I_{OY} = \Sigma a x^2$

Moment of inertia of the plane area about OX

$$= I_{OX} = a_1 y_1^{\,2} + a_2 y_2^{\,2} + a_3 y^2{}_3 + \ldots$$

$$I_{OX} = \Sigma a y^2$$

In the integration form it may be written as

$$I_{OY} = \int_A dA \cdot x^2 \quad \text{and} \quad I_{OX} = \int_A dA \cdot y^2$$

where dA is the elemental area and (x, y) is the coordinate of the elemental area from the reference axis.

DO YOU KNOW?

The moment of inertia is additive so, for a rigid body consisting of N point masses m_i with distances r_i to the rotation axis, the total M.I equals to the sum of the point-mass moment of inertia

$$I = \sum_{i=1}^{N} m_i r_i^2$$

5.3 UNITS OF MOMENT OF INERTIA

The units of moment of inertia depend upon both the factors, *i.e.*, units of the area and perpendicular distance as mentioned below :

(*a*) If the area is in mm^2 and distance in mm, then M.I. is given as mm^4.

(*b*) If the area is in m^2 and distance in m, then M.I. is given as m^4.

(*c*) If instead of area, mass of the body is considered and the mass is in kg, distance is in cm^2 then M.I. is given as kg cm^2.

5.4 RADIUS OF GYRATION (RADIUS OF INERTIA)

Consider an area A concentrated at distance K from a given axis. Let moment of inertia of the area about this axis be I. Then M.I. can be written as

$$I = AK^2$$

or

$$K = \sqrt{\frac{I}{A}}$$

This distance K is called radius of gyration (or radius of inertia)

The unit of K is the unit of length.

Considering axis of reference as XX and YY, we have

$$K_{XX} = \sqrt{\frac{I_{XX}}{A}} \text{ and } K_{YY} = \sqrt{\frac{I_{YY}}{A}}$$

Thus the radius of gyration can be defined as the distance of the point, at which the whole area is supposed to be concentrated without altering the M.I. from an axis and is obtained by the square root of the ratio of M.I. and the area.

5.5 MODULUS OF SECTION

The ratio of moment of inertia of a section about its centroidal axis and the distance of the extreme fibre from the centroidal axis is called Modulus of *section or section modulus*. It is generally denoted by Z.

Consider a rectangular section as shown in figure 5.3.

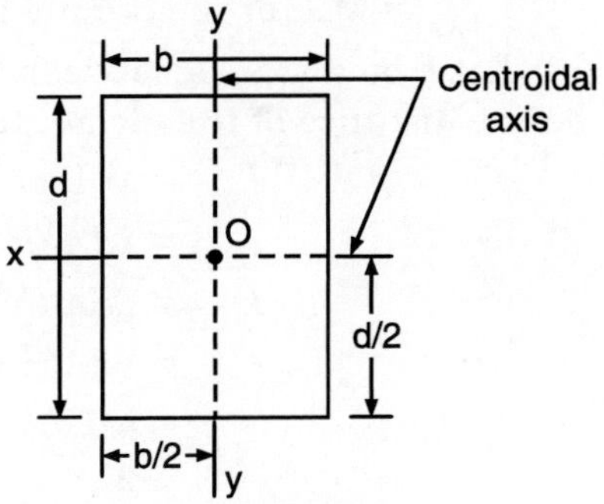

Fig. 5.3.

Let I_{XX} = M.I. of the section about XX axis

and I_{YY} = M.I. of the section about YY axis

$\frac{d}{2}$ = Distance of the extreme fibre from XX axis

$\frac{b}{2}$ = Distance of the extreme fibre from YY axis

$$\therefore \text{ Modulus of section about } XX = Z_{XX} = \frac{I_{XX}}{d/2}$$

$$\text{and, Modulus of section about } YY = Z_{YY} = \frac{I_{YY}}{b/2}$$

5.6 THEOREM OF PERPENDICULAR AXIS

If I_{OX} and I_{OY} be the M.I. of a lamina about two perpendicular axes OX and OY in the plane of the lamina and I_{OZ} be the M.I. of the lamina about an axis (OZ) perpendicular to the lamina and passing through the intersection of the axes OX and OY then $I_{OZ} = I_{OX} + I_{OY}$.

Consider a lamina as shown in the figure 5.4.

Let OX and OY be the two axes lying in the plane of lamina. The third axis OZ is perpendicular to the plane of the lamina and passing through point O.

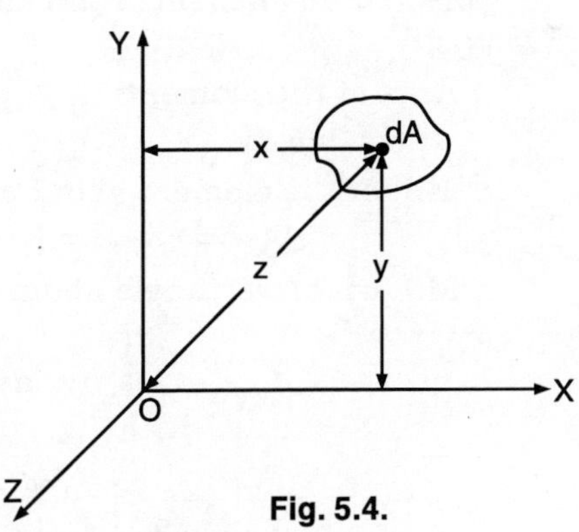

Fig. 5.4.

Then $Z^2 = x^2 + y^2$

Now, $I_{OX} = \Sigma dAy^2$

$I_{OY} = \Sigma dAx^2$

and $I_{OZ} = \Sigma dAZ^2$

Putting the value of Z^2 in I_{OZ} we get

$$I_{OZ} = \Sigma dA(x^2 + y^2)$$

or $$I_{OZ} = \Sigma dAx^2 + \Sigma dAy^2$$

or $$I_{OZ} = I_{OY} + I_{OX}$$

$$\therefore \quad I_{OZ} = I_{OX} + I_{OY}$$

The axis oz is called polar axis and hence I_{OZ} is called polar moment of inertia.

5.7 THEOREM OF PARALLEL AXIS

"The moment of inertia of a lamina about any axis in the plane of the lamina equals the sum of moment of inertia about a parallel centroidal axis in the plane of the lamina and the product of the area of the lamina with the square of the distance between the two axes."

Consider a lamina of area A shown in the figure 5.5. Let MN be an axis in the plane of the lamina which is parallel to the centroidal axis *XX*. We have to calculate the M.I. of the lamina about MN axis.

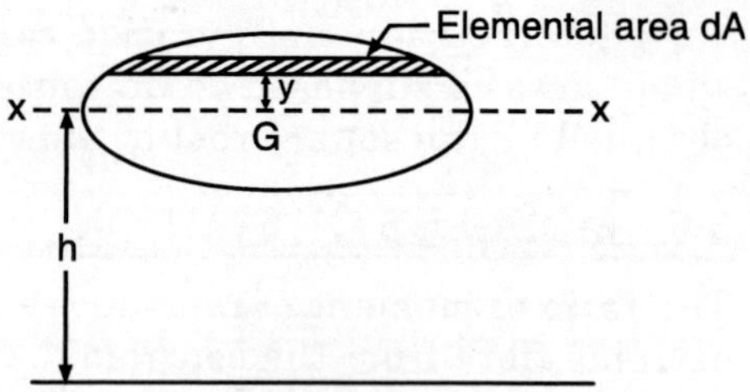

Fig. 5.5

Consider an elemental area dA at distance *y* from *XX* axis.

I_{XX} = M.I. of the lamina about *XX* axis = ΣdAy^2

Let *h* be distance between the two axes and $(h + y)$ be the distance of the elemental area from MN axis.

$\therefore$ I_{MN} = M.I. of the lamina about MN axis

or $I_{MN} = \Sigma dA\,(h + y)^2$

or $I_{MN} = \Sigma dA\,(h^2 + y^2 + 2hy)$

or $I_{MN} = \Sigma dAh^2 + \Sigma dAy^2 + \Sigma dA 2hy$

$= h^2\Sigma dA + \Sigma dAy^2 + 2h\Sigma dAy$

$= Ah^2 + I_{XX} + 2h\Sigma dAy$ ...(1)

DO YOU KNOW?

The theorem of parallel axes is also known as parallel axes rule or Huygens-Steiner's theorem.

ΣdAy represents the algebraic sum of moment of all the elemental area about XX axis. As the distance of the total area from *XX* is zero.

$$\Sigma dAy = 0$$

Substituting the value of ΣdAy in equation (1), we get

$$I_{MN} = Ah^2 + I_{XX} + 0$$

$$\therefore \quad I_{MN} = I_{XX} + Ah^2$$

5.8 MOMENT OF INERTIA OF PLANE FIGURES (LAMINAS)

1. Rectangular Lamina

Consider a rectangular lamina of breadth '*b*' and depth '*d*' as shown in figure 5.5.

DO YOU KNOW?

M.I. has two forms, a scalar form (when the axis of rotation is known) and a more general tersor form that does not require knowing the axis of rotation. Tersor M.I. is a convenient way to summerize all M.I. of a object with one quantity.

(*a*) M.I. about the centroidal axis *XX*

Consider an element of thickness *dy* at a distance y from the *XX* axis.

$\therefore$ Area of the element

$dA = b\,dy$

$\therefore$ M.I. of the element about axis XX

$= dAy^2 = bdyy^2 = by^2dy$

$\therefore$ M.I. of the rectangle about *XX*

$$I_{XX} = 2\int_0^{d/2} by^2\,dy = 2b\int_0^{d/2} y^2\,dy$$

$$I_{XX} = 2\,b\left[\frac{y^3}{2}\right]_0^{d/2}$$

$$= \frac{2\,b}{3}\left(\frac{d^3}{8} - 0\right) = \frac{2\,b\,d^3}{24}$$

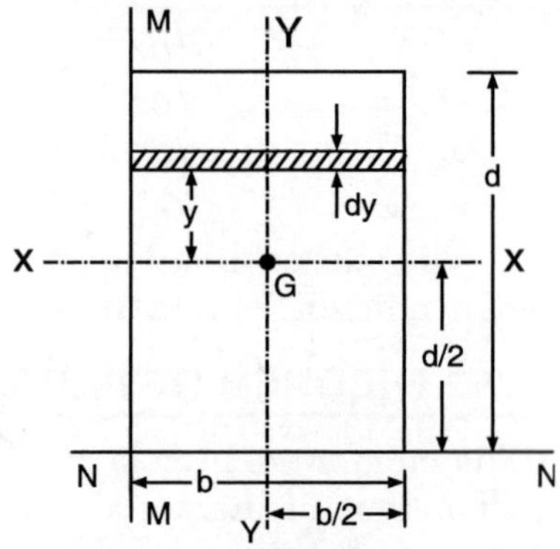

Fig. 5.6.

$$\therefore \qquad I_{XX} = \frac{b\,d^3}{12}$$

Similarly M.I. of the rectangle about the centroidal axis YY

$$I_{YY} = \frac{d\,b^3}{12}$$

(b) M.I. about an axis MM parallel to YY

From the theorem of parallel axis, we know

$$I_{MM} = I_{YY} + Ah^2$$

Here $\qquad h = \frac{b}{2}$

$$A = b.d$$

and $\qquad I_{YY} = \text{M.I. of the section about } YY = \frac{d\,b^3}{12}$

$$\therefore \qquad I_{MM} = \frac{db^3}{12} + bd \times \left(\frac{b}{2}\right)^2 = \frac{db^3}{12} + bd \times \left(\frac{b}{2}\right)^2$$

$$I_{MM} = \frac{db^3 + 3\,db^3}{12} = \frac{4\,db^3}{12} = \frac{db^3}{3}$$

Similarly the M.I. about the axis NN is given by

$$I_{NN} = \frac{bd^3}{3}$$

2. Rectangular Lamina with a Centrally Situated Rectangular Hole

Let I_{xx} be the M.I. of the hollow rectangle and I_{xx1} and I_{xx2} be the M.I. of the bigger rectangle and smaller rectangle respectively.

$$I_{XX_1} = \frac{BD^3}{12} \text{ and } I_{XX_2} = \frac{bd^3}{12}$$

$$\therefore \qquad I_{XX} = I_{XX_1} - I_{XX_2}$$

$$= \frac{BD^3}{12} - \frac{bd^3}{12}$$

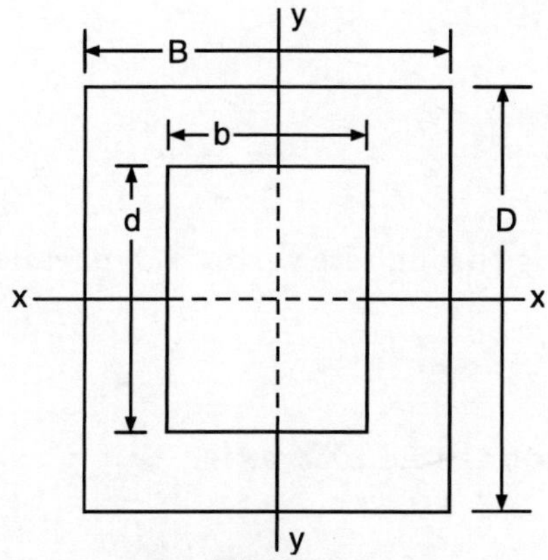

Fig. 5.7.

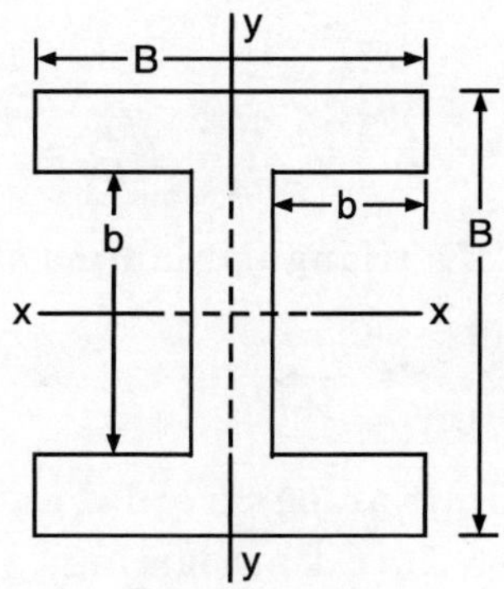

Fig. 5.8.

Similarly the M.I. of the hollow rectangular lamina about YY axis is

$$I_{YY} = \frac{DB^3}{12} - \frac{db^3}{12}$$

3. Moment of Inertia of an I Section

Consider an I section as shown in Fig. 5.8. The dimensions of the section are also shown. Then M.I. of the section about the centroidal axis XX

$$= I_{XX} = \frac{BD^3}{12} - 2 \times \frac{bd^3}{12}$$

5.9 MOMENT OF INERTIA OF TRIANGULAR LAMINA

Let ABC be a triangular lamina of base 'b' and height 'h'.

(a) M.I. about an axis MN passing through the vertex

Consider an elemental strip of thickness dy at a distance y from the axis MN.

Now from ΔADE and ΔABC

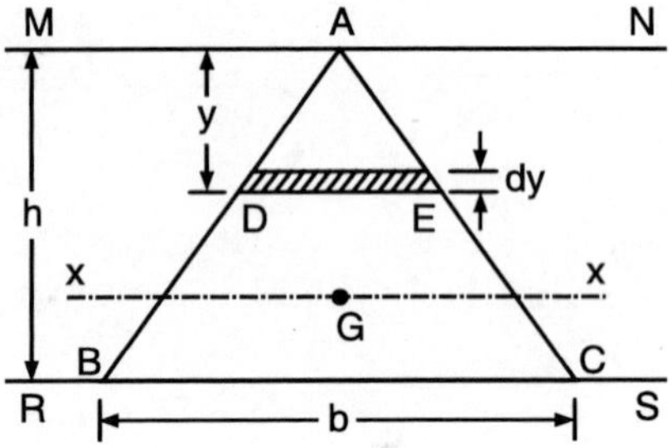

Fig. 5.9.

$$\frac{DE}{BC} = \frac{y}{h}$$

or $$DE = \frac{y}{h} \times BC = \frac{y}{h} \times b$$

$\therefore$ $$DE = \frac{by}{h}$$

$\therefore$ Area of the strip,

$$dA = DE \times dy = \frac{by}{h} dy$$

Now M.I. of the strip about MN axis = $dA.y^2$

$\therefore$ M.I. of the whole lamina about MN axis

$$= I_{MN} = \int_0^h dA \, . \, y^2 = \int_0^h \frac{by}{h} dy \, . \, y^2 = \frac{b}{h} \int_0^h y^3 \, . \, dy$$

Integrating we get,

$$I_{MN} = \frac{b}{h}\left[\frac{y^4}{4}\right]_0^h = \frac{b}{h} \times \frac{h^4}{4}$$

$\therefore$ $$I_{MN} = \frac{bh^3}{4}$$

$\therefore$ M.I. of a triangular lamina about an axis passing through its vertex and parallel to the centroidal axis $= \dfrac{bh^3}{4}$.

(*b*) M.I. about an centroidal axis and parallel to base i.e., XX axis.

Let G be the C.G. of lamina

$\therefore$ Distance of C.G. from the vertex

$$= h - \frac{h}{3} = \frac{2h}{3}$$

Let M.I. of the lamina about XX axis be I_{XX}.

$$\therefore \quad I_{MN} = I_{XX} + A \cdot \left(\frac{2h}{3}\right)^2 \text{ (from the theorem of parallel axes)}$$

$$\text{or} \quad I_{XX} = I_{MN} - A \times \left(\frac{2h}{3}\right)^2$$

$$I_{XX} = \frac{bh^3}{4} - \frac{1}{2} \times b \times h \times \frac{4 \times h^2}{9}$$

$$\text{or} \quad I_{XX} = \frac{bh^3}{4} - \frac{2\,bh^3}{9} = \frac{9\,bh^3 - 8\,bh^3}{36}$$

$$\therefore \quad I_{XX} = \frac{bh^3}{36}$$

(c) M.I about an axis passing through its base and parallel to the centroidal axis *i.e.*, axis RS.

$$I_{RS} = I_{XX} + A\left(\frac{h}{3}\right)^2$$

$$\text{or} \quad I_{RS} = \frac{bh^3}{36} + \frac{bh}{2} \times \frac{h^2}{9} = \frac{bh^3 + 2\,bh^3}{36}$$

$$\text{or} \quad I_{RS} = \frac{3\,bh^3}{36} = \frac{bh^3}{12}$$

5.10 MOMENT OF INERTIA OF CIRCULAR LAMINA

Let us consider a circular lamina of radius R.

Let XX and YY be centroidal axis.

Let ZZ be the polar axis perpendicular to the plane of the paper and passing through O.

Consider an elemental circular ring of radius 'r' and thickness 'dr'.

$\therefore$ Area of the elemental ring $(dA) = 2\,\pi r \times dr$

Now, M.I. of the elemental ring about ZZ axis $= dA\; r^2$

$\therefore$ M.I. of the circular lamina about ZZ axis

$$= I_{ZZ} = \int_0^R dA \,.\, r^2$$

$$= \int_0^R 2\,\pi r \,.\, dr \,.\, r^2 = 2\,\pi \int_0^R r^3\, dr$$

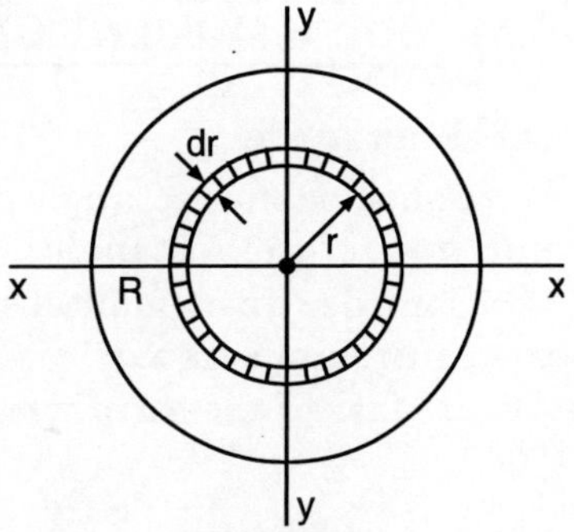

Fig. 5.10.

By integrating this we get

$$I_{ZZ} = 2\,\pi\, X \left[\frac{r^4}{4}\right]_0^R = 2\,\pi \times \frac{R^4}{4}$$

$$\therefore \qquad I_{ZZ} = \frac{\pi R^4}{2}$$

If D be the diameter of the circular lamina then

$$R = \frac{D}{2}$$

$$\therefore \qquad I_{ZZ} = \frac{\pi}{2} \times \left(\frac{D}{2}\right)^4 = \frac{p}{2} \times \frac{D^4}{16} = \frac{\pi D^4}{32}$$

By theorem of perpendicular axes we get

$$I_{ZZ} = I_{XX} + I_{YY}$$

For circular lamina

$$I_{XX} = I_{YY}$$

since the whole area is symmetrical with respect to XX and YY axis.

$$\therefore \qquad I_{ZZ} = I_{XX} + I_{XX}$$

$$\text{or} \qquad I_{ZZ} = 2I_{XX} \qquad (\text{As } I_{XX} = I_{YY})$$

$$\text{or} \qquad I_{XX} = I_{YY} = \frac{I_{ZZ}}{2}$$

$$\text{or} \qquad I_{XX} = I_{YY} = \frac{\pi D^2}{32 \times 2}$$

$$\therefore \qquad I_{XX} = I_{YY} = \frac{\pi D^2}{64}$$

5.10.1. M.I. of a Hollow Circular Section

Consider a hollow section of outer dia 'D' and inner dia 'd'. The M.I. of the section about any centroidal axis *i.e.*, $I_{XX} = I_{YY}$

= M.I. of the outer circle abo ut the centroidal axis – M.I. of the inner circle about the centroidal axis

$$= \frac{\pi}{64} D^4 - \frac{\pi}{64} D^4$$

$$\therefore \qquad I_{XX} = I_{YY} = \frac{\pi}{64} (D^4 - d^4)$$

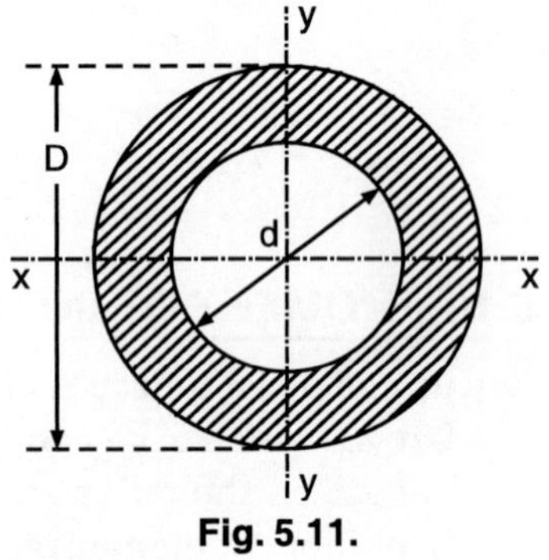

Fig. 5.11.

5.11 MASS MOMENT OF INERTIA

(a) Rectangle

Consider the rectangle $ABCD$ of mass M. Let m be the mass per unit area of the rectangle.

Consider an elemental strip of thickness dx at a distance x from the centroidal axis xx.

$\therefore$ Mass of the strip = $m \times bdx$

$$= \frac{M}{bd} bdx = \frac{Mdx}{d}$$

M.I. of the elemental strip about XX

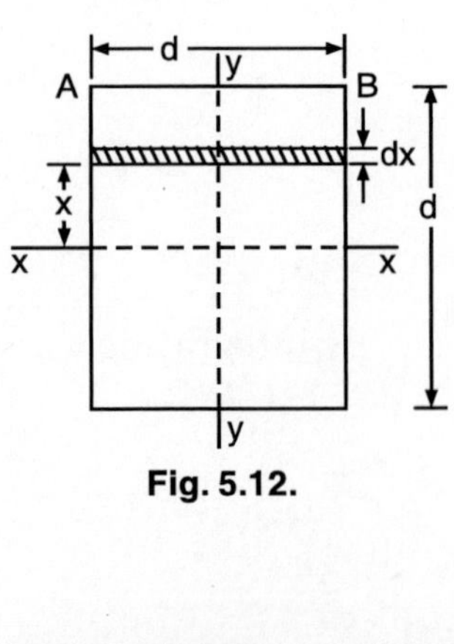

Fig. 5.12.

$$= \frac{M.dx}{d} \times x^2$$

$\therefore$ M.I. of the whole rectangle about XX

$$= I_{XX} = 2\int_0^{d/2} \frac{Mdx}{d}.x^2 = \frac{2M}{d}\int_0^{d/2} x^2 dx$$

Integrating, we gate

$$I_{XX} = \frac{2M}{d}\left[\frac{x^3}{3}\right]_0^{d/2} = \frac{2M}{d} \times \frac{1}{3} \times \frac{d^3}{8}$$

$$I_{XX} = \frac{Md^2}{12}$$

Similarly $$I_{YY} = \frac{Mb^2}{12}$$

M.I. of the rectangle about side DC

$$I_{DC} = I_{XX} + M\left(\frac{d}{2}\right)^2 = \frac{Md^2}{12} + M\frac{d^2}{4}$$

$$I_{DC} = \frac{Md^2 + 3Md^2}{12} = \frac{Md^2}{3}$$

5.12 STEPS FOR CALCULATING THE M.I. OF A COMPOSITE SECTION

(*a*) Divide the section into regular sections (like rectangle, triangle, circle etc.)
(*b*) Find out the C.G. of the composite section.
(*c*) Find out the M.I. of the regular sections about their centroidal axis.
(*d*) Then transfer these M.I. about the required axis using theorem of parallel axis.
(*e*) The M.I. of the composite section can be calculated by taking the algebraic sum of the M.I. of the regular section about the required axis.

Moment of inertia of some regular figures

(*i*) Rectangle

$$I_{XX} = \frac{bd^3}{12}$$

$$I_{YY} = \frac{db^3}{12}$$

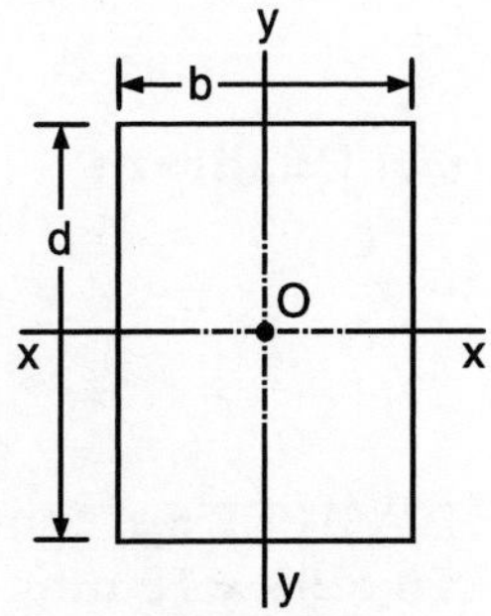

Fig. 5.13.

(*ii*) Hollow rectangle

$$I_{XX} = \frac{BD^3 - bd^3}{12}$$

(*iii*) Triangle

$$I_{XX} = \frac{bh^3}{36}$$

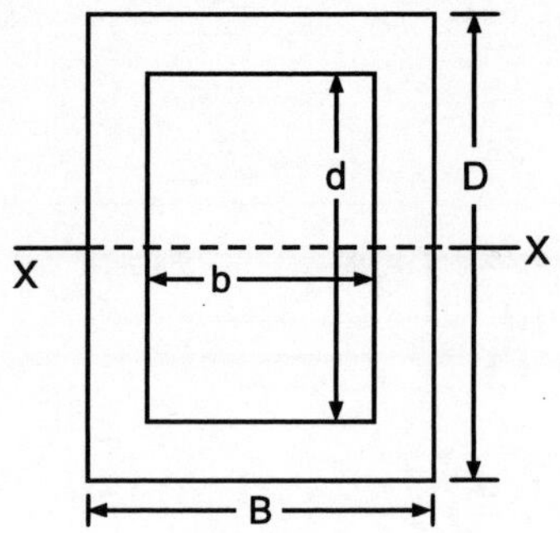

Fig. 5.14

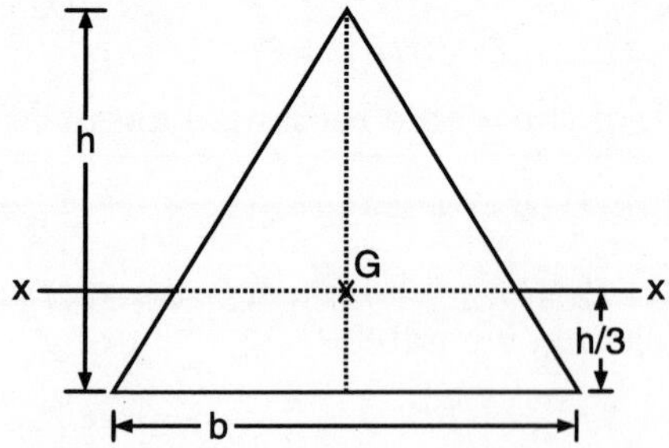

Fig. 5.15

$$I_{\text{base}} = \frac{bh^3}{12}$$

$$I_{\text{vertex}} = \frac{bh^3}{4}$$

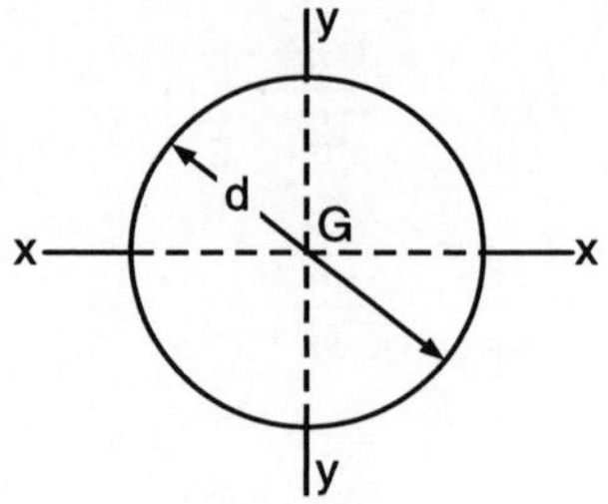

Fig. 5.16.

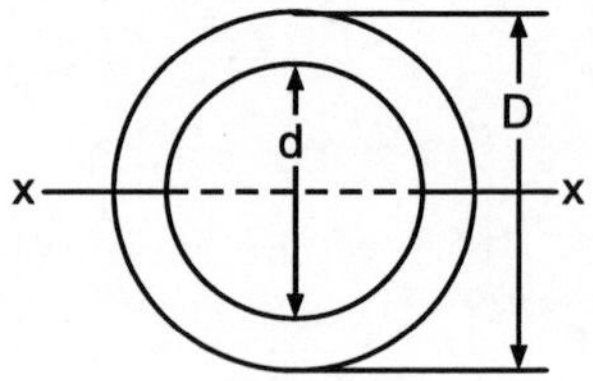

Fig. 5.17.

***(iv)* Circle**

$$I_{XX} = I_{YY} = \frac{\pi d^4}{64}$$

***(v)* Hollow Circle**

$$I_{XX} = I_{YY} = \frac{\pi}{64}(D^4 - d^4)$$

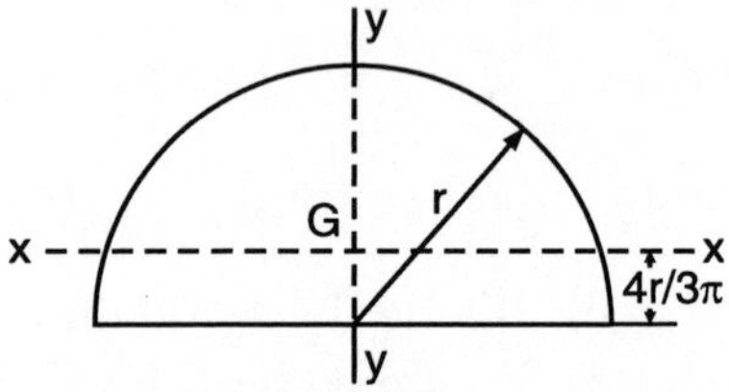

Fig. 5.18.

***(vi)* Semicircle**

$$I_{XX} = 0.11r^4$$

$$I_{YY} = \frac{\pi r^4}{8}$$

$$I_{OX} = \frac{1}{2} \times \frac{\pi d^4}{64} = \frac{\pi d^4}{128}$$

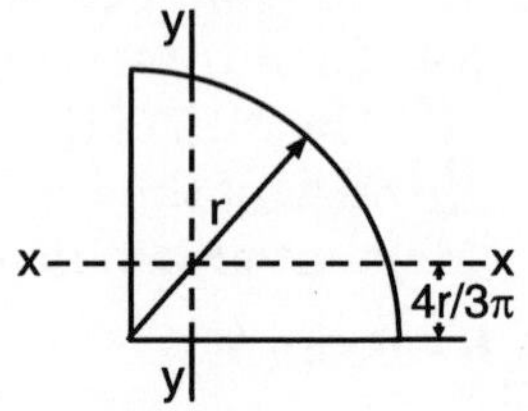

Fig. 5.19.

***(vii)* Quadrant**

$$I_{XX} = I_{YY} = 0.055\, r^4$$

SOLVED EXAMPLES

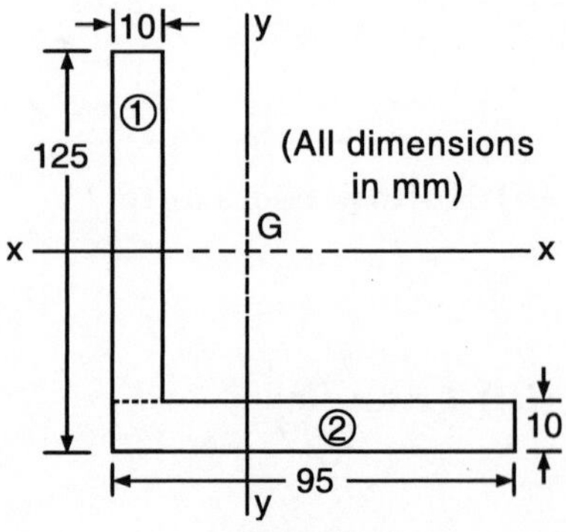

Fig. 5.20.

EXAMPLE 5.1. Find I_{XX} and I_{YY} for unequal angle section 125 × 95 × 10 mm. Details shown in the figure 5.20.

Solution: Let G be the C.G. of the L section

whose co-ordinate is $(\bar{X}, \bar{Y})$

SI. No.	Components	Area (a) (mm²)	Dist. of C.G from OL 'x' (mm)	Dist. of C.G from OM 'y' (mm)	ax (mm³)	ay (mm³)
1.	Rectangle (1)	115 × 10 = 1150	$\frac{10}{2}$ = 5	$10 + \frac{115}{2}$ = 67.5	5750	77625
2.	Rectangle (2)	95 × 10 = 950	$\frac{95}{2}$ = 47.5	$\frac{10}{2}$ = 5	45125	4750
		Sa = 2100			Sax = 50875	Say = 82375

$$\therefore \qquad \bar{X} = \frac{\Sigma ax}{\Sigma a} = \frac{50875}{2100} = 24.23 \text{ mm}$$

$$\therefore \qquad \bar{Y} = \frac{\Sigma ay}{\Sigma a} = \frac{82375}{2100} = 39.23 \text{ mm}$$

Let I_{XX_1} and I_{XX_2} be the M.I. of the rectangles (1) and (2) about XX axis then

$$I_{XX} = I_{XX_1} + I_{XX_2}$$

$$= \left[\frac{10 \times 115^3}{12} + 10 \times 115(67.5 - 39.23)^2\right] + \left[\frac{95 \times 10^3}{12} + 95 \times 10 \times (39.23 - 5)^2\right]$$

$$= 2186467.7 + 1121024.9$$

$$= 3307492.6 \text{ mm}^4 \qquad \textbf{Ans.}$$

Let I_{YY_1} and I_{YY_2} be the M.I. of the rectangles (1) and (2) about YY.

$$I_{YY} = I_{YY1} + I_{YY2}$$

$$= \left[\frac{115 \times 10^3}{12} + 115 \times 10(24.23 - 5)^2\right] + \left[\frac{10 \times 95^3}{12} + 10 \times 95 \times (47.5 - 24.23)^2\right]$$

$$= 434845.17 + 1228897.4$$

$$= 1663742.6 \text{ mm}^4 \qquad \textbf{Ans.}$$

EXAMPLE 5.2. Calculate the M.I. of the T-section about centroidal axes as shown in the figure 5.21.

Solution: Let G be the C.G. of the section having coordinates $(\bar{X}, \bar{Y})$ from O. Let OX and OY be the reference axis as shown in figure 5.22.

S.No.	Components	Area (a) (mm^2)	Dist. of C.G. from OY 'x' (mm)	Dist. of C.G. from ox 'y' (mm)	ax (mm^3)	ay (mm^3)
1.	Rectangle (1)	60×10 = 600	$\frac{60}{2}$ = 30	$100 + \frac{10}{2}$ = 105	18000	63000
2.	Rectangle (2)	100×10 = 1000	$25 + \frac{10}{2}$ = 30	$\frac{100}{2}$ = 50	30000	50000
		Sa = 1600			Sax = 48000	Say = 113000

$$\therefore \quad \bar{X} = \frac{\Sigma ax}{\Sigma a} = \frac{48000}{1600} = 30 \text{ mm}$$

$$\therefore \quad \bar{Y} = \frac{\Sigma ay}{\Sigma a} = \frac{113000}{1600} = 70.62 \text{ mm}$$

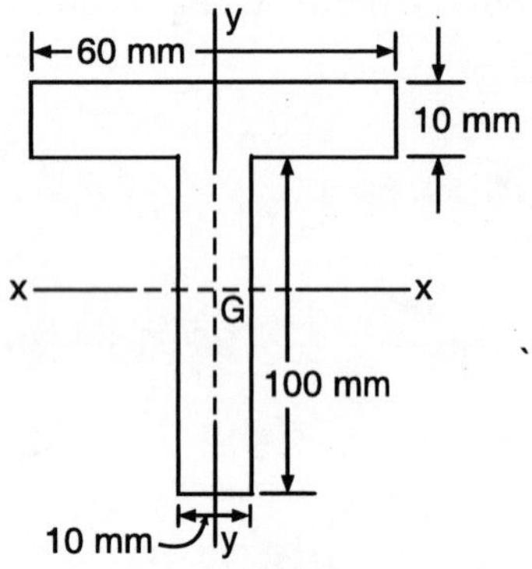

Fig. 5.21.

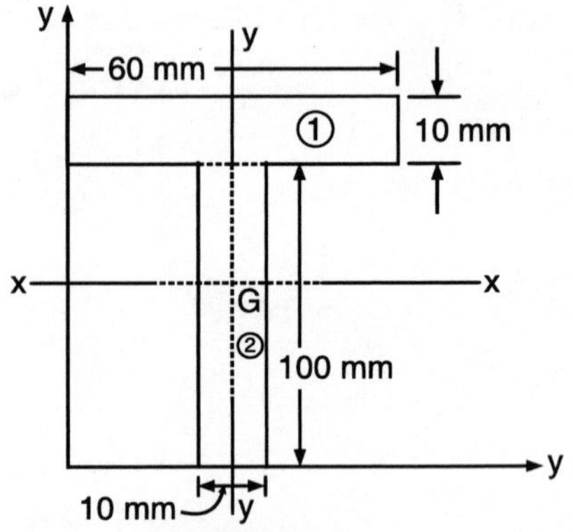

Fig. 5.22.

[As the section is symmetrical about YY axis, so the C.G. will lie on the YY and hence $\bar{X}\left\{=\frac{60}{2}=30 \text{ mm}\right\}$ can be calculated directly without making the table.]

Let I_{XX1} I_{XX2} be the M.I. of the rectangles (1) and (2) about XX axis.

$$\therefore \quad I_{XX} = I_{XX1} + I_{XX2}$$

$$= \left[\frac{60 \times 10^3}{12} + 60 \times 10(105 - 70.62)^2\right]$$

$$+ \left[\frac{10 \times 100^3}{12} + 10 \times 100\,(70.62 - 50)^2\right]$$

$$= 714190.64 + 1258517.7 = 1972708.4 \text{ mm}^4 \quad \textbf{Ans.}$$

Let I_{YY1} and I_{YY2} be the M.I. of the rectangles (1) and (2) about YY axis.

$$\therefore \quad I_{YY} = I_{YY1} + I_{YY2}$$

$$= \left[\frac{10 \times 60^3}{12} + 0\right] + \left[\frac{100 \times 10^3}{12} + 0\right]$$

$$= 188333.33 \text{ mm}^4 \quad \textbf{Ans.}$$

EXAMPLE 5.3. Find the moment of inertia about the horizontal axis through the C.G. of the section shown in the figure 5.23.

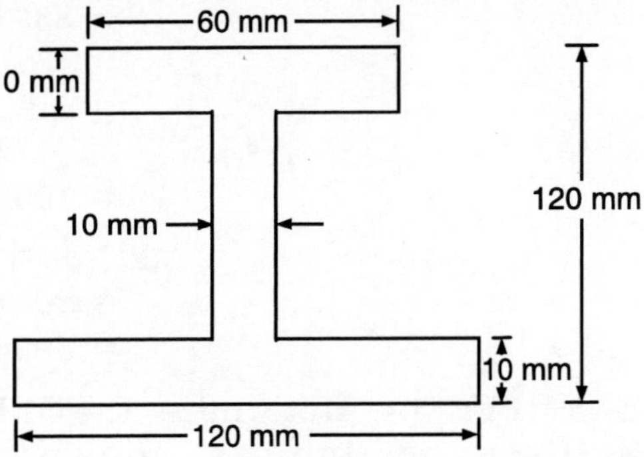

Fig. 5.23.

Solution: Divide the I section in three rectangles as shown in figure 5.23(*i*). The section is symmetrical about *Y* axis.

Let G be the C.G. of the I-section and *OX* and *OY* be the reference axis.

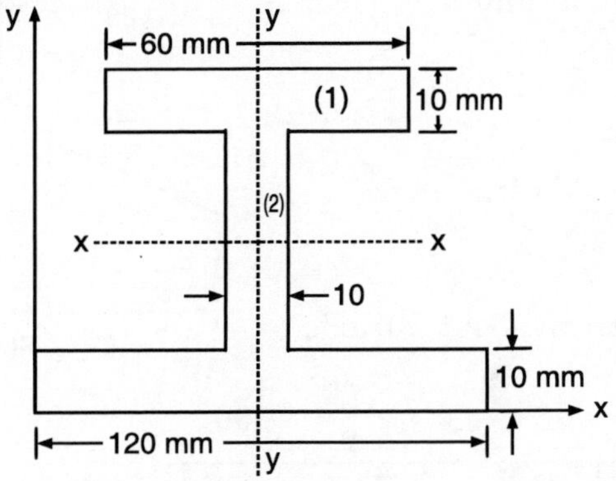

Fig. 5.23 (i).

S.No.	Components	Area (*a*) (mm²)	Dist. of C.G. from *OX* (*y*) (mm)	*ay* (mm³)
1.	Rectangle (1)	60 × 10 = 600	110 + 5 = 115	69000
2.	Rectangle (2)	100 × 10 = 1000	10 + 50 = 60	60000
3.	Rectangle (3)	120 × 10 = 1200	5	6000
		Sa = 2800		Say = 1,35,000

$$\therefore \qquad \bar{Y} = \frac{\Sigma ay}{\Sigma a} = \frac{135000}{2800} = 48.21 \text{ mm}$$

Let I_{XX1}, I_{XX2} and I_{XX3} be the M.I. of the rectangles (1), (2) and (3) respectively about *XX* axis and I_{XX} be the M.I. of I-section about *XX* axis.

$$I_{XX1} = \frac{60 \times 10^3}{12} + 60 \times 10 \times (115 - 48.21)^2$$

$$= 5000 + 2676542.5 = 2681542.5 \text{ mm}^4$$

$$I_{XX2} = \frac{10 \times 100^3}{12} + 100 \times 10 \times (60 - 48.21)^2$$

$$= 833333.33 + 139004.1 = 972337.4 \text{ mm}^4$$

$$I_{XX3} = \frac{120 \times 10^3}{12} + 120 \times 10(48.21 - 5)^2$$

$$= 10000 + 2240524.9 = 2250524.9 \text{ mm}^4$$

$$\therefore \quad I_{XX} = I_{XX1} + I_{XX2} + I_{XX3}$$

$$= 2681542.5 + 972337.4 + 2250524.9$$

$$= 5904404.8 \text{ mm}^4 = 5.904 \times 10^6 \text{ mm}^4 \quad \textbf{Ans.}$$

EXAMPLE 5.4. Determine the M.I. of the built-up section as shown in the figure 5.24(i) about its centroidal axes *X-X* and *Y-Y*.

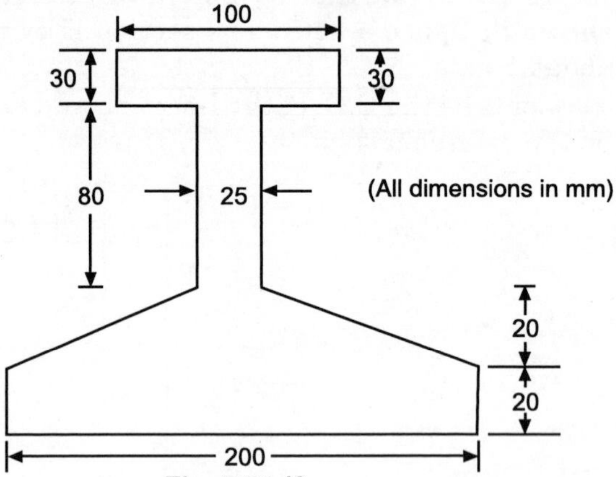

Fig. 5.24 (*i*).

Solution: Divide the section into five parts as shown in the figure 5.24(*ii*).

(*i*) Rectangle (1)
(*ii*) Rectangle (2)
(*iii*) Triangle (3)
(*iv*) Triangle (4)
(*v*) Rectangle (5)

The section is symmetrical about *YY* axis

$$\overline{X} = \frac{200}{2} = 100$$

S.No.	Components	area (*a*)	dist of C.G from OX axis (*y*)	*ay*
1.	Rectangle (1)	100 × 30 = 3000	135	405000
2.	Rectangle (2)	100 × 25 = 2500	70	175000
3.	Triangle (3)	$\frac{1}{2} \times 87.5 \times 20 = 875$	$20 + \frac{20}{3} = 25.67$	23335.25
4.	Triangle (4)			23335.25
5.	Rectangle (5)	20 × 200 = 4000	10	40000
		Sa = 11250		Say = 666672.5

$$\overline{Y} = \frac{\Sigma ay}{\Sigma a}$$

$$\overline{Y} = \frac{666672.5}{11250} = 59.26 \text{ mm.}$$

Let I_{xx1}, I_{xx2}, I_{xx3}, I_{xx4} and I_{xx5} be the M.I. of parts (1), (2), (3), (4) and (5) respectively about xx-axis and I_{xx} be the M.I. of the section about *xx*-axis.

$$I_{xx1} = \frac{100 \times 30^3}{12} + 100 \times 30 \times (135 - 59.26)^2 = 17434643 \text{ mm}^4$$

$$I_{xx2} = \frac{25 \times 100^3}{12} + 25 \times 100 \times (70 - 59.26)^2$$
$$= 2371702.30 \text{ mm}^4$$

$$I_{xx3} = \frac{87.5 \times 20^3}{36} + 87.5 \times 20 \times (59.26 - 25.67)^2$$
$$= 187133.6 \text{ mm}^4$$
$$I_{xx4} = I_{xx3} = 1878133.6 \text{ mm}^4$$

$$I_{xx5} = \frac{200 \times 20^3}{12} + 200 \times 20 \times (59.26 - 10)^2$$
$$= 9839523.7 \text{ mm}^4$$

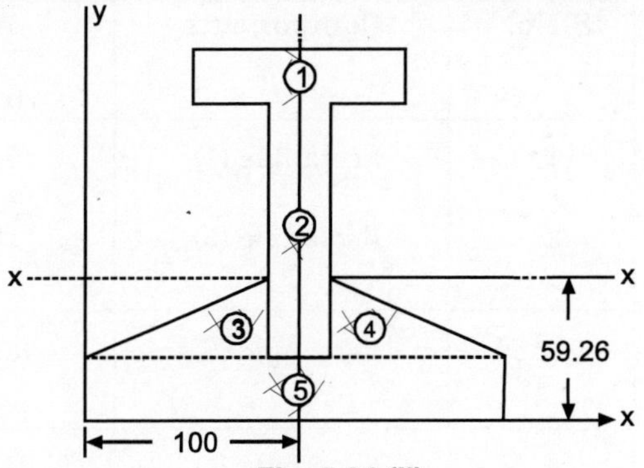

Fig. 5.24 (ii).

$$I_{xx} = I_{xx1} + I_{xx2} + I_{xx3} + I_{xx4} + I_{xx5}$$
$$= 17434643 + 2371702.30 + 1878133.6 + 1878133.6 + 9839523.7$$
$$= 33402136 \text{ mm}^4.$$

Let I_{yy1}, I_{yy2}, I_{yy3}, I_{yy4}, I_{yy5} and I_{yy} be the M.I. of parts (1), (2), (3), (4), (5) and the whole section about *yy*-axis.

$$I_{yy1} = \frac{30 \times 100^3}{12} = 2500000 \text{ mm}^4$$

$$I_{yy2} = \frac{100 \times 25^3}{12} = 13020.833 \text{ mm}^4$$

$$I_{yy3} = \frac{20 \times 87.5^3}{12} + 20 \times 87.5 \times 12.5^2 = 1389974 \text{ mm}^4$$

$$I_{yy4} = \frac{20 \times 87.5^3}{12} + 20 \times 87.5 \times 12.5^2 = 1389974 \text{ mm}^4$$

$$I_{yy5} = \frac{20 \times 200^3}{12} = 13333333 \text{ mm}^4$$

$$I_{yy} = 2500000 + 13020.83 + 1389974 + 1389974 + 13333333$$
$$= 18626302 \text{ mm}^4.$$

EXAMPLE 5.5. A T-beam is made up of two plates 150 × 10 mm and two angles 50 × 50 × 10 mm as shown in figure 5.25(i). Determine the M.I. of T-section about an axis passing through the centroid of the section and parallel to the top plate.

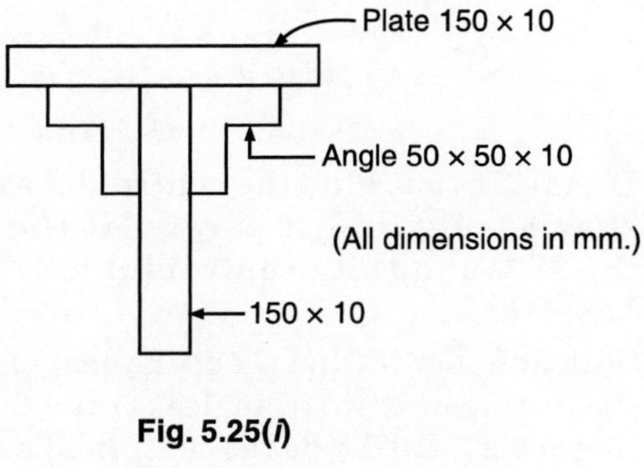

Fig. 5.25(*i*)

Solution: Let *OX* and *OY* be the reference axis. Divide the T-section into four components as shown in the figure 5.25 (*i*) Let *XX* be the centroidal axis parallel to the top plate.

S.No.	Components	Area (a) (mm^2)	Dist. of C.G from OX axis (y) (mm)	ay (mm^3)
1.	Rectangle (1)	150 × 10 = 1500	5	7500
2.	Rectangle (2)	110 × 10 = 1100	10 + 5 = 15	16500
3.	Rectangle (3)	30 × 40 = 1200	10 + 10 + 20	48000
4.	Rectangle (4)	10 × 10 = 1000	10 + 10 + 40 + 50 = 110	1,10,000
		Sa = 4800		Say = 1,82,000

$$\therefore \qquad \bar{Y} = \frac{\Sigma ay}{\Sigma a} = \frac{182000}{4800} = 37.92 \text{ mm.}$$

Let I_{XX1}, I_{XX2}, I_{XX3} and I_{XX4} be the M.I. of the rectangles (1), (2) and (3) and (4) about XX axis and I_{XX} be the M.I. of the T-Section about XX axis.

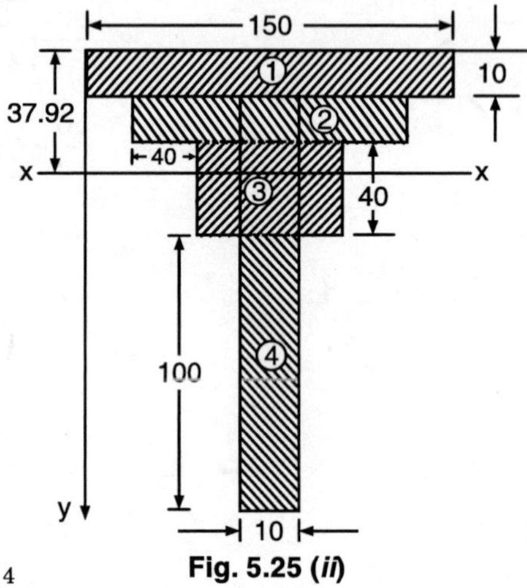

Fig. 5.25 (*ii*)

$$I_{XX1} = \frac{150 \times 10^3}{12} + 150 \times 10 \times (37.92 - 5)^2$$

$$= 12500 + 1625589.6 = 1638089.6 \text{ mm}^4$$

$$I_{XX2} = \frac{110 \times 10^3}{12} + 110 \times 10 \times (37.92 - 15)^2$$

$$= 9165.67 + 577859.04 = 587025.71 \text{ mm}^4$$

$$I_{XX3} = \frac{30 \times 40^3}{12} + 40 \times 30 \times (40 - 37.92)^2$$

$$= 160000 + 5191.68 = 165191.68 \text{ mm}^4$$

$$I_{XX4} = \frac{10 \times 100^3}{12} + 100 \times 10 \times (100 - 37.92)^2$$

$$= 833333.3 + 5195525.4 = 6028859.7 \text{ mm}^4$$

$$\therefore \quad I_{XX} = I_{XX1} + I_{XX2} + I_{XX3} + I_{XX4}$$

$$= 1638089.6 + 587025.71 + 165191.68 + 6028859.7$$

$$= 8419165.7 \text{ mm}^4 \quad \textbf{Ans.}$$

EXAMPLE 5.6. Find the centroidal axis of the lamina shown in figure 5.26 parallel to the base. Find also the M.I. about its centroidal axis parallel to the base.

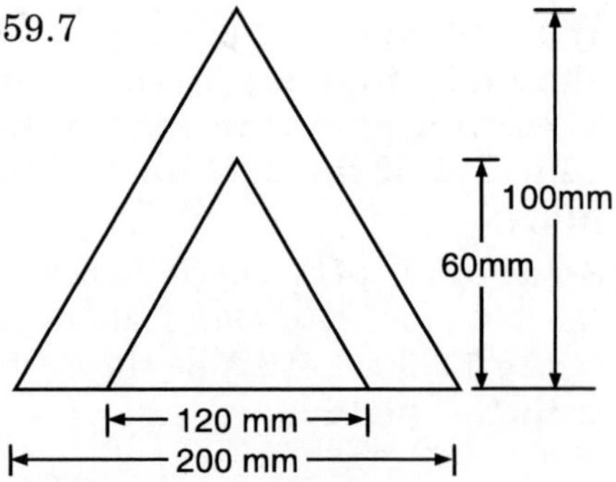

Fig. 5.26.

Solution: Let OX and OY be the axes of reference. Divide the lamina into two triangles (1) and (2) as shown in the figure 5.27. Let XX be the centroidal axis parallel to the base.

S.No.	Components	Area (a) (mm^2)	Dist. of C.G. from OX axis (y) (mm)	ay (mm^3)
1.	Triangle (1)	$\frac{1}{2} \times 200 \times 100$ = 10000(+)	$\frac{100}{3}$ = 33.33	333300 (+)
2.	Triangle (2)	$\frac{1}{2} \times 120 \times 60$ = 3600(–)	$\frac{60}{3}$ = 20	72000(–)
		Sa = 6400		Say = 261300

$$\bar{Y} = \frac{\Sigma ay}{\Sigma a} = \frac{261300}{6400} = 40.83 \text{ mm}$$

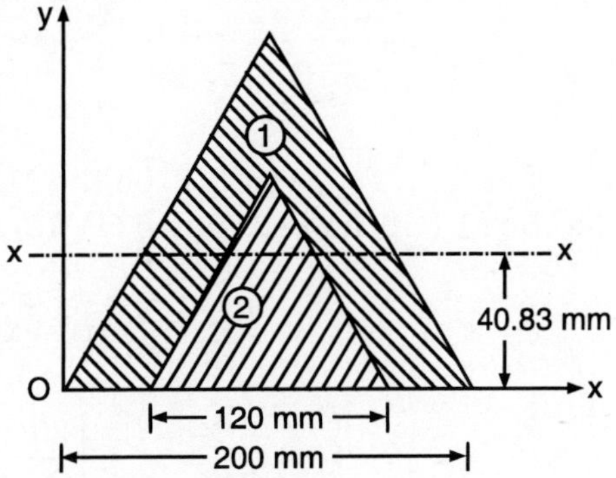

Fig. 5.27.

Let I_{XX1}, I_{XX2} and I_{XX} be the M.I. of the triangles (1), (2) and lamina respectively about XX axis.

$$\therefore \quad I_{XX1} = \frac{200 \times 100^3}{36} + \frac{1}{2} \times 200 \times 100\,(40.83 - 33.33)^2$$

$$= 5555555.6 + 562500$$
$$= 6118055.6 \text{ mm}^4 \text{ (+)}$$

$$\therefore \quad I_{XX2} = \frac{120 \times 60^3}{36} + \frac{1}{2} \times 120 \times 60 \times (40.83 - 21)^2$$

$$= 720000 + 1562000 = 2282000 \text{ mm}^4 (-)$$

$$\therefore \quad I_{XX} = I_{XX1} - I_{XX2} = 6118055.6 - 2282000 = 3836055.6 \text{ mm}^4 \quad \textbf{Ans.}$$

EXAMPLE 5.7. A channel section 160 × 80 × 20 mm stands with flanges horizontal. Determine moment of inertia about XX and YY axis passing through the centroid of the section.

Solution: Figure 5.28 shows the channel section stands with flanges horizontal. Let OX and OY be the axis of reference. Divide the channel section into three rectangles (1), (2) and (3) as shown in figure 5.28.

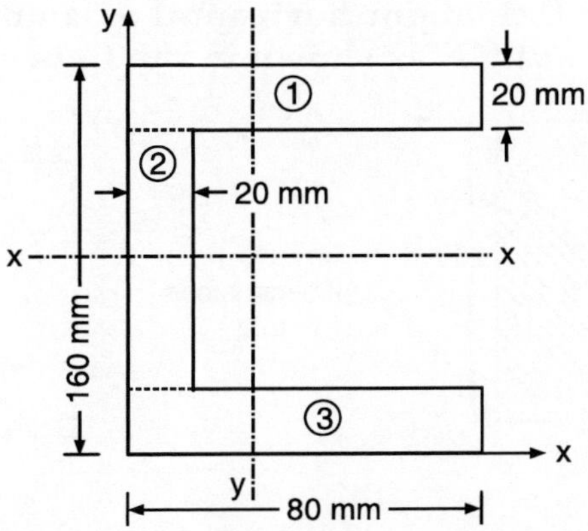

Fig. 5.28.

Components	Area (a) (mm²)	Dist. of C.G. from OX (y) (mm)	Dist of C.G. from oy (x) (mm)	ay (mm³)	ax (mm³)
Rectangle (1)	80 × 20 = 1600	20 + 120 + 10 = 150	40	240000	64000
Rectangle	120 × 20 = 2400	20 + 60 = 80	10	192000	24000
Rectangle (3)	80 × 20 = 1600	10	40	16000	64000
	Sa = 5600			Say = 448000	Sax = 152000

$$\bar{X} = \frac{\Sigma ax}{\Sigma a} = \frac{152000}{5600} = 27.14 \text{ mm}$$

$$\bar{Y} = \frac{\Sigma ay}{\Sigma a} = \frac{448000}{5600} = 80 \text{ mm}$$

Let I_{XX1}, I_{XX2} and I_{XX3} be the M.I. of the rectangles about XX axis and I_{YY1}, I_{YY2} and I_{YY3} be the M.I of the rectangles (1), (2) and (3) about YY axis.

$$I_{XX} = I_{XX1} + I_{XX2} + I_{XX3} = \left\{\frac{80 \times 20^3}{12} + 80 \times 20 \times (150 - 80)^2\right\}$$

$$+ \left\{\frac{20 \times 120^3}{12} + 120 \times 20 \times (0)\right\} + \left\{\frac{80 \times 20^3}{12} + 80 \times 20 \times (80 - 10)^2\right\}$$

$$= 7893333.3 + 2880000 + 7893333.3 = 18666667 \text{ mm}^4 \quad \textbf{Ans.}$$

$$\therefore \quad I_{YY} = I_{YY1} + I_{YY2} + I_{YY3} = \left\{\frac{20 \times 80^3}{12} + 80 \times 20 \times (40 - 27.14)^2\right\}$$

$$+ \left\{\frac{120 \times 20^3}{12} + 120 \times 20 \times (27.14 - 10)^2\right\} + \left\{\frac{20 \times 80^3}{12} + 80 \times 20 \times (40 - 27.14)^2\right\}$$

$$= 1117940.7 + 785071.04 + 1117940.7$$

$$= 3020952.4 \text{ mm}^4 \quad \textbf{Ans.}$$

EXAMPLE 5.8. Find the C.G. about horizontal axis and moment of inertia of section, shown hatched, about MN as shown in the figure 5.29.

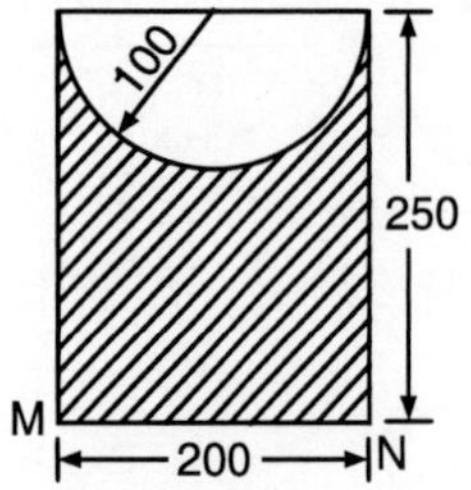

(All dimensions in mm.)

Fig. 5.29.

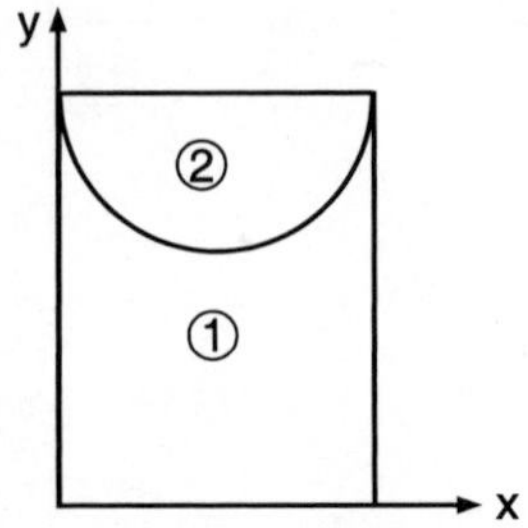

Fig. 5.30.

Solution: Let OX and OY be the reference axes. Divide the section into a rectangle (1) and semicircle (2) as shown in the figure 5.30.

Components	Area (a) (mm)	Dist. of C.G. from OX (y)	ay (mm²)
Rectangle (1)	250 × 200 = 50000(+)	125	6250000(+)
Semicircle (2)	$\frac{\pi \times 100^2}{2}$	$250 - \frac{4r}{3\pi}$	3260344.2 (–)
	= 15707.96 (–)	$= 250 - \frac{4 \times 100}{3\ \pi} = 207.56$	
	Sa = 34292.04		Say = 2989655.8

$$\bar{Y} = \frac{\Sigma ay}{\Sigma a} = \frac{2989655.8}{34292.04} = 87.18 \text{ mm}$$

Let I_{XX1} and I_{XX2} be the M.I. of the rectangle and the semicircle respectively about OX axis. Let I_{MN} be the M.I. of the hatched portion about MN.

$$I_{MN} = I_{XX1} - I_{XX2}$$

$$= \left\{\frac{200 \times 250^3}{12} + 200 \times 250 \times 125^2\right\}$$

$$- \left\{0.11 \times (100)^4 + \frac{\pi \times 100^2}{2} \times (207.56)^2\right\}$$

$$= 1.04 \times 10^9 - 5.87 \times 10^8 = 3.53 \times 10^8 \text{ mm}^4 \quad \textbf{Ans.}$$

EXAMPLE 5.9. Find the position of the C.G. and calculate the moment of inertia about the horizontal centroidal axis of the section in the figure 5.31.

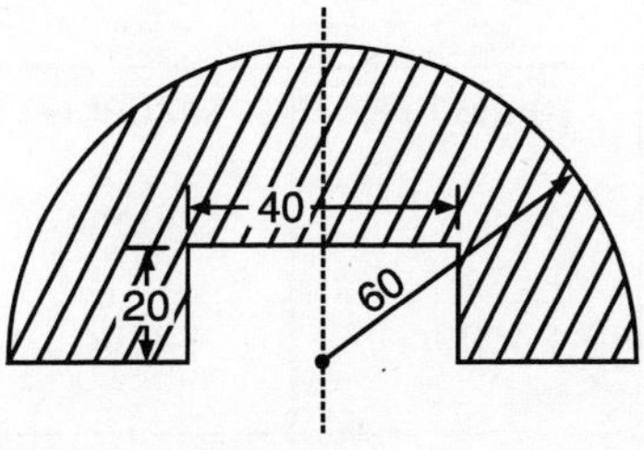

All dimensions in mm.

Fig. 5.31.

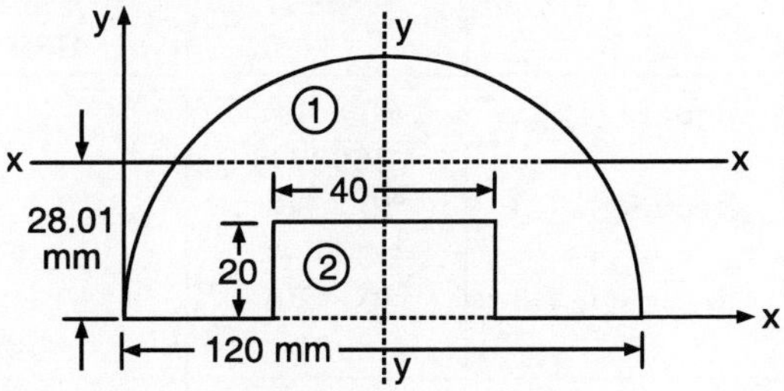

Fig. 5.32.

Solution: Let OX and OY be the reference axes. Divide the section into a semicircle (1) and rectangle (2).

Components	Area (a)	Dist of C.G. from OX (y)	ay
Semicircle (1)	$\frac{\pi \times 60^2}{2}$ = 5654.86 (+)	25.46	143972.74 (+)
Rectangle (2)	40 × 20 = 800(–)	10	8000(–)
	Sa = 4854.86		Say = 135972.74

$$\bar{Y} = \frac{\Sigma ay}{\Sigma a} = \frac{135972.74}{4854.86} = 28.01 \text{ mm}$$

Let I_{XX1}, I_{XX2} and I_{XX} be the M.I. of the semicircle, rectangle and hatched portion of the section about XX axis respectively.

$$I_{XX} = I_{XX1} - I_{XX2}$$

$$= \left\{0.11 \times (60)^4 + \frac{\pi \times 60^2}{2} \times (28.01 - 25.46)^2\right\} - \left\{\frac{40 \times 20^3}{12} + 40 \times 20 \times (28.01 - 10)^2\right\}$$

$$= 1462370.8 - 286154.75$$

$$= 1176215.1 \text{ mm}^4 \quad \textbf{Ans.}$$

EXAMPLE 5.10. Locate the centroid of the lamina as shown in the figure 5.33. Also calculate the moment of inertia about the centroidal axis *XX*. (All dimensions in mm).

Solution: Let OX and OY be the reference axes. Divide the lamina into three components

(1) Square of size 60 × 60

(2) Rectangle of size 40 × 20

(3) Rectangle of size 40 × 20

as shown in the figure 5.33.

The C.G. of the lamina can be found out from the table shown below :

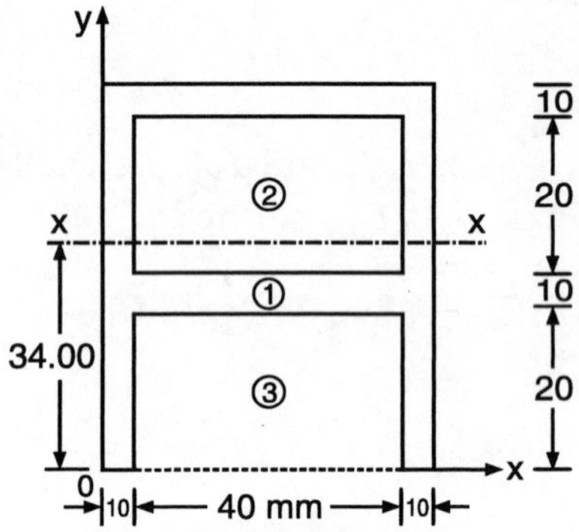

Fig. 5.33.

Components	Area (a) (mm²)	Dist of C.G. from oy (x) (mm)	Dist of C.G from ox(y) (mm)	ax (mm³)	ay (mm³)
Square (1)	60 × 60 = 3600 (+)	30	30	108000 (+)	108000(+)
Rectangle (2)	40 × 20 = 800(–)	10 + 20 = 30	20 + 10 + 10 = 40	24000 (–)	32000(–)
Rectangle (3)	40 × 20 = 800(–)	10 + 20 = 30	10	24000(–)	8000(–)
	$\Sigma a = 2000$			$\Sigma ax = 60000$	$\Sigma ay = 68000$

$$\bar{X} = \frac{\Sigma ax}{\Sigma a} = \frac{60000}{2000} = 30 \text{ mm}$$

$$\bar{Y} = \frac{\Sigma ay}{\Sigma a} = \frac{68000}{2000} = 34 \text{ mm}$$

Let I_{XX1}, I_{XX2} and I_{XX3} be the moment of inertia of the square (1) and rectangles (2) and (3) about the centroidal axis XX.

The M.I. of lamina (I_{XX}) about XX axis can be found out by substracting the moment of inertia of the square and rectangles *i.e.*,

$$I_{XX} = I_{XX1} - I_{XX2} - I_{XX3}$$

$$I_{XX1} = \frac{60\times 60^3}{12} + 60 \times 60 \times (34 - 30)^2$$
$$= 1080000 + 57600 = 1137600 \text{ mm}^4$$

$$I_{XX2} = \frac{40\times 20^3}{12} + 40 \times 20 \times (40 - 34)^2$$
$$= 26665.67 + 28800 = 55465.67 \text{ mm}^4$$

$$I_{XX3} = \frac{40\times 20^3}{12} + 40 \times 20 \times (34 - 10)^2$$
$$= 26665.67 + 460800 = 487465.67 \text{ mm}^4$$

$$\therefore \quad I_{XX} = 1137600 - 55465.67 - 487465.67$$
$$= 594665.66 \text{ mm}^4 \quad \textbf{Ans.}$$

EXAMPLE 5.11. Find the moment of the lamina with a circular hole of 30 mm diameter about the axis LM as shown in the figure 5.34.

Solution: Let I_{XX1}, I_{XX2} and I_{XX3} be the M.I. of the triangle LMN, Semicircle and the circle respectively about LM axis.

The M.I. of the lamina about LM axis is found out by substracting the M.I. of the circle with the M.I. of triangle and Semicircle, *i.e.*,

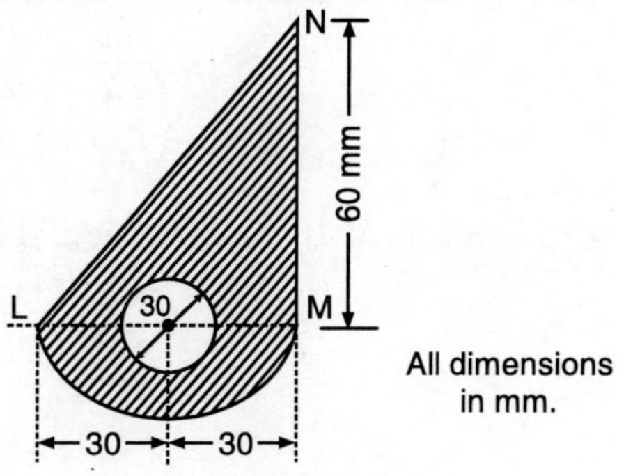

Fig. 5.34.

$$I_{XX} = I_{XX1} + I_{XX2} - I_{XX3}$$

$$I_{XX1} = \frac{bd^3}{12} = \frac{60 \times 60^3}{12}$$
$$= 1080000 \text{ mm}^4$$

$$I_{XX2} = \frac{1}{2} \times \frac{\pi d^4}{64} = \frac{1}{2} \times \frac{\pi \times (60)^4}{64} = 318085.26 \text{ mm}^4$$

$$I_{XX3} = \frac{\pi d_1^4}{64} = \frac{\pi \times (30)^4}{64} = 39760.78 \text{ mm}^4$$

$$\therefore \quad I_{XX} = 1080000 + 318085.26 - 39760.78$$
$$= 1358325.5 \text{ mm}^4 \quad \textbf{Ans.}$$

EXAMPLE 5.12. Calculate the moment of inertia of the section shown in the Fig. 5.35 about the axis passing through the C.G. and parallel to the base.

Solution: Let *OX* and *OY* be the reference axes. Divide the lamina into four regular components as shown in the figure 5.35.

The C.G of the section can be calculated as below :

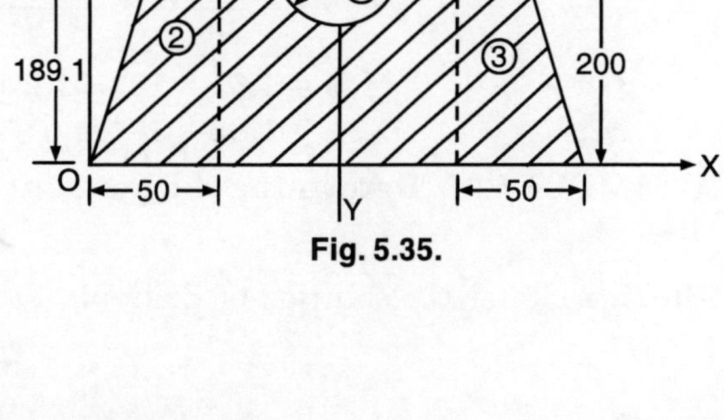

Fig. 5.35.

Components	Area (a) mm^2	Dist. of C.G. from OX (y) mm	ay mm^3
Rectangle (1)	300×400 $= 120000$	200	24000000
Triangle (2)	$\frac{1}{2} \times 50 \times 400$ $= 10000$	$\frac{400}{3}$ $= 133.33$	1333300
Triangle (3)	$\frac{1}{2} \times 50 \times 400$ $= 10000$	$\frac{400}{3}$ $= 133.33$	1333300
Circle (4)	$\frac{\pi \times 150^2}{4}$ $= 17671.46$	200	3534292 (–)
	$\Sigma a = 122328.54$		$\Sigma ay = 23132308$

$$\bar{Y} = \frac{\Sigma ay}{\Sigma a} = \frac{23132308}{122328.54} = 189.1 \text{ mm}$$

Let I_{XX1}, I_{XX2}, I_{XX3} and I_{XX4} be the M.I. of the components (1), (2), (3) and (4) respectively about xx axis.

Then, the M.I. of the section about the xx axis is given by

$$I_{XX} = I_{XX1} + I_{XX2} + I_{XX3} - I_{XX4}$$

$$I_{XX1} = \frac{300 \times 400^3}{12} + 300 \times 400 \times (200 - 189.1)^2$$

$$= 1.61 \times 10^9 \text{ mm}^4.$$

$$I_{XX2} = \frac{bh^3}{36} + \frac{1}{2} bh \text{ (distance of C.G. of the section}$$

$$- \text{distance of C.G. of the triangle)}^2$$

$$= \frac{50 \times 400^3}{36} + \frac{1}{2} \times 50 \times 400 \times (189.1 - 133.33)^2$$

$$= 1.2 \times 10^8 \text{ mm}^4$$

$$I_{XX3} = I_{XX2} = 1.2 \times 10^8 \text{ mm}^4$$

$$I_{XX4} = \frac{\pi d^4}{64} + \frac{\pi d^2}{4} (200 - 189.1)^2$$

$$= \frac{\pi \times 150^4}{64} + \frac{\pi \times 150^2}{4} (10.9)^2 = 26950035 \text{ mm}^4$$

$$\therefore \quad I_{XX} = 1.61 \times 10^9 + 1.2 \times 10^8 + 1.2 \times 10^8 - 26950035$$

$$= 1.82 \times 10^9 \text{ mm}^4 \quad \textbf{Ans.}$$

EXAMPLE 5.13. Determine the moment of inertia of the shaded area shown in figure 4.35.

Slution: Refer the solution of Example 17 of Chapter 4 for $\bar{X}$ and $\bar{Y}$

$\bar{X} = 200$ mm and $\bar{Y} = 113.83$ mm

Refer figure 5.36

Let I_{XX1}, I_{XX2}, I_{XX3} and I_{XX4} be the moment of inertia of triangle (1), Rectangle (2), triangle (3), and Semicircle (4) about *XX* axis and I_{XX} be the moment of inertia of the shaded area about *XX* axis.

$$I_{XX1} = \frac{100 \times 190^3}{36} + \frac{1}{2} \times 100 \times 190 \left(113.83 - \frac{190}{3}\right)^2$$
$$= 43280153 \text{ mm}^4$$

$$I_{XX2} = \frac{200 \times 190^3}{12} + 200 \times 190 \times \left(113.83 - \frac{190}{2}\right)^2$$
$$= 1.28 \times 10^8 \text{ mm}^4$$

$$I_{XX3} = I_{XX1} = 43280153 \text{ mm}^4$$
$$I_{XX4} = 0.11 \times (140)^4 = 42257600 \text{ mm}^4$$
$$I_{XX} = I_{XX1} + I_{XX2} + I_{XX3} - I_{XX4}$$
$$= 43280153 + 1.28 \times 10^8 + 43280153 - 42257600$$
$$= 1.72 \times 10^8 \text{ mm}^4 \quad \textbf{Ans.}$$

Let I_{YY1}, I_{YY2}, I_{YY3} and I_{YY4} be the moment of inertia of triangle (1), rectangle (2, triangle (3) and semicircle (4) about *YY* axis and I_{YY} be the moment of inertia of the shaded area bout YY axis.

$$I_{YY1} = \frac{190 \times 100^3}{36} + \frac{1}{2} \times 190 \times 100\ (200 - 66.67)^2$$
$$= 1.74 \times 10^8 \text{ mm}^4$$

$$I_{YY2} = \frac{190 \times 200^3}{12} = 1.27 \times 10^8 \text{ mm}^4$$

$$I_{YY3} = \frac{190 \times 100^3}{36} + \frac{1}{2} \times 190 \times 100\ (333.33 - 200)^2 = 1.74 \times 10^8 \text{ mm}^4$$

$$I_{YY4} = \frac{\pi \times (140)^4}{8} = 1.51 \times 10^8 \text{ mm}^4$$
$$I_{YY} = I_{YY1} + I_{YY2} + I_{YY3} - I_{YY4}$$
$$= 1.74 \times 10^8 + 1.27 \times 10^8 + 1.74 \times 10^8 - 1.51 \times 10^8$$
$$= 3.24 \times 10^8 \text{ mm}^4 \quad \textbf{Ans.}$$

EXAMPLE 5.14. Determine the moment of inertia of the built up section shown in the figure 5.36 about an axis *AB* passing through the top most fibre of the section.

Solution: Let I_{XX1}, I_{XX2}, I_{XX3} and I_{XX4} be the moment and inertia of rectangles (1), (2), (3) and (4) respectively about the axis *AB*.

$$I_{XX1} = \frac{400 \times 20^3}{12} + 400 \times 20 \times 10^2 = 1066665.67 \text{ mm}^4$$

$$I_{XX2} = 2 \times \left[\frac{100 \times 10^3}{12} + 100 \times 10 \times (20 + 5)^2\right]$$

[∵ There are two rectangles of no. 2]

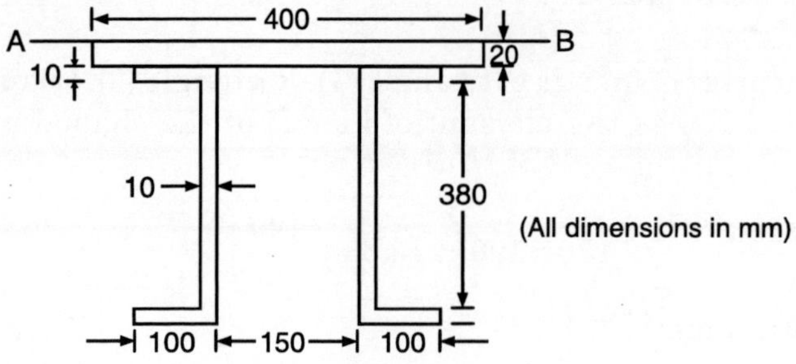

(All dimensions in mm)

Fig. 5.36.

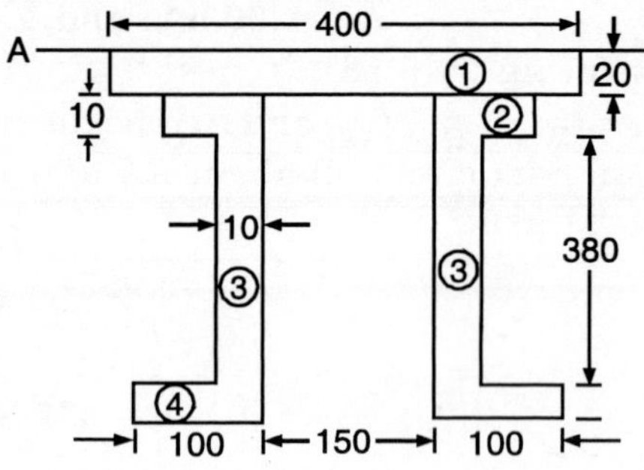

Fig. 5.37

$$= 1266665.67 \text{ mm}^4$$

$$I_{XX3} = 2 \times \left[\frac{100 \times 380^3}{12} + 380 \times 10 \times (20 + 10 + 190)^2\right]$$

$$= 459293333.2 \text{ mm}^4$$

$$I_{XX4} = 2 \times \left[\frac{100 \times 10^3}{12} + 100 \times 10 \times (20 + 10 + 380 + 5)^2\right]$$

$$= 344466665.6 \text{ mm}^4$$

I_{XX} = M.I. of the built up section ab1out AB

$$= I_{XX1} + I_{XX2} + I_{XX3} + I_{XX4}$$

$$= 1066665.67 + 1266665.67 + 459293333.2 + 344466665.6$$

$$= 806093333 = 8.06 \times 10^8 \text{ mm}^4$$ **Ans.**

SUMMARY

1. **Moment of inertia** of a body is defined as the product of the mass and the square of the perpendicular distance of its C.G. from the axis.
2. **Radius of gyration** is the square root of the ratio of moment of inertia and the area of the body

$$K = \sqrt{\frac{I}{A}}$$

3. The ratio of moment of inertia about the centroidal axis and the distance of the extreme fibre from the centroidal axis is called modulus of section.
4. If I_{OX} and I_{OY} be the M.I. of a lamina about two perpendicular axis OX and OY in the plane of the lamina and I_{OZ} be the M.I. of the lamina about the OZ axis, Which is perpendicular to the lamina and passing through the intersection of OX and OY, then according to the theorem of perpendicular axis.

$$I_{OZ} = I_{OX} + I_{OY}$$

5. The M.I. of lamina about any axis equals the sum of M.I. about a parallel centroidal axis in the plane of the lamina and the product of the area of the lamina with the square of the distance between the two axis. This is the theorem of parallel axis.

OBJECTIVE TYPE QUESTIONS

I. Multiple Choice Questions

1. M.I. of a square, about xx axis, as shown in figure 5.38 is

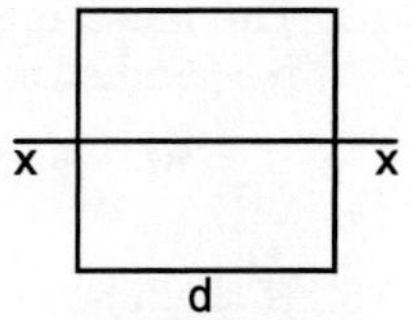

Fig. 5.38.

(a) $\frac{d^4}{12}$ (b) $\frac{d^3}{12}$

(c) $\frac{d^4}{24}$ (d) $\frac{d^3}{6}$

2. M.I. of the rectangle about xx axis is

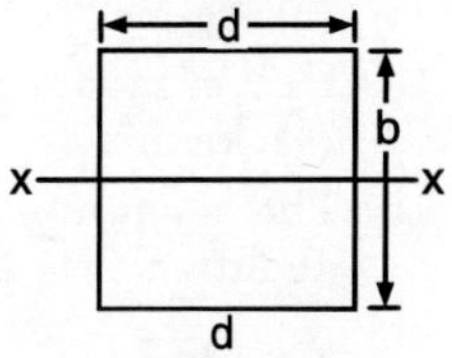

Fig. 5.39

(a) $\frac{bd^3}{12}$ (b) $\frac{db^3}{12}$

(c) $\frac{db^3}{24}$ (d) $\frac{bd^3}{12}$

3. M.I. of the triangle about LM axis is

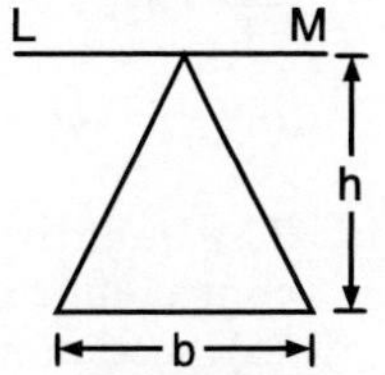

Fig. 5.40.

(a) $\frac{bh^3}{36}$ (b) $\frac{bh^3}{12}$

(c) $\frac{bh^3}{4}$ (d) $\frac{bh^3}{24}$

4. M.I. of the semicircle about xx axis is

(a) $0.11\ r^4$, (b) $0.39\ r^4$,
(c) $0.02\ r^4$, (d) $0.15\ r^4$.

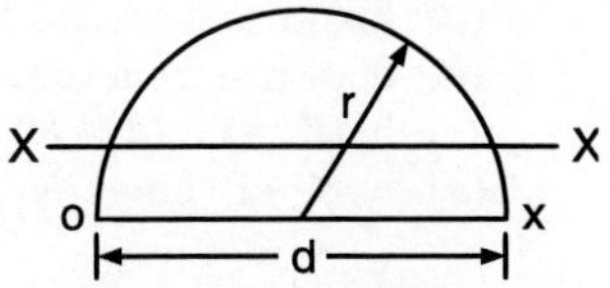

Fig. 5.41.

5. M.I. of the semicircle about ox axis is

(a) $\frac{\pi d^4}{128}$ (b) $\frac{\pi d^4}{128}$

(c) $\frac{\pi d^4}{32}$ (d) None

6. The radius of gyration of lamina of M.I. (I) and area A is given by

(a) $\sqrt{\frac{I}{A'}}$ (b) $\frac{I}{A^2}$ (c) $\sqrt{\frac{A}{I'}}$ (d) $\sqrt{IA}$

7. M.I. of a lamina is also called as

(a) First moment of area (b) Second moment of area
(c) Third moment of area (d) None

8. Unit of M.I. of a lamina having area in m^2 and distance in m is

(a) m^3 (b) m^4 (c) N-m^2 (d) N-m^3

9. The theorem of perpendicular axis is used to obtain the M.I. of

(a) Rectangular lamina (b) Square lamina
(c) Triangular lamina (d) Circular lamina.

10. The theorem of parallel axes cannot be used in obtaining the M.I. of

(a) Triangular lamina (b) Circular lamina
(c) Square (d) Rectangular

11. The moment of inertia of a spherical shell of mass 'm' and radius 'r' about a diameter is

(a) $\frac{1}{3}mr^2$ (b) $\frac{2}{3}mr^2$ (c) $\frac{2}{5}mr^2$ (d) $\frac{2}{7}mr^2$.

12. The mass moment of inertia of rectangular plate of mass 'M' and sides 'a' and 'b' about an axis perpendicular to plate and passing through its centre is
    (a) $\frac{M}{12}(a^2+b^2)$ (b) $\frac{M}{6}(a^2+b^2)$ (c) $\frac{Ma^2}{12}+\frac{Mb^2}{4}$ (d) None.
13. If the mass of a body is measured in kilograms and the distance in metres, then the moment of inertia of body will be measured in
    (a) kg-m (b) kg-m^2 (c) kg/m (d) kg/m^3.
14. Unit of modulus of section is
    (a) kg/m^2 (b) m^3 or cm^3 (c) cm^4 or m^4 (d) cm^2 or m^2.
15. The moment of inertia of a body about its C.G. is I_{CG}. The value of moment of inertia of the body about an axis AB parallel to C.G. will be given by
    (a) $I_{AB} = I_{CG} + Ah^2$ (b) $I_{CG} = I_{AB} + Ah^2$ (c) $I_{AB} = I_{CG} - Ah^2$ (d) $I_{CG} = I_{AB} - Ah^2$
16. I_{xx} or I_{yy} of a hollow circular section of internal diameter 'd' and external diameter '*D*' will be
    (a) $\frac{\pi}{32}(D-d)^2$ (b) $\frac{\pi}{32}(d^4-d^4)$ (c) $\frac{\pi}{64}(D^4-d^4)$ (d) $\frac{\pi}{64}(D-d)^4$
17. The Quantity obtained by dividing the moment of inertia the cross-section of a structure about its centroidal axis by the distance of the extreme fibre from centroidal axis is known as
    (a) section modulus (b) radius of gyration (c) second moment (d) None
18. Radius of gyration may be defined as the
    (a) distance from the axis of reference where the whole mass of the body is assumed to be concentrated.
    (b) distance from the axis of reference where the whole area of the body is assumed to be concentrated.
    (c) both (a) and (b)
    (d) none of the above

Answers

1. (a)	2. (b)	3. (c)	4. (a)	5. (a)	5. (a)
7. (b)	8. (b)	9. (d)	10. (b)	11. (b)	12. (a)
13. (b)	14. (b)	15. (a)	15. (c)	17. (a)	18. (a)

THEORETICAL QUESTIONS

1. (a) Define moment of inertia.
   (b) What is radius of gyration?
   (c) Define modulus of section.
2. State and prove theorem of parallel axes.
3. State and prove theorem of perpendicular axes.
4. Derive the expression for moment of inertia of the following sections:
   (a) Rectangular section about its base parallel to the centroidal axis.
   (b) Triangular section about its vertex.
   (c) Circular section about its centroidal axis.
   (d) Rectangular section with a centrally situated rectangular hole, about its horizontal centroidal axis.

NUMERICALS

1. Find I_{XX} and I_{YY} for the unequal angle section 100 × 80 × 20 mm.
   (I_{XX} = 2906667 mm^4, I_{YY} = 1626667 mm^4)

2. Find the M.I of the T section about an axis passing the top fibre of the flange having flange = 150 × 50 mm and web = 100 × 50 mm. (6041667 mm^4)
3. Find the M.I of a channel section of 75 × 175 × 10 mm about *XX* and *YY* axis passing through C.G. (13329103 mm^4, 1521217 mm^4)
4. Calculate the M.I about the centroidal axis x-x for lamina shown in the figure 5.42. (I_{XX} = 40176677 mm^4)

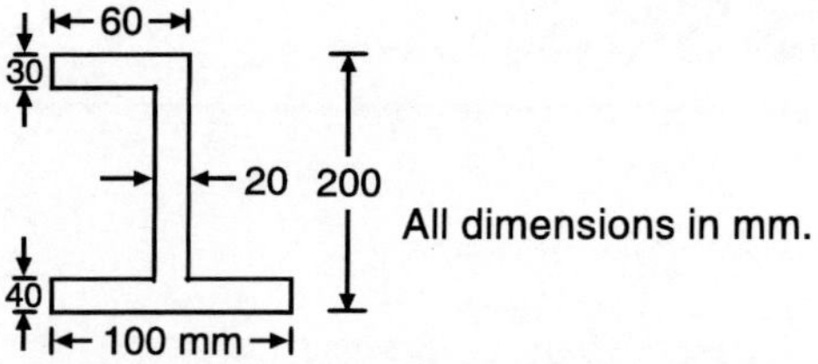

Fig. 5.42.

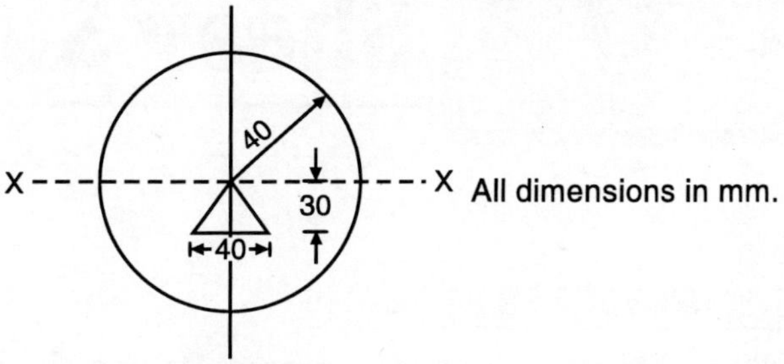

Fig. 5.43.

5. Find the M.I of the section shown in the figure 5.43 about *xx* axis. (I_{XX} = 1740619 mm^4)
6. A rectangular hole is made in a triangular section as shown in figure 5.44. Determine the M.I. of the section about *xx* axis passing through the C.G and the base *BC*. (I_{XX} = 1.824 × 10^6 mm^4; I_{BC} = 4.815 × 10^6 mm^4)

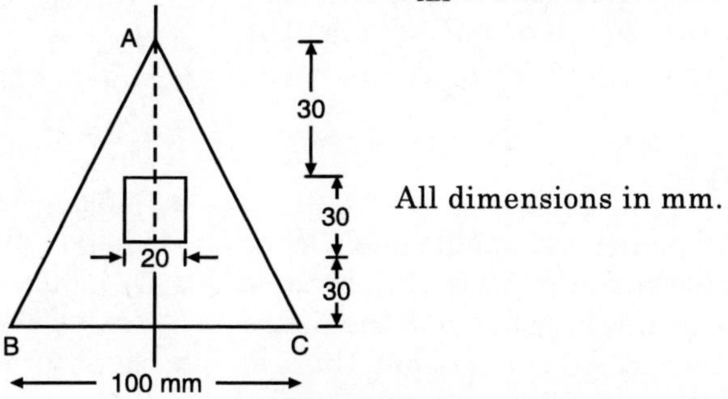

Fig. 5.44.

7. Determine the M.I. of a built up section shown in the figure 5.45 about *xx*-axis passing through the C.G. of the section. (I_{XX} = 5.74 × 10^7 mm^4)

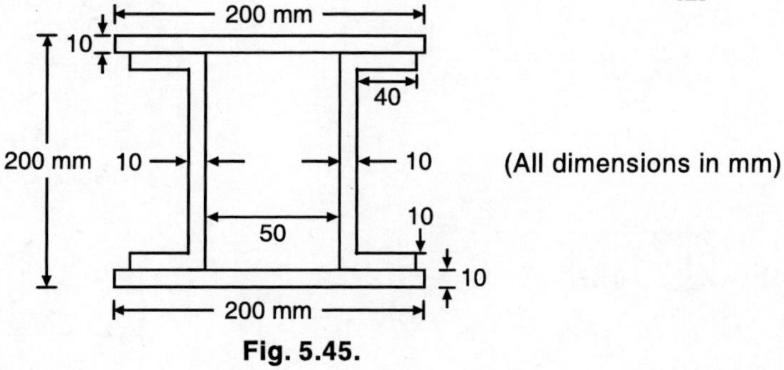

Fig. 5.45.

CHAPTER 6

Bending and Shear Stresses

6.1 STRESSES IN BEAMS

When a beam is subjected to external loading, the shear force and bending moment are developed at all sections of the beam. Due to this shear force and bending moment, the beam undergoes certain deformation. The material of the member will offer resistance or stresses. The stresses produced to counter bending moment are known as *bending stresses* or *longitudinal stresses*. The stresses produced to counter shear force are known as *shear stresses*.

6.2 PURE BENDING

Consider a beam CABD loaded with point-loads W at point C and D as shown in the figure 6.1 (*i*). Shear force and bending moment diagrams are shown in the figure 6.1 (*ii*) and 6.1 (*iii*). In the part AB of the beam, there is no shear force and is subjected to bending moment only. This condition of the beam between A and B is called ***pure bending*** or ***simple bending***.

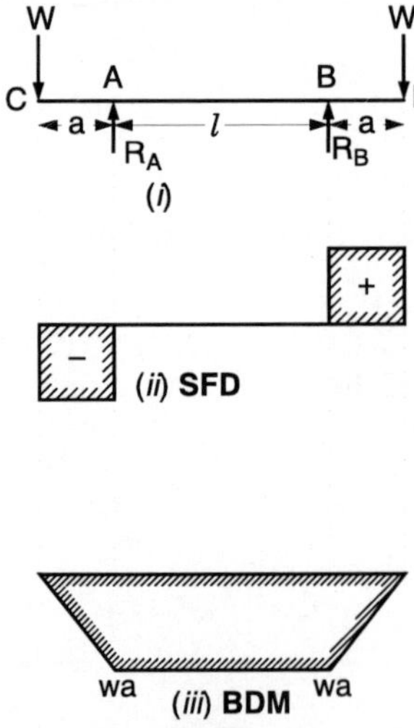

Fig. 6.1

Chapter Outline

- Stresses in Beams
- Pure Bending
- Assumptions in the Theory of Simple or Pure Bending
- Theory of Simple Bending (Bending Equation)
- Neutral Axis
- Moment of Resistance
- Modulus of Section (Section Modulers)
- Section Modulus for Various Shapes of Beam Section
- Distribution of Bending Stress Across the Beam Section
- Steps for Finding out the Bending Stress
- Beams of Heterogeneous Materials (Flitched Beam)
- Shear Stress
- Shear Stress at a Section of a Beam
- Shear Stress Distribution for Beam Section of Various Shapes Rectangular Section

DO YOU KNOW?

In case of simple bending, the beam is subjected to pure moments and the shearing force is absent.

The bending of the beam not accompanied by shear force is known as pure bending.

6.3 ASSUMPTIONS IN THE THEORY OF SIMPLE OR PURE BENDING

The following assumptions have been made in the theory of pure bending:

(*i*) The material of the beam is homogeneous* and isotropic #.

(*ii*) The value of Young's modulus of elasticity (E) is the same in tension and compression.

(*iii*) The transverse section of the beam which were plane before bending remain plane after bending.

(*iv*) The material of the beam obey's Hooke's law and is stressed within its elastic limit.

(*v*) The resultant push or pull on the transverse section of the beam is zero.

(*vi*) The loads are applied in the plane of bending.

(*vii*) The transverse section of the beam is symmetrical about an axis passing through the centroid and parallel to the plane of bending.

(*viii*) The radius of curvature of the beam before bending is very large in comparison to the transverse dimensions of the beam.

(*ix*) Each layer of the beam is free to expand or contract independently.

6.4 THEORY OF SIMPLE BENDING (BENDING EQUATION)

Figure 6.2 (*i*) shows a part of beam subjected to bending moment M. Consider small length '*dx*' between two sections AB and CD. Due to this bending moment 'M' the portion of the beam bends like an arc of radius R and centre O as shown in figure 6.2 (*ii*).

Consider a layer **G H** at a distance '*y*' from N. A.

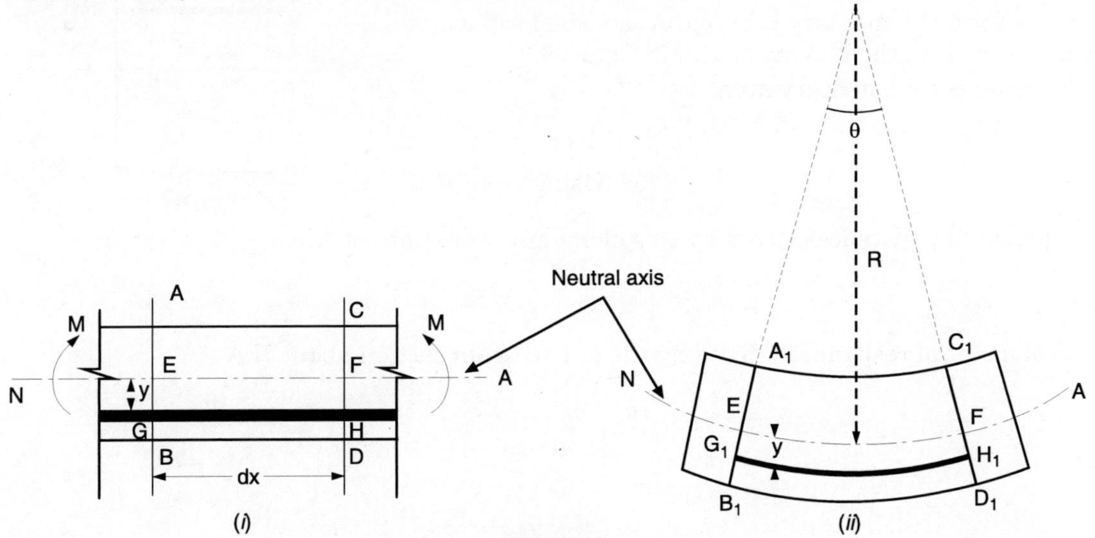

Fig. 6.2

Original length of this fibre *GH* = *dx*

After deformation the length of fibre GH increases and will take the position G_1H_1.

Thus, will be no change in length of the fibre EF on the N.A. after the deformation of the beam.

Length of fibre $GH = EF = dx = R\theta$ [Refer figure 6.2 (*ii*)]

---

* Homogeneous means the material is of same kind throughout.

\# Isotropic means having the same elastic properties in all directions.

$$\left[\begin{array}{l}\because \text{lenght of arc} = \text{Radius of curvatuve} \times \text{angle subtended} \\ \qquad dx = R \times \theta\end{array}\right]$$

Length of fibre $G_1H_1 = (R + y)\,\theta$

Change in length of fibre $GH = G_1H_1 - GH = (R + y)\,\theta - R\,\theta = y\theta$

Strain in the fibre GH, $e = \dfrac{\text{change in length}}{\text{original length}}$

$$e = \frac{y\theta}{R\theta} = \frac{y}{R} \qquad ...(i)$$

But strain, $e = \dfrac{\text{stress}}{E} = \dfrac{\sigma_b}{E}$ ...(ii)

From (*i*) and (*ii*), we get,

$$\frac{\sigma_b}{E} = \frac{y}{R}$$

or
$$\frac{\sigma_b}{y} = \frac{E}{R}$$

or,
$$\sigma_b = \frac{E}{R}y \qquad ...(iii)$$

DO YOU KNOW?

In ordinary bending, an engineer has to deal with a beam cross-section which has to resist both bending and shearing force.

∴ *The stress intensity at any fibre is proportional to the distance of the fibre from N.A.*

Consider an elementary area 'δa' of the beam section at a distance 'y' from the N.A. as shown in figure 6.3.

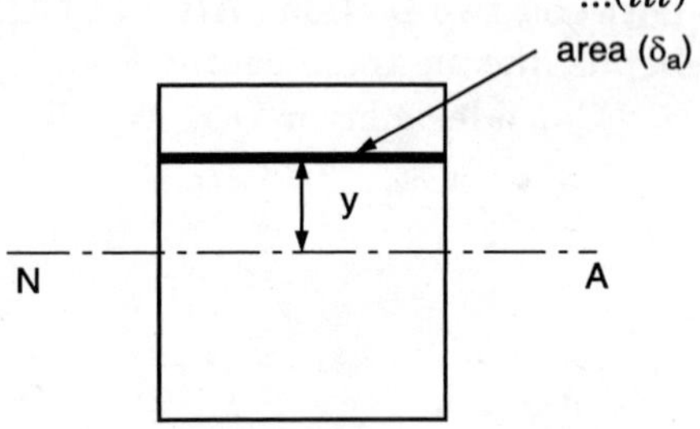

Fig 6.3

Force on the elementary area

$$= \sigma_b \times \delta_a$$

$$= \left(\frac{E}{R}y\right)\delta a \text{ [Using eqn. (iii)]}$$

Moment of resistance offered by this elementary area about N.A.

$$= \left(\frac{E}{R}y\right)\delta a \times y = \frac{E}{R}y^2\delta a$$

∴ Moment of resistance offered by the entire beam section about N.A.

$$M = \int \frac{E}{R}y^2\delta a$$

$$M = \frac{E}{R}\int y^2\delta a$$

$$= \frac{E}{R}I \qquad \left[\because \int y^2\delta a = I = M.I. \text{ of the beam section about N.A}\right]$$

or
$$\frac{M}{I} = \frac{E}{R} \qquad ...(iv)$$

From (*iii*) and (*iv*), we get

$$\boxed{\frac{M}{I} = \frac{\sigma_b}{y} = \frac{E}{R}}$$

This is called *bending equation.*

DO YOU KNOW?

The term EI is known as flexural rigidity of a beam section.

Where M = Moment of resistance of the section or bending moment
I = Moment of inertia of the section about N.A.
σ_b = Bending stress at a distance 'y' from N.A.
y = Distance of fibre under consideration from N.A.
R = Radius of curvature of N.A.
E = Young's molulus of elasticity

DO YOU KNOW?

Since $\frac{1}{R} = \frac{M}{EI}$, the curvature of the axis of beam in bending is directly proprotional to bending moment and inversely proportional to flexural rigidity.

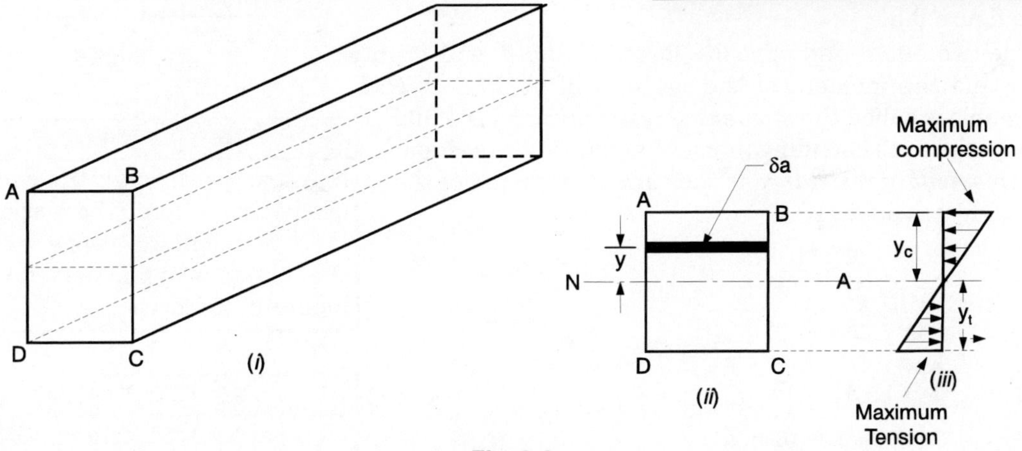

Fig. 6.4

It a section is subjected to pure sagging moment, all layers above N.A. will be under compressive stress varying from zero at the neutral axis to maximum at the top most layer as shown in figure 6.4(*iii*). The layers below N.A. will be under tensile stress varying from zero at N.A. to maximum at the bottom most layer.

Position of neutral axis

Consider an elemental (small) area δa in a section ABCD of a beam [Fig 6.4 (*ii*)] at a distances 'y' from N.A.

DO YOU KNOW?

Neutral axis is the axis at which the stress is zero.

Stress on elemental area $= \frac{E}{R}y$

Force on elemental area $= \left(\frac{E}{R}\right)y \times \delta a$

Total force on the beam section

$$= \int \frac{E}{R} y\delta a = \frac{E}{R}\int y\delta a$$

Since no axial load has been applied, the total force on the beam section is zero.

$$\therefore \qquad \frac{E}{R}\int y\,\delta a = 0$$

As E and R are constants $\therefore$ $\frac{E}{R}$ cannot be zero.

and so $\qquad \int y\,\delta a = 0$

$\int y\,\delta a = 0$, is possible if $y = 0$, *i.e.*, if neutral axis and centroidal axis concides. Thus neutral axis is a centroidal axis.

6.5 NEUTRAL AXIS

Neutral axis of a beam is the axis at which stress is zero. Neutral axis always passes through the centroid of the cross-section of a beam and is parallel to the base of the cross-section. It is abbreviated as N.A.

6.6 MOMENT OF RESISTANCE

When a beam is subjected to B.M, internal forces or stresses are developed in the section above and below neutral axis (N.A.). These forces or stresses are of opposite nature as shown in the figure 6.5.

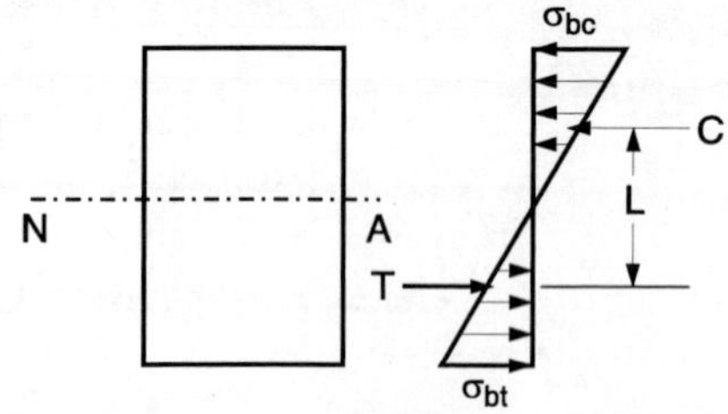

Fig.6.5

The two equal and opposite forces C and T will form a couple and the moment of the couple will be $T \times L$ or $C \times L$. This couple is called the *moment of resistance* and is equal and opposite to the bending moment acting on the section. Thus, *moment of resistance* is the sum of moments of the internal forces about N.A.

> **DO YOU KNOW?**
> N.A. passes through the centroid of a beam section if the material of the beam follows hook's law and the section is not subjected to any longitudinal force.

We have

$$\frac{M}{I} = \frac{\sigma_b}{y}$$

or, $$M = \sigma_b \times \frac{I}{y}$$

∴ $$M = \sigma_b \times Z$$

> **DO YOU KNOW?**
> Neutral axis coincides with the centroidal axis.

where Z = Section moleculus = $\frac{I}{y}$

Thus, moment of resistance depends upon the section modulus of the section.

6.7 MODULUS OF SECTION (SECTION MODULUS)

Modulus of section or section modulus is defined as the ratio of moment of intertia of a beam section about the neutral axis (N.A.) to the distance of extreme (outermost) fibre from the N.A. It is denoted by Z.

We have the bending equation.

> **DO YOU KNOW?**
> The strength of beam section depends mainly on section modulus.

$$\frac{M}{I} = \frac{\sigma_b}{y}$$

or, $$M = \sigma_b \times \frac{I}{y} = \sigma_b \times z$$

where z = Section modulus = $\frac{I}{y}$

I = Moment of intertia of the section about N.A.

y = Distance of outermost fibre from N.A.

6.8 SECTION MODULUS FOR VARIOUS SHAPES OF BEAM SECTION

(*i*) Rectangular section

Consider a rectangular section of width '*b*' and depth '*d*'. Moment of inertia of the section about N.A.

$$I = \frac{1}{12}bd^3$$

Distance of outermost fibre from N.A.

$$= y_{max} = \frac{d}{2}$$

Section modulus, $$z = \frac{I}{y_{max}} = \frac{\frac{1}{12}bd^3}{d/2}$$

$$= \frac{1}{6}bd^2$$

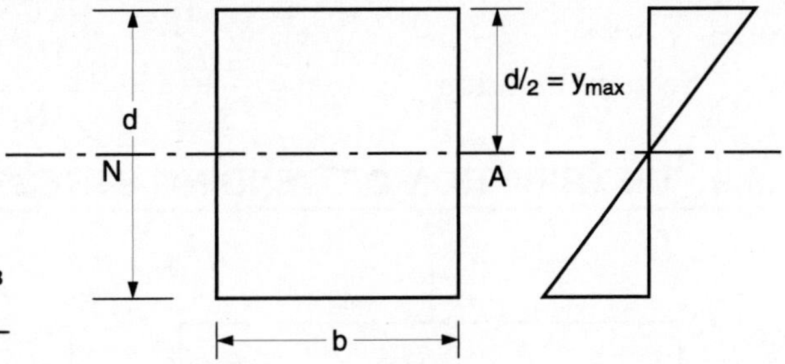

Fig. 6.6

(ii) Hollow rectangular section

Moment of inertia of the section about N.A.

$$= \frac{1}{12}(BD^3 - bd^3)$$

Distance of outermost fibre from N.A.

$$= y_{max} = D/2$$

Sections modules= $$Z = \frac{\frac{1}{12}(BD^3 - bd^3)}{D/2}$$

$$= \frac{1}{6}\frac{(BD^3 - bd^3)}{D}$$

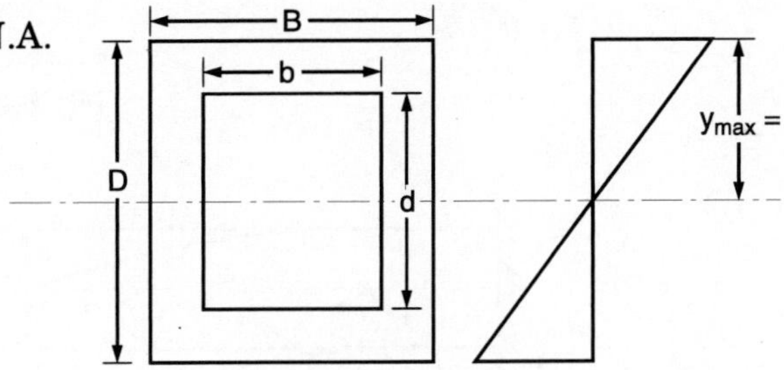

Fig. 6.7

> **DO YOU KNOW?**
> Greater the value of *z*, stronger will be the section.

(iii) Circular section

Moment of interia of the section of diameter '*d*' about N.A.

$$= \frac{\pi}{64}d^4$$

Distance of outermost fibre from N.A.

$$= y_{max} = d/2$$

Section modulus $$= z = \frac{I}{y_{max}}$$

$$= \frac{\frac{\pi}{64}d^4}{\frac{d}{2}} = \frac{\pi d^3}{32}$$

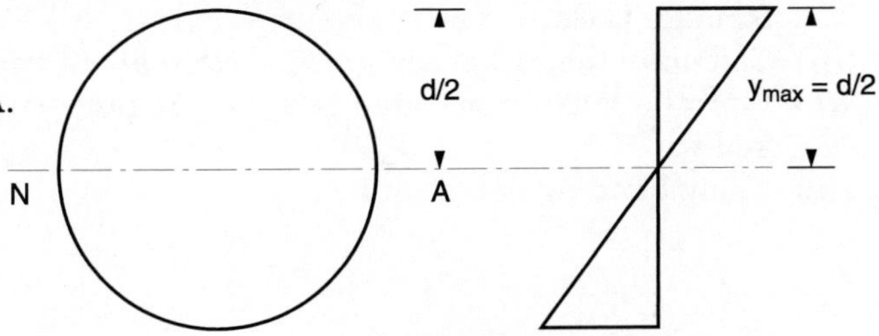

Fig. 6.8

(iv) Hollow circular section

Moment of interia of hollow section about N.A.

$$I = \frac{\pi}{64}(D^4 - d^4)$$

Distance of outer most fibre from N.A.

$$= y_{max} = D/2$$

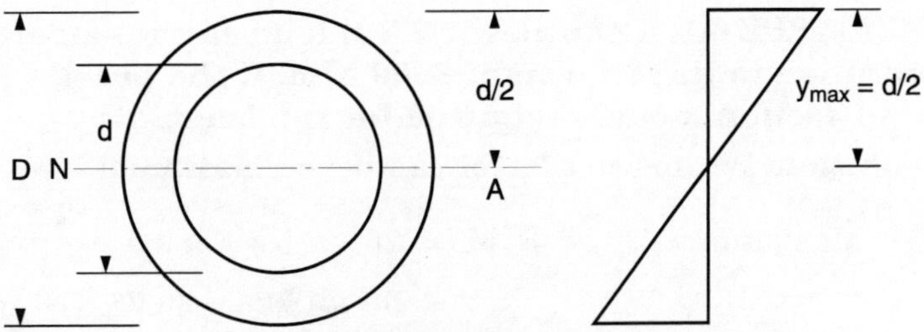

Fig. 6.8 (*a*)

Section modulus, $z = \dfrac{I}{y_{max}} = \dfrac{\frac{\pi}{64}(D^4 - d^4)}{D/2} = \dfrac{\pi(D^4 - d^4)}{32\,D}$

6.9 DISTRIBUTION OF BENDING STRESS ACROSS THE BEAM SECTION

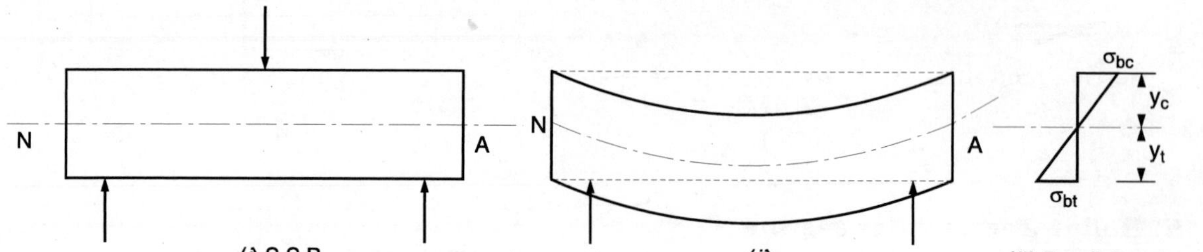

Fig. 6.9

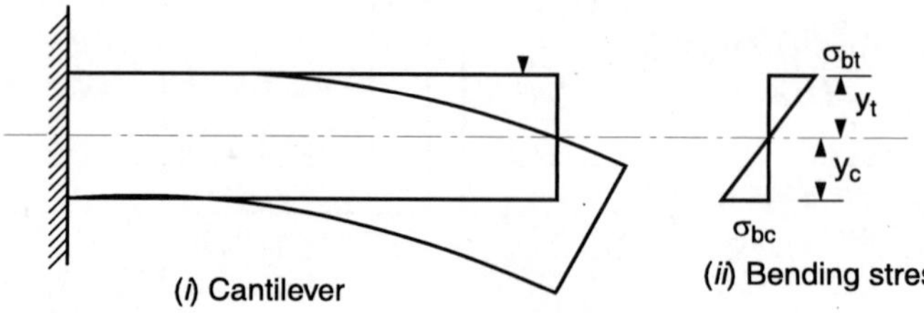

Fig. 6.10

6.10 STEPS FOR FINDING OUT THE BENDING STRESS

(*i*) Calculate, the maximum B.M. acting on the beam for the given load.

(*ii*) Calculate, the distance of C.G. of the section from the bottom. *i.e.*, $\overline{Y}$. The axis passing through C.G., will be the neutral axis (N.A.).

(*iii*) Calculate the M.I of the given section about the N.A.

(*iv*) Find the maximum value of y (*i.e.*, distance of N.A. from the outermost fibre), out of y_c and y_t.

(*v*) Using bending equation *i.e*,.

$$\frac{M}{I} = \frac{\sigma_b}{y} = \frac{E}{R}$$

Calculate the bending stress, σ_b.

DO YOU KNOW?

The flexure formula does not give accurate results in the regions close to the supports or close to a concentreted load.

SOLVED EXAMPLES

EXAMPLE 6.1. A cantilever 2.5 m long carries a point load of 20 KN at the free end in addition to its self weight 6000 N/m. If the permissible bending stress is 120 N/mm^2, find section moduls required for the beam.

Solution: We known that for cantilever, maximum B.M. will be at the fixed end.

$$\text{Maximum B. M.} = 20 \times 2.5 + 6 \times 2.5 \times \frac{2.5}{2}$$

$$= 50 + 18.75 = 68.75 \text{ KN-m} = 68.75 \times 1000 \times 1000$$

$$= 68.75 \times 10^6 \text{ N-mm}$$

Using the relation

$$\frac{M}{I} = \frac{\sigma_b}{y}$$

$$M = \sigma_b . \frac{I}{y} = \sigma_b \times z \qquad \text{[where } z = \text{section modulus} = I/y]$$

$$68.75 \times 10^6 = 120 \times z$$

$$\therefore \quad z = \frac{68.75 \times 10^6}{120} = 0.5729 \times 10^6 = 57.29 \times 10^4 \text{ mm}^3 \quad \textbf{Ans.}$$

EXAMPLE 6.2. A steel plate of width 120 mm and of thickness 20 mm is bent into a circular arc of radius 10 m. Determine the maximum stress induced and the bending moment which will produce the maximum stress. Take $E = 2 \times 10^5$ N/mm².

Solution: Given

Width of plate, $b = 120$ mm

Thickness of plate, $t = 20$ mm

Radius of arc, $R = 10$ m $= 10000$ mm

$E = 2 \times 10^5$ N/mm²

REMEMBER
M.I of a rectangular section $= \frac{1}{12}$ bd³

Moment of intertia of the plate,

$$I = \frac{1}{12} bt^3 = \frac{1}{12} \times 120 \times (20)^3 = 80000 \text{ mm}^3$$

Using the relation

$$\frac{\sigma_b}{y} = \frac{E}{R}$$

$$\sigma_b = \frac{E}{R} \times y \qquad ...(i)$$

From (*i*) it is clear that σ_b will be maximum if y is maximum. y will be maximum at the top or bottom fibre.

$$\therefore \quad y = \frac{t}{2} = \frac{20}{2} = 10 \text{ mm}$$

$$\sigma_b = \frac{2 \times 10^5 \times 10}{10000} = 200 \text{ N/mm}^2$$

Again

$$\frac{M}{I} = \frac{E}{R}$$

$$\therefore \quad M = \frac{E}{R} \times I = \frac{2 \times 10^5 \times 80000}{10000}$$

$$= 16 \times 10^5 \text{ N-mm} = \frac{16 \times 10^5}{1000 \times 1000} \text{ KN-m}$$

$$= 1.6 \text{ KN.m} \qquad \textbf{Ans.}$$

EXAMPLE 6.3. A rectangular beam 300 mm deep is simply supported over a span of 4m; what uniformly distributed load per metre the beam may carry, if the bending stress is not to exceed 120 N/mm² ? Take $I = 8 \times 10^6$ mm⁴.

Solution: Given

Thickness / Depth, $t = 300$ mm $\therefore y = \frac{300}{2} = 150\text{mm}$

Span, $l = 4 \text{ m} = 4000$ mm

Moment of intertia , $I = 8 \times 10^6 \text{ mm}^4$

Bending stress, $\sigma_b = 120 \text{ N/mm}^2$

Using the relation

$$\frac{M}{I} = \frac{\sigma_b}{y}$$

$$M = \sigma_b \times \frac{I}{y} = \sigma_b \times z$$

where z = sectional modulus = $\frac{8\times10^6}{150} = 5.33 \times 10^4 \text{ mm}^3$

$$M = \sigma_b \times z = 120 \times 5.33 \times 10^4$$
$$= 639.6 \times 10^4 \text{ N-mm}$$

Maximum B.M $= \frac{wl^2}{8}$, where $w = UDL$

$$639.6 \times 10^4 = \frac{wl^2}{8}$$

$$\frac{639.6\times10^4\times8}{l^2} = w$$

$$w = \frac{639.6\times10^4\times8}{4000\times4000} = 3.2 \text{ N/mm}$$
$$= 3.2 \times 1000 = 3200 \text{ N/m}$$ **Ans.**

REMEMBER

$$\frac{M}{I} = \frac{\sigma_b}{y} = \frac{E}{R}$$

EXAMPLE 6.4. A symmetrical section 200 mm deep has a moment of inertia of 2.26 × 10^{-5} m^4 about its neutral axis. Determine the longest span over which, when simply supported, the beam would carry a *UDL* of 4KN/m run without the stress due of bending exceeding 125 N/mm^2.

Solution: Given depth , $d = 200$ mm $\therefore y = \frac{200}{2} = 100$ mm

Moment of interia, $I = 2.26 \times 10^{-5} \text{ m}^4 = 2.26 \times 10^{-5} \times (10^3)^4$
$= 2.26 \times 10^7 \text{ mm}^4$.

$$UDL, w = 4 \text{ KN/m} = \frac{4000}{1000} = 4\text{N/mm}$$

Let the span be l mm.

Maximum B.M., $M = \frac{wl^2}{8} = \frac{4l^2}{8} = \frac{l^2}{2} N - \text{mm}$

Using the relation

$$\frac{M}{I} = \frac{\sigma_b}{y}$$

REMEMBER

Maximum B.M with udl in case of S.S.Beam = $\frac{wl^2}{8}$

$$\text{or,} \quad \frac{l^2}{2 \times 2.26 \times 10^7} = \frac{125}{100}$$

$$\text{or,} \quad l^2 = \frac{125}{100} \times 2 \times 2.26 \times 10^7$$

$$\text{or,} \quad l^2 = 5.65 \times 10^7$$

$$\therefore \quad l = (5.67 \times 10^7)^{1/2} = 7529.94 \text{ mm}$$

$$= 7.53 \text{ m} \quad \textbf{Ans.}$$

EXAMPLE 6.5. A timber beam of rectangular section is to support a load of 20 KN uniformly distributed over a span of 3.6 m when beam is simply supported. If the depth of section is to be twice the breadth and the stress in the timber, is not to exceed 7 N/mm², find the dimensions of the cross-section.

Solution: Given

Load, $W = 20$ KN

Span, $l = 3.6$ m

$$\therefore \quad \text{w} = \text{UDL} = \frac{w}{l} = \frac{20}{3.6} = 5.55 \text{ KN/m}$$

Let the breath of the section be b

$\therefore$ depth, $d = 2\,b$

Stress, $\sigma_b = 7$ N/mm²

$$\text{Maximum B.M.} \quad M = \frac{wl^2}{8} = \frac{5.55 \times (3.6)^2}{8} = 9 \text{ KN-m}$$

$$= 9 \times 10^6 \text{ N-mm}$$

Moment of inertia of beam,

$$I = \frac{bd^3}{12} = \frac{b \times (2b)^3}{12} = \frac{8b^4}{12} = \frac{2b^4}{3} \text{mm}^4$$

$$y = \frac{d}{2} = \frac{2b}{2} = b$$

$$\therefore \text{ Section modulus} \quad z = \frac{I}{y} = \frac{2b^4}{3b} = \frac{2b^3}{3}$$

Using the relation

$$M = \sigma_b \times z$$

$$9 \times 10^6 = 7 \times \frac{2b^3}{3}$$

$$\frac{9 \times 10^6 \times 3}{14} = b^3.$$

$$b^3 = 1.93 \times 10^6$$

$$\therefore \quad b = \sqrt[3]{1.93 \times 10^6} = 1.245 \times 100 \text{ mm} = 124.5 \text{ mm} \quad \textbf{Ans.}$$

$$\text{and} \quad d = 2\,b = 2 \times 124.5 = 249 \text{ mm} \quad \textbf{Ans.}$$

EXAMPLE 6.6. The cross-section of a cast iron beam is shown in the figure 6.11. The beam is simply supported at the ends and carries a udl of 20KN/m. If the span of the beam is 3 m, determine the maximum tensile and compressive stresses in the beam.

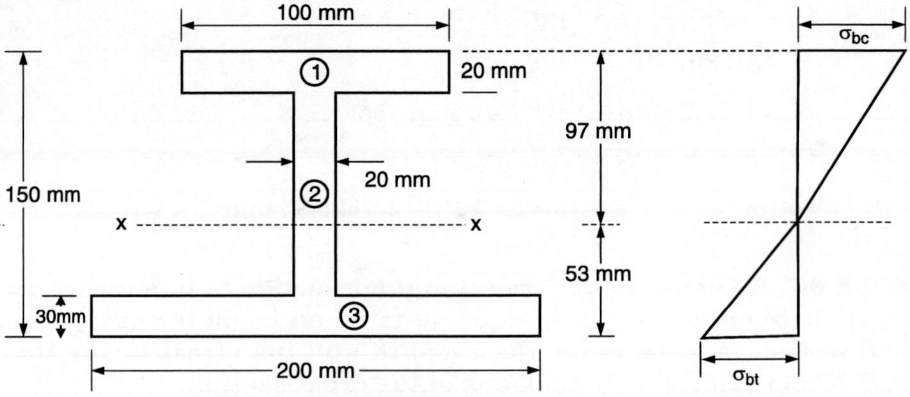

Fig. 6.11

Solution: Given

$$w = 20 \text{ KN/m}$$
$$l = 3 \text{ m}$$

Calculation of C.G. and M.I. $\bar{X} = \dfrac{200}{2} = 100$ mm

S.No.	Particulars	*a*	*y*	*ay*
1.	Rect (1)	2000	$30 + 100 + \dfrac{20}{2} = 140$	280000
2.	Rect (2)	2000	$30 + \dfrac{100}{2} = 80$	160000
3.	Rect (3)	6000	15	90000
		$\Sigma a = 10000$		$\Sigma ay = 530000$

$$\bar{Y} = \frac{\Sigma ay}{\Sigma a} = \frac{530000}{10000} = 53\,\text{mm}$$

$$I_{XX1} = \frac{1}{12} \times 100 \times 20^3 + 100 \times 20 \times (140 - 53)^2$$
$$= 66666.67 + 2000 \times 7569 = 66666.67 + 15138000$$
$$= 15204667 \text{ mm}^4$$

$$I_{XX2} = \frac{1}{12} \times 20 \times (100)^3 + 100 \times 20 \times (80 - 53)^2$$
$$= 1666666.7 + 1458000 = 3124666.7 \text{ mm}^4$$

$$I_{XX3} = \frac{1}{12} \times 200 \times (30)^3 + 200 \times 30 \times (53 - 15)^2$$
$$= 450000 + 8664000 = 9114000 \text{ mm}^4$$

$$I_{XX} = 15204667 + 3124666.7 + 9114000$$
$$= 27443334 \text{ mm}^4$$

Maximum $\quad BM = \dfrac{wl^2}{8} = \dfrac{20 \times 3^2}{8} = \dfrac{20 \times 9}{8}$

$= 22.5 \text{ KN-m} = 22.5 \times 10^6 \text{ N-mm}$

Maximum compressive bonding stress

y_c = Distance of extreme top fibre from NA = 150 – 53
= 97 mm

$$\frac{M}{I} = \frac{\sigma_{bc}}{y_c}$$

$$\sigma_{bc} = \frac{M \times y_c}{I} = \frac{22.5 \times 10^6 \times 97}{27443334} = 79.53 \text{ N/mm}^2. \quad \textbf{Ans.}$$

Maximum tensile bending stress

y_t = 53 mm

$$\frac{M}{I} = \frac{\sigma_{bt}}{y_t}$$

$$\therefore \quad \sigma_{bt} = \frac{M \times y_t}{I} = \frac{22.5 \times 10^6 \times 53}{27443334} = 43.45 \text{ N/mm}^2 \quad \textbf{Ans.}$$

EXAMPLE 6.7. A floor has to carry a load of 15 KN/m². The floor is supported on floor joists 300 mm × 100 mm and 5 m long. Determine the spacing of the joists from centre to centre such that the bending stress does not exceed 8.5 N/mm².

Solution:

Given

load, = 15 KN/m²
= 15000 N/m²

depth, d = 300 mm
width, b = 100 mm
span, l = 5 m
bending stress, σ_b = 8.5 N/mm²

Let c/c spacing of joists be 'x' m.

Total load on each joist ,

$W = 15000 \times 5 \times x$ N
$= 75000\, x\, N$

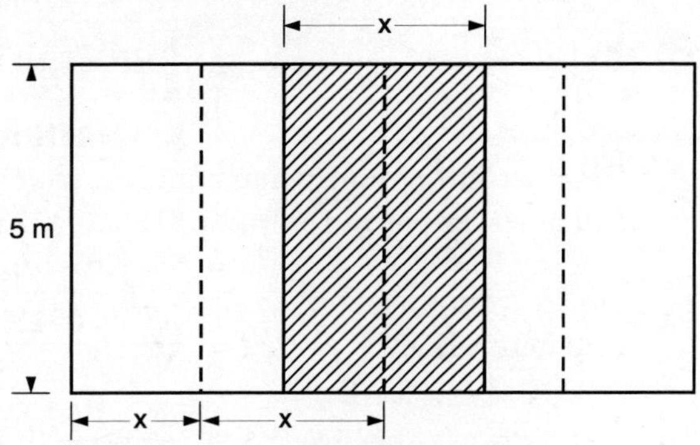

Fig. 6.12

Maximum B.M. $\quad M = \dfrac{Wl}{8} = \dfrac{75000x \times 5}{8} = 46875x$ N-m

$= 46875\, x \times 1000$ N-mm

Moment of inertia $\quad I = \dfrac{bd^3}{12} = \dfrac{100 \times (300)^3}{12} = 2.25 \times 10^8 \text{ mm}^4$

$$y = \frac{d}{2} = 150 \text{ mm}$$

Using relation

$$\frac{M}{I} = \frac{\sigma_b}{y}$$

$$= \frac{46875x \times 1000}{2.25 \times 10^8} = \frac{8.5}{150}$$

$$\therefore \quad x = \frac{8.5 \times 2.25 \times 10^8}{46875 \times 150 \times 1000} = 0.272 \text{ m} \quad \textbf{Ans.}$$

EXAMPLE 6.8. A cast iron water main 1 m long 550 mm inside diameter and 25 mm wall thickness runs full of water and is supported at its ends. Calculate the maximum stress in the metal if density of cast iron is 70.6 KN/m^3 and that of water is 9.81 KN/m^3.

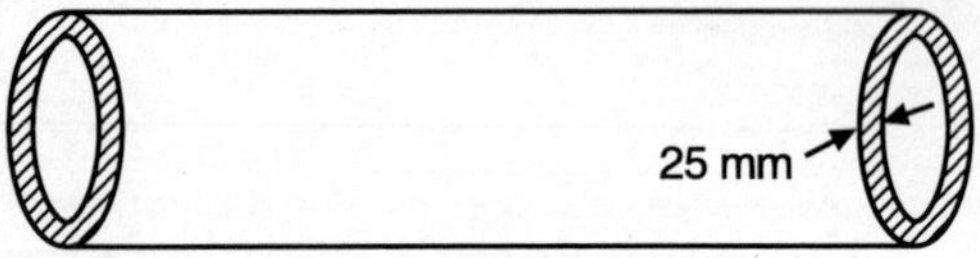

Fig. 6.13

Solution: Given

Length, l = 1m

Inner dimeter, d_i = 550 mm

Wall thickness, t = 25 mm

Outer diameter, $d_o = di + 2t = 550 + 2 \times 25$

$= 600$ mm

Density of cast iron = ρ_c = 70.6 KN/m^3 = 70.6 × 10^3 = 70.6 × 10^3 N/m^3

Density of water = ρ_w = 9.81 KN/m^3 = 9.81 × 10^3 = 9.81 × 10^3 N/m^3

Weight of pipe per meter length

$$= \frac{\pi}{4}(d_o^2 - d_i^2) \times 1 \times 70.6 \times 10^3 = \frac{\pi}{4}\left[(6.0)^2 - (0.55)^2\right] \times 70.6 \times 10^3$$

$= 3188.32$ N/m

Weight of water per metre length of pipe

$$= \frac{\pi}{4} di^2 \times 1 \times 9.81 \times 10^3 = \frac{\pi}{4} \times (0.55)^2 \times 9.81 \times 10^3$$

$= 2330.69$ N/m

Total weight of pipe when running full of water per m run $= w = 3188.32 + 2330.69$

$= 5519.01$ N/m

Maximum B.M. $$M = \frac{wl^2}{8} = \frac{5519.01 \times 1^2}{8} = 689.88 \text{ N-m}$$

$$y = \frac{d_o}{2} = \frac{600}{2} = 300 \text{ mm} = 0.3 \text{ m}$$

M.I. of the pipe, $$I = \frac{\pi}{64}(d_o^4 - d_i^4) = \frac{\pi}{64}\{(0.6)^4 - (0.55)^4\}$$

$$= \frac{\pi}{64}\{0.1296 - 0.0915\} = 1.87 \times 10^{-3} \text{ m}^4$$

Using relation

$$\frac{M}{I} = \frac{\sigma_b}{y}$$

$$M = \sigma_b \times \frac{I}{y}$$

or, $$689.88 = \sigma_b \times \frac{1.87 \times 10^{-3}}{0.3}$$

REMEMBER
M.I of a hollow circular section $= \frac{\pi}{64}(d_0^4 - d_i^4)$

or, $\dfrac{689.88 \times 0.3 \times 10^3}{1.87} = \sigma_b = 110675.94 \text{ N/m}^2$ **Ans.**

EXAMPLE 6.9. A cast iron beam is of T-section as shown in the figure 6.14. The beam is simply supported on a span of 8 m. The beam carries a *UDL* of 5KN/m length on the entire span. Determine the maximum stress.

Solution: Given

Span, $l = 8$ m

UDL, $w = 5$ KN/m = 5000 N/m

Calculation of M.I.

S.No.	Particulars	a	y	ay
1.	Rect (1)	200 × 40 = 8000	280	2240000
2.	Rect (2)	260 × 40 = 10400	130	1352000
		Σa = 18400		Σay = 3592000

$$\therefore \quad \bar{Y} = \frac{\Sigma ay}{\Sigma a} = \frac{3592000}{18400} = 195.22 \text{ mm}$$

$$\bar{X} = \frac{200}{2} = 100 \text{ mm}$$

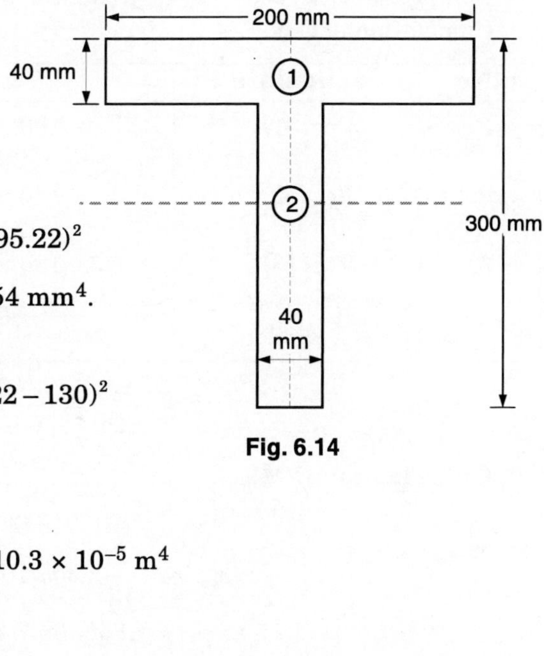

Fig. 6.14

$$I_{XX} = I_{XX1} + I_{XX2}$$

$$I_{XX1} = \frac{1}{12} \times 200 \times 40^3 + 200 + 40\,(280 - 195.22)^2$$

$$= 1066666.7 + 57501187 = 58567854 \text{ mm}^4.$$

$$= 5.86 \times 10^{-5} \text{ m}^4.$$

$$I_{XX2} = \frac{1}{12} \times 40 \times (260)^3 + 40 \times 260 \times (195.22 - 130)^2$$

$$= 58586667 + 44237943$$

$$= 1.03 \times 10^8 \text{ mm}^4$$

$$= \frac{1.03 \times 10^8}{10^{12}} = \frac{1.03 \times 10^{-4} \times 10}{10} = 10.3 \times 10^{-5} \text{ m}^4$$

$$I_{xx} = 5.86 \times 10^{-5} + 10.3 \times 10^{-5}$$

$$= 16.16 \times 10^{-5} \text{ m}^4$$

$$y = 195.22 \text{ mm} = 0.195 \text{ m}$$

Maximum B.M. $\quad M = \dfrac{wl^2}{8} = \dfrac{5000 \times 8^2}{8} = 40000$ N-m

Using relation

$$\frac{M}{I} = \frac{\sigma_b}{y}$$

$$\therefore \quad \sigma_b = \frac{M \times y}{I} = \frac{40000 \times 0.195}{16.16 \times 10^{-5}} = 482.67 \text{ N/m}^2 = 48.27 \text{ N/mm}^2$$

EXAMPLE 6.10. The horizontal beam of section shown in figure 6.15 is 4 m long and is simply supported at the ends. Calculate the maximum *UDL* it can carry if the tensile and compressive stresses must not exceed 25 MN/m² and 45 MN/m² respectively.

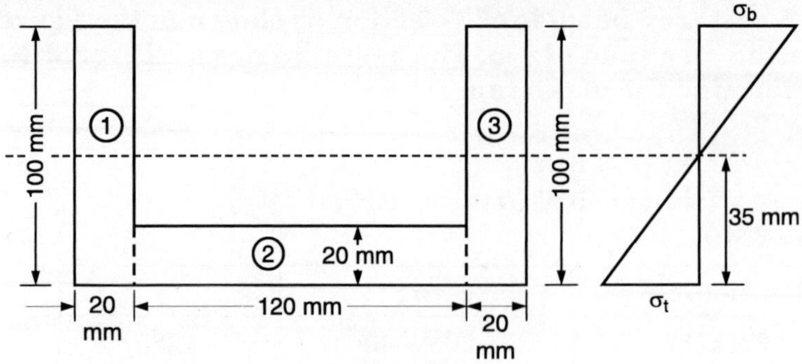

Fig. 6.15

Solution: Given

span, $l = 4$ m

Tensible stress, $\sigma_t = 25$ MN/m^2 = 25×10^6 N/m^2

Compressive.stress, $\sigma_c = 45$ MN/m^2 = 45×10^6 N/m^2

Calculation of C.G.

S.No.	Particulars	*a*	*y*	*ay*
1.	Rect (1)	100 × 20 = 2000	50	100000
2.	Rect (2)	120 × 20 = 2400	10	24000
3.	Rect (3)	100 × 20 = 2000	50	100000
		Σa = 6400 mm^2		Σay = 224000

$$\bar{Y} = \frac{\Sigma ay}{\Sigma a} = \frac{224000}{6400} = 35mm$$

Calculation of M.I.

$$I_{XX} = I_{XX1} + I_{XX2} + I_{XX3}$$

$$I_{XX1} = \frac{1}{12} \times 20 \times 100^3 + 20 \times 100\,(50 - 35)^2$$

$$= 1666666.7 + 450000 = 2116666.7 \text{ mm}^4$$

$$I_{XX2} = \frac{1}{12} \times 120 \times 20^3 + 120 \times 20 \times (35 - 10)^2$$

$$= 80000 + 1500000 = 1580000 \text{ mm}^4.$$

$$I_{XX3} = \frac{1}{12} \times 20 \times 100^3 + 20 \times 100 \times (50 - 35)^2$$

$$= 2116666.7 \text{ mm}^4.$$

$$I_{XX} = 2116666.7 + 1580000 + 2116666.7$$

$$= 5813333.4 \text{ mm}^4 = \frac{5813333.4}{10^{12}} = 5.81 \times 10^{-6} \text{ m}^4$$

$$y = 35 \text{ mm}$$

Maximum. B.M. $M = \frac{wl^2}{8} = \frac{w \times 4^2}{8} = 2w$ N-m ; where w = UDL

Suppose the maximum tensile stress be 25×10^6 N/m². Then the corresponding compressive stress

$$= \frac{25 \times 10^6}{35} \times 65 = 46.43 \times 10^6 \text{ N/m}^2$$

But the compressive stress should not exceed 45×10^6 N/m².

If the compressive stress is 45×10^6 N/m² then corresponding tensile stress

$$\sigma_t = \frac{45 \times 10^6}{65} \times 35 = 24.23 \times 10^6 \text{ N/m}^2$$

Using relation

$$\frac{M}{I} = \frac{\sigma_t}{y}$$

$$\frac{2w}{5.81 \times 10^{-6}} = \frac{24.23 \times 10^6}{0.035}$$

$$w = \frac{24.23 \times 10^6 \times 5.81 \times 10^{-6}}{2 \times 0.35} = 2011.09 \text{ N/m} = 2.01 \text{ KN/m} \quad \textbf{Ans.}$$

EXAMPLE 6.11. Calculate the dimensions of the strongest section that can be cut of a circular log of wood 200 mm in diameter.

Solution: Given

Diameter = 200 mm

Let a strongest rectangular section of dimensions b and d be cut-out of the circular log.

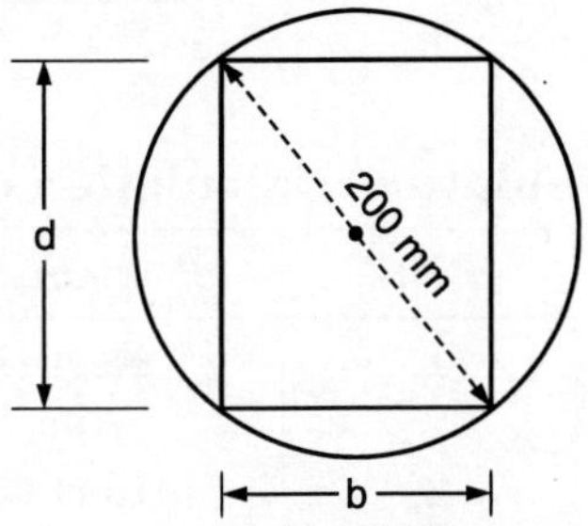

Fig. 6.16

For minimum wastage

$$(dia)^2 = b^2 + d^2$$

$$(200)^2 = b^2 + d^2$$

or $$d^2 = (200)^2 - b^2 = 40000 - b^2 \quad ...(i)$$

Section modulus of the strongest section,

$$z = \frac{1}{6} bd^2$$

$$z = \frac{1}{6} \times b\,(40000 - b^2) = \frac{40000\,b - b^3}{6}$$

For a beam to be strongest the section modulus has to be maximum and for z to be maximum

$$\frac{dz}{db} = 0$$

$$\frac{d}{db}\frac{(40000\,b - b^3)}{6} = 0$$

$$\frac{1}{6}\frac{d}{db}(40000\,b - b^3) = 0$$

$$\frac{1}{6}(40000 - 3b^2) = 0$$

or $$40000 = 3b^2$$

or, $$b^2 = \frac{40000}{3} = 13333.33$$

$\therefore$ $$b = 115.47 \text{ mm}$$

Putting the value of b in (i), we get

$$d^2 = 40000 - (115.47)^2 = 40000 - 13333.33$$

$\therefore$ $$d = 163.3 \text{ mm} \quad \textbf{Ans.}$$

EXAMPLE 6.12. A cast iron bracket subjected to bending has the cross-section of I-form with unequal flanges. Find:

(*i*) The position of N.A. and moment of inertia of the section about N.A.

(*ii*) The maximum B.M. on the section if tensile stress in the top flange is not to exceed 40 N/m². What is then the value of the compressive stress in the bottom flange?

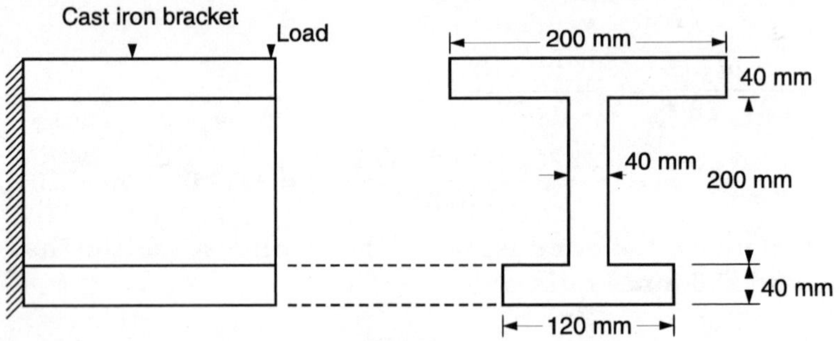

Fig. 6.17

Solution: Calculation of C.G.

S.No.	Particulars	a(mm²)	y (mm)	ay (mm³)
1.	Rect (1)	200 × 40 = 8000	40 + 200 + 20 = 260	2080000
2.	Rect (2)	200 × 40 = 8000	40 + 100 = 140	1120000
3.	Rect (3)	120 × 40 = 4800	20	96000
		Σa = 20800		Σay = 3296000

$$\bar{Y} = \frac{\Sigma ay}{\Sigma a} = \frac{3296000}{20800} = 158.46 \text{ mm from bottom fibre}$$

Calculation of M.I.

$$I_{XX1} = \text{M.I. of rect (1) about centroidal axis.}$$

$$= \frac{1}{12} \times 200 \times (40)^3 + 200 + 40\,(260 - 158.46)^2$$

$$= 1066666.7 + 82482973 = 83549640 \text{ mm}^4$$

$$I_{XX2} = \frac{1}{12}\times 40\times(200)^3 + 40\times 200\,(158.46-140)^2$$
$$= 2666666.7 + 2726172.8$$
$$= 5392839.5 \text{ mm}^4.$$
$$I_{XX3} = \frac{1}{12}\times 120\times(40)^3 + 120\times 40\times(158.46-20)^2$$
$$= 640000 + 92021624 = 92661624 \text{ mm}^4.$$
$$I_{XX} = 83549640 + 5392839.5 + 92661624$$
$$= 1.82\times 10^8 \text{ mm}^4 = \frac{1.82\times10^8}{10^{12}} = 1.82\times10^{-4} \text{ m}^4.$$
$$y_c = 158.46 \text{ mm}$$
$$y_t = 280 - 158.46 = 121.54 \text{ mm}$$

For maximum tensile stress

$$\sigma_t = 40\text{N/mm}^2$$
$$\frac{M}{I} = \frac{\sigma_t}{y_t}$$

or $$M = \sigma_t \times \frac{I}{y_t} = 40\times\frac{1.82\times10^8}{121.54}$$
$$= 59897976 \text{ N-mm} = 59897.98 \text{ N-m} \quad \textbf{Ans.}$$
$$\frac{M}{I} = \frac{\sigma_c}{y_c}$$
$$\sigma_c = \frac{M\times y_c}{I}$$
$$= \frac{59897976\times158.46}{1.82\times10^8} = 52.15 \text{ N/mm}^2. \quad \textbf{Ans.}$$

DO YOU KNOW?

Flitched beams are statically indeterminate.

6.11 BEAMS OF HETEROGENEOUS MATERIALS (FLITCHED BEAM)

A flitched beam is a beam made up of two or more different materials to act as a single beam.

Filched beams are used in places where space available is less to accommodate the cross-section of a beam. To overcome this situation, a material is strengthened with some other material of higher strength, hereby to reduce the cross-section, figure 6.18 shows a flitched beam.

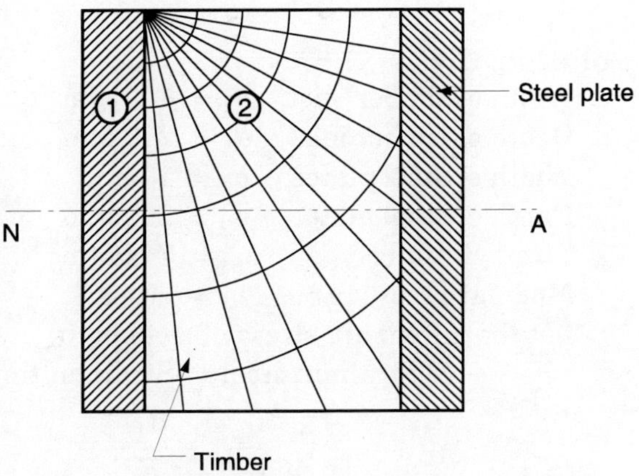

Fig. 6.18 Flitched beam

In flitched beams:

(*i*) Strain at the common surface will be same in both the materials.

$$\frac{\sigma_{bs}}{E_s} = \frac{\sigma_{bt}}{E_t}$$

or, $$\sigma_{bs} = \frac{E_s}{E_t}\times\sigma_{bt}$$

$\therefore \qquad \sigma_{bs} = m\,\sigma_{bt}$

where $\qquad m =$ modular ratio $= \dfrac{E_s}{E_t}$

(*ii*) The moment of resistance of flitched beam is the summation of moment of resistance of individual materials

$$M = M_1 + M_2 = \sigma_{bs} Z_s + \sigma_{bt} Z_t$$

The moment of resistance of the flitched beam is determined by considering the equivalent section. This is obtained by multiplying the dimensions of steel plate in the direction parallel to N.A. by the modular ratio (m).

EXAMPLE 6.13. A flitched beam consists of a timber joist 100 mm wide and 200 mm deep strengthened by two steel plates 10 mm thick and 200 mm deep as shown in figure 6.19. If the maximum stress in the timber joist is 7 N/mm², find the corresponding maximum stress attaind in steel. Find also the moment of resistance of the composite section. Take modular ratio = 20.

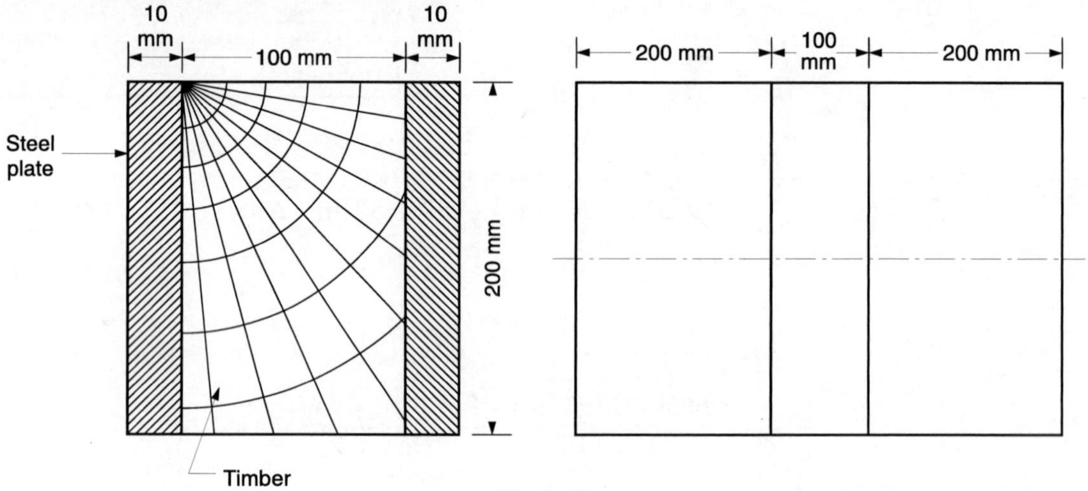

Fig. 6.19

Solution: Given

Width of timber joist, $b_1 = 100$ mm

Depth of timber joist, $d_1 = 200$ mm

Width of steel plates, $b_2 = 10$ mm

Depth of steel plate, $d_2 = 200$ mm

Stress in timber, $\sigma_{b1} = 7$ N/mm²

Modular ratio, $m = 20$

Let the maximum stress of steel be σ_{b2}.

Strain in steel = Strain in timber

$$\frac{\sigma_{b_2}}{E_2} = \frac{\sigma_{b_1}}{E_1}$$

$$\sigma_{b2} = \frac{E_2}{E_1}\sigma_{b_1} = \sigma_{b1} \times m \text{ where } m = \text{modular ratio} = \frac{E_2}{E_1}$$

$$= 7 \times 20 = 140 \text{ N/mm}^2$$

Equivalent timber section $= m\,b_2$

$= 20 \times 10 = 200$ mm

Equivalent moment of inertia,

$$I = \frac{bd^3}{12} = \frac{500 \times (200)^3}{12} = 3.33 \times 10^8 \text{ mm}^4$$

$$y = \frac{d}{2} = \frac{200}{2} = 100 \text{ mm}$$

∴ Total moment of resistance,

M = Moment of resistance × equivalent timber section

$= \sigma_b \times Z$

$$= 7 \times \frac{3.33 \times 10^8}{100} \qquad \left[\because z = \frac{I}{y}\right]$$

$= 23333333$ N-mm

$= 23333.33$ N-m

EXAMPLE 6.14. A timber beam 100 mm wide and 200 mm deep is to be reinforced by bolting on two steel flitches each 150 mm × 12.5 mm in section. Calculate the moment of resistance in the following cases:

(*i*) Flitches attached symmetrically at the top and bottom

(*ii*) Flitches attached symmetrically at the sides.

Allowable stress in timber is 6 N/mm².

Take modular ratio = 20

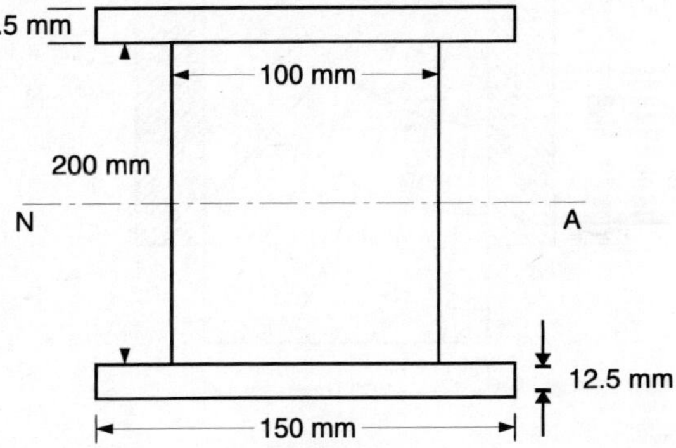

Fig. 6.20

Solution: (*i*) Flitches attached symmetrically at the top and bottom

Equivalent timber width for steel flitches

$= 20 \times 150$

[150 mm is the dimension of steel flitches parallel to N.A.]

$= 3000$ mm

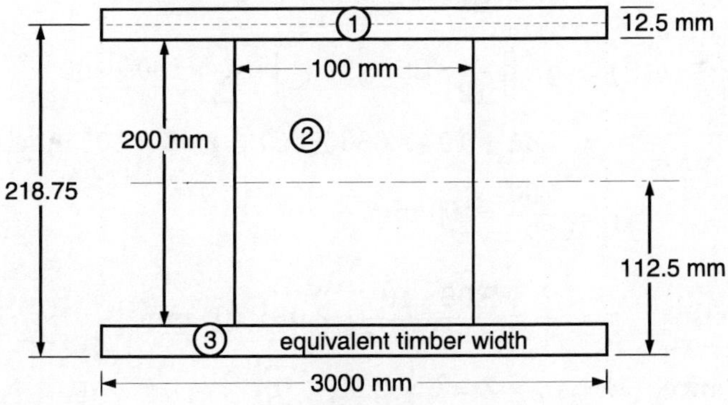

Fig. 6.21

Equivalent M.I.,

$$I = 2 \times \left[\frac{1}{12} \times 3000 \times (12.5)^3\right] + 3000 \times 12.5 \times (218.75 - 112.5)^2 + \frac{1}{12} \times 100 \times (200)^3$$

[$\because$ dist of C.G. of rect (1) from bottom most fibre = 12.5 + 200 + 12.5/2 = 218.75 mm]

$$= 2 \times (488281.25 + 4.23 \times 10^8) + 66666667$$

$$= 8.51 \times 10^8 + 66666667 = 9.18 \times 10^8 \text{ mm}^4$$

$$y = \frac{225}{2} = 112.5 \text{ mm}$$

Section modulus, $Z = \frac{I}{Y} = \frac{9.18 \times 10^8}{112.5} = 8.16 \times 10^6 \text{ mm}^3$

Moment of resistance, $M = \sigma_b \times z = 6 \times 8.16 \times 10^6$

$$= 48.96 \times 10^6 \text{ N-mm} \qquad \textbf{Ans.}$$

(*ii*) **Flitches attached symmetrically at the sides**

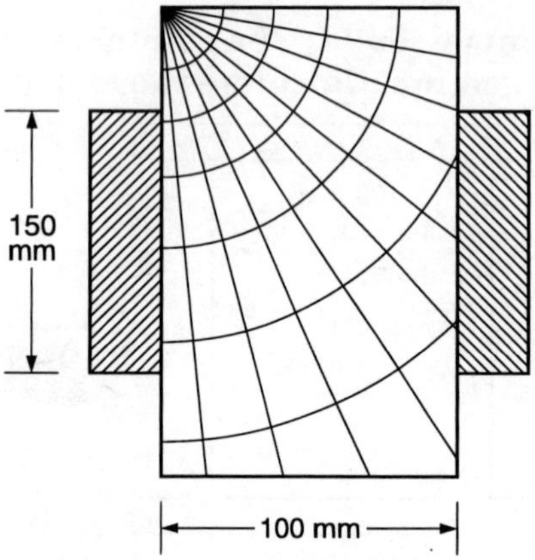

Fig. 6.22

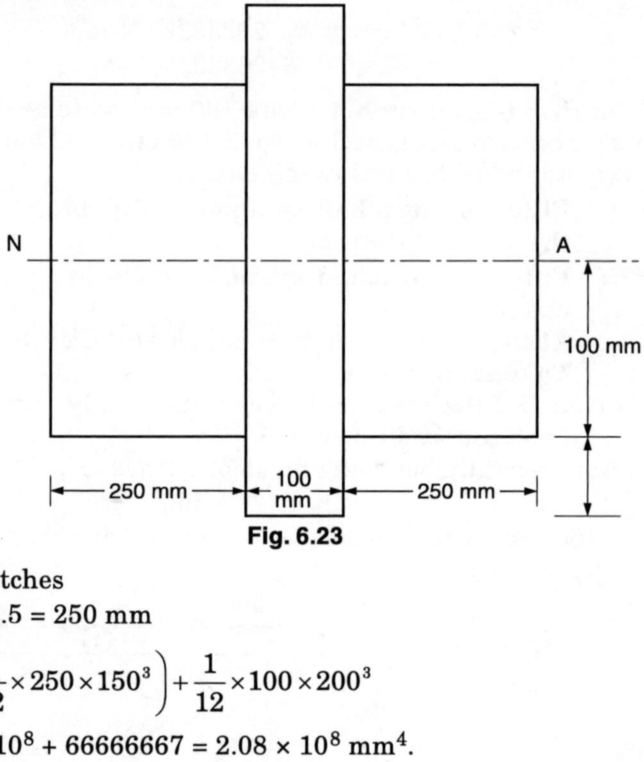

Fig. 6.23

Equivalent timber width for steel flitches

$$= 20 \times 12.5 = 250 \text{ mm}$$

Equivalent M.I., $I = 2 \times \left(\frac{1}{12} \times 250 \times 150^3\right) + \frac{1}{12} \times 100 \times 200^3$

$$= 1.41 \times 10^8 + 66666667 = 2.08 \times 10^8 \text{ mm}^4.$$

$$y = \frac{200}{2} = 100 \text{ mm}$$

Modules of section, $z = \frac{I}{y} = \frac{2.08 \times 10^8}{100} = 2.08 \times 10^6 \text{ mm}^3$

Moment of resistance, $M = \sigma_b \times Z = 6 \times 2.08 \times 10^6$

$$= 12.48 \times 10^6 \text{ N-mm} \qquad \textbf{Ans.}$$

EXAMPLE 6.15. A wooden beam 160 mm wide and 180 mm deep is reinforced with steel plates 6 mm thick along its longer sides. If the bending stresses in the composite beam are to be limited to 120 N/mm² in steel and 10 N/mm² in the timber, estimate the permissible bending moment in the beam. Assume modular ratio = 20.

Solution: Given,

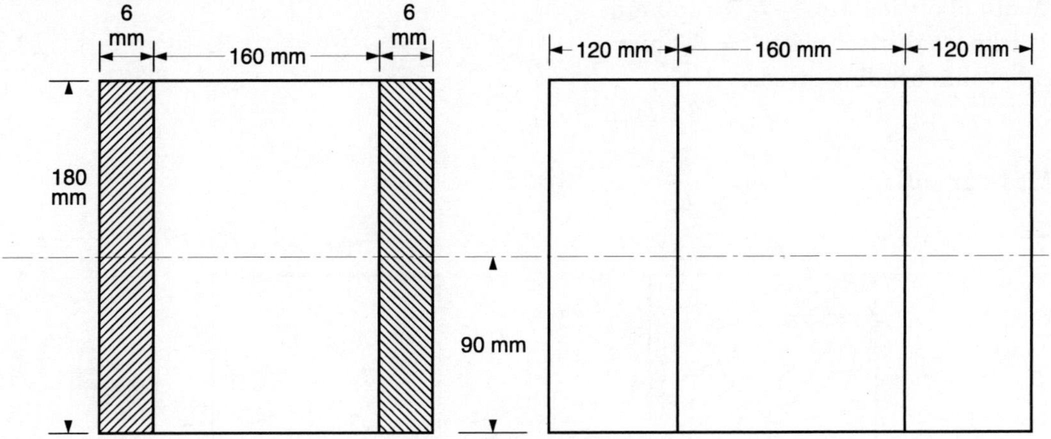

Fig. 6.24

Modular ratio, $m = 20$

Stress in timber, $\sigma_w = 10 \text{ N/mm}^2$

Stress in steel, $\sigma_s = 120 \text{ N/mm}^2$

Equivalent timber width for steel flitches = 20 × 6 = 120 mm

Equivalent M.I., $I = \dfrac{1}{12} \times 400 \times 180^3 = 1.94 \times 10^8 \text{ mm}^4.$

Strain in steel = Strain in timber

$$\frac{\sigma_s}{E_s} = \frac{\sigma_w}{E_w}$$

$$\Rightarrow \quad \sigma_s = \sigma_w \times \frac{E_s}{E_w} = 10 \times 20 = 200 \text{ N/mm}^2 \ (> 120 \text{ N/mm}^2)$$

Stress in steel is more than the permissible value.

$$\therefore \quad \frac{120}{E_s} = \frac{\sigma_w}{E_w}$$

$$\Rightarrow \quad \sigma_w = 120 \times \frac{E_w}{E_s} = \frac{120}{20} = 6 \text{ N/mm}^2$$

$$y = \frac{180}{2} = 90 \text{ mm}$$

Section modulus, $Z = \dfrac{I}{y} = \dfrac{1.94 \times 10^8}{90} = 216 \times 10^4 \text{ mm}^3$

Moment of resistance, $M = \sigma_w \times z = 6 \times 216 \times 10^4$

$= 12.96 \times 10^6$ N-mm **Ans.**

EXAMPLE 6.16. A wooden beam 150 mm wide and 200 mm deep is reinforced at the bottom by a steel plate 150 mm wide and 10 mm thick. If the allowable stress in timber is 6N/mm²; find the moment of resistance of the beam. Take $E_s = 15 E_w$.

Solution: Given

Width of wooden beam, $b = 150$ mm

Depth of wooden beam, d = 200 mm
Width of steel plate, b_s = 150 mm
Thickness of steel plate, t = 10 mm
Allowable bending stress,

$$\sigma_b = 6 \text{ N/mm}^2$$

Modular ratio,

$$m = \frac{E_s}{E_w} = \frac{15E_w}{E_w} = 15$$

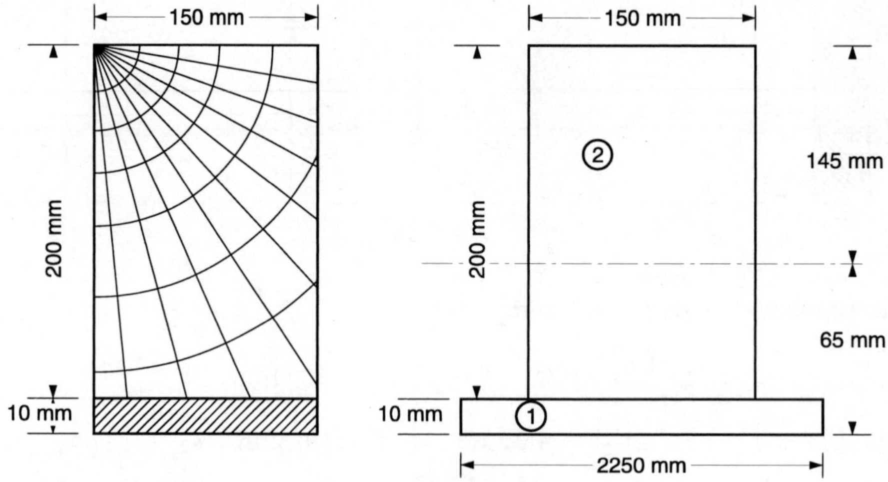

Fig. 6.25

Equivalent timber width for steel plate

$$= 15 \times 150 = 2250 \text{ mm}$$

Calculation of C.G.

S.No.	Particulars	*a*	*y*	*ay*
1.	Rect (1)	22500	5	112500
2.	Rect (2)	30000	110	3300000
		$\Sigma a = 52500$		$\Sigma ay = 3412500$

$$\bar{Y} = \frac{\Sigma ay}{\Sigma a} = \frac{3412500}{52500} = 65mm$$

Equivalent M.I.,

$$I = \left[\frac{1}{12} \times 2250 \times 10^3 + 2250 + 10 \times (65 - 5)^2\right]$$

$$= +\left[\frac{1}{12} \times 150 \times (200)^3 + 150 \times 200\ (110 - 65)^2\right]$$

$$= 187500 + 81000000 + 1 \times 10^8 + 60750000$$

$$= 2.42 \times 10^8 \text{ mm}^4.$$

y = 145 mm

Moment of resistance, $M = \sigma_b \times z$

$$= \sigma_b \times \frac{I}{y}$$

$$= \frac{6 \times 2.42 \times 10^8}{145} = 10013793 \text{ N-mm} \quad \textbf{Ans.}$$

EXAMPLE 6.17. A timber beam 150 mm wide and 100 mm deep is to be reinforced by two steel flitches each 150 mm × 10 mm is section. Calculate the ratio of the moment of resistances in the two mentioned cases:-

(*i*) Flitches attached at top and bottom.

Take *m* = 20

(*ii*) Flitches attached symmetrically on the sides

Solution: (*i*) Flitches attached at top and bottom

Let the modular ratio be 20.

Equivalent timber width for steel flitches = 20 × 150 = 3000 mm

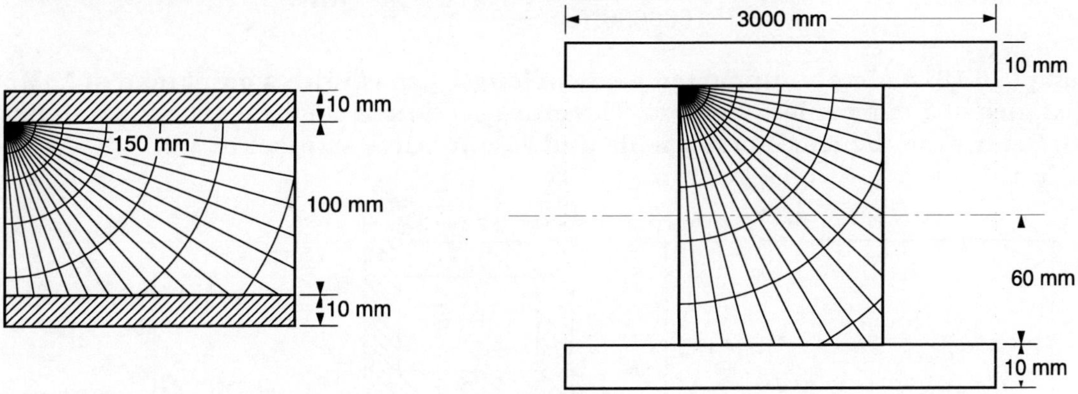

Fig. 6.26

Equivalent moment of intertia

$$I = 2 \times \left[\frac{1}{12} \times 3000 \times 10^3 + 3000 \times 10 \times (55)^2\right] + \frac{1}{12} \times 150 \times 100^3$$

$$= 2\,[250000 + 9075000] + 1.25 \times 10^7 = 3.11 \times 10^7 \text{ mm}^4$$

$$y = 60 \text{ mm}$$

Moment of resistance,

$$M_{r1} = \sigma_b \times \frac{3.11 \times 10^7}{60} = 518333.33\ \sigma_b \text{ N-mm}$$

(*ii*) Flitches attached symmetrically on the sides

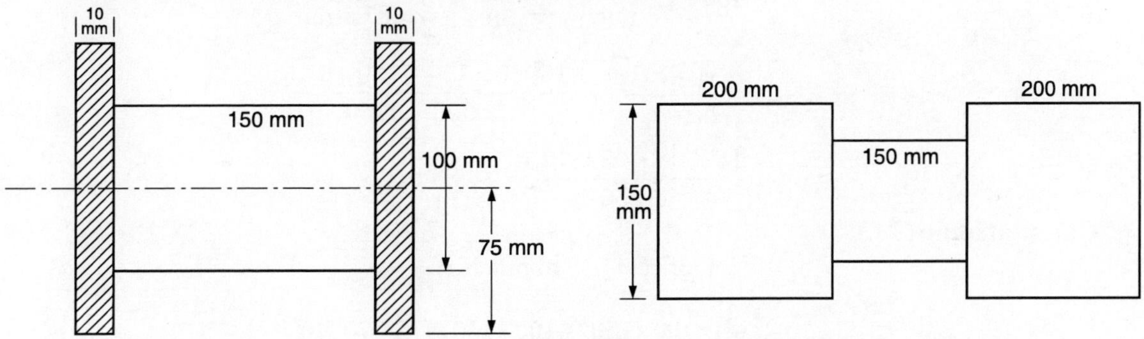

Fig. 6.27

Equivalent timber width for steel flitches

$$= 20 \times 10 = 200 \text{ mm}$$

Equivalent M.I., $I = 2 \times \frac{1}{12} \times 200 \times 150^3 + \frac{1}{12} + 150 \times 100^3$

$$= 1.125 \times 10^8 + 0.125 \times 10^8 = 1.25 \times 10^8 \text{ mm}^4.$$

$$y = 75 \text{ mm}$$

Moment of resistance ,

$$M_{r2} = \sigma_b \times \frac{1.25 \times 10^8}{75} = 1666666.7\ \sigma_b \text{ N-mm}$$

The ratio of $\frac{M_{r_1}}{M_{r_2}} = \frac{518333.33\ \sigma_b}{1666666.7\sigma_b} = 0.311$ **Ans.**

Example 6.18. A simply supported beam of length 3 m carries a point load of 12 KN at a distance of 2 m from left support. The cross-section of the beam is shown in figure 6.28. Determine the maximum tensile and compressive stresses at *x-x*.

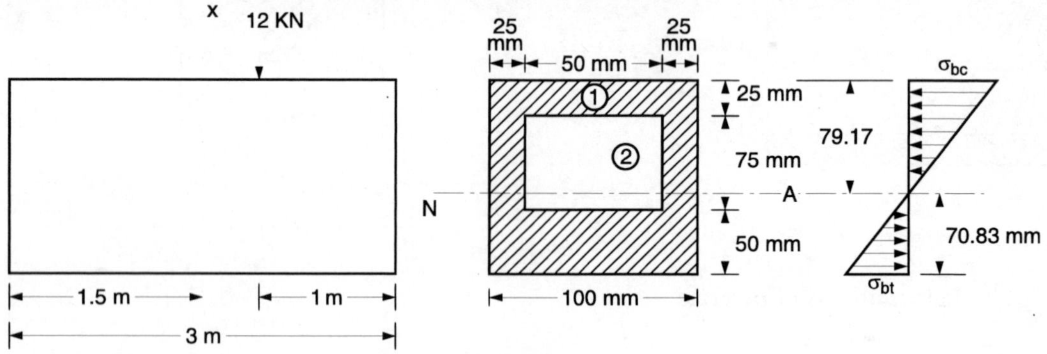

Fig. 6.28

Solution: Given, $W = 12 \text{ KN} = 12000 \text{ N}$

Span, $l = 3\text{m} = 3000 \text{ mm}$

Calculation of C.G.

$$a_1 = \text{area of rect (1)} = 100 \times 150 = 15000 \text{ mm}^2$$

$$a_2 = \text{area of react (2)} = 75 \times 50 = 3750 \text{ mm}^2.$$

$$y_1 = \frac{150}{2} = 75 \text{ mm} ;\ y_2 = 50 + \frac{75}{2} = 87.5 \text{ mm}$$

$$\bar{Y} = \frac{a_1 y_1 - a_2 y_2}{a_1 - a_2} = \frac{15000 \times 75 - 3750 \times 87.5}{15000 - 3750}$$

$$= \frac{1125000 - 328125}{11250} = 70.83 \text{ mm}$$

Calculation of M.I.

$$I_{xx1} = \text{M.I. of rect. (1) about N.A.}$$

$$= \frac{1}{12} \times 100 \times 150^3 + 100 \times 150 \times (75 - 70.83)^2$$

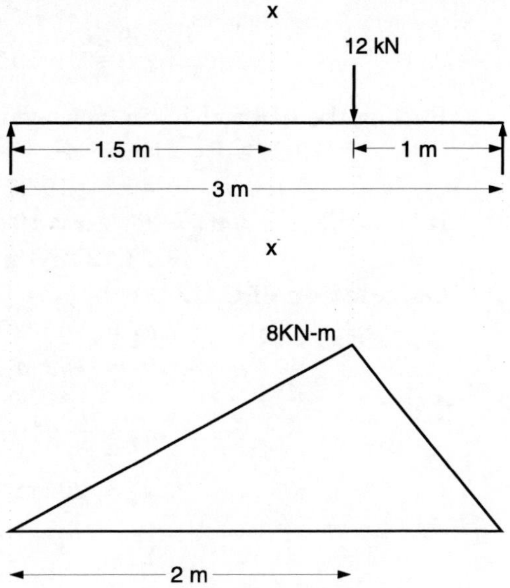

Fig. 6.29

$= 28125000 + 260833.5$
$= 28385833.5 \text{ mm}^4.$

I_{xx2} = M.I. of rect. 2 about N.A.

$$= \frac{1}{12} \times 50 \times (75)^3 + 50 \times 75 \times (87.5 - 70.8)^2$$

$= 2803650 \text{ mm}^4$

$I_{xx} = 28385833.5 - 2803650$
$= 25582183.5 \text{ mm}^4$

$y_t = 70.83$ mm, $y_c = 150 - 70.83 = 79.17$ mm

Maximum B.M. $= \dfrac{12 \times 2 \times 1}{3}$

$= 8$ KN-m

$\therefore$ B.M. at x-x, $M_x = \dfrac{8}{2} \times 1.5 = 6$ KN-m

$= 6 \times 10^6$ N-mm

Using relation

$$\frac{\sigma_{bc}}{y_c} = \frac{M_x}{I}$$

$$\sigma_{ab} = \frac{6 \times 10^6 \times 79.17}{25582183.5} = 18.57 \text{ N/mm}^2 \quad \textbf{Ans.}$$

and

$$\frac{\sigma_{bt}}{y_t} = \frac{M_x}{I}$$

$$\sigma_{bt} = \frac{6 \times 10^6 \times 70.83}{25582183.5} = 16.61 \text{ N/mm}^2 \quad \textbf{Ans.}$$

EXAMPLE 6.19. A simply supported beam and its cross-section are as shown in figure 6.30. The beam carries a load 20 KN. Its self weight is 7 KN/m. Calculate the maximum bending stress at *x*-*x*.

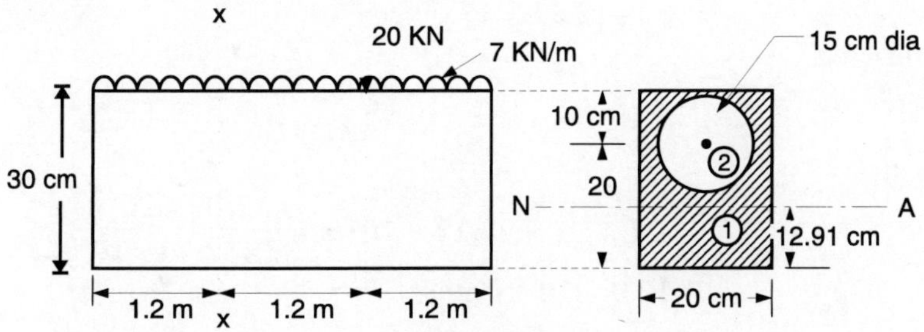

Fig. 6.30

Solution: $R_A + R_B = 20 + 7 \times 3.6 = 45.2$ KN ... (*i*)

Taking moment of all the forces about B *i.e.*, $\Sigma M_B = 0$

$R_A \times 3.6 = 20 \times 1.2 + 7 \times 3.6 \times 1.8$

$R_A \times 3.6 = 69.36$

$$\therefore \quad R_A = \frac{69.36}{3.6} = 19.27 \text{ KN}$$

Putting the value of R_A in (*i*) we get

$$19.27 + R_B = 45.2$$

$$\therefore \quad R_B = 45.2 - 19.27 = 25.93 \text{ KN}$$

B.M. at XX, $M_X = R_A \times x = 19.27 \times 1.2$

$$= 23.12 \text{ KN-m} = 23.12 \times 10^3 \text{ N-m}$$

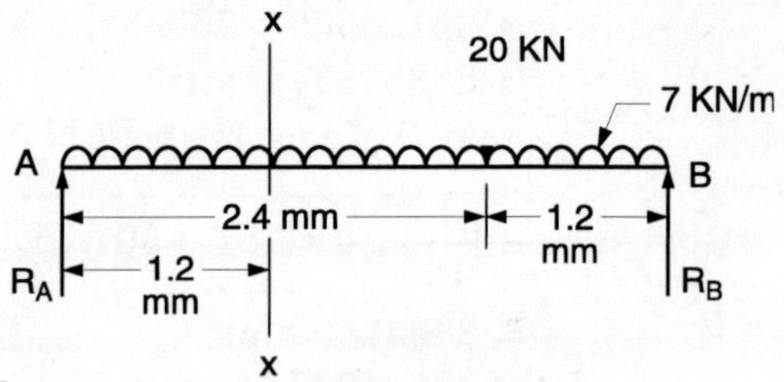

Fig. 6.31

Calculation of C.G.

$$a_1 = 30 \times 20 = 600 \text{ cm}^2$$

$$y_1 = 15 \text{ cm}$$

$$a_2 = \frac{\pi}{4} d^2 = \frac{\pi}{4} \times (15)^2 = 176.71 \text{ cm}^2$$

$$y_2 = 20 \text{ cm}$$

$$\bar{Y} = \frac{a_1 y_1 - a_2 y_2}{a_1 - a_2} = \frac{600 \times 15 - 176.71 \times 20}{600 - 176.71} = \frac{9000 - 3534.2}{423.29}$$

$$= 12.91 \text{ cm.}$$

Calculation of M.I.

I_{XX1} = M.I. of react (1) about *XX* axis

$$= \frac{1}{12} \times 20 \times 30^3 + 20 \times 30 \times (15 - 12.91)^2$$

$$= 47620.86 \text{ cm}^4$$

I_{XX2} = M.I. of circle (2) about *XX* axis

$$= \frac{\pi}{64} \times (15)^4 + \frac{\pi}{4} (15)^2 \times (20 - 12.91)^2$$

$$= 11368.15 \text{ cm}^4$$

$$I_{XX} = I_{XX1} - I_{XX2}$$

$$= 47620.86 - 11368.15 = 36252.71 \text{ cm}^4$$

$$= 36252.71 \times 10^{-8} \text{ m}^4$$

$$y_{max} = 30 - 12.91 = 17.09 \text{ cm}$$

Using relation

$$\frac{M}{I} = \frac{\sigma_b}{y}$$

$$\sigma_b = \text{sM} \times \frac{y}{I} = 23.12 \times 10^3 \times \frac{17.09 \times 10^8}{36252.71 \times 10^2} \text{ N/m}^2$$

$$= 10.9 \times 10^6 \text{ N/m}^2$$ **Ans.**

6.12 SHEAR STRESS

When a beam is loaded, the shear force along with bending moment will be set up at any section of the beam. Shear stress is the stress caused by the shear force at a section of the beam. Shear stress varies along the depth of the section. As compared to the bending stress, the effect of shear stress is quite small.

6.13 SHEAR STRESS AT A SECTION OF A BEAM

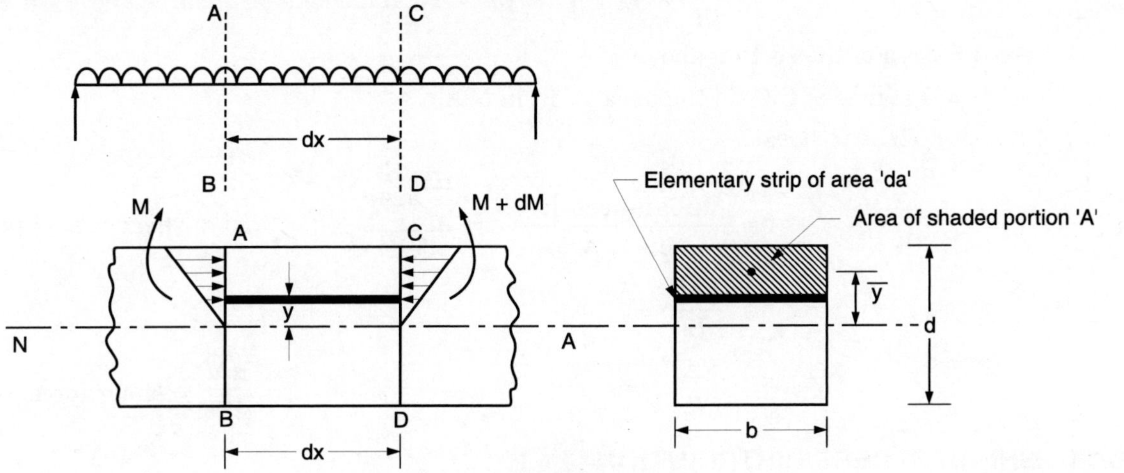

Fig. 6.32

Consider section AB and CD a simply supported beam loaded with UDL, as shown in figure 6.32.

Let M = Bending moment at AB

V = Shear force at AB

$M + dM$ = Bending moment at CD

$V + d\text{v}$ = Shear force at CD.

I = moment of intertia of the section about neutral axis.

Consider an elementary strip of area 'da' at a distance y from N.A.

The intensity of bending stress across AB at a distance y from N.A.

$$= \sigma_b = \frac{M}{I}y \qquad \left[\because \frac{M}{I} = \frac{\sigma_b}{y}\right]$$

Similarly

Intensity of bending stress across CD at a distance y from N.A.

$$= \sigma_b + d\sigma_b = \frac{M + dM}{I} \times y$$

Force acting across AB = stress × area

$$= \frac{M}{I} y \times da \qquad \left[\because \sigma_b = \frac{M}{I} y\right]$$

Force acting across CD at a distance

$$y \text{ from N.A.} = \frac{M + dM}{I} \times y \times da \qquad \left[\because \sigma_b + d\sigma_b = \frac{M + dM}{I} \times y\right]$$

Net unbalanced force on the elemental

$$\text{strip} = \frac{M + dM}{I} \times yda - \frac{M}{I} \times yda = \frac{dM}{I} yda$$

∴ Total unbalanced force in the section above y between the section AB and CD.

$$= \int \frac{dM}{I} yda = \frac{dM}{I} \int yda$$

$$= \frac{dM}{I} \times A\bar{y} \quad [\because \int y da = \text{Total moment of the area above } y = A\bar{y}]$$

Where A = Area of the section above 'y'

$\bar{y}$ = Distance of C.G. of the area 'A' from N.A.

$\therefore$ Intensity of shear stress,

$$\tau = \frac{\text{unbalanced force}}{\text{shear area}} = \frac{\frac{dM}{I}.A\bar{y}}{bdx} \qquad [\because \text{shear area} = bdx]$$

$$= \frac{dM}{dx} \times \frac{A\bar{y}}{Ib}$$

$$\therefore \qquad \tau = \frac{VA\bar{y}}{Ib} \qquad [\because \frac{dM}{dx} = \text{shear force} = V]$$

6.14 SHEAR STRESS DISTRIBUTION FOR BEAM SECTIONS OF VARIOUS SHAPES

(*i*) Rectangular Section

Figure 6.33 shows a rectangular section of width 'b' and depth 'd'. Consider a layer 'CD' at a distance y from N.A.

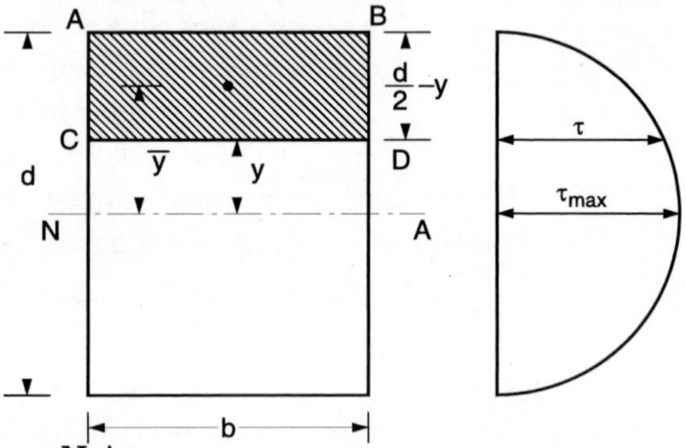

Fig. 6.33

Shear stress at this layer,

$$\tau = \frac{VA\bar{y}}{Ib} \qquad ...(i)$$

where $A\bar{y}$ is the moment of the area above CD, about the N.A.

A = Area of section above

$$\text{CD} = b\left(\frac{d}{2} - y\right)$$

$\bar{y}$ = Distance of C.G. of area 'A' from N.A.

$$= y + \frac{1}{2}\left(\frac{d}{2} - y\right) = \frac{y}{2} + \frac{d}{4}$$

$$= \frac{1}{2}\left(y + \frac{d}{2}\right) = \frac{1}{2}\left(\frac{d}{2} + y\right)$$

$$\therefore \qquad A\bar{y} = b\left(\frac{d}{2} - y\right) \times \frac{1}{2}\left(\frac{d}{2} + y\right) = \frac{1}{2}(b)\left(\frac{d^2}{4} - y^2\right)$$

$$[\because (a + b)(a - b) = a^2 - b^2]$$

Putting the value of $A\bar{y}$ in (*i*), we get

$$\tau = \frac{V + \frac{1}{2}b\left(\frac{d^2}{4} - y^2\right)}{Ib} = \frac{V}{2I}\left(\frac{d^2}{4} - y^2\right)$$

At the top edge *i.e.*, at $y = \frac{d}{2}$

$$\tau = \frac{V}{2I}\left(\frac{d^2}{4} - \frac{d^2}{4}\right) = 0$$

At the neutral axis, *i.e.*, at $y = 0$

$$\tau = \frac{V}{2I}\left(\frac{d^2}{4} - 0\right) = \frac{Vd^2}{8I} = \frac{Vd^2 \times 12}{8 \times bd^3} \qquad \left[\because I = \frac{1}{12}bd^3\right]$$

$$= \frac{3}{2}\frac{V}{bd} = 1.5\frac{V}{bd} \qquad ...(ii)$$

$$\text{Average shear stress} = \frac{\text{shear force}}{\text{cross-sectional area}}$$

$$\tau_{av} = \frac{V}{bd} \qquad ...(iii)$$

From (*ii*) and (*iii*), we get

$$\tau = 1.5\ \tau_{av} \qquad ...(iv)$$

The, shear stress at the neutral axis is maximum as $y = 0$

∴ Equation (*iv*) becomes

$$\tau_{max} = 1.5\ \tau_{av}$$

Hence maximum shear stress for a rectangular beam section is 1.5 times the average shear stress.

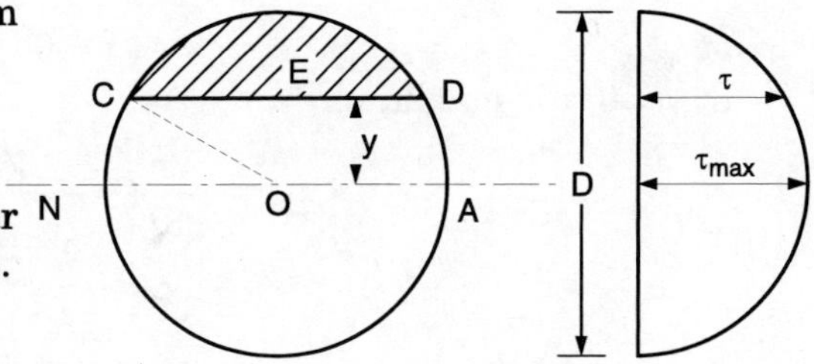

Fig. 6.34

(*ii*) Circular section

Figure 6.34 shows a solid circular section of diameter *d*. Consider a layer CD at a distance *y* from N.A.

Width of section at CD, $b = 2\sqrt{r^2 - y^2}$.

consider a strip of thickness '*dy*' at a distance *y* from N.A.

∴ Area of strip $= b\ dy$

$$= 2\sqrt{r^2 - y^2} \times dy \qquad \left[\begin{array}{l}\because \text{In right angle } \Delta\ OEC\ OC^2 = OE^2 + CE^2 \\ r^2 = y^2 + CE^2 \\ \therefore CE = \sqrt{r^2 - y^2} \quad \text{and} \\ CD = 2CE\end{array}\right]$$

Moment of the this area about N.A.

$$= \left(2\sqrt{r^2 - y^2}\ dy\right)y = 2y\sqrt{r^2 - y^2}\,dy$$

∴ Moment of area of the section above CD about N.A.

$$A\bar{y} = \int_y^r 2y\sqrt{r^2 - y^2}\,dy$$

Let $\quad 2\sqrt{r^2 - y^2} = u$

∴ $\quad 4(r^2 - y^2) = u^2$

Differentiating, we get

$$-8y\ dy = 2\ udu$$

$$\therefore \quad y\,dy = -\frac{1}{4}u\,du$$

at $\quad y = r,\ u = 0$

$\quad y = y,\ u = u$

$$\therefore \quad A\bar{y} = \int_u^0 -\frac{1}{4}u\,du \times u = \int_0^u \frac{1}{4}u^2 du$$

$$= \frac{1}{4}\int_0^u u^2 du = \frac{1}{12}u^3 = \frac{1}{12}b^3 \qquad \left[\because u = 2\sqrt{r^2 - y^2} = b\right]$$

Shear stress at the level CD,

$$\tau = \frac{VA\bar{Y}}{Ib} = V \times \frac{u^3}{12} \times \frac{1}{uI} \qquad [\because u = b]$$

$$= \frac{Vu^2}{12\,I} = \frac{V \times 4(r^2 - y^2)}{12\,I} = \frac{V(r^2 - y^2)}{3\,I}$$

at $\quad y = r,\ \tau = 0$

at $y = 0$, τ is maximum

$$\therefore \quad \tau_{max} = \frac{Vr^2}{3I} = \frac{Vr^2}{3 \times \frac{\pi}{64}D^4} = \frac{Vr^2 \times 64}{3 \times \pi \times (2r)^4}$$

$$= \frac{Vr^2 \times 64}{3\pi \times 16r^4} = \frac{4}{3}\frac{V}{\pi r^2} \qquad ...(i)$$

$$\text{Average shear stress, } \tau_{av} = \frac{\text{shear force}}{\text{cross-sectional area}}$$

$$= \frac{V}{\pi r^2} \qquad ...(ii)$$

From (*i*) and (*iii*), we get

$$\tau_{max} = \frac{4}{3}\tau_{av}$$

(*iii*) I-section

Figure 6.35 shows an I-section of overall depth, '*D*' and flange width '*B*'. Let *b* and *d* be the width and depth of web respectively.

Shear stress on a layer at a distance *y* from the neutral axis,

$$\tau = \frac{VA\bar{y}}{IB} \qquad ...(i)$$

Shear stress distribution in the web and in the flange are to be calculated separately.

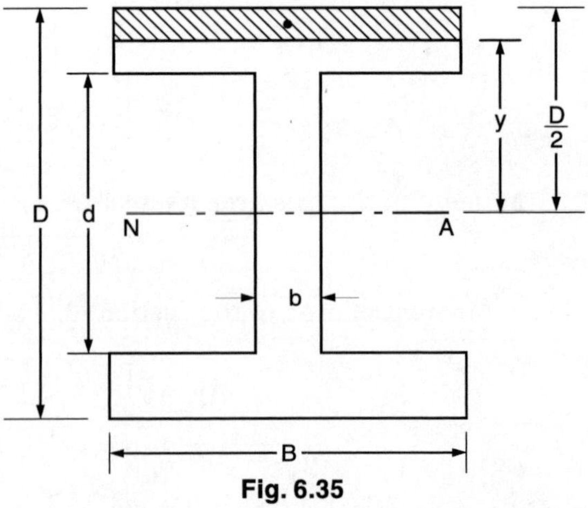

Fig. 6.35

Shear stress distribution in the flange

Consider a section of flange at a distance *y* from the neutral axis as shown in figure 6.35.

Width of section = B
Area of shaded portion

$$= \left(\frac{D}{2} - y\right) \times B$$

Distance of C.G. of the shaded area from N.A $\bar{y} = y + \frac{1}{2}\left(\frac{D}{2} - y\right) = y + \frac{D}{4} - \frac{y}{2}$

$$\bar{y} = \left(\frac{D}{4} + \frac{y}{2}\right) = \frac{1}{2}\left(\frac{D}{2} + y\right)$$

$$A\,\bar{y} = \left(\frac{D}{2} - y\right) B \times \frac{1}{2}\left(\frac{D}{2} + y\right) = \frac{1}{2} B \left(\frac{D^2}{4} - y^2\right)$$

Putting the value of $A\,\bar{y}$ in (*i*), we get

$$\tau = \frac{V \times B\left(\frac{D^2}{4} - y^2\right)}{2\,I\,B} = \frac{V}{2I}\left(\frac{D^2}{4} - y^2\right)$$

Hence in the flange, shear stress follows a parabolic law.

At $y = \frac{D}{2}$ *i.e.*, the upper edge of the flange

$$\tau = \frac{V}{2I}\left(\frac{D^2}{4} - \frac{D^2}{4}\right) = 0$$

At $y = \frac{d}{2}$ *i.e.*, the lower edge of the flange

$$\tau = \frac{V}{2I}\left(\frac{D^2}{4} - \frac{d^2}{4}\right) = \frac{V}{8I}(D^2 - d^2)$$

Shear stress distribution in the web

Consider a section in the web at a distance y from N.A. as shown in figure 6.36.

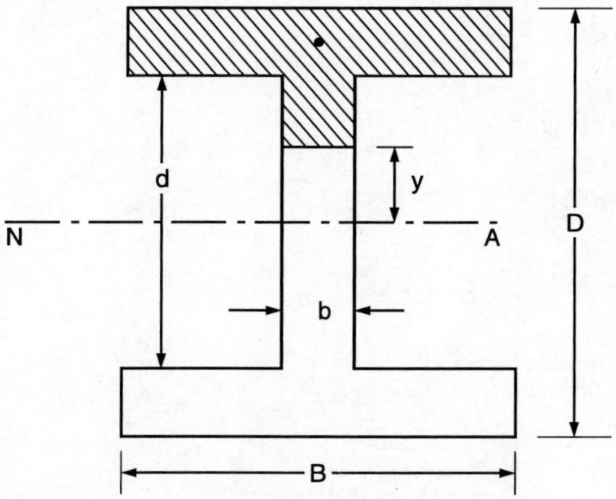

Fig. 6.36

DO YOU KNOW?

Failure due to shear stress in the web usually takes the form of bucking. For this reason deep webs are often supported by vertical stiffeners.

Width of the section = b

$$A\bar{y} = \text{Moment of shaded area about N.A.}$$

$= \text{moment of the flange area about N.A.} + \text{moment of shaded area of web about N.A.}$

$$= B \times \left(\frac{D-d}{2}\right) \times \frac{1}{2}\left(\frac{D+d}{2}\right) + b\left(\frac{d}{2}-y\right) \times \frac{1}{2}\left(\frac{d}{2}+y\right)$$

Distance of C.G. of flange from N.A.

$$= \frac{D}{2} - \frac{1}{2}\left(\frac{D-d}{2}\right) = \frac{D}{2} - \frac{D}{4} + \frac{d}{4} = \frac{D}{4} + \frac{d}{4}$$

$$= \frac{1}{4}(D+d) = \frac{1}{2}\left(\frac{D+d}{2}\right)$$

Distance of C.G. of the shaded portion of webs from N.A.

$$= \frac{1}{2}\left(\frac{d}{2}-y\right) + y = \frac{d}{4} - \frac{y}{2} + y = \frac{d}{4} + \frac{y}{2}$$

$$= \frac{1}{2}\left(\frac{d}{2}+y\right)$$

$$= \frac{1}{2}B\left(\frac{D^2-d^2}{4}\right) + \frac{1}{2}b\left(\frac{d^2}{4}-y^2\right)$$

$$= \frac{B}{8}\left(D^2-d^2\right) + \frac{b}{2}\left(\frac{d^2}{4}-y^2\right)$$

$$\text{Shear stress in the webs} = \frac{VA\bar{y}}{Ib} = \frac{V}{Ib}\left[\frac{B}{8}\left(D^2-d^2\right) + \frac{b}{2}\left(\frac{d^2}{4}-y^2\right)\right]$$

Hence, from the above equation it is clear that in the web, the shear stress follows a parabolic law.

At $y = 0$, *i.e.*, at the N.A,

$$\text{Shear stress} = \frac{V}{Ib}\left[\frac{B}{8}(D^2-d^2) + \frac{bd^2}{8}\right]$$

At $y = \frac{d}{2}$, *i.e.*, at the junction of flange and web

$$\text{shear stress} = \frac{V}{Ib}\left[\frac{B}{8}(D^2-d^2)\right]$$

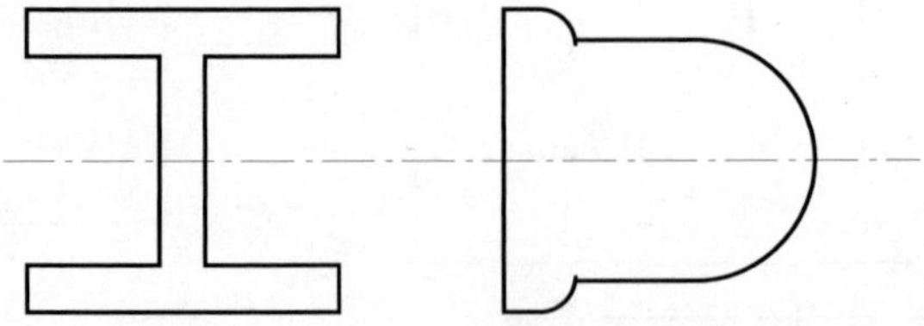

(*i*) (*ii*) Shear stress distribution

Fig. 6.37

(*iv*) Channel section: Figure 6.38 (*ii*) shows shear stress distribution.

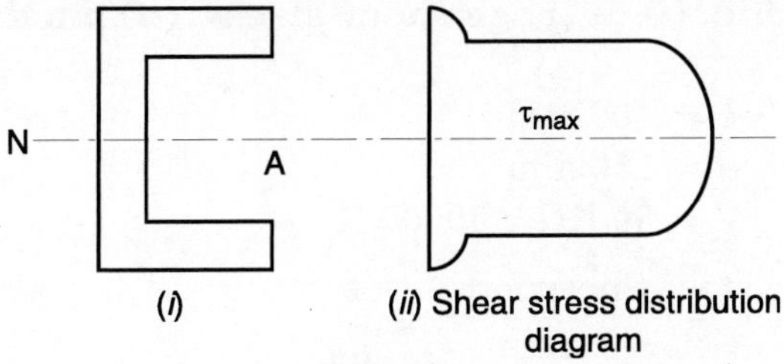

Fig. 6.38

(*v*) Angle section: Figure 6.39 (*ii*) shows shear stress distribution.

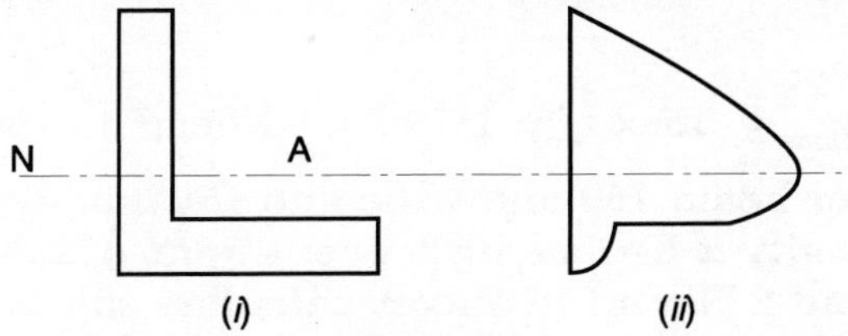

Fig. 6.39

(*vi*) T-section: Figure 6.40 (*ii*) shows shear stress distribution of T-section

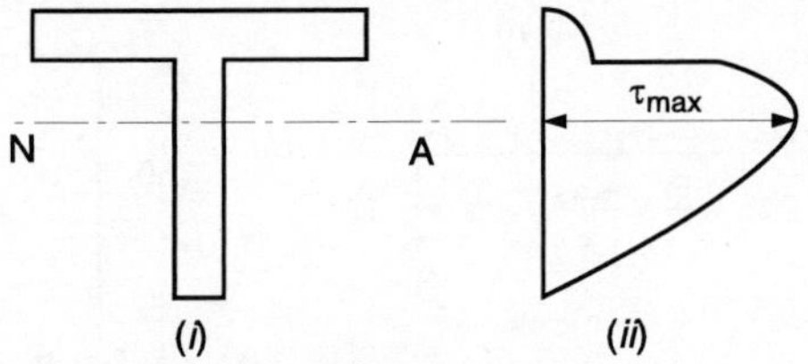

Fig. 6.40

EXAMPLE 6.20. Find the depth of a beam 100 mm wide subjected to a maximum shear force of 50 KN having the maximum shear stress as 3 N/mm².

Solution: Given

Width, $b = 100$ mm

Maximum shear force, $V = 50$ KN

Maximum shear stress, $\tau_{max} = 3$ N/mm^2

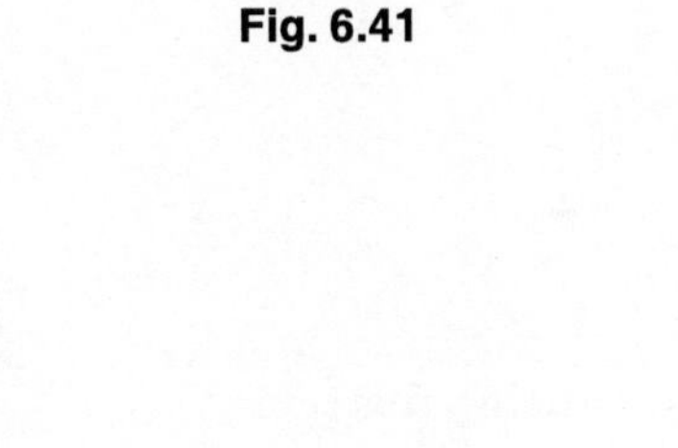

Fig. 6.41

Maximum shear stress,

$$\tau_{max} = 1.5\,\tau_{av}$$

$$3 = 1.5\,\tau_{av}$$

$$\frac{3}{1.5} = \tau_{av}$$

$$\therefore \quad \tau_{av} = 2 \text{ N/mm}^2$$

$$\tau_{av} = \frac{\text{Shear force}}{\text{Area}} = \frac{V}{bd}$$

$$2 = \frac{50 \times 1000}{100 \times d}$$

$$\therefore \quad d = \frac{50 \times 1000}{100 \times 2} = 250 \text{ mm} \qquad \textbf{Ans.}$$

EXAMPLE 6.21. A rectangular beam 100 mm wide and 150 mm deep is subjected to shear of 30 KN. Determine. (*i*) average shear stress, (*ii*) maximum shear stress.

Solution: Given

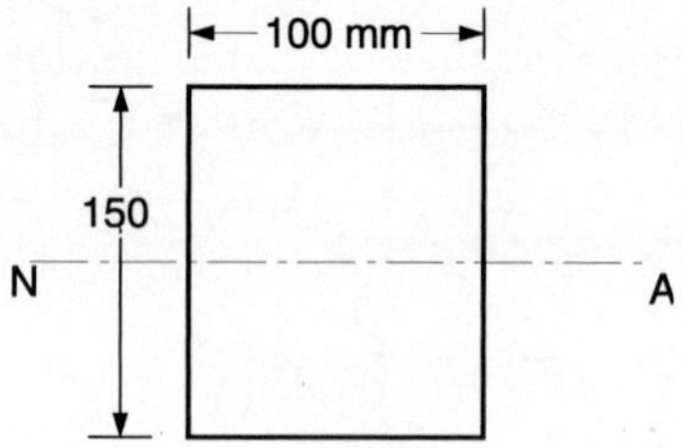

Fig. 6.42

Width, $b = 100$ mm

Depth, $d = 150$ mm

Shear force, $V = 30 \text{ KN} = 30000 \text{ N}$

Average shear stress $\tau_{av} = \dfrac{\text{Shear force}}{\text{Area}} = \dfrac{V}{bd}$

$$= \frac{30000}{100 \times 150} = 2 \text{ N/mm}^2$$

Maximum shear stress,

$$\tau_{max} = 1.5 \times \tau_{av} = 1.5 \times 2 = 3 \text{ N/mm}^2 \qquad \textbf{Ans.}$$

EXAMPLE 6.22. A timber beam 100 mm wide and 150 mm deep support a uniformly distributed load of intensity *w* KN/m length over a span of 2 m. If the safe stresses are 28 N/mm² in bending and 2 N/mm² in shear, calculate the safe intensity of the load which can be supported by the beam.

Solution: Given,

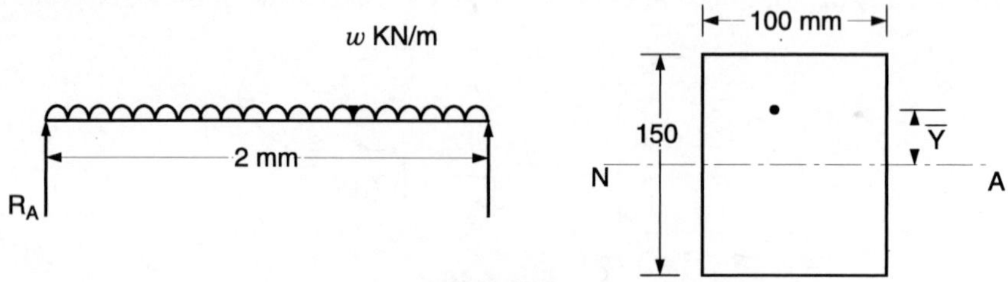

Fig. 6.43

Width, $b = 100$ mm

Depth, $d = 150$ mm

UDL $= w$ KN/m

Span, $l = 2$ m

Bending stress, $\sigma_b = 28 \text{ N/mm}^2$

Shear stress, $\tau = 2 \text{ N/mm}^2$

$$M = \frac{wl^2}{8} = \frac{w \times (2)^2}{8} = \frac{w}{2} \text{KN-m} = \frac{1000 \times 1000\, w}{2}$$

$$= 5 \times 10^5\, w \text{ N-mm}$$

$$I = \frac{bd^3}{12} = \frac{100 \times (150)^3}{12} = 28125000 \text{ mm}^2$$

$$y = 75 \text{ mm}$$

using relation

$$\frac{M}{I} = \frac{\sigma_b}{y}$$

$$\frac{5\times10^5 w}{28125000} = \frac{28}{75}$$

$$w = \frac{28\times28125000}{5\times10^5\times75} = 21 \text{ KN/m} \quad ...(i)$$

Maximum shear force, $V = \frac{wl}{2} = \frac{w\times2}{2} = wKN = 1000\ w\text{N}$

Maximum shear stress will be at the N.A., *i.e*,. at $y = 0$

Area above $NA = 100 \times 75 = 7500 \text{ mm}^2$

Distance of C.G. of the area A from neutral axis,

$$\bar{y} = \frac{75}{2} = 37.5 \text{ mm}$$

$$\tau_{max} = \frac{VA\bar{y}}{I_b}$$

$$z = \frac{1000\ w\times7500\times37.5}{28125000\times100}$$

$$\frac{2\times28125000\times100}{7500\times37.5\times1000} = w$$

$\therefore$ $w = 20 \text{ N/mm}$ ...(*ii*)

$\therefore$ The safe load intensity will be smaller of (*i*) and (*ii*), *i.e*., 20 N/mm. **Ans.**

EXAMPLE 6.23. A beam of span 'L' meters simply supported by the ends, carries a central load W. The beam section has an overall depth of 290 mm, with horizontal flanges each 150 mm × 20 mm and a vertical web 250 mm × 10 mm. If the maximum shear stress is to be 45 N/mm² then the maximum bending stress is 150 N/mm². Calculate the value of the centrally applied point load 'W' and the span 'L'.

Solution: Given

Maximum shear stress,

$$\tau_{max} = 45 \text{ N/mm}^2$$

Maximum bending stress,

$$\sigma_{bmax} = 150 \text{ N/mm}^2$$

Maximum B.M. $M = \frac{WL}{4}$ N-m

$$= \frac{WL}{4}\times1000 \text{ N-mm}$$

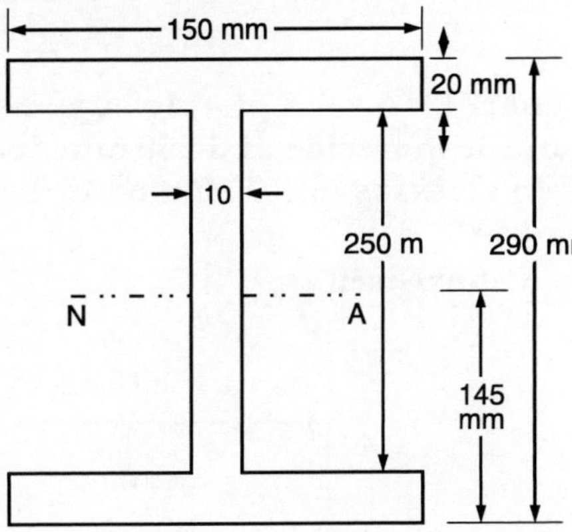

Fig. 6.44

M.I. $I = \frac{1}{12}\times150\times(290)^3 - \frac{1}{12}\times140\times(250)^3$

$= 3.05 \times 10^8 - 1.82 \times 10^8 = 1.23 \times 10^8 \text{ mm}^4.$

$$y = \frac{290}{2} = 145 \text{ mm}$$

Using relation

$$\frac{M}{I} = \frac{\sigma_b}{y}$$

$$\frac{WL \times 1000}{4 \times 1.23 \times 10^8} = \frac{150}{145}$$

$$WL = \frac{150 \times 4 \times 1.23 \times 10^8}{145 \times 1000} = 508965.52 \text{ N-mm} \qquad ...(i)$$

Maximum shear stress will be at the neutral axis i.e., at $y = 0$

$$\text{Maximum shear force}, V = \frac{W}{2} \text{N}$$

$$A\bar{y} = \text{Moment of total area (above N.A.) about N.A.}$$

$= $ moment of area of upper flange about N.A. + moment of area of webs about N.A.

$$= 150 \times 20 \times \left(145 - \frac{20}{2}\right) + 125 \times 10 \times \left(\frac{125}{2}\right)$$

$$= 150 \times 20 \times 135 + 125 \times 10 \times 62.5 = 483125 \text{ mm}^3$$

$$\tau_{max} = \frac{Va\bar{y}}{I_b}$$

$$45 = \frac{W \times 483125}{2 \times 1.23 \times 10^8 \times 10} = \frac{45 \times 2 \times 1.23 \times 10^8 \times 10}{483125} = W$$

$\therefore$ $W = 229133.25 \text{ N} = 229.13 \text{ KN}$ **Ans.**

Putting the value of w in (1), we get

$$229133.25 \times L = 508965.52$$

$$\therefore L = \frac{508965.52}{229133.25} = 2.22 \text{ m}$$ **Ans.**

EXAMPLE 6.24. A simply supported wooden beam of span 1.3 m having a cross-section 150 mm wide and 250 mm deep carries a point load 'W' at the centre. The permissible stresses are 7 N/mm² in bending and 1N/mm² in shearing. Calculate the safe load 'W'.

Solution: Given

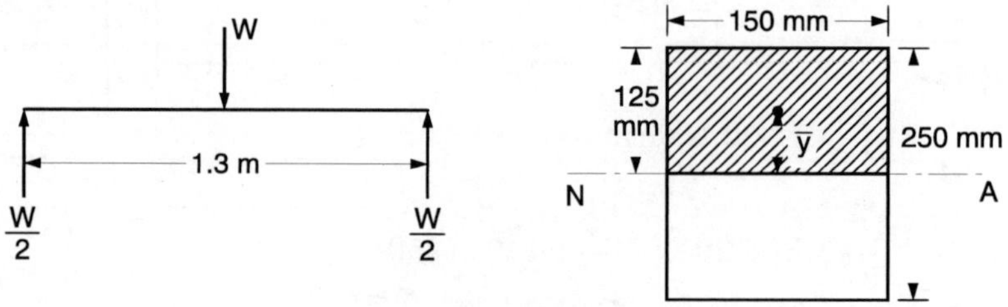

Fig. 6.45

Depth of beam,	$d =$	250 mm
Width	$b =$	150 mm
Span,	$l =$	1.3 m
Bending stress,	$\sigma_{b\,max} =$	7 N/mm²

Shearing stress, $\tau_{max} = 1 \text{ N/mm}^2$

Maximum shear force, $V = \dfrac{W}{2}$

Maximum bending moment, $M = \dfrac{Wl}{4} = \dfrac{W \times 1.3}{4} \text{ N-m} = \dfrac{W \times 1.3 \times 1000}{4} \text{ N-mm}$

$= 325 \text{ W N-mm}$

$$y = \frac{250}{2} = 125 \text{ mm}$$

$$\bar{y} = \frac{125}{2} = 62.5 \text{ mm}$$

$$I = \frac{1}{12}bd^3 = \frac{1}{12} \times 150 \times (250)^3 = 1.95 \times 10^8 \text{ mm}^4$$

A = Area of beam above N.A = 150 × 125 = 18750 mm^2

Using relation $\dfrac{M}{I} = \dfrac{\sigma_{b\max}}{y}$

$$\frac{325W}{1.95 \times 10^8} = \frac{7}{125}$$

or, $$W = \frac{7 \times 1.95 \times 10^8}{325 \times 125} = 33600 \text{ N} \quad ...(i)$$

We known

$$\tau_{max} = \frac{VA\bar{y}}{I_b}$$

$$l = \frac{W \times 18750 \times 62.5}{2 \times 1.95 \times 10^8 \times 150}$$

$$\frac{1 \times 2 \times 1.95 \times 10^8 \times 150}{18750 \times 62.5} = W$$

∴ $W = 49920$ N ...(*ii*)

The safe load will be smaller of (*i*) and (*ii*) *i.e.*, 33600 N. **Ans.**

EXAMPLE 6.25. An I-section has the following dimensions:

Flanges : 150 mm × 20 mm

Web : 300 mm × 10 mm

Find the maximum shear stress developed in the beam for a shear force of 50 KN.

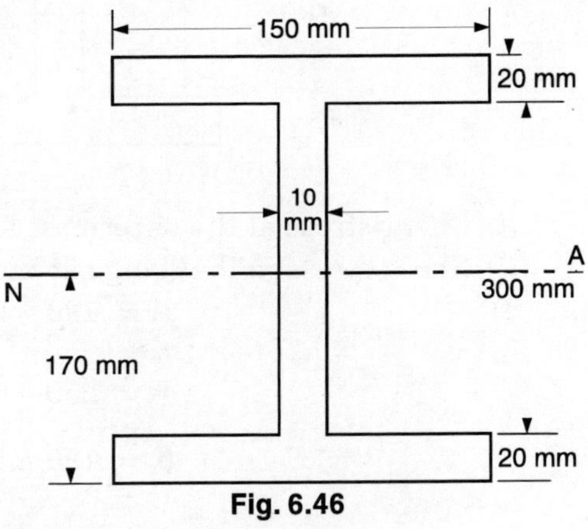

Fig. 6.46

Solution: Given

Width of flange, $b_f = 150$ mm

Depth, $d_f = 20$ mm

Width of web, $b_w = 10$ mm

Depth, $d_w = 300$ mm

Shear force, $V = 50 \text{ KN} = 50000 \text{ N}$

M.I. $$I = \frac{1}{12}\times 150\times(340)^3 - \frac{1}{12}\times 140\times(300)^3$$

$$= 4.91 \times 10^8 - 3.15 \times 10^8 = 1.76 \times 10^8 \text{ mm}^4$$

Shear stress will be maximum at the N.A., *i.e.*, $y = 0$

$A\bar{y}$ = Moment of area (above N.A) about N.A.

= Moment of area of flange about N.A. + moment of area of web about N.A.

$$= 150 \times 20 \times (170 - \frac{20}{2}) + 150 \times 10 \times \frac{150}{2}$$

$$= 480000 + 112500 = 592500 \text{ mm}^3$$

Maximum shear stress,

$$\tau_{max} = \frac{VA\bar{y}}{I\,b_w} = \frac{50000\times 592500}{1.76\times 10^8\times 10}$$

$$= 16.83 \text{ N/mm}^2$$ **Ans.**

EXAMPLE 6.26. A beam of I-section is having overall depth as 500 mm and overall width as 190 mm. The thickness of flanges is 25 mm where as the thickness of the web is 15 mm. The moment of intertia about N.A. is given as 6.45 × 10⁸ mm⁴. If the section carries a shear force of 40 KN, calculate the maximum shear stress. Also sketch the shear stress distribution across the section.

Solution: Given

Width of flange, b_f = 190 mm

Depth, d_f = 25 mm

Width of web, b_w = 15 mm

Depth, d_w = 450 mm

M.I., $I = 6.45 \times 10^8$ mm⁴.

Shear force, V = 40 KN = 40000 N

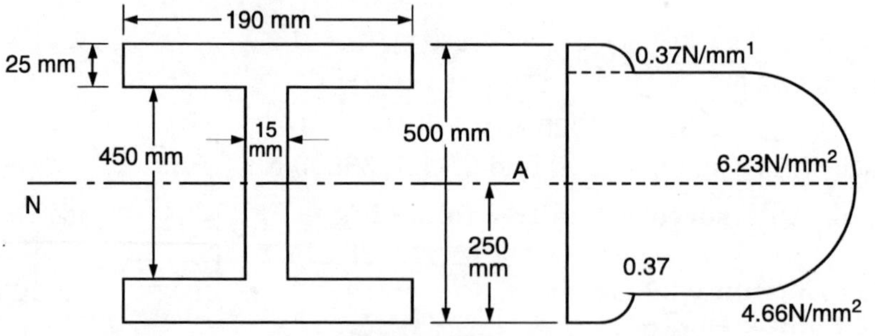

Fig. 6.47

(*i*) Shear stress at the outermost fibre of the flange = 0 [∵ A = 0]

(*ii*) Shear stress in the flange at the junction of flange and web

$$A = 190 \times 25 = 4750 \text{ mm}^2$$

$$\bar{y} = 250 - \frac{25}{2} = 237.5 \text{ mm}$$

$$b_f = 190 \text{ mm}$$

Shear stress, $\tau = \frac{VA\bar{y}}{I \times b_f} = \frac{40000 \times 4750 \times 237.5}{6.45 \times 10^8 \times 190}$

$= 0.37 \text{ N/mm}^2$

(*iii*) Shear stress in the web at the junction of flange and web

$A = 190 \times 25 = 4750 \text{ mm}^2$

$\bar{y} = 250 - \frac{25}{2} = 237.5 \text{ mm}$

$b_w = 15 \text{ mm}$

Shear stress, $\tau = \frac{VA\bar{y}}{Ib_w} = \frac{40000 \times 4750 \times 237.5}{6.45 \times 10^8 \times 15}$

$= 4.66 \text{ N/mm}^2$

(*iv*) Shear stress in the web at the neutral axis

$A\bar{y} = 190 \times 25 \times 237.5 + 15 \times 225 \times \frac{225}{2}$

$= 1507812.5 \text{ mm}^3$

$b_w = 15 \text{ mm}$

Shear stress, $\tau = \frac{VA\bar{y}}{Ib_w} = \frac{40000 \times 1507812.5}{6.45 \times 10^8 \times 15} = 6.23 \text{ N/mm}^2$

Maximum shear stress will be at the N.A.

$\therefore \quad \tau_{max} = 6.23 \text{N/mm}^2$ **Ans.**

EXAMPLE 6.27. A T-section has the flange of 80 mm × 20 mm size and the web of 120 mm × 20 mm size. The section is used as a beam with its web vertical and its flange above the web. It is subjected to a shear force of 100 KN at a certain cross section. Plot the shear stress distribution diagram indicating all important values. Also find the ratio of maximum shear stress to average shear stress.

Solution: Given

Width of flange, $b_f = 80 \text{ mm}$

Depth, $d_f = 20 \text{ mm}$

Width of web, $b_w = 20 \text{ mm}$

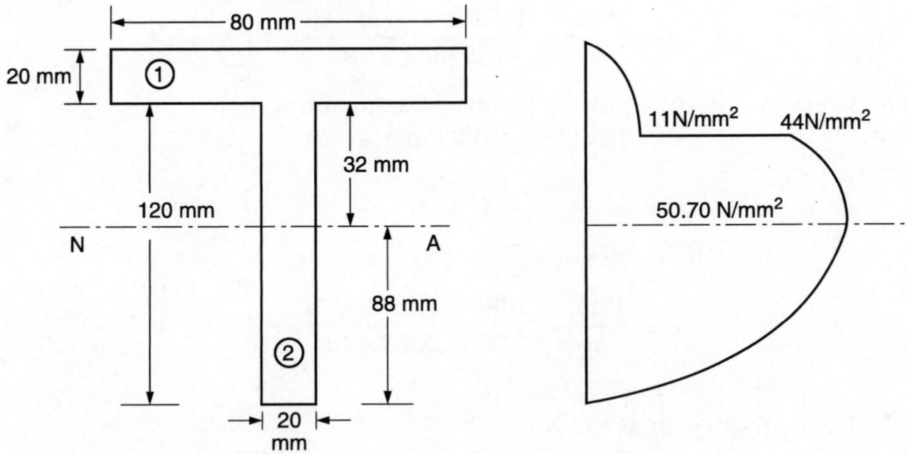

Fig. 6.48

Depth, $d_w = 120$ mm
Shear force, $V = 100$ KN = 100000 N

Calculation of C.G

$$a_1 = 80 \times 20 = 1600 \text{ mm}$$

$$y_1 = 120 + \frac{20}{2} = 130 \text{ mm}$$

$$a_2 = 120 \times 20 = 2400 \text{ mm}$$

$$y_2 = \frac{120}{2} = 60 \text{ mm}$$

$$\bar{Y} = \frac{a_1 y_1 + a_2 y_2}{a_1 + a_2} = \frac{1600 \times 130 + 2400 \times 60}{1600 + 2400} = 88 \text{ mm}$$

Calculation of M.I.

$$I_{XX1} = \text{M.I. of flange about N.A.} = \frac{1}{12} \times 80 \times 20^3 + 80 \times 20 \times (130 - 88)^2$$

$$= 53333.33 + 2822400 = 2875733.3 \text{ mm}^4.$$

$$I_{XX2} = \text{M.I. of web about N.A.}$$

$$= \frac{1}{12} \times 20 \times (120)^3 + 20 \times 120 \times (88 - 60)^2$$

$$= 2880000 + 1881600 = 4761600 \text{ mm}^4.$$

$$I_{XX} = \text{M.I. of T-section about- N.A.}$$

$$= Ixx_1 + Ixx_2 = 2875733.3 + 4761600$$

$$= 7637333.3 \text{ mm}^4.$$

Shear stress at the outermost fibre of the flange = 0 [$\because A = 0$]

(*i*) Shear stress in the flange at the junction of flange and web.

$$A = 80 \times 20 = 1600 \text{ mm}^2$$

$$\bar{y} = 32 + \frac{20}{2} = 42 \text{ mm}$$

$$b_f = 80 \text{ mm}$$

shear stress, $$\tau = \frac{VA\bar{y}}{Ib_f} = \frac{100000 \times 1600 \times 42}{7637333.3 \times 80} = 11 \text{ N/mm}^2$$

(*ii*) Shear stress in the web at the junction of flange and web

$$A = 80 \times 20 = 1600 \text{ mm}^2$$

$$\bar{y} = 32 + \frac{20}{2} = 42 \text{ mm}$$

$$b_w = 20 \text{ mm}$$

Shear stress, $$\tau = \frac{VA\bar{y}}{Ib_w} = \frac{100000 \times 1600 \times 42}{7637333.3 \times 20}$$

$$= 44 \text{ N/mm}^2$$

(*iv*) Shear stress in the web at the N.A.

$$A\bar{y} = 80 \times 20 \times 42 + 32 \times 20 \times 16$$

$$= 67200 + 10240 = 77440 \text{ mm}^3.$$

$$b_w = 20 \text{ mm}$$

Maximum shear stress,

$$\tau_{max} = \frac{VA\bar{y}}{Ib_w} \qquad [\because \text{ Shear stress is maximum at the N.A.}]$$

$$= \frac{100000 \times 77440}{7637333.3 \times 20} = 50.70 \text{ N/mm}^2.$$

(v) Shear stress at the bottom most fibre of web = 0 [$\because$ A = 0]

Area of T-section $= 80 \times 20 + 120 \times 20$

$= 4000 \text{ mm}^2$

$$\text{Average shear stress, } \tau_{av} = \frac{V}{A} = \frac{100000}{4000} = 25 \text{ N/mm}^2$$

$$\text{Ratio} = \frac{\tau_{max}}{\tau_{av}} = \frac{50.70}{25} = 2.03 \qquad \textbf{Ans.}$$

EXAMPLE 6.28. The shear force acting on a beam of an I-section with unequal flanges is 80 KN. The dimensions of the I-section are:

Top flange : 250 mm × 50 mm

Web : 250 mm × 50 mm

Bottom flange : 150 × 50 mm

The moment of inertia of the section is 50196 × 10⁴ mm⁴. Draw the shear stress distribution over the depth of the section.

Solution: Given

Shear force, $V = 80 \text{ KN} = 80000 \text{ N}.$

M.I., $I = 50196 \times 10^4 \text{ mm}^4$

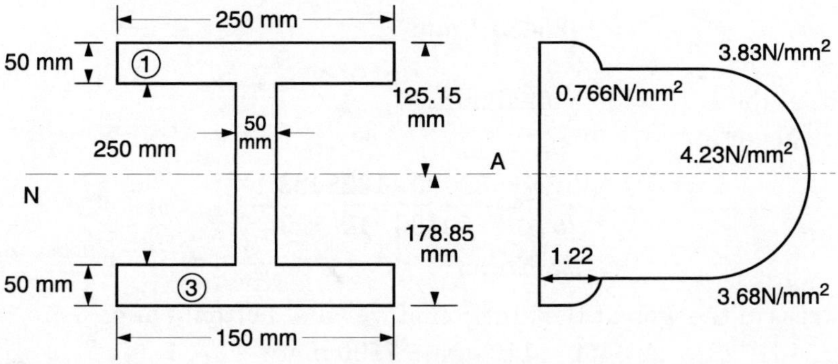

Fig. 6.49

Calculation of C.G.

$$a_1 = 250 \times 50 = 12500 \text{ mm}^2$$

$$y_1 = 250 + \frac{50}{2} = 275 \text{ mm}$$

$$a_2 = 250 \times 50 = 12500 \text{ mm}^2$$

$$y_2 = 50 + \frac{250}{2} = 175 \text{ mm}$$

$$a_3 = 50 \times 150 = 7500$$

$$y_3 = \frac{50}{2} = 25$$

$$\bar{y} = \frac{a_1y_1 + a_2y_2 + a_2y_3}{a_1 + a_2 + a_3} = \frac{12500 \times 275 + 12500 \times 175 + 7500 \times 25}{12500 + 12500 + 7500}$$

$$= \frac{3437500 + 2187500 + 187500}{32500} = 178.85 \text{ mm}$$

(*i*) Shear stress at the outermost fibre of top flange = 0 [$\because A = 0$]

(*ii*) Shear stress in the flange at the junction of top flange and web

$$V = 80000 \text{ N}$$

$$A = 250 \times 50 = 12500 \text{ mm}^2.$$

$$\bar{y} = 121.15 - \frac{50}{2} = 96.15 \text{ mm}, b = 250 \text{ mm}$$

Shear stress, $\tau = \frac{VA\bar{y}}{Ib} = \frac{80000 \times 12500 \times 96.15}{50196 \times 10^4 \times 250}$

$$= 0.766 \text{ N/mm}^2$$

(*iii*) Shear stress in the web at the junction of top flange and web

$$b = 50 \text{ mm}$$

Shear stress, $\tau = \frac{VA\bar{y}}{Ib} = \frac{80000 \times 12500 \times 96.15}{50196 \times 10^4 \times 50}$

$$= 3.83 \text{ N mm}^2$$

(*iv*) Shear stress at the N.A.

$$A\bar{y} = 50 \times 250 \times 96.15 + 71.15 \times 50 \times \frac{71.15}{2} \quad [\because 121.15 - 50 = 71.15]$$

$$= 1328433.1 \text{ mm}^3$$

$$b = 50 \text{ mm}$$

Shear stress at the N.A. will be maximum

∴ Maximum shear stress,

$$\tau_{max} = \frac{VA\bar{y}}{Ib} = \frac{80000 \times 1328433.1}{50196 \times 10^4 \times 50}$$

$$= 4.23 \text{ N/mm}^2$$

(*v*) Shear stress in the web at the junction of web and bottom flange

$$A = 50 \times 150 \text{ mm} = 7500 \text{ mm}^2$$

$$\bar{y} = 178.85 - \frac{50}{2} = 153.85 \text{ mm}$$

$$b = 150 \text{ mm}$$

Shear stress, $\tau = \frac{VA\bar{y}}{Ib} = \frac{80000 \times 7500 \times 153.85}{50196 \times 50}$

$$= 3.68 \text{ N/mm}^2$$

(*vi*) Shear stress in the flange at the junction of web and bottom flange

$$b = 150 \text{ mm}$$

Shear stress, $\tau = \dfrac{VA\bar{y}}{Ib} = \dfrac{80000 \times 7500 \times 153.85}{50196 \times 10^4 \times 150}$

$= 1.22 \text{ N/mm}^2$

(*vii*) shear stress at the bottom most fibre of bottom flange = 0 [$\because A = 0$]

EXAMPLE 6.29. A beam of inverted T-section with 60 mm × 20 mm flange and 60 mm × 20 mm web is subjected to a shear force of 12 kN at a particular section. Draw the shear stress distribution diagram across the depth of the section. Also find the ratio of maximum shear stress to the average shear stress.

Solution: Given

Shear force, $V = 12 \text{ KN} = 12000 \text{ N}$

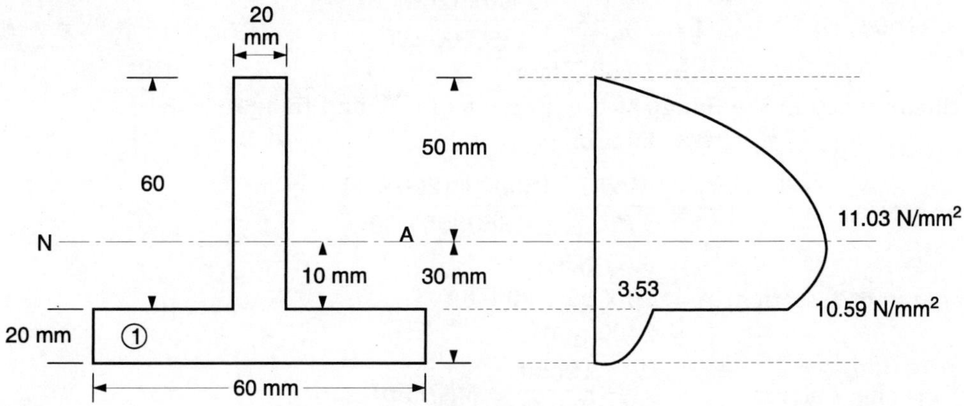

Fig. 6.50

Calculation of C.G.

$$a_1 = 60 \times 20 = 1200 \text{ mm}^2$$

$$y_1 = \frac{20}{2} = 10 \text{ mm}$$

$$a_2 = 60 \times 20 = 1200 \text{ mm}^2$$

$$y_2 = 20 + \frac{60}{2} = 50 \text{ mm}$$

$$\bar{y} = \frac{a_1 y_1 + a_2 y_2}{a_1 + a_2} = \frac{1200 \times 10 + 1200 \times 50}{1200 + 1200}$$

$$= 30 \text{ mm from base}$$

M.I. $$I = \left[\frac{1}{12} \times 60 \times 20^3 + 60 \times 20 \times (30 - 10)^2\right]$$

$$= + \left[\frac{1}{12} \times 20 \times 60^3 + 20 \times 60 \ (50 - 30)^2\right]$$

$$= 520000 + 840000 = 1360000 \text{ mm}^4$$

(*i*) Shear stress at the topmost fibre = 0 [$\because A = 0$]

(*ii*) Shear stress at the N.A.

Shear stress will be maximum at N.A.

$b = 20$ mm; $A = 20 \times 50 = 1000 \text{ mm}^2$; $\bar{y} = \dfrac{50}{2} = 25$ mm

Maximum shear stress,

$$\tau_{max} = \frac{VA\bar{y}}{I\,b} = \frac{12000 \times 1000 \times 25}{1360000 \times 20}$$

$$= 11.03 \text{ N/mm}^2$$

(*iii*) Shear stress in the web at the junction of web and flange

$$A = 20 \times 60 = 1200 \text{ mm}^2$$

$$\bar{y} = 10 + \frac{20}{2} = 20 \text{ mm}$$

$$b = 20 \text{ mm}$$

Shear stress, $$\tau = \frac{VA\bar{y}}{Ib} = \frac{12000 \times 1200 \times 20}{1360000 \times 20}$$

$$= 10.59 \text{ N/mm}^2$$

(*iv*) Shear stress in the flange at the junction of web and flange.

$$b = 60 \text{ mm}$$

Shear stress, $$\tau = \frac{VA\bar{y}}{Ib} = \frac{12000 \times 1200 \times 20}{1360000 \times 60}$$

$$= 3.53 \text{ N/mm}^2$$

Total area of T-section, $A = 20 \times 60 + 20 \times 60$

$$= 2400 \text{ mm}^2$$

Average shear stress $= \dfrac{V}{A} = \dfrac{12000}{2400} = 5 \text{ N/mm}^2$

Ratio $= \dfrac{11.03}{5} = 2.21$

EXAMPLE 6.30. A beam of channel section is 120 mm × 60 mm has uniform thickness of 15 mm. Draw diagram showing the distribution of shear stress for a vertical section where shear force is 50 KN. Find the ratio between maximum shear stress and mean shear stress.

Solution: Given

Shear force, $V = 50 \text{ KN} = 50000 \text{ N}$

$$\bar{y} = \frac{120}{2} = 60 \text{ mm}$$

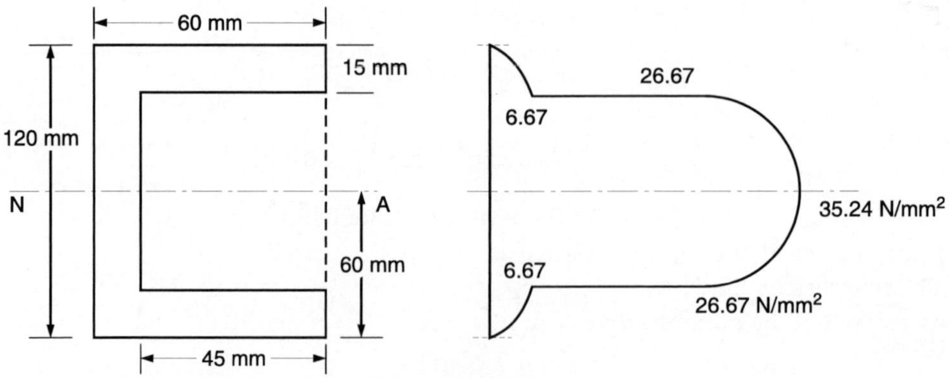

Fig. 6.51

M.I. about N.A.

$$I = \frac{1}{12}\times 60\times(120)^3 - \frac{1}{12}\times 45\times(90)^3$$

$$= 8640000 - 2733750 = 5906250 \text{ mm}^4$$

(*i*) Shear stress at the outermost fibre of the flange = 0 [$\because$ A = 0]

(*ii*) Shear stress in the flange at the junction of flange and web.

$$A = 60 \times 15 = 900 \text{ mm}$$

$$\bar{y} = 60 - \frac{15}{2} = 52.5 \text{ mm}$$

$$b = 60 \text{ mm}^2$$

shear stress, $$\tau = \frac{VA\bar{y}}{Ib} = \frac{50000\times 900\times 52.5}{5906250\times 60} = 6.67 \text{ N/mm}^2$$

(*iii*) Shear stress in the web at the junction of flange and web

$$A = 60 \times 15 = 900 \text{ mm}^2$$

$$\bar{y} = 52.5 \text{ mm}$$

$$b = 15mm$$

Shear stress, $$\tau = \frac{Va\bar{y}}{Ib} = \frac{50000\times 900\times 52.5}{5906250\times 15} = 26.67 \text{ N/mm}^2$$

(*iv*) Shear stress at the neutral axis

Shear stress will be maximum at the neutral axis.

Maximum shear stress,

$$\tau_{max} = \frac{VA\bar{y}}{Ib}$$

$$A\,\bar{y} = \text{Moment of area (above N.A.) about the N.A.}$$

$$= 60 \times 15 \times 52.5 + 45 \times 15 \times \frac{45}{2}$$

$$= 47250 + 15187.5 = 62437.5 \text{ mm}^3$$

$$\tau_{max} = \frac{50000\times 62437.5}{5906250\times 15} = 35.24 \text{ N/mm}^2$$

Total area of channel section,

$$A = 60 \times 15 + 90 \times 15 + 60 \times 15 = 3150 \text{ mm}^2$$

Average shear stress, $$\tau_{av} = \frac{50000}{3150} = 15.87 \text{ N/mm}^2$$

Ratio $$= \frac{35.24}{15.87} = 2.22$$

SUMMARY

1. The bending of a beam not accompanied by shear force is known as pure bending.
2. The stresses produced to counter bending moment are known as bending stresses.
3. Bending equation is given by

$$\frac{M}{I} = \frac{\sigma_b}{y} = \frac{E}{R}$$

Where M = Moment of resistance or bending moment.

σ_b = Bending stress

I = moment of intertia of the section.

y = Distance of fibre from N.A.

E = Young's modulus of elasticity

R = Radius of curvature of N.A.

4. Neutral axis of a beam is the axis at which stress is zero.
5. Moment of resistance is the sum of moments of the internal forces about N.A.
6. Moment of resistance depends upon the section modulus of the section.

$$\text{M.R.} = \sigma_b \times z$$

7. Section modulus is the ratio of moment of inertia about N.A. to the distance of extreme fibre from the N.A.

$$Z = \frac{I}{y_{max}}$$

8. Shear stress is the stress caused by the shear force at a section of the beam.
9. Shear stress at any layer across the transverse section of a beam

$$\tau = \frac{VA\bar{y}}{Ib}$$

where V = Shear force at a section

A = Area of section above the layer.

$\bar{y}$ = Distance of C.G. of the section of area A from N.A.

I = Moment of inertia of the beam section about N.A.

b = width of the beam at that layer

10. The maximum shear stress in a rectangular section is 1.5 times of average shear stress.
11. The maximum shear stress in circular section is 4/3 times the average shear stress.
12. The maximum shear stress is at the N.A.
13. The shear stress distribution diagram is drawn by calculating the shear stress at important points.
14. Shear stress distribution diagram across a rectangular and circular section is parabolic.

OBJECTIVE TYPE QUESTIONS

(*i*) Fill in the Blanks

1. The strength of the beam is proportional to its..
2. The stresses induced in the beam due to bending moments are known as...........................
3. Bending stresses are also known as ....
4. Bending equation is $\frac{M}{I}$ = ...
5. The flexural strength of a section means the .... offered by it.
6. In case of a circular section of dimeter 'd' the section modulus is ....
7. The stress caused by the shearing force at a section of a beam is called ....
8. Shear stress at any section. $\tau = \frac{VA\bar{y}}{......}$
9. Shear stress varies along the .... of the beam.
10. Shear stress is .... at the N.A.
11. Moment of resistance = ... × ....

Answers

1. Section modulus
2. bending stress
3. longitudinal stress
4. $\frac{M}{I} = \frac{\sigma_b}{y} = \frac{E}{R}$
5. moment of resistance
6. $\frac{\pi d^3}{32}$
7. shear stress
8. Ib
9. Depth
10. Maximum
11. bending stress, modulus of section.

THEORETICAL QUESTIONS

1. What do you mean by simple bending?
2. What are the assumptions made in theory of simple bending?
3. Derive the bending equation; $\frac{M}{I} = \frac{\sigma_b}{y} = \frac{E}{R}$
4. What do you mean by modulus of section?
5. Define (*i*) Neutral axis(*ii*) Moment of resistance.
6. What is a flitched beam? What is the necessity of these beam?
7. What is shear stress?
8. Derive the expression

$$\tau = \frac{VA\bar{y}}{Ib}$$

where
V = Shear froce at a layer
A = Area of section above the layer
$\bar{y}$ = Distance of C.G. of area A from N.A.
I = M.I. of the beam section.
b = width of the layer.

NUMERICALS

1. A beam 250 mm wide and 300 mm deep is used to carry a uniformly distributed load of 1000 N/m over a span of 5 m. Find the maximum stress developed in the beam. (0.833 N/mm^2)
2. A cantilever of 3 m span carries a point load of 25 KN at the free end in addition to its self weight of 8000 N/m. If the permissible bending stress is 120 N/mm^2, find section modulus required for the beam section. (925000 mm^3)
3. A steel plate of width 60 mm and of thickness 10 mm is bent into a circular arc of radius 10 m. Determine the maximum stress induced and the bending moment which will produce the maximum stress. Assume $E = 2 \times 10^5$ N/mm^2 (100 N/mm^2 ; 100 Nm)
4. What is the least internal radius to which a bar of steel 100 mm wide and 10 mm thick can be bent so that the maximum stress will not exceed 500 N/mm^2. Assume E = 2×10^5 N/mm^2 (1.995 m)
5. A rectangular beam 300 mm deep and 150 mm wide is simply supported over a span of 8 m. Find the maximum UDL that the beam can carry, if the bending stress is not to exceed 120 N/mm^2. (33.75 KN/m)
6. A cast iron test beam 20 mm × 20 mm in section and 1m long and supported at the ends fails when a central load of 640 N is applied. What uniformly distributed load will break a cantilever of the same material 50 mm wide, 100 mm deep and 2 meters long? (5000 N/m)
7. A rectangular beam 100 mm deep and 75 mm wide in simply supported over a span of 6 m. It carries a uniformly distributed load of 4000 N/m over the whole span. Calculate the maximum bending stress developed in the section at: (*i*) supports, (*ii*) 1.5 m from the support, (*iii*) mid span of the beam. (*O*; 108 N/mm^2; 144 N/mm^2)
8. The moment of intertia of a beam section 500 mm deep is 69.49×10^7 mm^4. Find the longest span over which a beam of this section, when simply supported, could carry a UDL of 50 KN/m. The bending stress is not to exceed 110 N/mm^2. (7 m)

9. A timber beam of rectangular section, simply supported over a span of 4 m carrying a UDL of 5.05 KN/m over the entire span. Calculate the width and depth of the beam if the bending stress is not to exceed 8N/mm^2. Take ratio of depth to width of beam as 1.5 (150 mm; 225 mm)

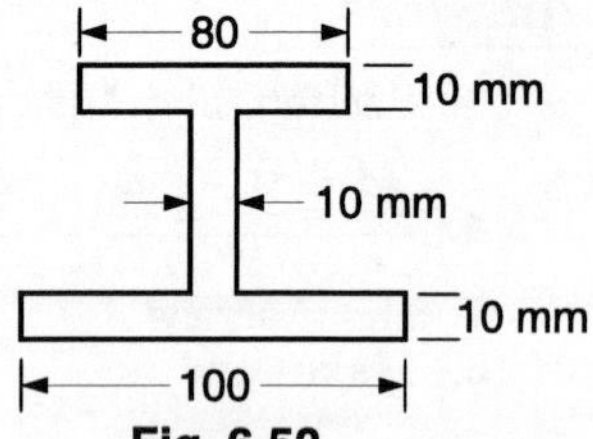

Fig. 6.52

10. Figure 6.52 shows a cast iron beam. The beam is simply supported at the ends and carries a UDL of 20 KN/m. If the span of the beam is 3 m, determine:
    (*i*) the maximum tensile and compressive stresses in the beam.
    (*ii*) the position of N.A. of the section.
    (*iii*) Stress at a distance of 60 mm from the top of beam.

(75.3 mm.; 154.23 N/mm^2; 137.12 N/mm^2; 109.25 N/mm^2)

11. A cast iron water main 12 m long of 500 mm inside diameter and 25 mm wall thickness runs full of water and is supported at its ends. Calculate the maximum stress in the metal if density of cast iron is 7200 kg/m^3 and that of water is 1000 kg/m^3. (16.82 MN/m^2)
12. Figure 6.53 shows a T-beam section. The beam is simply supported over a span of 5 m. The beam carries a concentrated load of 1500 N at mid span. Calculate the safe uniformly distributed load which the beam can carry if the maximum allowable stress is 160 N/mm^2. (2.4 KN/m)

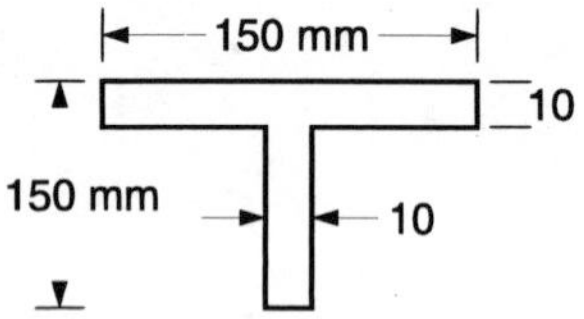

Fig. 6.53

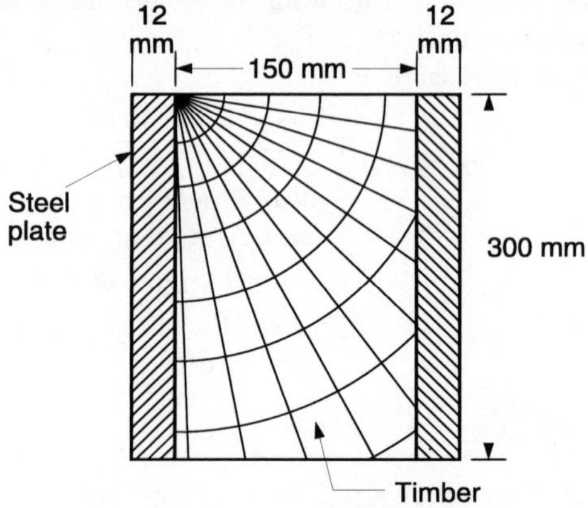

Fig. 6.54

13. A flitched timber beam made up of steel and timber had a section as shown in figure 6.54 If the maximum stresses in the timber beam is 7 N/mm^2, find the corresponding stress in steel. Find also the moment of resistance of the composite section. Take E for steel = 2 × 10^5 N/mm^2 and for timber = 1 × 10^4 N/mm^2. (66150 Nm)
14. A timber beam 110 mm wide and 220 mm deep is strengthened by two steel plates 110 mm wide and 20 mm thick at top and bottom of the timber section. Find the moment of resistance of the section. Allowable stress in timber = 7N/mm^2. Take $E_{steel} = 20\ E_{wood}$. (87035.78 Nm)
15. A beam of rectangular cross-section is 200 mm wide and 350 mm deep. If the section is subjected to a maximum shear force 30 KN, find the maximum shear stress and draw the shear stress distribution along the depth of the beam. (0.642 N/mm^2)
16. A rectangular beam 100 mm wide is subjected to a maximum shear force of 100 KN. Find the depth of the beam if the maximum shear stress is 6N/mm^2. (250 mm)
17. A wooden beam 100 mm wide and 150 mm deep is simply supported over a span of 4 m. If shear force at a section of the beam is 4500N, determine, (*i*) average shear stress, (*ii*) maximum shear stress (*iii*) shear stress at a distance of 25 mm above the neutral axis: (0.3N/mm^2; 0.45N/mm^2, 0.4 N/mm^2)
18. A timber beam 150 mm × 300 mm in cross-section is simply supported at its ends and has a span of 4 m. The maximum safe allowable stress in bending is 8N/mm^2. Find the maximum safe UDL which the beam can carry. What is the maximum shear stress in this beam for the calculated UDL?

(9 KN/m; 0.6 N/mm^2)

□□□

CHAPTER 7

Slope and Deflection

7.1 SLOPE AND DEFLECTION

Figure 7.1(*i*) shows a beam without any load. When the beam is loaded with some load it deflects or sag as shown in figure 7.1(*ii*). Due to this deflection or sag. N.A. becomes a curved line and is called elastic curve. Consider a point *M* on the bent beam. Let T_M be the tangent at point *M*. The angle 'θ' made by the tangent at point *M* with the horizontal is slope at point *M*. The vertical distance between the N.A. and elastic curve at point *M* is known as deflection at M as shown in figure 7.1 (*iii*).

Slope at any section of a beam is the angle which the tangent at the section makes with the horizontal. It is measured in radian and is denoted by θ.

Chapter Outline

- Slope and Deflection
- Strength and Stiffness of a Beam
- Necessity of Calculating Deflection
- Methods of Finding Slope and Deflection
- Moment Area Method (Mohr's Theorem)
- Determination of Maximum Slope and Deflection
- Macaulay's Method

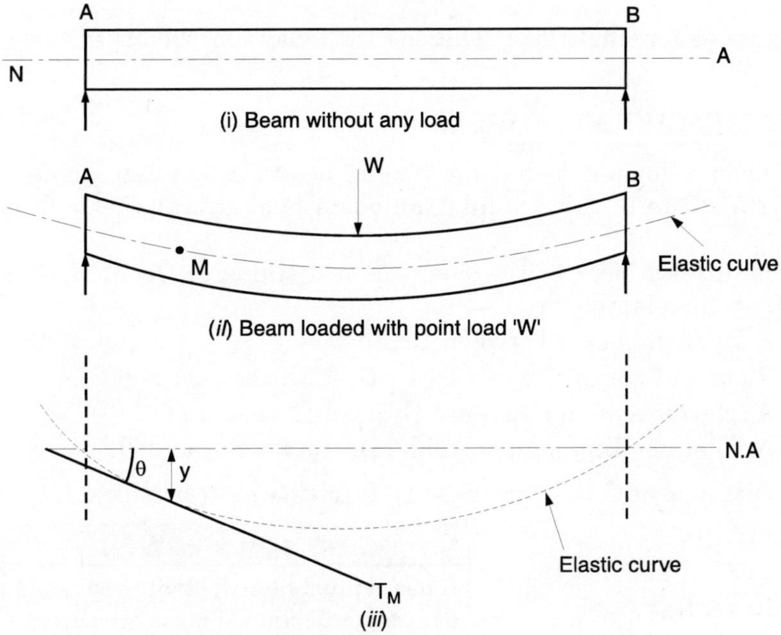

Fig. 7.1.

DO YOU KNOW?

The cross-section of a beam must be strong enough to resist the bending and shearing stresses. This is one criterion of beam design, known as **strength criterion**.

Deflection at any point on a beam is the vertical distance between the elastic curve of the loaded beam and neutral axis of the unloaded beam. The amount of deflection at any point of a beam depends upon its cross-sectional area and the bending moment.

7.2 STRENGTH AND STIFFNESS OF A BEAM

(*i*) *Strength* of any section is also known as *flexural strength* and is equal to moment of resistance offered by it.

Moment of resistance,

$$M = \sigma_b \times Z$$

where σ_b = bending stress

z = section modulus

Stiffness of a beam is a measure of its resistance to deflection. It is the ratio of maximum deflection to span of the beam. It is inversely proportional to the deflection.

7.3 NECESSITY OF CALCULATING DEFLECTION

Two basic criteria are to be satisfied in the design of a beam. These are : (*i*) Strength, (*ii*) Stiffness.

As per strength criteria, the beam should be strong enough to resist the shear force and bending moment developed due to loading.

As per stiffness criteria, the beam should be stiff enough so that it does not deflect more than the allowable limit due to loading.

7.4 METHODS OF FINDING SLOPE AND DEFLECTION

These are various methods for finding slope and deflections. Important methods are:

(*i*) Moment area method
(*ii*) Double integration method
(*iii*) Macaulay's method

First two methods are suitable for single load, whereas Macaulay's method is suitable for several loads.

7.5 MOMENT AREA METHOD (MOHR'S THEOREM)

Let us consider a beam *AB* of span '*l*' loaded with some type of load. Its bending moment diagram is shown in figure 7.2(*ii*). Due to this loading the beam bent into $A\ C_1\ D_1\ B$ as shown in figure 7.2(*iii*).

Take an element *CD* of small length *dx* of the beam at a distance '*x*' from *B*. The corresponding points are $C_1\ D_1$ on the elastic curve.

Let M = B.M. of the beam between *C* and *D*

R = Radius of curvature of part $C_1\ D_1$ of the elastic curve

$d\theta$ = Angle subtended by an arc $C_1\ D_1$ at the centre *O*

A = Area of bending moment diagram between *A* and *B*.

$\bar{x}$ = Distance of C.G. of the B.M.D. from the reference line.

From geometry of elastic curve

$$C_1D_1 = R\,d\theta$$

Since the elastic curve is almost flat

$$C_1D_1 = dx$$

> **DO YOU KNOW?**
>
> A beam must be stiff enough to resist against deflection. This is the second criterion for beam design, known as stiffness criterion.

$$\therefore \quad dx = Rd\theta$$

$$\text{or} \quad \frac{1}{R} = \frac{d\theta}{dx} \qquad ...(i)$$

From bending equation,

$$\frac{M}{I} = \frac{E}{R}$$

$$\frac{1}{R} = \frac{M}{EI} \qquad ...(ii)$$

From (i) and (ii) we get,

$$\frac{d\theta}{dx} = \frac{M}{EI}$$

$$\therefore \quad d\theta = \frac{M}{EI}dx$$

Total change in slope of the elastic curve is obtained by inte grating $d\theta$ between the limits 0 to l.

$$\therefore \quad \theta = \int_0^l \frac{M}{EI}dx = \frac{1}{EI}\int_0^l Mdx$$

But $\quad Mdx$ = Area of BMD between C and D

$$\therefore \quad \int_0^l Mdx = \text{Area of BMD between } A \text{ and } B$$

$$= A$$

$$\text{So} \quad \theta = \frac{1}{EI} \times A$$

$$\therefore \quad \theta = \frac{A}{EI} \qquad ...(iii)$$

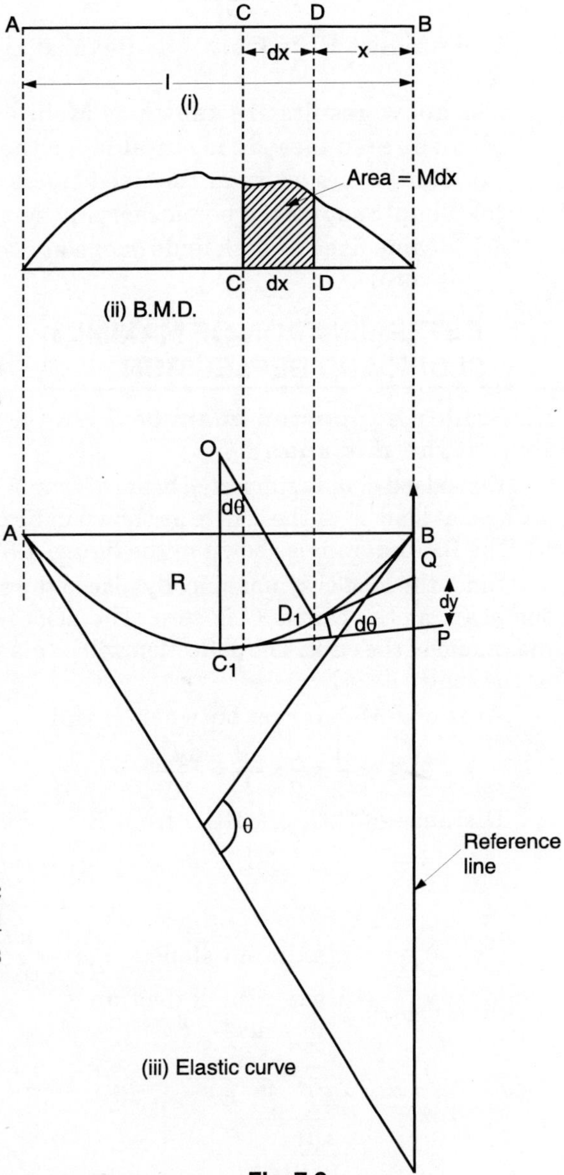

Fig. 7.2.

Now draw tangents at C, and D, on elastic curve. These two tangents intersect the vertical line (reference line) drawn through B at points P and Q as shown in figure 7.2 (iii).

From geometry of figure,

$$PQ = xd\theta$$

$$\text{or} \quad dy = xd\theta \qquad [\because PQ = dy] \;...iv)$$

Putting the value of dθ in (iv), we get

$$dy = x\frac{M}{EI}dx = \frac{M}{EI}xdx$$

Total intercept between A and B on vertical line

$$y = \int_0^l \frac{M}{EI}xdx = \frac{1}{EI}\int_o^l Mdx \times x$$

$Mdx \times x$ = moment of area of BMD between C and D about the reference line.

$$\therefore \quad \int_0^l M dx \times x = \text{Moment of area of BMD between A and B about the reference line.}$$

$$= A\overline{X}$$

$$\therefore \quad y = \frac{A\overline{X}}{EI}$$

> **DO YOU KNOW?**
>
> Differential equation of deflected beam is given by EI $\frac{d^2y}{dx^2} = M$

The product EI is known as flexural rigidity of the beam.

The above results are known as Mohr's theorems.

Moment area method is suitable for the following types of problems:

(*a*) Cantilevers (slope at the fixed end is zero).

(*b*) Simply supported beam carrying symmetrical loading (slope at mid span is zero).

(*c*) Beams fixed at both ends (slope at each end is zero)

7.6 DETERMINATION OF MAXIMUM SLOPE AND DEFLECTION

(*i*) Simply supported beam carrying a point load at the mid span :

Consider a simply supported beam of span '*l*' loaded with point load '*w*' at the centre as shown in figure 7.3 (*i*). The BM diagram is shown in the figure 7.3 (*ii*).

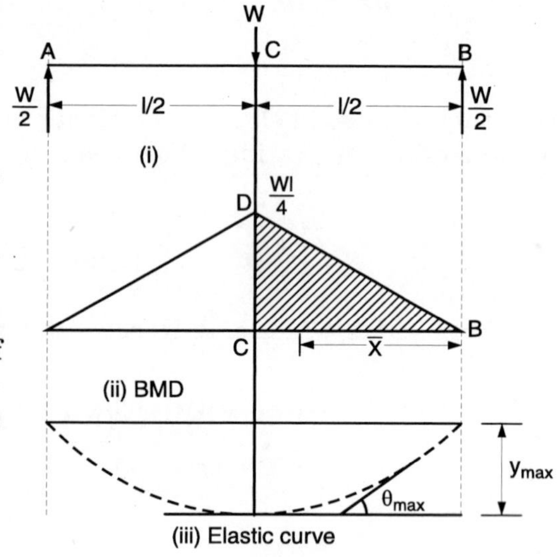

Fig. 7.3.

Since the load is symmetrically placed the slope of the beam at the centre *C* is zero. The slope will be maximum at the ends. The deflection will be maximum at the centre *i.e.*, *C*.

Area of *BM* diagram between *B* and *C*,

$$A = \frac{1}{2} \times \frac{l}{2} \times \frac{wl}{4} = \frac{wl^2}{16}$$

Distance of C.G. of ΔBCD from B

$$\bar{x} = \frac{2}{3} \times \frac{l}{2} = \frac{l}{3}$$

$$\theta_{max} = \text{Maximum slope} = \frac{A}{EI} = \frac{wl^2}{16EI}$$

$$y_{max} = \text{Maximum deflection}$$

$$y_{max} = \frac{A\overline{X}}{EI} = \frac{\frac{wl^2}{16} \times \frac{l}{3}}{EI}$$

$$y_{max} = \frac{wl^3}{48EI}$$

(*ii*) Simply Supported Beam Carrying a U.D.L.:

Consider a simply supported beam of length '*l*' loaded with *udl* of *w*/unit run.

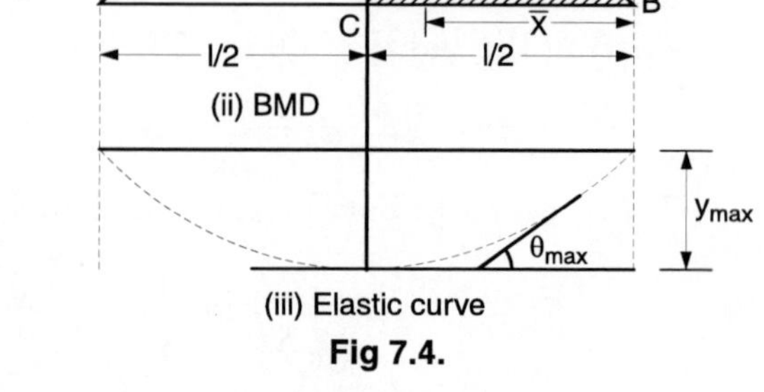

Fig 7.4.

As the beam is symmetrically loaded the slope at the centre will be zero. Maximum deflection will be at the centre.

The shape of BMD is parabolic.

Area of BM diagram between B and C,

$$A = \frac{2}{3}\times\frac{l}{2}\times\frac{wl^2}{8} = \frac{wl^3}{24}$$

Distance of C.G of area BCD from B

$$\bar{x} = \frac{5}{8}\times\frac{l}{2} = \frac{5l}{16}$$

$$\theta_{max} = \text{Maximum slope at } A \text{ and } B = \frac{A}{EI} = \frac{wl^3}{24EI}$$

$$y_{max} = \text{Maximum deflection, } = \frac{A\overline{X}}{EI} = \frac{\frac{wl^3}{24}\times\frac{5l}{16}}{EI} = \frac{5wl^4}{384EI}$$

DO YOU KNOW?

Double integration method for finding deflection is given by

$EIy = \iint M$

(*iii*) **Cantilever beam carrying a point load at the free end**

Consider a cantilever of length 'l' loaded with point load 'w' at the free end. Figure 7.5(*ii*) shows BM diagram.

At the fixed end A, the slope and deflection will be zero. The slope and deflection will be maximum at the free end 'B'.

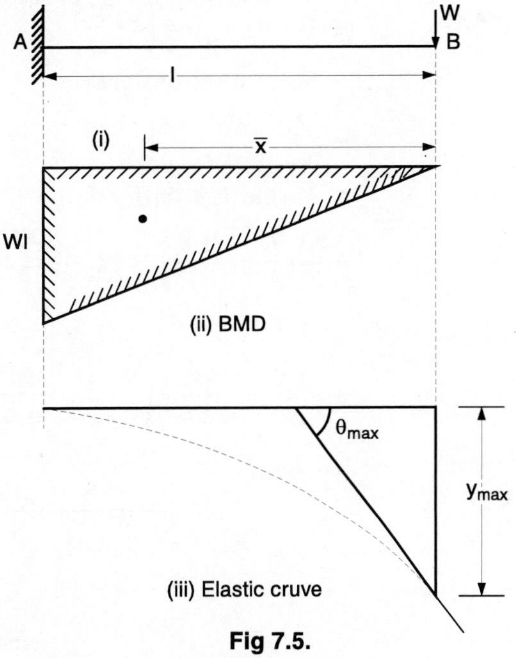

Fig 7.5.

Area of BM diagram, $\quad A = \frac{1}{2}\times l\times wl = \frac{wl^2}{2}$

Distance of C.G. of area (A) from B,

$$\bar{x} = \frac{2}{3}l$$

$$\theta_{max} = \text{Maximum slope} = \frac{A}{EI} = \frac{wl^2}{2EI}$$

y_{max} = Maximum deflection

$$= \frac{A\overline{X}}{EI} = \frac{\frac{wl^2}{2}\times\frac{2}{3}l}{EI} = \frac{wl^3}{3EI}$$

(*iv*) **Cantilever beam carrying a *u.d.l.*:**

Consider a cantilever of length 'l' loaded with *u.d.l* of *w*/unit run. Figure 7.6 (*ii*) shows BM diagram. The shape of BMD is parabolic curve.

The slope and deflection is maximum at the free end.

Area of BM diagram between A and B,

$$A = \frac{1}{3}\times l\times\frac{wl^2}{2} = \frac{wl^3}{6}$$

Distance of C.G. of BM diagram between A and B from B

$$\bar{x} = \frac{3l}{4}$$

$$\theta_{max} = \text{Maximum slope} = \frac{A}{EI} = \frac{wl^3}{6EI}$$

$$y_{max} = \text{Maximum deflection}$$

$$= \frac{A\overline{X}}{EI} = \frac{Wl^3 \times 3l}{6EI \times 4} = \frac{Wl^4}{8EI}$$

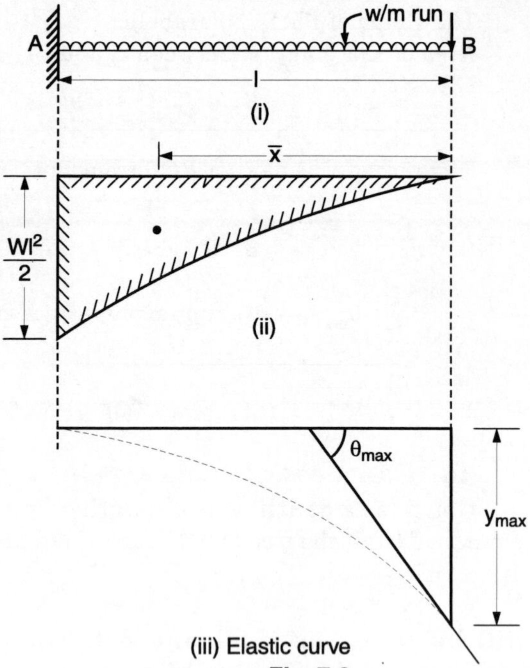

Fig. 7.6.

(*v*) Cantilevers beam carrying a point load at some distance from the free end

Consider a cantilevers of length '*l*' loaded with a point load '*W*' at a distance '*a*' from fixed end. The slope and deflection will be maximum at the free end.

Area of BM diagram between *A* and *C*,

$$A = \frac{1}{2} \times a \times Wa = \frac{Wa^2}{2}$$

Distance of C.G. *of* BM diagram from B,

$$\bar{x} = (l - a) + \frac{2}{3}a = \left(l - \frac{a}{3}\right)$$

$$\theta_{max} = \text{Maximum deflection}$$

$$= \frac{A}{EI} = \frac{Wa^2}{2EI}$$

$$y_{max} = \text{Maximum deflection}$$

$$= \frac{A\overline{X}}{EI} = \frac{Wa^2}{2EI} \times \left(l - \frac{a}{3}\right)$$

$$y_c = \text{Deflection at } C$$

$$= \frac{A\bar{X}}{EI} = \frac{wa^2}{2EI} \times \left(\frac{2}{3}a\right) = \frac{wa^2}{3\ EI}$$

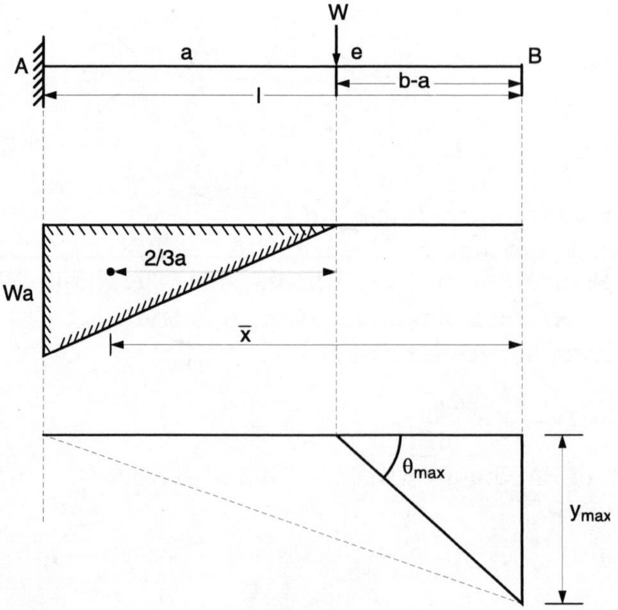

Fig. 7.7.

Table 7.1. Maximum slope and deflection

S.No.	Loading conditions	Slope	Deflection
1.	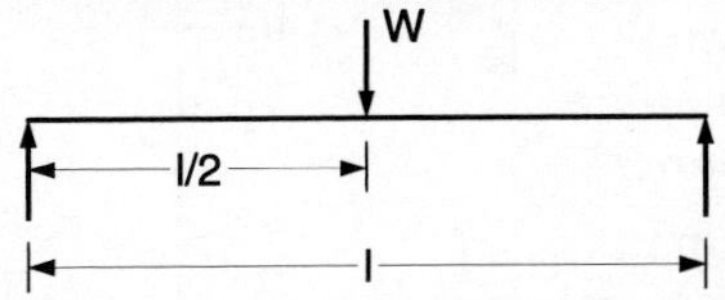	$\theta_{max} = \frac{Wl^2}{16EI}$ at supports	$y_{max} = \frac{Wl^3}{48EI}$ at mid span
2.	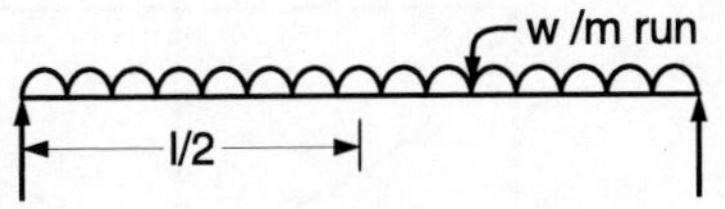	$\theta_{max} = \frac{Wl^3}{24EI}$ at supports	$y_{max} = \frac{5}{384}\frac{Wl^4}{EI}$ at mid span
3.	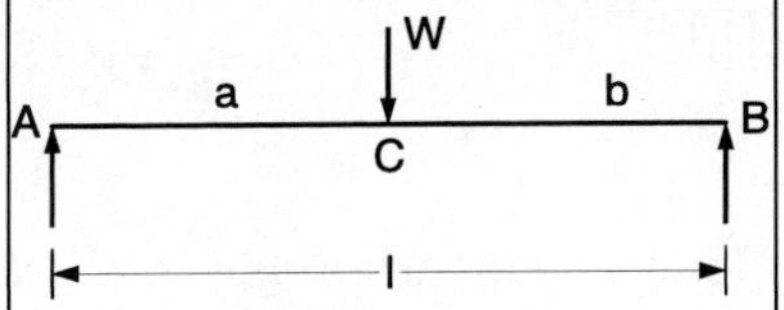	$\theta_A = \frac{Wb}{6EIl}(l^2 - b^2)$ $\theta_B = \frac{Wa(l^2 - a^2)}{6EIl}$	$y_c = \frac{Wa}{6EIl}(l^2 - a^2 - b^2)$ $y_{max} = \frac{Wa}{9\sqrt{3}EIl}(l^2 - a^2)^{3/2}$
4.	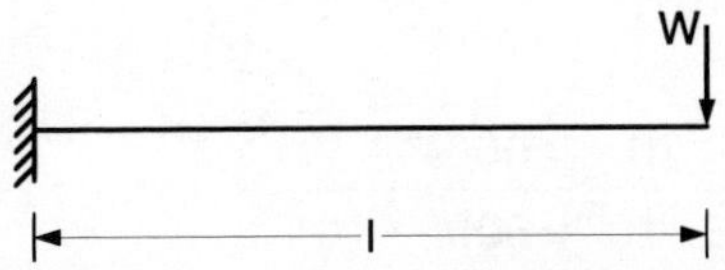	$\theta_{max} = \frac{Wl^2}{2EI}$ at the free end	$y_{max} = \frac{Wl^3}{3EI}$ at free end
5.	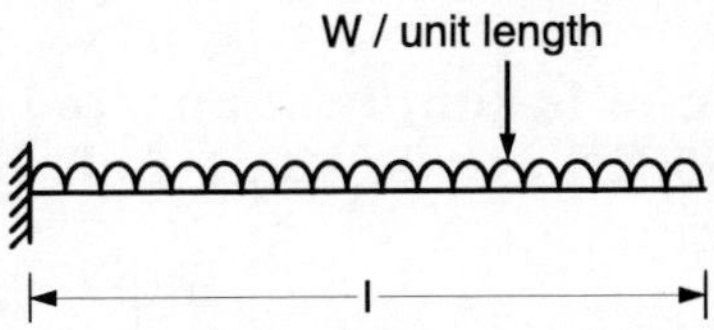	$\theta_{max} = \frac{Wl^3}{6EI}$ at free end	$y_{max} = \frac{Wl^4}{8EI}$ at free end
6.	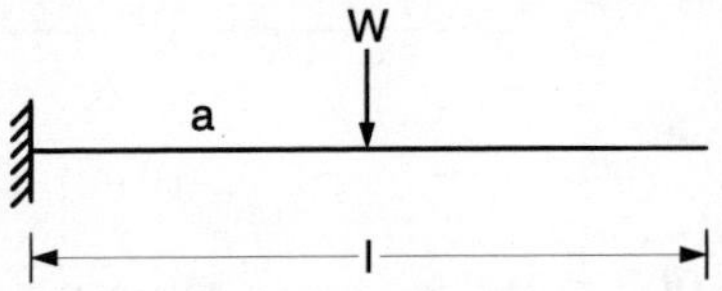	$\theta_{max} = \frac{Wa^2}{2El}$ at free end	$y_{max} = \frac{Wa^2}{2EI}\left(l - \frac{a}{3}\right)$ at free end
7.	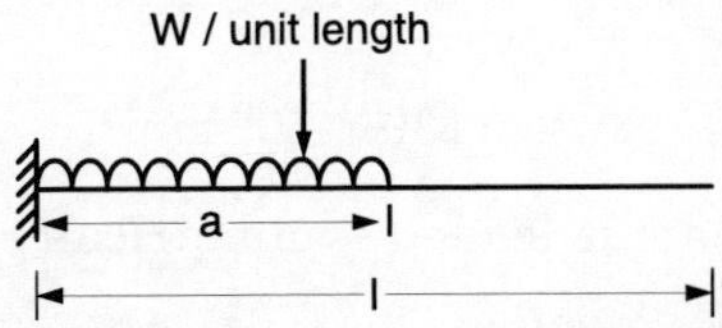	$\theta_{max} = \frac{Wa^3}{6EI}$ at free end	$y_{max} = \frac{Wa^4}{8EI} + \frac{Wa^3}{6EI}(l - a)$ at free end

SOLVED EXAMPLES

EXAMPLE 7.1. A simply supported beam 4m span carries *a u.d.l.* of 20 KN/m on the whole span and in addition it carries a point load of 40 KN at the centre of span. Calculate the maximum slope and deflection of the beam.

Take $E = 2 \times 10^5$ N/mm² and $I = 5000 \times 10^4$ mm⁴

Solution: Given :

Point load, $W =$ 40 KN = 40000 N

U.D.L. $w =$ 20 KN/m

Span, $l = 4\text{ m} = 4000\text{ mm}$

$E = 2 \times 10^5\text{ N/mm}^2$

$I = 5000 \times 10^4\text{ mm}^4$

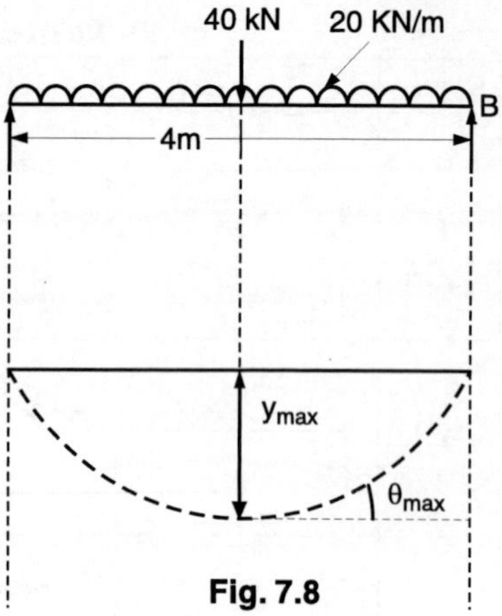

Fig. 7.8

As the beam is loaded symmetrically the maximum slope will be at the ends and maximum deflection will at the centre.

Maximum slope, θ_{max} = Slope due to point load + Slope due to *u.d.l.*

$$= \frac{Wl^2}{16EI} + \frac{wl^3}{24EI}$$

$$= \frac{40000 \times (4000)^2}{16 \times 2 \times 10^5 \times 5000 \times 10^4} + \frac{20 \times (4000)^3}{24 \times 2 \times 10^5 \times 5000 \times 10^4}$$

$= 4 \times 10^{-3} + 5.33 \times 10^{-3}$

$= 9.33 \times 10^{-3} = 0.0093$ radian **Ans.**

Maximum deflection, y_{max} = Deflection due to point load + Deflection due to *u.d.l.*

$$= \frac{Wl^3}{48\,EI} + \frac{5\,wl^4}{384\,EI}$$

$$= \frac{40000 \times (4000)^3}{48 \times 2 \times 10^5 \times 5000 \times 10^4} + \frac{5 \times 20 \times (4000)^4}{384 \times 2 \times 10^5 \times 5000 \times 10^4}$$

$= 5.33 + 6.67 = 12$ mm **Ans.**

EXAMPLE 7.2. A rolled steel joist 150 mm × 450 mm is simply supported at ends and 5.3 m span, carries a concentrated load of 100 KN at mid span. Calculate the deflection at the centre.

Take $E = 2 \times 10^5$ N/mm²

Solution: Given :

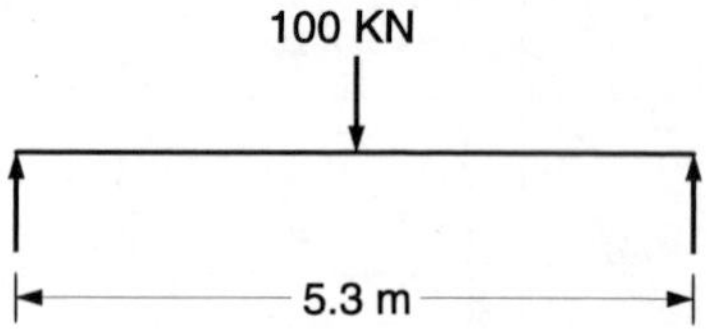

Fig. 7.9

Width, $b = 150$ mm

Depth, $d = 450$ mm

Span, $l = 5.3\text{ m} = 5300$ mm

$W = 100\text{ KN} = 100 \times 1000 = 10^5$ N

$E = 2 \times 10^5\text{ N/mm}^2$

Moment of intertia, $I = \frac{1}{12}bd^3 = \frac{1}{2} \times 150(450)^3 = 1.139 \times 10^9\text{ mm}^4.$

Maximum deflection will be at the centre as the beam is loaded symmetrically

Maximum deflection, $y_{max} = \frac{Wl^3}{48EI} = \frac{10^5 \times (5300)^3}{48 \times 2 \times 10^5 \times 1.139 \times 10^9} = 1.36$ mm. **Ans.**

EXMAPLE 7.3. A simply supported beam is 4.2 m long and is of circular cross-section 100 mm in diameter. If the maximum permissible deflection is 5 mm, determine the maximum value of the central load *W*. Take $E = 2 \times 10^5$ N/mm². Also calculate the slope at the supports.

Solution: Given :

span, $l = 4.2\text{ m} = 4200$ mm

$$\text{diameter, } d = 100 \text{ mm}$$
$$y_{max} = 5 \text{ mm}$$
$$E = 2 \times 10^5 \text{ N/mm}^2$$

$$\text{Moment of intertia I} = \frac{\pi}{64} \times (100)^4 = 4.91 \times 10^6 \text{ mm}^4$$

Maximum deflection in case of point load,

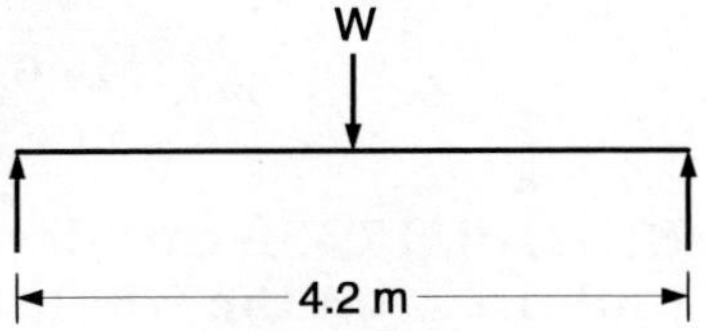

Fig. 7.10

$$y_{max} = \frac{Wl^3}{48EI}$$

$$5 = \frac{W \times (4200)^3}{48 \times 2 \times 10^5 \times 4.91 \times 10^6}$$

or $$\frac{5 \times 48 \times 2 \times 10^5 \times 4.91 \times 10^6}{(4200)^3} = W$$

$\therefore$ $W = 3181.08$ N **Ans.**

Slope will be maximum at the supports

$$\therefore \text{ Maximum slope, } \quad \theta_{max} = \frac{Wl^2}{16EI} = \frac{3181.08 \times (4200)^2}{16 \times 2 \times 10^5 \times 4.91 \times 10^6} = 0.00357 \text{ radian. } \textbf{Ans.}$$

EXAMPLE 7.4. When a simply supported beam 4m long carries a load of 200 KN at its mid span, the maximum deflection is 4 mm. What would be the deflection at free end if a cantilever of the same material, length and cross-section carries a load of 100 KN at a point 1.8m from free end?

Solution. Case I :

Span, $l = 4 \text{ m} = 4000 \text{ mm}$

Load, $W = 200 \text{ KN} = 200 \times 1000 = 2 \times 10^5 \text{ N}$

$y_{max} = 4 \text{ mm}$

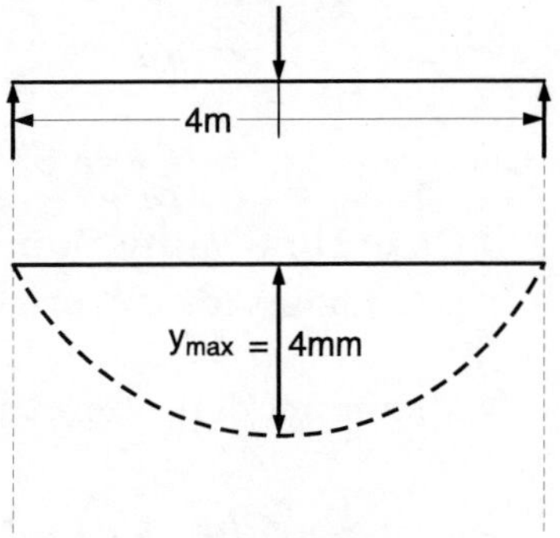

Fig. 7.11

Let the M.I. be I and modulus of elasticity be E.

Maximum deflection in case of point load

$$y_{max} = \frac{Wl^3}{48EI}$$

$$4 = \frac{2 \times 10^5 \times (4000)^3}{48\ EI}$$

$$\therefore \quad EI = \frac{2 \times 10^5 \times 64 \times 10^9}{48 \times 4}$$
$$= 6.67 \times 10^{13} \text{ N mm}^2 \quad \quad ...(i)$$

Case II :

$W = 100 \text{ KN} = 100 \times 1000 = 10^5 \text{ N}$

$l = 4000$ mm,

$a = 2.2 \text{ m} = 2200 \text{ mm}$

As the cantilever is of same material and cross-section to that of simply supported beam the value of I and E will be same for both the beams.

Maximum deflection at the free end,

$$y_{\max} = \frac{Wa^2}{2EI}\left[l - \frac{a}{3}\right]$$

$$= \frac{10^5 \times (2200)^2}{2 \times 6.67 \times 10^{13}}\left[4000 - \frac{2200}{3}\right]$$

$$= 11.85 \text{ mm.} \quad \textbf{Ans.}$$

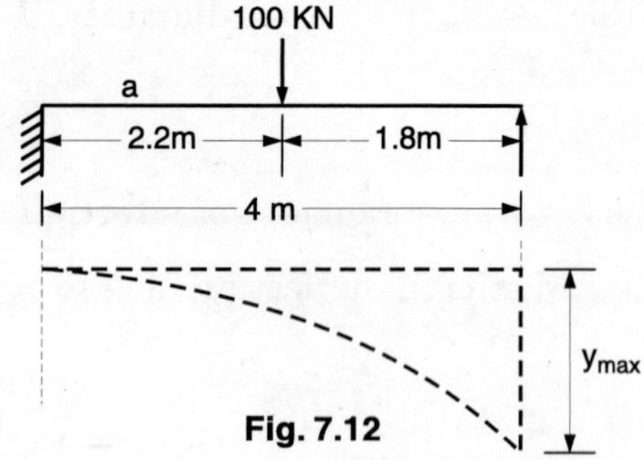

Fig. 7.12

EXAMPLE 7.5. A cantilever beam 3 m span is loaded uniformly on the whole span with 20 KN/m and a point load of 20 KN at the free end. If the same beam is simply supported at the two ends, what point load at the centre, it should carry to have the same deflection as for the cantilever.

Solution: Case I : Cantilever beam

Given :

Span, l = 3 × 1000 = 3000 mm

$u.d.l.$, w = 20 KN/m = 20 N/mm

Point load, W = 20 KN = 20000 N

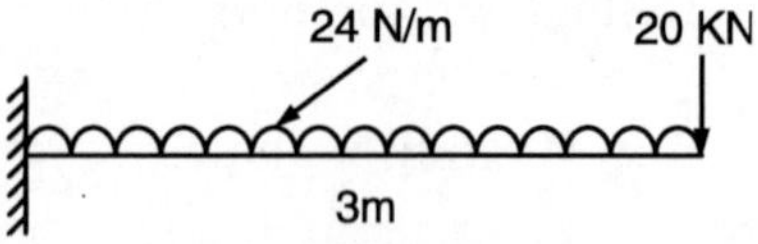

Fig. 7.13

Maximum deflection,

$y_{\max}$ = Deflection due to point-load + Deflection due to $u.d.l.$

$$= \frac{Wl^3}{3EI} + \frac{wl^4}{8EI} = \frac{20.000 \times (3000)^3}{3\ EI} + \frac{20 \times (3000)^4}{8\ EI}$$

$$= \frac{1}{EI}(1.8 \times 10^{14} + 2.02 \times 10^{14})$$

$$= \frac{1}{EI} \times 3.82 \times 10^{14} \text{mm}$$

Case II : Simply Supported beam

Let a point load W produces same deflection in S.S. beam as for the cantilever.

$$\text{Maximum deflection, } y_{\max} = \frac{Wl^3}{48EI}$$

$$\frac{1}{EI} \times 3.82 \times 10^{14} = \frac{W \times (3000)^3}{48\ EI}$$

$$3.82 \times 10^{14} = \frac{W \times (3000)^3}{48 \times EI}$$

$$W = \frac{3.82 \times 10^{14} \times 48}{(3000)^3} 1$$

$$\therefore \quad W = 679111.11 \text{ N.} \quad \textbf{Ans.}$$

EXAMPLE 7.6. A cantilever of 4 m span carries a u.d.l. of 5 KN/m. Find the value of concentrated load 'W' acting at the free end so that the maximum deflection does not exceed 10 mm anywhere in the cantilever. Also find the maximum slope.

Take $E = 2 \times 10^5$ N/mm² and $I = 2 \times 10^8$ mm⁴.

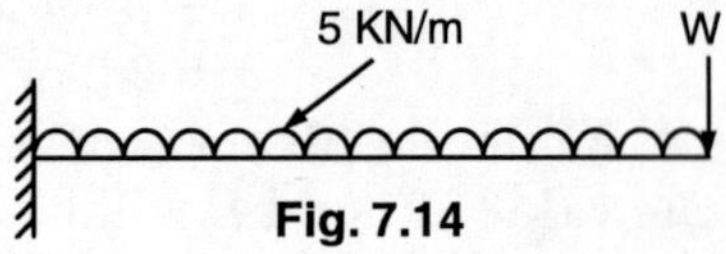

Fig. 7.14

Solution. Given :

Span, $l = 4 \text{ m} = 4000 \text{ mm}$

u.d.l., $w = 5 \text{ KN/m} = 5 \text{ N/mm}$

$y_{max} = 10 \text{ mm}$

$E = 2 \times 10^5 \text{ N/mm}^2, \; I = 2 \times 10^8 \text{ mm}^4$

Maximum deflection, y_{max} = Deflection due to point load + Deflection due to *u.d.l.*

$$= \frac{Wl^3}{3EI} + \frac{wl^4}{8EI}$$

$$10 = \frac{W \times (4000)^3}{3 \times 2 \times 10^5 \times 2 \times 10^8} + \frac{5 \times (4000)^4}{8 \times 2 \times 10^5 \times 2 \times 10^8}$$

$$10 = \frac{W \times 64 \times 10^9}{12 \times 10^{13}} + 4$$

$$10 - 4 = \frac{W \times 64 \times 10^9}{12 \times 10^{13}}$$

$$\frac{6 \times 12 \times 10^{13}}{64 \times 10^9} = W$$

$\therefore$ $W = 11250 \text{ N}$

Maximum slope, $$\theta_{max} = \frac{Wl^2}{2EI} + \frac{wl^3}{6EI} = \frac{11250 \times (4000)^2}{2 \times 2 \times 10^5 \times 2 \times 10^8} + \frac{5 \times (4000)^3}{6 \times 2 \times 10^5 \times 2 \times 10^8}$$

$$= 2.25 \times 10^{-3} + 1.33 \times 10^{-3} = 3.58 \times 10^{-3}$$

$= 0.00358$ radius **Ans.**

EXAMPLE 7.7. A timber cantilevers beam is 3 m long. It is loaded with u.d.l. of 200 KN (total). It is desired that the end of cantilever be raised to its original level as before loading. To achieve this, an upward force '*P*' is applied at the end. Find the value of P.

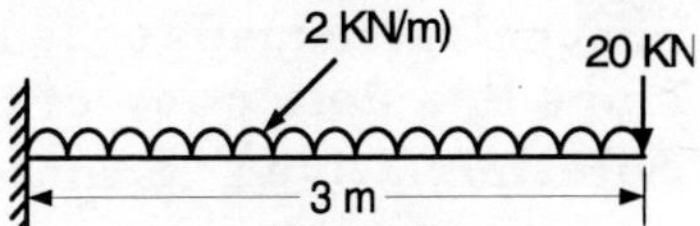

Fig. 7.15

Solution: Given :

Span, $l = 3 \text{ m} = 3000 \text{ mm}$

udl (total), $wl = 200 \text{ KN} = 200000 \text{ N}$

Maximum deflection, $$y_{max} = \frac{wl^4}{8EI} \quad \text{.....}(i)$$

An upward force '*P*' is applied at the end of the cantilever to raise the end of the cantilever.

Maximum deflection due to P $$= \frac{Pl^3}{3EI} \quad \text{......}(ii)$$

Both the deflection must be same so that the end of cantilever be raised to its original level.

$$\therefore \quad \frac{Pl^3}{3EI} = \frac{wl^4}{8EI}$$

or $$\frac{P}{3} = \frac{wl}{8} \quad \text{[On simplification]}$$

or $$P = \frac{3wl}{8} = \frac{3 \times 200000}{8} = 75000 \text{ N}$$ **Ans.**

EXAMPLE 7.8. A 4.5 m long pole is standing vertically on the ground. It is circular in cross-section with 250 mm diameter. The wind is blowing at the rate of 0.020 N/mm^2. Find the deflection of the pole at its top point in the direction of wind. Take E = 2.1 × 10^5 N/mm^2.

Solution. Given :

Span, l = 4.5 m = 4500 mm

Diameter, d = 250 mm

wind pressure = 0.020 N/mm^2

E = 2.1 × 10^5 N/mm^2

u.d.l. due to wind, w = wind pressure × diameter

$= 0.020 \times 250$

$= 5$ N/mm

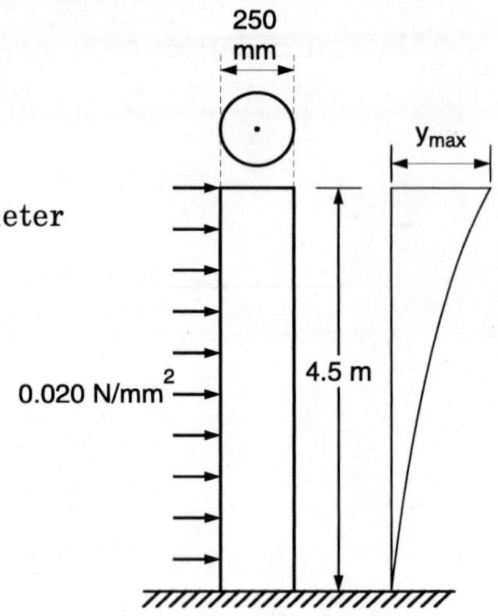

Fig. 7.16

M.I., $$I = \frac{\pi d^4}{64} = \frac{\pi \times (250)^4}{64}$$

$= 1.917 \times 10^8$ mm^4

Deflection at the top point of the pole,

$$y_{max} = \frac{wl^4}{8EI}$$

$$= \frac{5 \times (4500)^4}{8 \times 2.1 \times 10^5 \times 1.917 \times 10^8}$$

$$= \frac{2.05 \times 10^{15}}{32.2 \times 10^{13}} = 6.37 \text{ mm.} \quad \textbf{Ans.}$$

EXAMPLE 7.9. A beam 8 m long is simply supported at its ends and carries concentrated loads of 20 KN each at points 2m from the ends. Calculate the maximum slope and deflection of the beam. Take EI = 50 × 10^{12} N/mm^2.

Solution: Given : Span, l = 8 m

EI = 50 × 10^{12} N/mm^2

Point loads, W = 20 KN

$R_A + R_B = 20 + 20 = 40$ .....(*i*)

Taking moment about B, $\Sigma M_B = 0$

$R_A \times 8 = 20 \times 6 + 20 \times 2 = 160$

$\therefore$ $$R_A = \frac{160}{8} = 20 \text{ KN}$$

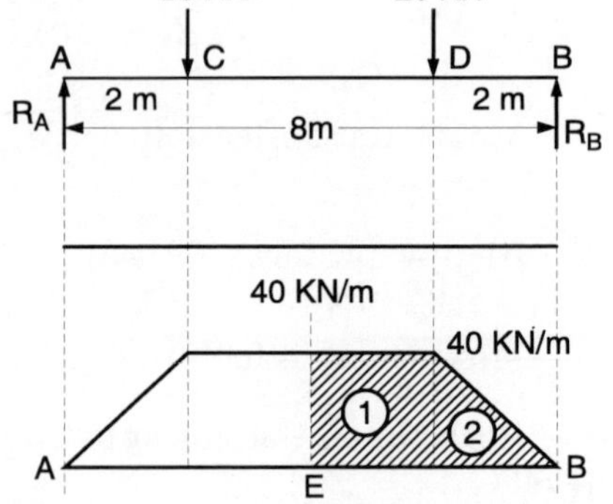

Fig. 7.17

BM at A = 0

BM at C = 20 × 2 = 40 KN-m

BM at D = 20 × 6 – 20 × 4 = 40 KN-m

BM at B = 0

Area of *BMD*, (hatched portion) $A = A_1 + A_2$

A_1 = Area of rectangle 1 = 2 × 40 = 80 KN m^2

$$A_2 = \text{Area of triangle } 2 = \frac{1}{2} \times 40 \times 2 = 40 \text{ KN m}^2$$

$A = A_1 + A_2$

$= 80 + 40 = 120$ KN m^2

Moment of area (A) of BMD between B and E about B

$$A\overline{X} = A_1 x_1 + A_2 x_2$$

(x_1 and x_2 are distance of C.G. of rectangle (1) and trianlge (2) in BMD from B)

$$= 80\ (1 + 2) + 40 \times \left(\frac{2}{3} \times 2\right) = 240 + 53.33$$

$$= 293.33 \text{ KN m}^3$$

Maximum deflection, $y_{max} = \dfrac{A\overline{X}}{EI} = \dfrac{293.33 \times 1000 \times (1000)^3}{50 \times 10^{12}} = 5.87$ mm **Ans.**

The slope will be maximum at the supports A and B.

Maximum slope, $\theta_{max} = \dfrac{A}{EI} = \dfrac{120 \times 1000 \times (1000)^2}{50 \times 10^{12}} = 0.0024$ radian. **Ans.**

EXAMPLE 7.10. A rolled steel joist having depth equal to 1.5 times width is simply supported at its ends. It carries a u.d.l. of 3 KN/m over the entire span. The maximum central deflection is 15 mm and the slope at the end is 0.005 radian. Find the dimensions of the beam if maximum bending stress is 125 N/ mm².

Solution: Given :

Bending stress, $\sigma_b = 125 \text{ N/mm}^2$

Slope, $\theta_{max} = 0.005$ radian

Deflection, $y_{max} = 15$ mm

Let 'b' be width of joist of span 'l'

$\therefore$ depth, $d = 1.5b$

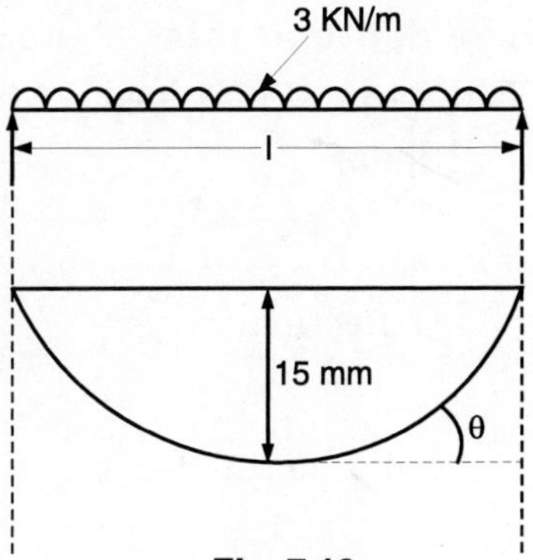

Fig. 7.18

For simply supported beam with *u.d.l.*, maximum slope is given by

$$\theta_{max} = \frac{wl^3}{24EI} = 0.005 \text{ (given)} \quad ...(i)$$

and maximum deflection, $y_{max} = \dfrac{5}{384}\dfrac{wl^4}{EI} = 15$ (given)

$$\frac{5l}{16 \times 24} \times \frac{wl^3}{EI} = 15 \qquad [\because 384 = 16 \times 24]$$

$$\frac{5l}{16} \times \frac{wl^3}{24EI} = 15$$

$$\frac{5l}{16} \times 0.005 = 15 \qquad \text{[using } (i)]$$

$$\therefore \quad l = \frac{15 \times 16}{5 \times 0.005} = 9600 \text{ mm}$$

We have, $\dfrac{M}{I} = \dfrac{\sigma_b}{y}$

or, $M = \dfrac{\sigma_b}{y} \times I = \dfrac{125}{\frac{d}{2}} \times \dfrac{1}{12}bd^3 \qquad \left[\because y = \dfrac{d}{2} \text{ and } I = \dfrac{1}{12}bd^3\right]$

$$= \frac{250 \times bd^3}{d \times 12} = \frac{125}{6}bd^2 \text{ Nmm} \quad ...(ii)$$

But $$M = \frac{wl^2}{8} = \frac{3\times(9600)^2}{8} = 34560000 \text{ Nmm} \quad ...(iii)$$

From (*ii*) and (*iii*), we get

$$\frac{125}{6}bd^2 = 34560000$$

$$\frac{125}{6}\times b\times(1.5b)^2 = 34560000$$

$$b^3 = \frac{34560000\times 6}{125\times 1.5\times 1.5} = 737280 \text{ mm}^3$$

$$\therefore \quad b = 90.34 \text{ mm}$$

$$d = 1.5 \times 90.34 = 135.51 \text{ mm.} \quad \textbf{Ans.}$$

EXAMPLE 7.11. A rolled steel joist rests freely on supports 6 m apart with a point load of 1500 N at mid span. If the maximum permissible bending stress is not to exceed 150 N/mm² and the central deflection not to exceed $\frac{1}{325}$ of span, determine the depth of joist. Take $E = 2 \times 10^5$ N/mm²

Solution: Point load, $W = 1500$ N

Span, $l = 6$ m

$$\sigma_{bmax} = 150 \text{ N/mm}^2$$

Deflection, $$y_{max} = \frac{1}{325}\times \text{span} = \frac{6000}{325} = 18.46 \text{ mm}$$

$$E = 2 \times 10^5 \text{ N/mm}^2$$

Let d be the depth of joist

$$y_{max} = \frac{wl^3}{48EI}$$

or $$I = \frac{wl^3}{48E\times y_{max}} = \frac{1500\times(6000)^3}{48\times 2\times 10^5\times 18.46} = 1828277.4 \text{ mm}^4$$

$$M = \text{Maximum BM at the mid span} = \frac{Wl}{4} = \frac{1500\times 6000}{4} = 2250000 \text{ Nmm}$$

We know $$\frac{M}{I} = \frac{\sigma_b}{y}$$

or, $$y = \frac{\sigma_b\times I}{M} = \frac{150\times 1828277.4}{2250000} = 121.88 \text{ mm} \simeq 122 \text{ mm}$$

$\therefore$ Depth of joist, $$d = 2y = 2\times 122 = 244 \text{ mm.} \quad \textbf{Ans.}$$

EXAMPLE 7.12. A beam simply supported at the two ends having a span 4 m is carrying a point load of 40 KN at a distance of 3m from the left end. Calculate the slope at the two supports and deflection under the load. Also calculate the maximum deflection.

Take $EI = 2.6 \times 10^7$ Nm².

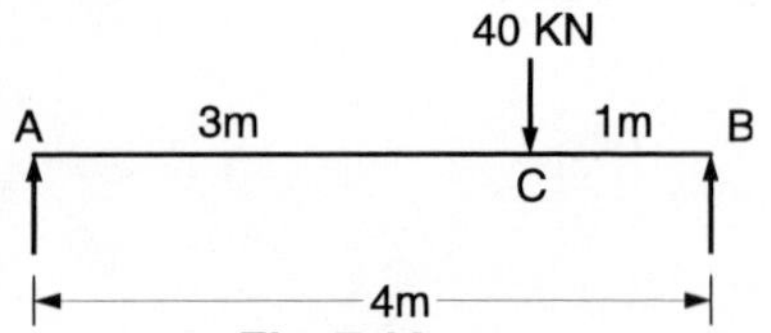

Fig. 7.19

Solution:

$$W = 40 \text{ KN} = 40000 \text{ N}$$
$$l = 4 \text{ m}$$
$$a = 3 \text{ m}$$
$$b = 1 \text{ m}$$
$$EI = 2.6 \times 10^7 \text{ Nm}^2$$

Slope at A,
$$\theta_A = \frac{Wb}{6EIl}(l^2 - b^2) = \frac{40000 \times 1}{6 \times 2.7 \times 10^7 \times 4}\left(4^2 - 1^2\right)$$
$$= 0.000926 \text{ radian} \quad \textbf{Ans.}$$

Slope at B,
$$\theta_B = \frac{Wa(l^2 - a^2)}{6EIl}$$
$$= \frac{40000 \times 3(4^2 - 3^2)}{6 \times 2.6 \times 10^7 \times 4} = \frac{120000 \times 7}{62.4 \times 10^7}$$
$$= 0.001346 \text{ radian} \quad \textbf{Ans.}$$

Deflection under the load,
$$y_c = \frac{Wab}{6EIl}(l^2 - a^2 - b^2)$$
$$= \frac{4000 \times 3 \times 1}{6 \times 2.6 \times 10^7 \times 4} \times (4^2 - 3^2 - 1^2) = 0.0115 \text{ m} \quad \textbf{Ans.}$$

Maximum deflection,
$$y_{max} = \frac{Wa}{9\sqrt{3}EIl}(l^2 - a^2)^{3/2} = \frac{40000 \times 3 \times (4^2 - 3^2)^{3/2}}{9\sqrt{3} \times 2.6 \times 10^7 \times 4}$$
$$= 0.00137 \text{ m. } \textbf{Ans.}$$

EXAMPLE 7.13. A cantilever 3 m long carries two point loads 60 KN each, at a distance of 0.75 m and 1.75 m respectively from the fixed end. Determine the deflection at the free end. Take $E = 200$ GN/m² and $I = 12689400$ cm⁴.

Solution: Method-I : Given :

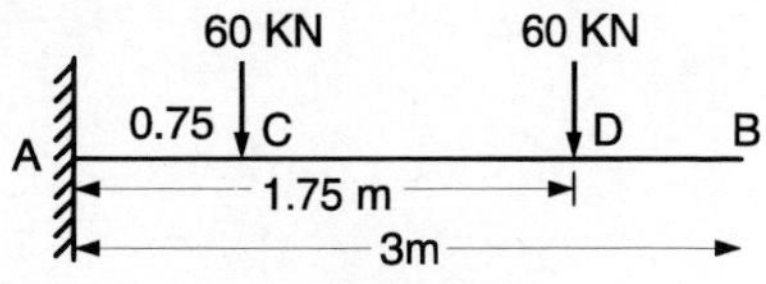

Fig. 7.20

load, $W_1 = 60$ KN

load $W_2 = 60$ KN

$$a = AC = 0.75 \text{ m}$$
$$b = AD = 1.75 \text{ m}$$
$$E = 200 \text{ GN/m}^2 = 200 \times 10^6 \text{ KN/m}^2$$
$$I = 12689400 \text{ cm}^4$$
$$= 12689400 \times 10^{-8} \text{m.}$$

Span, $l = 3$ m

Maximum deflection at free end

y_{max} = Deflection due to W_1 at free end + Deflection due to W_2 at free end

$$= \frac{W_1 a^2}{2EI}\left(l - \frac{a}{3}\right) + \frac{W_2 b^2}{2EI}\left(l - \frac{b}{3}\right)$$

$$= \frac{60 \times (0.75)^2 \times \left(3 - \frac{0.75}{3}\right)}{2 \times 200 \times 10^6 \times 12689400 \times 10^{-8}} + \frac{60 \times (1.75)^2 \times \left(3 - \frac{1.75}{3}\right)}{2 \times 200 \times 10^6 \times 12689400 \times 10^{-8}}$$

$$= \frac{92.81}{50757600} + \frac{444.06}{50757600}$$

$$= 1.05771 \times 10^{-5} \text{ m} = 0.0106 \text{ mm.}$$

Method-II : Refer figure 7.21

Area of BM diagram, $A = A_1 + A_2 + A_3$

$$A_1 = \frac{1}{2} \times 60 \times 1 = 30 \text{ KN m}^2$$

$$A_2 = 0.75 \times 60 = 45 \text{ KN m}^2$$

$$A_3 = \frac{1}{2} \times 0.75 \times 90 = 33.75 \text{ KN m}^2$$

$$A = 30 + 45 + 33.75$$

$$= 108.75 \text{ KN m}^2$$

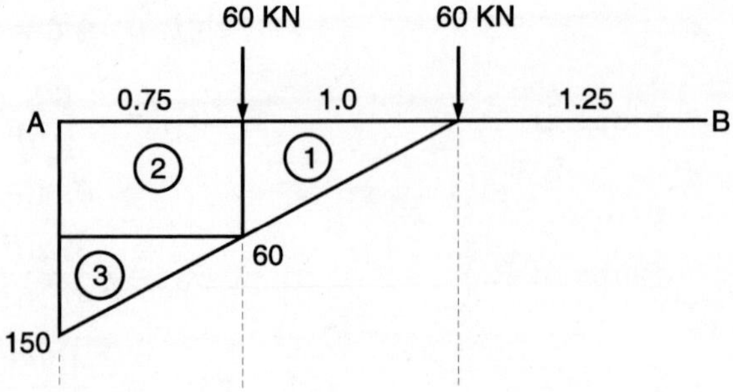

Fig. 7.21 BM Diagram

Moment of area of BMD about B

$$A\overline{X} = A_1x_1 + A_2x_2 + A_3x_3$$

$$= 30 \times \left(1.25 + \frac{2}{3} \times 1\right) + 45\left(1 + 1.25 + \frac{0.75}{2}\right)$$

$$+ 33.75 \times \left(1 + 1.25 + \frac{2}{3} \times 0.75\right)$$

$$A\overline{X} = 30 \times 1.92 + 45 \times 2.625 + 33.75 \times 2.75$$

$$= 57.6 + 118.12 + 92.81$$

$$= 268.53 \text{ KN m}^3$$

Maximum Deflection at the free end,

$$y_{max} = \frac{A\overline{X}}{EI}$$

$$= \frac{268.53}{200 \times 10^6 \times 12689400 \times 10^{-8}}$$

$$= 1.05808 \times 10^{-5} \text{m}$$

$$= 0.0106 \text{ mm} \quad \textbf{Ans.}$$

EXAMPLE 7.14. A cantilever beam with a span of 3 m carries a point load 30 KN at distance 2 m from the fixed end. Determine the slope and deflection at the free end and at the point where load is applied.

Take, $I = 11924 \times 10^{-8} \text{m}^4$ and $E = 2 \times 10^8 \text{ KN/m}^2$.

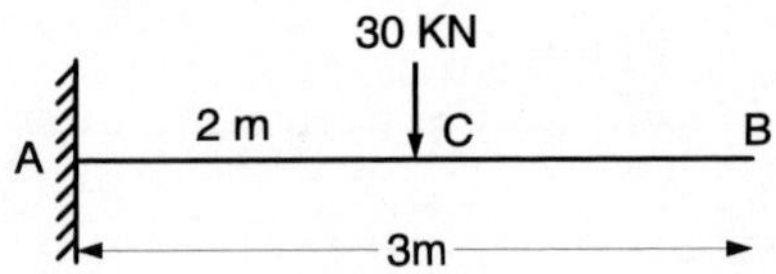

Fig. 7.22

Soluton: Method-I :

Span, $l = 3$ m

Point load, $W = 30$ KN

M.I. $I = 11924 \times 10^{-8} \text{ m}^4$

$E = 2 \times 10^8 \text{ KN/m}^2$

$AC = a = 2$ m

$$\text{Slope at point } B = \text{Slope at point } C = \theta_{max} = \frac{Wa^2}{2EI}$$

$$\therefore \qquad \theta_{max} = \frac{30 \times 2^2}{2 \times 2 \times 10^8 \times 11924 \times 10^{-8}} = 0.002516 \text{ radian} \quad \textbf{Ans.}$$

Deflection at C, $$y_c = \frac{Wa^3}{3EI} = \frac{30 \times 2^3}{3 \times 2 \times 10^8 \times 11924 \times 10^{-8}}$$

$$= 3.354 \times 10^{-3} \text{m} = 3.35 \text{ mm}$$

Maximum deflection, $$y_{max} = \frac{wa^2}{2EI}\left(l - \frac{a}{3}\right)$$

$$= \frac{30 \times 2^2 \left(3 - \dfrac{2}{3}\right)}{2 \times 2 \times 10^8 \times 11924 \times 10^{-8}} = \frac{279.6}{47696}$$

$$= 5.862 \times 10^{-3} = 5.862 \text{ mm}$$

Method-II : Refer figure 7.23

Area of BMD, $A = \frac{1}{2} \times 2 \times 60$

$= 60 \text{ KNm}^2$

Slope at B = Slope at C = θ_{max}

[$\because$ CB remains straight]

$$= \frac{A}{EI} = \frac{60}{2 \times 10^8 \times 11924 \times 10^{-8}}$$

$$= 2.51593 \times 10^{-3} = 0.002516 \text{ radian}$$

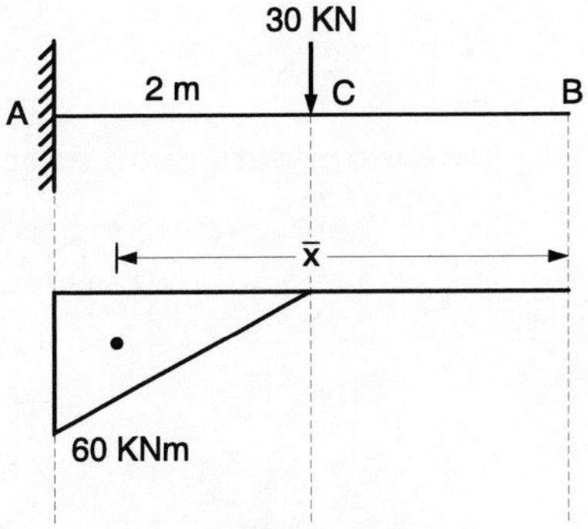

Fig. 7.23 BMD

Deflection at c, $$y_c = \frac{A\overline{X}}{EI}$$

where $\overline{X}$ = Distance of C.G. of BMD from C

$$= \frac{2}{3} \times 2 = \frac{4}{3} = 1.33 \text{ m}$$

$$\therefore \qquad y_c = \frac{60 \times 1.33}{2 \times 10^8 \times 11924 \times 10^{-8}}$$

$$= 3.346 \times 10^{-3} \text{ m} = 3.346 \text{ mm}$$

Maximum deflection at free end,

$$y_{max} = \frac{A\overline{X}}{EI}$$

where $\overline{X}$ = Distance of C.G. of BMD from B

$$= 1 + \frac{4}{3} = \frac{7}{3} = 2.33 \text{ m}$$

$$\therefore \qquad y_{max} = \frac{60 \times 2.33}{2 \times 10^8 \times 11924 \times 10^{-8}}$$

$$= 5.862 \times 10^{-3} \text{ m} = 5.862 \text{ mm. } \textbf{Ans.}$$

EXAMPLE 7.15. A cantilever 2.5 m long is loaded with UDL of 10 KN per m over a length of 1.5 m from the fixed end. Determine the slope and deflection at the free end of the cantilever. Take, I = 9500 cm^4 and $E = 2.1 \times 10^8$ KN/m^2.

Solution: Span, $l = 2.5$ m

u.d.l, $w = 10$ KN/m

$AC = a = 1.5$ m

$I = 9500 \text{ cm}^4 = 9500 \times 10^{-8} \text{ m}^4$

$E = 2.1 \times 10^8 \text{ KN/m}^2$

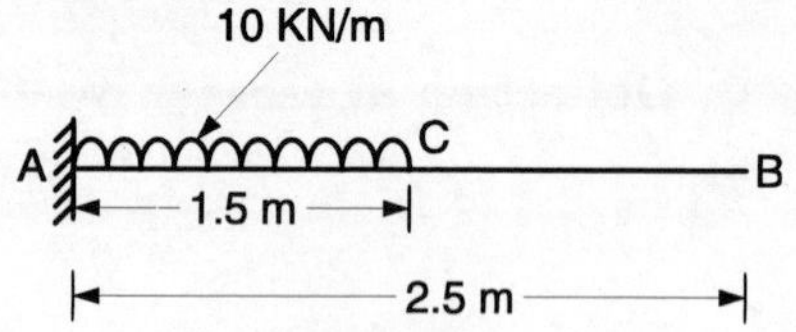

Fig. 7.24

Maximum slope at the free end

$$\theta_{max} = \frac{wa^3}{6EI}$$

$$= \frac{10 \times (1.5)^3}{6 \times 2.1 \times 10^8 \times 9500 \times 10^{-8}}$$

$$= \frac{33.75}{119700} = 2.8195 \times 10^{-4}$$

$$= 0.000282 \text{ radian.}$$

Maximum deflection at the free end

$$y_{max} = \frac{wa^4}{8EI} + \frac{wa^3}{6EI}(l - a)$$

$$= \frac{10 \times (1.5)^4}{8 \times 2.1 \times 10^8 \times 9500 \times 10^{-8}} + \frac{10 \times (1.5)^3 (2.5 - 1.5)}{6 \times 2.1 \times 10^8 \times 9500 \times 10^{-8}}$$

$$= \frac{50.625}{159600} + \frac{33.75}{119700} = 5.99 \times 10^{-4}$$

$$= 0.599 \text{ mm.} \qquad \textbf{Ans.}$$

EXAMPLE 7.16. Determine slope and deflection at A for the beam shown in figure 7.25. ***(Delhi, Dec., 09)***

Take $E = 2 \times 10^5$ **N/mm^2**

$I = 2 \times 10^8$ **mm^4.**

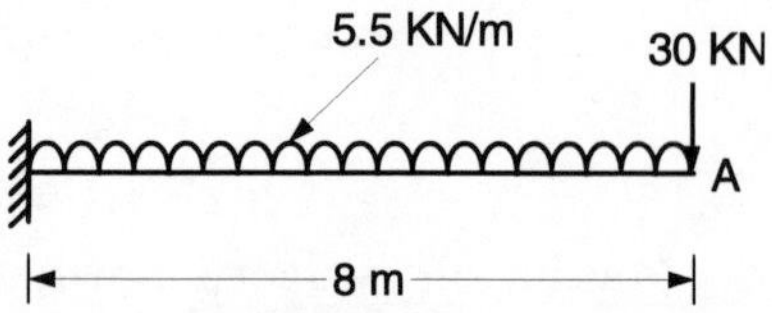

Fig. 7.25

Solution: Given :

Span, $l = 8 \text{ m} = 8000 \text{ mm}$

udl, $w = 5.5 \text{ KN/m} = 5.5 \text{ N/mm}$

Point load, $W = 30 \text{ KN} = 30000\text{N}$

$E = 2 \times 10^5 \text{ N/mm}^2$,

$I = 2 \times 10^8 \text{ mm}^4$

Maximum slope and deflection will be at the free end A.

$$\theta_A = \theta_{max} = \text{Slope due to point load + Slope due to } udl$$

$$= \frac{Wl^2}{2EI} + \frac{wl^3}{6EI} = \frac{30000 \times (8000)^2}{2 \times 2 \times 10^5 \times 2 \times 10^8} + \frac{5.5 \times (8000)^3}{6 \times 2 \times 10^5 \times 2 \times 10^8}$$

$$= \frac{24 \times 10^{10}}{8 \times 10^{13}} + \frac{2816 \times 10^9}{24 \times 10^{13}} = 0.003 + 0.0117$$

$$= 0.0147 \text{ radian} \qquad \textbf{Ans.}$$

$$y_A = y_{max} = \text{Deflection due to point load + Deflection due to } udl$$

$$= \frac{Wl^3}{3EI} + \frac{wl^4}{8EI} = \frac{30000 \times (8000)^3}{3 \times 2 \times 10^5 \times 2 \times 10^8} + \frac{5.5 \times (8000)^4}{8 \times 2 \times 10^5 \times 2 \times 10^8}$$

$$= \frac{1536 \times 10^{13}}{12 \times 10^{13}} + \frac{4096 \times 10^{12}}{32 \times 10^{13}} = 128 + 12.8 = 140.8 \text{ mm } \textbf{Ans.}$$

7.7 MACAULAY'S METHOD

Macaulay's method is suitable for a beam:

(*i*) having *a u.d.l* extended only on a part of the span.

(*ii*) carrying more than one point loads.

(*iii*) for overhanging beams.

The following rules are followed in macaulay's method:

(*i*) while integrating the terms within the brackets are not to be disturbed and should be kept as it is *i.e.,* integration of $(x - a)$ is written as $\frac{(x-a)^2}{2}$.

(*ii*) If for any value of x, the quantity within the square bracket in any term is negative, the term should be neglected.

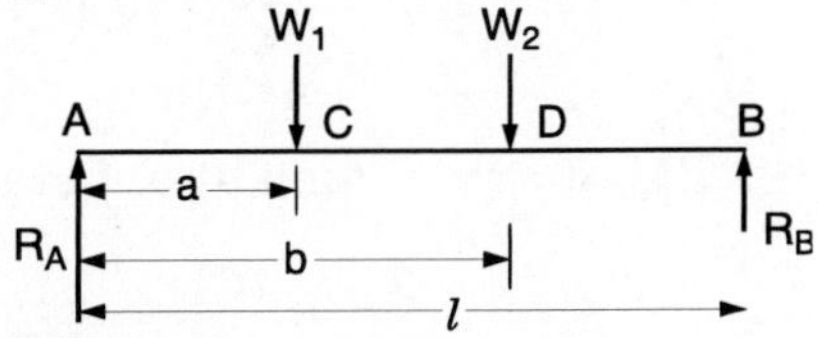

Fig. 7.26

Figure 7.26 shows a simply supported beam loaded with point loads w_1 and w_2 at C and D respectively.

Consider a section $x - x$ between A and C at a distance x from A

$$M_x = R_A x \quad ; x \text{ varies from } 0 \text{ to } a$$

Consider a section $x - x$ between CD at a distance x from A.

$$M_x = R_A x - W_1(x - a) \; ; x \text{ varies from } a \text{ to } b$$

Similarly ; $B.\ M$ at a distance x from A in portion DB

$$M_x = R_A x - W_1(x - a) - W_2(x - b) \; ; x \text{ varies from } b \text{ to } l$$

In general at any section the bending movement is given by

$$M_x = \frac{EId^2y}{dx^2} = R_A x \,|- w_1(x-a)\,| - w_2(x-b) \quad ..(i)$$

(For any section in portion AC, stop at the first partition line; for any section in portion CD, stop at the 2nd partition line and so on. In this manner the above expression is written)

Integrating eqn (*i*) we get the general expression for slope

$$EI\frac{dy}{dx} = R_A \frac{x^2}{2} + C_1 \left| - \frac{W_1(x-a)^2}{2} \right| - \frac{W_2(x-b)^2}{2} \quad ...(ii)$$

Further integrating we get general expression for deflection

$$EIy = R_A \frac{x^3}{6} + c_1 x + c_2 \left| - \frac{W_1(x-a)^3}{6} \right| - \frac{W_2(x-b)^3}{6} \quad ...(iii)$$

Constants C_1 and C_2 are written after the first term.

To find the values of C_1 and C_2 the following boundary conditions are used:

(*i*) The deflection at A and B is zero *i.e.,* at $x = 0$ and $x = l$, $y = 0$

Putting $x = 0$ and $y = 0$ in eqn. (*iii*) we get $C_2 = 0$

Putting $x = l$ and $y = 0$ in eqn. (*iii*) we get the value of C_1

By putting the value of C_1 and C_2 in slope and deflection equations we can calculate the slope and deflection at any point in the given beam.

EXAMPLE 7.17. A beam 4 m long is supported at its ends carries concentrated load of 10 KN, 15 KN and 20 KN at each 1 m from the LHS support. Calculate the slope and deflection under each point load. Take $EI = 2.6 \times 10^7$ N-m².

(Delhi, Jan. 2009)

Solution: Let R_A and R_B be the reactions at A and B

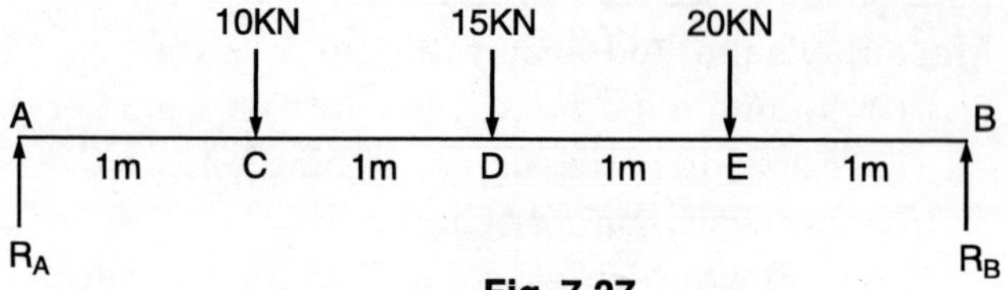

Fig. 7.27

$$\Sigma F_y = 0$$

$$R_A + R_B = 10 + 15 + 20$$

$$\therefore \quad R_A + R_B = 45\ KN \qquad ...(i)$$

$$\Sigma M_B = 0$$

$$R_A \times 4 = 10 \times 3 + 15 \times 2 + 20 \times 1 = 80$$

$$R_A = \frac{80}{4} = 20 \text{ KN}$$

$$\therefore \quad R_A = 25 \text{ KN}$$

BM at any section at a distance x from A is given by

$$EI\frac{d^2y}{dx^2} = 20 \times x \;|- 10\,(x-1)\,|\; - 15\,(x-2)\,|\; - 20\,(x-3)$$

Integrating, we get

$$EI\frac{dy}{dx} = \cdot\frac{20x^2}{x} + c_1 \left| -\frac{10(x-1)^2}{2} \right| - \frac{15(x-2)^2}{2} \left| - \frac{20(x-3)^2}{2} \right. \qquad ...(i)$$

Integrating again, we get

$$EIy = \frac{10x^3}{3} + c_1x + x_2 \left| -\frac{5}{3}(x-1)^3 \right| - \frac{5}{2}(x-2)^3 \left| - \frac{10}{3}(x-3)^3 \right. \qquad ...(ii)$$

At $\quad x = 0,\ y = 0$

From (*ii*), we get $\quad C_2 = 0$

At $\quad x = 4,\ y = 0$

From (*iii*), we get
$$0 = \frac{10 \times 4^3}{3} + c_1 \times 4 + 0 - \frac{5}{3}(4-1)^3 - \frac{5}{2}(4-2)^3 - \frac{10}{3}(4-3)^3$$

$$0 = \frac{640}{3} + 4c_1 - 45 - 20 - \frac{10}{3}$$

$$0 = \frac{630}{3} - 65 + 4c_1$$

$$0 = 145 + 4c_1$$

$$\therefore \quad 1c_1 = \frac{-145}{4}\cdot$$

Slope at ($x = 1$ m) c, θ_c

$$EI\frac{dy}{dx} = 10x^2 + C_1 = 10 \times 1^2 - \frac{145}{4} \qquad EI\frac{dy}{dx} = \frac{-105}{4}$$

$$\theta_c = \frac{dy}{dx} = \frac{-26.25}{EI} = \frac{-26.25}{2.6 \times 10^7} = -1.0096 \times 10^{-6} \text{ radian.}$$

Deflection at C,
$$y_c = \frac{1}{EI}\left[\frac{10}{3}(1)^3 - \frac{145}{4}\right]$$

$$= \frac{1}{2.6 \times 10^7}\left[\frac{10}{3} - \frac{145}{4}\right] = \frac{1}{2.6 \times 10^7}[3.33 - 36.25]$$

$$= \frac{-32.92}{2.6 \times 10^7} = -1.2661 \times 10^{-6} \text{ m}$$

Slope at D,
$$\theta_D = \frac{1}{EI}\left[10(2)^2 - \frac{145}{4} - 5(2-1)^2\right]$$

$$= \frac{1}{2.6 \times 10^7}[40 - 36.25 - 5] = \frac{-1.25}{2.6 \times 10^7}$$

$$= -4.808 \times 10^{-8}$$

Deflection at D,
$$y_D = \frac{1}{EI}\left[\frac{10}{3} \times (2)^3 - \frac{145}{4} \times 2 - \frac{5}{3}(2-1)^3\right]$$

$$= \frac{1}{2.6 \times 10^7}\left[\frac{80}{3} - \frac{145}{2} - \frac{5}{3}\right]$$

$$= \frac{1}{2.6 \times 10^7}[26.67 - 72.5 - 1.67]$$

$$= \frac{-47.5}{2.6 \times 10^7} = -1.8269 \times 10^{-6} \text{ m}$$

Slope at E,
$$\theta_E = \frac{1}{EI}\left[10(3)^2 - \frac{145}{4} - 5(3-1)^2 - \frac{15}{2}(3-2)^2\right]$$

$$= \frac{1}{2.6 \times 10^7}\left[90 - \frac{145}{4} - 20 - \frac{15}{2}\right]$$

$$= \frac{1}{2.6 \times 10^7}(90 - 36.25 - 20 - 7.5)$$

$$= \frac{1}{2.6 \times 10^7}(90 - 63.75) = \frac{26.25}{2.6 \times 10^7} = 1.0096 \times 10^{-6}$$

Deflection at E,
$$y_E = \frac{1}{2.6 \times 10^7}\left[\frac{10}{3} \times 3^3 - \frac{145}{4} \times 3 - \frac{5}{3}(3-1)^2 - \frac{5}{2}(3-2)^3\right]$$

$$= \frac{1}{2.6 \times 10^7}[90 - 108.75 - 6.67 - 2.5]$$

$$= \frac{-27.92}{2.6 \times 10^7} = -1.0738 \times 10^{-6} \text{ m}$$

SUMMARY

1. Slope at any section of a beam is the angle which the tangent at the section makes with the horizontal.
2. Deflection at any point on a beam is the vertical distance between elastic curve of the loaded beam and the N.A at that point.
3. In moment area method (Mohr's theorem)

$$\theta_{max} = \frac{A}{EI} \cdot \qquad \text{(Theorem I)}$$

$$y_{max} = \frac{A\bar{x}}{EI} \cdot \qquad \text{(Theorem II)}$$

4. In simply supported beam
   (*i*) For point load at mid span

$$y_{max} = \frac{Wl^4}{48EI}, \theta_{max} = \frac{Wl^2}{16EI} \cdot$$

   (*ii*) For U.D.L over the entire span

$$\theta_{max} = \frac{Wl^3}{24EI}, y_{max} = \frac{5}{384}\frac{Wl^4}{EI} \cdot$$

5. In cantilevers
   (*i*) For point load at free end

$$\theta_{max} = \frac{Wl^2}{2EI}, y_{max} = \frac{Wl^3}{3EI} \cdot$$

   (*ii*) For U.D.L over entire span

$$\theta_{max} = \frac{Wl^3}{6EI}, y_{max} = \frac{Wl^4}{8EI} \cdot$$

6. Macaulay's method is suitable for determining the deflections of a beam subjected to point loads or discontinuous loads. It is the modification of double integration method.
7. According Macaulay's method, slope at any section is given by

$$\theta_x = \frac{1}{EI}\left[R_A\frac{x^2}{2} + c_1 \left| - \frac{W_1(x-a)^2}{2} \right| - \frac{W_2(x-b)^2}{2}\right]$$

Deflection at x

$$y_x = \frac{1}{EI}\left[R_A\frac{x^3}{6} + c_1x + c_2 \left| - \frac{W_1(x-a)^3}{6} \right| - \frac{W_2(x-b)^3}{6}\right]$$

OBJECTIVE TYPE QUESTIONS

I. Fill in the Blanks:

(*i*) For simply supported beam with point load at the centre, the maximum slope will be at the...........

(*ii*) For simply supported beam with point load at the centre, the maximum deflection will be at the..........

(*iii*) For cantilevers, the maximum deflection will be at the..........

(*iv*) Slope and deflection are both zero in cantilever beam at..........end.

(*v*) In moment area method, the deflection is equal to ....

(*vi*) The amount of deflection of a beam subjected to some type of loading depends upon...........

(*vii*) The deflection at the free end of a cantilever of length *l*' carrying a point load '*W*' at its free end is given as.........

(*viii*) The ratio of maximum deflection of a beam to its span is known as..........

(*ix*) The angle which the tangent at the section makes with the horizontal is .... at any section of a beam.

(*x*) The vertical intercept between the bent and unbent beam at any point is called...........

Answers

(*i*) supports, (*ii*) centre, (*iii*) free end, (*iv*) fixed, (*v*) $\frac{A\bar{X}}{EI}$, (*vi*) bending moment and cross section, (*vii*) $\frac{wl^3}{3EI}$, (*viii*) stiffness, (*ix*) slope, (*x*) deflection.

THEORETICAL QUESTIONS

1. Define slope and deflection and state its importance. (*Delhi, December, 2009*)
2. What are the difficult methods for finding slope and deflection?
3. What is moment area method? Write its suitability.
4. Derive an expression for maximum deflection at the free end for a cantilever beam carrying point load '*W*' at the free end.
5. Define Mohr's theorem for finding slope and deflection.

NUMERICALS

1. A simply supported beam of length 5 m carries a u.d.l of 9 KN/m over the entire length. It also carries a concentrated load of 20 KN at the centre of span. Calculate the maximum slope and deflection of the beam.
Take $E = 2 \times 10^5$ N / mm^2 and $I = 85 \times 10^6$ mm^4 (0.0046 radian ; 7.37 mm)

2. A wooden beam 150 mm wide and 250 mm deep has a span of 4 metres. Determine the load, that can be placed at its centre to cause the beam a deflection of 12 mm. Also find the slope.
Take $E = 6 \times 10^6$ KN/m^2 (10. 55 KN ; 0.009 radian)

3. A beam is simply supported at ends on a span of 3 m and carries a u.d.l of 5 KN/m on the whole span. What point load at the centre it should carry so that the maximum deflection is doubled? (18750 N)

4. A simply supported beam of I-section 4 m long carries a total uniform load of 40 KN and a concentrated load of 70 KN at mid span.

(*i*) Find the maximum deflection of the beam.

(*ii*) If permissible deflection is limited to $\frac{1}{360}$ of the span, is this beam acceptable based on deflection?

(*iii*) Find slope at the ends

Take $E = 2.1 \times 10^8$ KN/m^2 and $I = 8.98 \times 10^{-5}$ m^4

(6.717 mm ; acceptable ; 0.005126 radian)

5. A beam of 4 m span is carrying a point load of 40 KN at a distance of 3 m from the left end. Calculate the slope at the two supports and deflection udner the load. Also calculate maximum deflection. Take $EI = 2.6 \times 10^7$ N/m^2. (0.0009615 rad ; 0.001346 rad ; 1.154 mm; 1.37 mm)

6. A simply supported beam of 5 m span has rectangular section 150 mm wide and 250 mm deep. The beam carries a point load of 10 KN at its centre and the deflection at the centre is 12.5 mm. Neglecting the self-weight of the beam calculate:
(*i*) Young's modulus (*ii*) slope at supports (1.06 × 10^8 KN / m^2 ; 0.000754 radian)

7. A simply supported beam 5 m long carries concentrated loads of 10 KN each a points 1 m from the ends. Calculate the maximum slope and deflection of the beam. Also calculate slope and deflection under each load.
   Take $EI = 1.2 \times 10^4$ KN m^2 (0.001666 rad ; 2.465 mm ; 0.001249 rad ; 1.5625 mm)
8. A cantilever 3 m long carries a point load of 50 KN at its free end. Calculate maximum slope and deflection.
   Take $I = 250 \times 10^6$ mm^4 and $E = 2 \times 10^5$ N / mm^2 (0.0045 radian ; 9 mm)
9. A steel cantilever projecting 3 m from a wall is loaded with a u.d.l of 20 KN / m over the entire span. Find the slope and deflection of the beam.
   Take $I = 7550$ cm^4 and $E = 210 \times 10^9$ N / m^2.
10. A simply supported R.C.C rectangular beam of span 4 m and cross-section 150 mm × 300 mm is loaded with u.d.l. of 10 KN/m over the entire span and a point load of 50 KN at the mid span. Find the maximum slope and deflection of the beam.
    Take $E = 2 \times 10^4$ N/ mm^2 (0.0114 radian ; 14.82 mm)
11. A cantilever of span 3 m is loaded with point load of 30 KN at a distance 2 m from the fixed end. b Calculate the maximum slope and deflection of the beam.
    Take $E = 2 \times 10^4$ N/mm^2 and $I = 2.25 \times 10^8$ mm^4 (0.013 radian ; 31.13 mm)
12. A cantilever of span 3 m is loaded with point load of 20 KN at the free end and a point of 10 KN at a distance 1 m from the fixed end. Calculate the maximum slope and deflection of the beam.
    Take $I = 3.375 \times 10^8$ mm^4 and $E = 2 \times 10^4$ N / mm^2 (0.014 radian ; 28.7 mm)
13. A cantilever beam of I - section projects 2.5 m out of the wall and supports a u.d.l of 20 KN/m for a length of 1.5 m from the fixed end. Calculate the maximum deflection and slope in the cantilevers.
    Take $E = 2 \times 10^5$ N / mm^2 and $I = 95 \times 10^6$ mm^4. (0.000592 rad ; 1.258 mm)

CHAPTER 8

Columns and Struts

8.0 STRUTS AND COLUMN

Any member of structure subjected to axial compressive force is known as strut. A strut may be horizontal or vertical. A strut inclined at 90° to the horizontal (*i.e.*, a vertical strut) is known as column, pillar or stanchion.

Chapter Outline

8.1 DEFINITIONS

(*a*) **Column :** A long vertical slender member or bar subjected to an axial compressive force is known as column.

(*b*) **Strut :** A slender member in any position other than vertical subjected to axial compressive force is known as strut.

(*c*) **Slenderness ratio :** It is the ratio of length of column to the least radius of gyration. Slenderness ratio has no unit.

Slenderness ratio =

$$\frac{\text{Length of column}}{\text{Least radius of gyration}} = \frac{L}{K}$$

It represents the extent to which the column is long and slender. As the slenderness ratio of a column increases its compressive strength decreases. A slenderness ratio of 200 is extremely large for a column.

(*d*) **Buckling* load :** The maximum axial compressive load at which the column starts buckling is known as buckling load or crippling load or critical load. Buckling always takes place about the axis having least moment of inertia. The value of buckling load is less than the crushing load.

*When a column bends and deflects laterally under axial compressive load, then we say that the column has buckled?

(*e*) **Safe load :** It is the load which a column can withstand safely without any buckling/ failure. It is the ratio of buckling load and factor of safety.

$$\text{Safe load} = \frac{\text{Buckling load}}{\text{Factor of safety}}$$

(*f*) **Buckling factor :** Buckling factor is the ratio of equivalent length to the least radius of gyration.

8.2 CLASSIFICATION OF COLUMNS

Depending upon the length to diameter ratio or slenderness ratio, a column can be classified as:

(*i*) **Short column :** If the length of a column is less than 8 times its least lateral dimension then the column is said to be a short column. If the slenderness ratio of a column is less than 32, then the column is also called a short column. In short columns, buckling is negligible and the column fails due to direct crushing only.

(*ii*) **Medium size column :** A column is said to be medium size column if its length varying from 8 times to 30 times of its least lateral dimension. If the slenderness ratio of a column lies between 32 to 120, then the column is also said to be medium size or intermediate column. In these columns, both buckling as well as direct stresses are of significant value. In the design of medium size column, both the stresses are taken into account.

(*iii*) **Long column :** A column is said to be long column if its length is more than 30 times the least lateral dimension or the slenderness ratio of the column is more than 120. In this type of column, direct compressive stress is very small as compared to buckling stress, *i.e.*, the failure is only due to buckling stress. Hence, long columns are designed to withstand buckling stresses.

8.3 FAILURE OF COLUMNS

The failure of a column takes place due to any of the following stresses:

(*i*) Direct compressive stress

(*ii*) Buckling stress

(*iii*) Combination of direct and buckling stresses.

(*i*) Failure of short column : Consider a short column of cross-sectional area '*A*' subjected axial compressive load *P*, then

$$\text{Compressive stress} = \frac{P}{A}$$

If the compressive load is gradually increased, a stage will be reached at which the column will be at a point of failure by crushing. The stress induced in the column corresponding to this load is called *crushing stress* and the load is known as *crushing load* (P_c).

$$\therefore \quad \text{Crushing stress} = \frac{\text{Crushing load}}{\text{Cross section area}} = \frac{P_c}{A}$$

(*ii*) Failure of a long column : Figure 8.1 shows a long column of cross-sectional area '*A*' subjected to an axial compressive load *P*. In case of long column, the failure is due to crushing as well as buckling (bending). The load at which the column just buckles is known as buckling load. The buckling load is less than the crushing load.

Let, P = Axial compressive load

l = length of the column

A = Cross-sectional area

e = Maximum bending at the centre of the column

z = Section modulus in the axis of bending

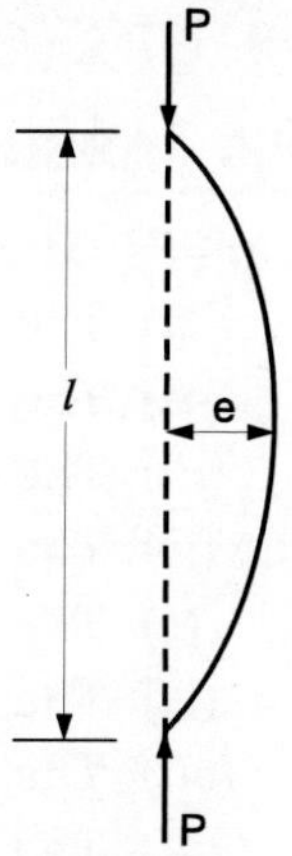

Fig. 8.1.

Direct stress, $\sigma_d = \dfrac{P}{A}$

Bending stress, $\sigma_b = \dfrac{P.e}{z}$

Stresses at the mid-section of the column are:

Maximum stress, $\sigma_{max} = \sigma_b + \sigma_d$

Minimum stress, $\sigma_{min} = \sigma_b - \sigma_d$.

If the maximum stress is more than crushing stress, the column will fail. In long columns, direct compressive stress is negligible as compared to buckling stress. Hence, very long columns are subjected to buckling stresses only.

8.4 END CONDITIONS OF COLUMN

The following end conditions of columns are important:

(*i*) Both the ends hinged [Figure 8.2(*i*)].

(*ii*) Both the ends fixed [Figure 8.2(*iii*)].

(*iii*) One end fixed and other end hinged [Figure 8.2 (*ii*)].

(*iv*) One end fixed and other free [Figure 8.2(*iii*)].

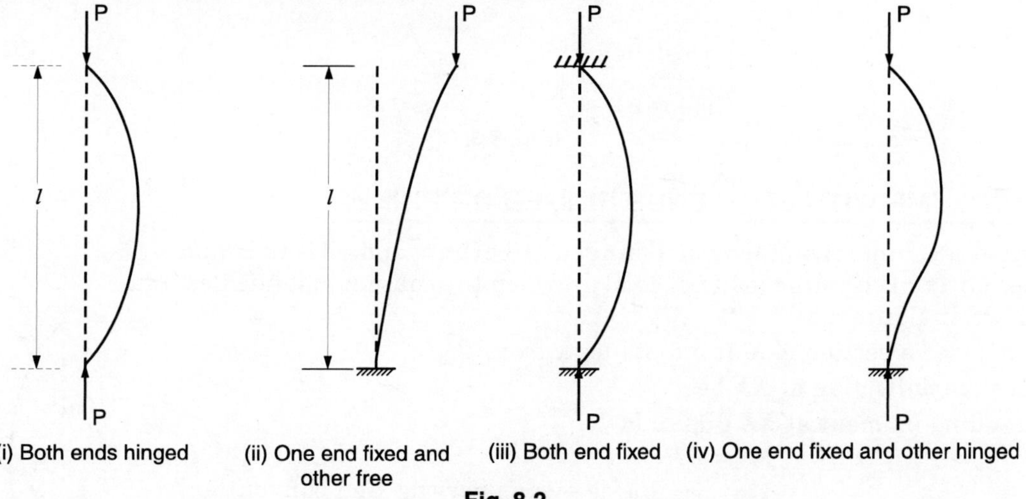

Fig. 8.2.

8.5 FORMULAE FOR FINDING BUCKLING LOAD IN COLUMNS

The following formulae are used to find out buckling load in columns:

(*i*) Euler's formula

(*ii*) Rankine Gordon formula

(*iii*) Johnson's parabolic formula

(*iv*) IS formula

(v) Straight line formula.

8.6 ASSUMPTIONS MADE IN THE EULER'S THEORY

Euler's formula for crippling load is based on the following assumptions:

(i) The section of the column is uniform through out its length.

(ii) The column is initially perfectly straight and axially loaded.

(iii) The column material is homogeneous and isotropic.

(iv) The column material is perfectly elastic and obeys Hooke's law.

(v) The length of the column is very large as compared to the lateral dimensions.

(vi) The self weight of the column is neglected.

(vii) The direct stress is very small as compared to bending strees.

(viii) The column will fail by buckling alone.

Sign Convention for Bending Moment : A bending moment is taken as positive if it bends the column with its convexity towards the actual centre line as shown in figure 8.3(i).

A bending moment is taken as negative if it bends the column with its concavity towards the actual centre line, as shown in figure 8.3 (ii).

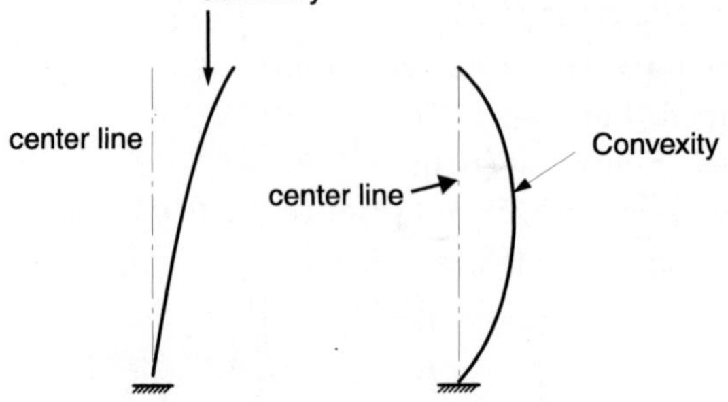

(i) + ve B.M. (ii) – ve B.M.

Fig. 8.3.

8.7 COLUMN WITH BOTH ENDS HINGED OR PINNED

Consider a column AB of length 'l' hinged at both its ends A and B and carries an axial crippling load *i.e.*, load at which the column just buckles, as shown in figure 8.4.

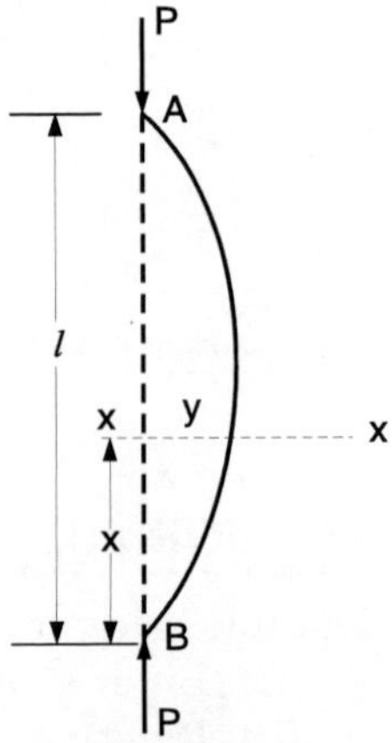

Fig. 8.4.

Consider a section X-X at a distance x from B.

Let the deflection at XX be y.

Bending moment at XX due to P,

$$M = -Py$$

[$\because$ BM is –ve as per the sign convention]

We know,
$$EI\frac{d^2y}{dx^2} = M = -Py$$

$$EI\frac{d^2y}{dx^2} = -Py$$

$$\frac{d^2y}{dx^2} = -\frac{Py}{EI} = -K^2y$$

where $$K^2 = \frac{P}{EI}\cdot$$

$$\therefore \quad \frac{d^2y}{dx^2} + K^2y = 0$$

Solution of this differential equation is given by

$$y = A\cos Kx + B\sin Kx$$

where A and B are constants

or $$y = A\cos\left(\sqrt{\frac{P}{EI}}\right) + B\sin\left(x\sqrt{\frac{P}{EI}}\right)$$

At B, $y = 0, \quad x = 0$

$\therefore \quad A = 0$

At A, $y = 0$, $x = l$

$$\therefore \quad 0 = B\sin l\sqrt{\frac{P}{EI}}$$

or $$\sin\left(l\sqrt{\frac{P}{EI}}\right) = 0$$

$$\therefore \quad l\sqrt{\frac{P}{EI}} = 0, \pi, 2\pi, 3\pi \ldots$$

Considering the least significant value, we get

$$l\sqrt{\frac{P}{EI}} = \pi$$

or $$\sqrt{\frac{P}{EI}} = \frac{\pi}{l}$$

Squaring both sides we get,

$$\frac{P}{EI} = \frac{\pi^2}{l^2}$$

$$\therefore \quad P = \frac{\pi^2 EI}{l^2}\cdot$$

8.8 COLUMN WITH ONE END FIXED AND OTHER END FREE

Consider a column AB of length 'l' with one of its end fixed and the other end free. The column is loaded with a crippling load 'P'. Due to the load, the deflection at A be 'a'.

Consider a section X-X at a distance 'x' from B.

Let the deflection at XX be y.

Bending moment at X-X due to load P,

$$M = +P(a - y)$$
[$\because$ BM is +ve as per sign convention]

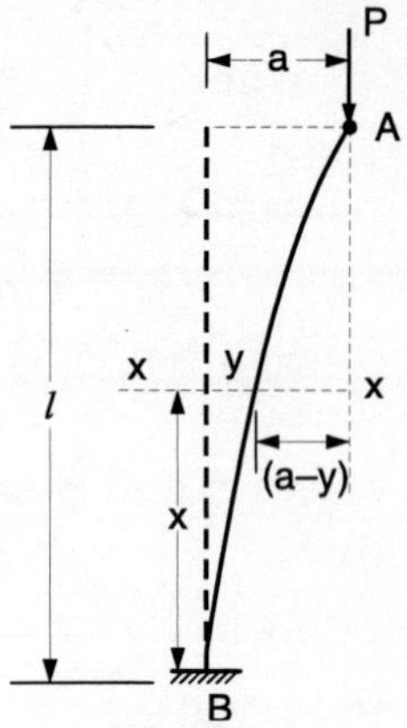

Fig. 8.5.

We know, $$EI \frac{d^2y}{dx^2} = M = P(a - y)$$

$$\frac{d^2y}{dx^2} = \frac{P}{EI}(a - y)$$

$$\frac{d^2y}{dx^2} = \frac{P}{EI}a - \frac{Py}{EI}$$

$$\therefore \quad \frac{d^2y}{dx^2} + \frac{Py}{EI} = \frac{Pa}{EI}$$

The solution of this differential equation is

$$y = A\cos(x\sqrt{P/EI}) + B\sin(x\sqrt{P/EI}) + a \quad \text{...(i)}$$

where A and B are constants

At B, $x = 0, \quad y = 0$

Equation (*i*) becomes

$$0 = A + 0 + a$$

$$\therefore \quad A = -a$$

Differentiating (*i*) w.r.t. x

$$\frac{dy}{dx} = -A\sqrt{P/EI}\sin(x\sqrt{P/EI}) + B\sqrt{P/EI}\cos(x\sqrt{P/EI})$$

At B, $x = 0$ and $\frac{dy}{dx} = 0$

$$0 = 0 + B\sqrt{P/EI}$$

$$\therefore \quad B = 0 \qquad \left[\because \sqrt{P/EI} \neq 0\right]$$

Putting the value of A and B in (*i*), we get

$$y = -a\cos(x\sqrt{P/EI}) + a \quad \text{...(ii)}$$

At B, $x = l, \quad y = a$

Equation (*ii*) becomes

$$a = -a\cos(l\sqrt{P/EI}) + a$$

$$\therefore \quad \cos(l\sqrt{P/EI}) = 0 = \cos\frac{\pi}{2}, \cos\frac{3\pi}{2}, \cos\frac{5\pi}{2}, \ldots$$

$$l\sqrt{P/EI} = \frac{\pi}{2}, \frac{3\pi}{2}, \frac{5\pi}{2}, \ldots$$

Considering the least significant value, we get

$$l\sqrt{P/EI} = \frac{\pi}{2}$$

$$\sqrt{P/EI} = \frac{\pi}{2l}$$

Squaring both sides, we get

$$\frac{P}{EI} = \frac{\pi^2}{4l^2}$$

$$\therefore \quad P = \frac{\pi^2 EI}{4l^2}.$$

8.9 COLUMN WITH BOTH ENDS FIXED

Consider a column AB of length 'l' fixed at both ends and carries an axial crippling load 'P'. Due to this crippling load 'P', the column deflects as shown in figure 8.6.

As the ends are fixed, fixed end moment M_0 is acting at both ends.

Consider a section X-X at a distance 'x' from B. Let the deflection at XX be y.

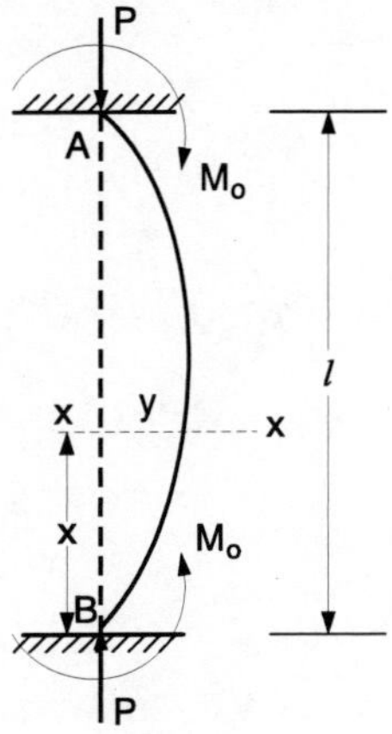

Fig. 8.6.

Bending moment at $XX = M_0 - Py$

[$\because$ Bending moment due to load 'P' is – ve as per sign convention]

We know,

$$EI \frac{d^2y}{dx^2} = M = M_0 - Py$$

or

$$\frac{d^2y}{dx^2} = \frac{M_0}{EI} - \frac{P}{EI} y$$

or

$$\frac{d^2y}{dx^2} + \frac{P}{EI} y = \frac{M_0}{EI}$$

[Multiplying and dividing RHS by P]

$$\frac{d^2y}{dx^2} + \frac{P}{EI} y = \frac{M_0}{EI} \times \frac{P}{P}$$

$$\frac{d^2y}{dx^2} + \frac{P}{EI} y = \frac{P}{EI} \times \frac{M_0}{P} \qquad ...(i)$$

The solution of differential equation is given by

$$y = A \cos(x\sqrt{P/EI}) + B \sin(x\sqrt{P/EI}) + \frac{M_0}{P}$$

where A and B are constants.

At B, $\quad x = 0, \quad y = 0$

Equation (i) becomes

$$0 = A + 0 + \frac{M_0}{P}$$

$$\therefore \quad A = -\frac{M_0}{P}$$

Differentiating equation (*i*) w.r.t. *x*, we get

$$\frac{dy}{dx} = -A\sqrt{P/EI}\sin x(\sqrt{P/EI}) + B\sqrt{P/EI}\cos(x\sqrt{P/EI})$$

At *B*, $x = 0, \quad \frac{dy}{dx} = 0$

$$0 = 0 + B\sqrt{P/EI}$$

$\therefore$ $B = 0$ $\qquad \left[\because \sqrt{P/EI} \neq 0\right]$

Putting the value of *A* and *B* in (*i*) we get

$$y = -\frac{M_0}{P}\cos(x\sqrt{P/EI}) + 0 + \frac{M_0}{P}$$

$$y = \frac{M_0}{P}\left[1 - \cos\left(x\sqrt{P/EI}\right)\right] \qquad ...(ii)$$

At, *A*, $x = l, \quad y = 0$

$$0 = \frac{M_0}{P}\left[1 - \cos(l\sqrt{P/EI})\right]$$

$$1 - \cos(l\sqrt{P/EI}) = 0$$

$$\cos(l\sqrt{P/EI}) = 1 = \cos 0°.2\pi, 4\pi, ...$$

$$l\sqrt{P/EI} = 0, 2\pi, 4\pi, 6\pi, ...$$

Considering the least significant value

$$l\sqrt{P/EI} = 2\pi$$

$$\sqrt{P/EI} = \frac{2\pi}{l}$$

Squaring both sides, we get

$$\frac{P}{EI} = \frac{4\pi^2}{l^2}$$

$$\therefore \quad P = \frac{4\pi^2 EI}{l^2}.$$

8.10 COLUMN IS FIXED AT ONE END AND HINGED AT THE OTHER END

Consider a column *AB* of length '*l*' whose upper end is hinged and the lower end is fixed. The column is subjected to crippling load '*P*'. As the column is fixed at *A*, a fixed end moment M_0 acts at *A*. In order to balance the fixed end moment at *B*, the hinge at *A* will exert a horizontal force '*H*' at *A*, as shown in Figure 8.7.

Consider a section *X-X* at a distance *x* from *B*. Let the deflection at *XX* be *y*.

Bending moment at *XX* = Moment due to *P* + Moment due to *H*.

$$M = -Py + H(l - x)$$

We know,

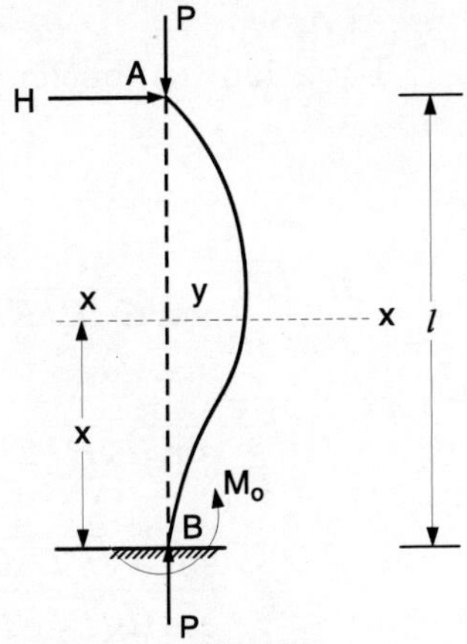

Fig. 8.7.

$$EI\frac{d^2y}{dx^2} = M = -Py + H(l - x)$$

$$EI\frac{d^2y}{dx^2} + Py = H(l - x)$$

Dividing both sides by EI, we get

$$\frac{d^2y}{dx^2} + \frac{P}{EI}y = \frac{H}{EI}(l - x)$$

$$\frac{d^2y}{dx^2} + \frac{P}{EI}y = \frac{P}{EI} \times \frac{H}{P}(l - x)$$

The solution of differential equation is given by

$$y = A\cos(x\sqrt{P/EI}) + B\sin(x\sqrt{P/EI}) + \frac{H}{P}(l - x) \quad ...(i)$$

where A and B are constants

At B, $x = 0, \quad y = 0$

The equation (i) becomes

$$0 = A + \frac{H}{P}(l - 0)$$

$$\therefore \quad A = -\frac{H}{P}l$$

Differentiating equation (i), w.r.t. x, we get

$$\frac{dy}{dx} = -A\sqrt{P/EI}\sin(x\sqrt{P/EI}) + B\sqrt{P/EI}\cos(x\sqrt{P/EI}) - \frac{H}{P} \quad ...(ii)$$

At, B, $x = 0, \quad \frac{dy}{dx} = 0$

Equation (ii) becomes

$$0 = 0 + B\sqrt{P/EI} - \frac{H}{P}$$

or

$$B\sqrt{\frac{P}{EI}} = \frac{H}{P}$$

$$\therefore \quad B = \frac{H}{P}\sqrt{\frac{EI}{P}}$$

Putting the value of A and B in equation (i), we get

$$y = \frac{-H}{P}l\cos(x\sqrt{P/EI}) + \frac{H}{P}\sqrt{\frac{EI}{P}}\sin(x\sqrt{P/EI}) + \frac{H}{P}(l - x) \quad ...(iii)$$

At A, $x = l,\ y = 0$

Equation (*iii*) becomes

$$0 = -\frac{H}{P} l \cos(l\sqrt{P/EI}) + \frac{H}{P}\sqrt{\frac{EI}{P}} \sin(l\sqrt{P/EI})$$

$$\frac{H}{P}\sqrt{\frac{EI}{P}} \sin(l\sqrt{P/EI}) = \frac{H}{P} l \cos(l\sqrt{P/EI})$$

$$\sqrt{\frac{EI}{P}} \sin(l\sqrt{P/EI}) = l \cos(l\sqrt{P/EI}) \qquad \left[\text{Eliminating } \frac{H}{P} \text{ from both sides}\right]$$

$$\frac{\sin(l\sqrt{P/EI})}{\cos(l\sqrt{P/EI})} = l \times \sqrt{\frac{P}{EI}}$$

$$\tan(l\sqrt{P/EI}) = l \times \sqrt{\frac{P}{EI}}$$

The value of $(l\sqrt{P/EI})$ in radians should be such that its tangent is equal to itself. The only angle whose tangent is equal to itself is 4.5 radians.

$$\therefore \quad l\sqrt{\frac{P}{EI}} = 4.5$$

Squaring both sides, we get

$$l^2 \times \frac{P}{EI} = (4.5)^2 = 20.25$$

$$P = \frac{20.25 \times EI}{l^2}$$

$$\therefore \quad P = \frac{2\pi^2 EI}{l^2} \qquad [\because \text{ The value of } 20.25 \text{ is approximately } 2\pi^2]$$

8.11 EQUIVALENT LENGTH OR EFFECTIVE LENGTH

Equivalent length is the length of the long column which is actually involved in bending.

Equivalent length of a column is also defined as the distance between adjacent point of inflexion*. The equivalent length of a column is obtained by multiplying it with some constant factor '*c*'. The constant factor '*c*' depends on end conditions of the column. If l is the actual length of a column, then its equivalent length, $L = c \times l$.

Hence, in case of column with:

(*i*) Both ends fixed, equivalent length, $L = l/2$

(*ii*) Both ends hinged, $L = l$

(*iii*) One end fixed and other end free, $L = 2l/\sqrt{2}$

(*iv*) One end fixed and other hinged, $L = l/\sqrt{2}$

*A point of inflexion is found at every column end that is free to rotate and at every point where there is a change of the axis.

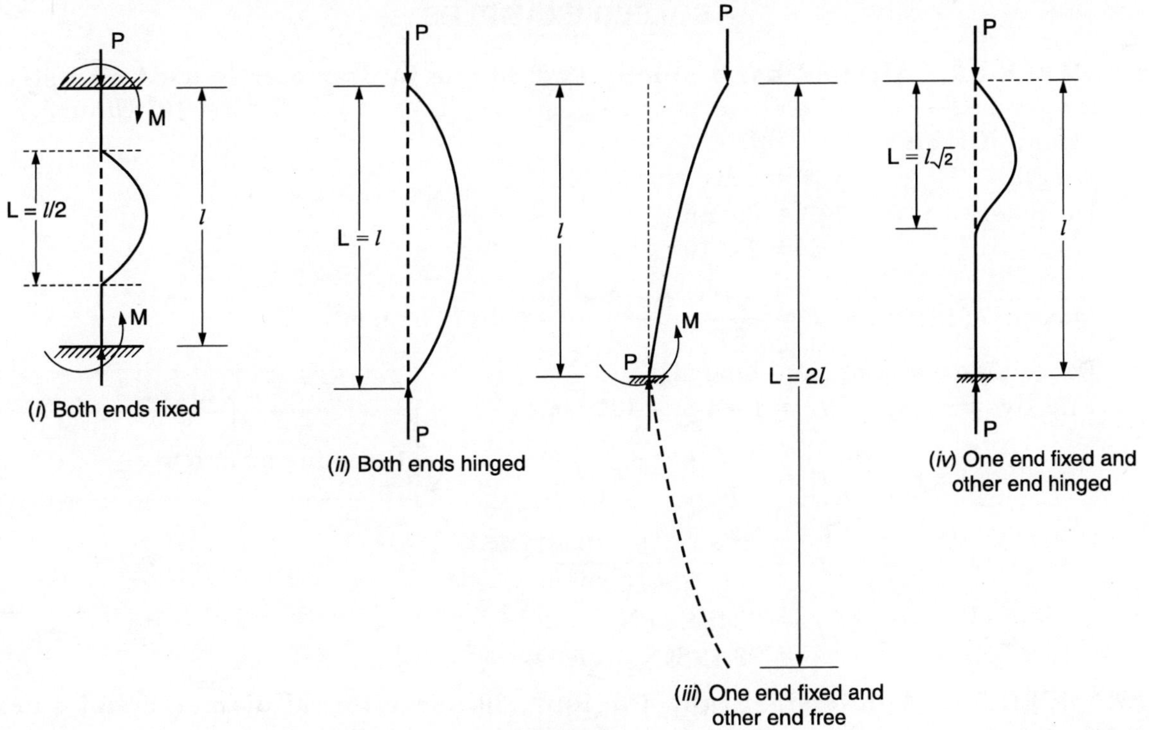

Fig. 8.8.

Table 8.1. Equivalent Length

S. No.	*End conditions of column*	*Relation between effective length (L) and actual length (l)*	*Value of factor 'C'*	*Crippling load in terms of*	
				Actual length	*Effective length*
1.	Both ends fixed	$L = l/2$	1/2	$P = \dfrac{4\pi^2 EI}{l^2}$	$P = \dfrac{\pi^2 EI}{L^2}$
2.	Both ends hinged	$L = l$	1	$P = \dfrac{\pi^2 EI}{l^2}$	$P = \dfrac{\pi^2 EI}{L^2}$
3.	One end fixed and other end free	$L = 2l$	2	$P = \dfrac{\pi^2 EI}{4l^2}$	$P = \dfrac{\pi^2 EI}{L^2}$
4.	One end fixed and and other hinged	$L = l/\sqrt{2}$	$1/\sqrt{2}$	$P = \dfrac{2\pi^2 EI}{l^2}$	$P = \dfrac{\pi^2 EI}{L^2}$

Note: It is to be noted that no factor of safety is included in the Euler's equations.

8.12 VALIDITY OF EULER'S FORMULA (LIMITATIONS FOR EULER'S FORMULA)

(*i*) Euler's formula is valid only if the assumptions made by Euler's are fulfilled.

(*ii*) Euler's formula is valid upto a certain value of slenderness ratio of column material. If the slenderness ratio is less than a certain limit, Euler's formula gives a value of the crippling load greater than the crushing load. Hence, if the slenderness ratio is less than 80 for mild steel column then Euler's formula will not be valid.

SOLVED EXAMPLES

EXAMPLE 8.1. A round bar 4 m long and 60 mm in diameter is used as a strut with both ends hinged. Determine the crippling load. Take **$E = 2 \times 10^5$ N/mm².**

Solution. Given :

Length, $l = 4$ m

Diameter, $d = 60$ mm

$E = 2 \times 10^5$ N/mm²

Moment of inertia, $I = \dfrac{\pi d^4}{64} = \dfrac{\pi \times (60)^4}{64} = 636172.51$ mm⁴

For column with both and hinged,

Effective length, $L = l = 4$ m $= 4000$ mm

Crippling load, $P = \dfrac{\pi^2 EI}{L^2}$

REMEMBER

Crippling load, $P = \dfrac{\pi^2 EI}{L^2}$

$$P = \frac{\pi^2 \times 2 \times 10^5 \times 636172.51}{(4000)^2}$$

$= 78484.64$ N

$= 78.48$ KN **Ans.**

EXAMPLE 8.2. A mild steel tube 4 m long, 30 mm internal diameter and 4 mm thick is used as a strut. Determine the safe compressive loads when this strut is used with the following end conditions :

(*i*) Both ends are hinged

(*ii*) Both ends are fixed

(*iii*) One end is fixed and the other is hinged

(*iv*) One end is fixed and the other is free.

Take factor of safety = 3 and $E = 2 \times 10^5$ N/mm².

Solution: Given :

Actual length, $l = 4$ m $= 4000$ mm

Internal diameter, $d_i = 30$ mm

Thickness, $t = 4$ mm

∴ Outer diameter, $d_0 =$ Internal diameter $+ 2t$

$= 30 + 2 \times 4 = 38$ mm

Moment of inertia, $I = \dfrac{\pi}{64}(d_0^4 - d_i^4)$

$= \dfrac{\pi}{64}(38^4 - 30^4)$

$= \dfrac{\pi}{64}(2085136 - 810000)$

$= 62593.09$ mm⁴

Let the crippling load be P.

(*i*) When both ends are hinged

Effective length, $L = l = 4000$ m

REMEMBER

When both ends are hinged, effective length of a column, L = actual length, l.

Crippling load, $P = \frac{\pi^2 EI}{L^2}$

$= \frac{\pi^2 \times 2 \times 10^5 \times 62593.09}{(4000)^2} = 7722.11$ N

Safe load $= \frac{\text{Crippling load}}{\text{Factor of safety}} = \frac{7722.11}{3}$

$= 2574.04$ N **Ans.**

***(ii)* When both ends are fixed**

Effective length, $L = \frac{l}{2} = \frac{4000}{2} = 2000$ mm

REMEBER

When both ends are fixed, effective length, $L = l/2$

Crippling load, $P = \frac{\pi^2 EI}{L^2}$

$= \frac{\pi^2 \times 2 \times 10^5 \times 62593.09}{(2000)^2} = \frac{1.235 \times 10^{11}}{4 \times 10^6}$

$= 30875$ N

Safe load $= \frac{\text{Crippling load}}{\text{FOS}} = \frac{30875}{3}$

$= 10291.67$ N **Ans.**

***(iii)* When one end is fixed and other end is hinged**

Effective length, $L = \frac{l}{\sqrt{2}} = \frac{4000}{\sqrt{2}}$

$= 2828.43$ mm

Crippling load, $P = \frac{\pi^2 EI}{L^2} = \frac{\pi^2 \times 2 \times 10^5 \times 62593.09}{(2828.43)^2}$

$= \frac{1.235 \times 10^{11}}{8 \times 10^6} = 15437.5$ N

DO YOU KNOW?

When one end fixed and other end hinged

Effective length, $L = \frac{l}{\sqrt{2}}$

Safe load $= \frac{\text{Cripping load}}{\text{FOS}} = \frac{15437.5}{3}$

$= 5145.83$ N **Ans.**

***(iv)* When one end is fixed and other end is free**

Effective length, $L = 2l$

$= 2 \times 4000 = 8000$ mm

Crippling load, $P = \frac{\pi^2 EI}{L^2} = \frac{\pi^2 \times 2 \times 10^5 \times 62593.09}{(8000)^2}$

$P = \frac{1.235 \times 10^{11}}{64 \times 10^6} = 1929.69$ N

$$\text{Safe load} = \frac{\text{Crippling load}}{\text{FOS}} = \frac{1929.69}{3}$$

$$= 643.23 \text{ N} \qquad \textbf{Ans.}$$

EXAMPLE 8.3. Find the maximum length of a solid mild steel rod having diameter 40 mm used as a column with both ends fixed to carry a load of 20 KN. Allow factor of safety = 3. Take $E = 2 \times 10^5$ N/mm².

Solution: Given :

Diameter, $d = 40$ mm

Safe load = 20 KN

FOS = 3

$E = 2 \times 10^5 \text{ N/mm}^2$

Moment of inertia, $I = \frac{\pi}{64} d^4 = \frac{\pi}{64} \times (40)^4$

$= 125663.71 \text{ mm}^4$

Crippling load, P = Safe load × FOS

$= 20 \times 3 = 60 \text{ KN} = 60000 \text{ N}$

Let the effective length be L and actual length be l.

$$P = \frac{\pi^2 EI}{L^2}$$

$$60000 = \frac{\pi^2 \times 2 \times 10^5 \times 125663.71}{L^2}$$

$$L^2 = \frac{\pi^2 \times 2 \times 10^5 \times 125663.71}{60000}$$

$$L^2 = 4134170.2$$

$$\therefore \quad L = \sqrt{4134170.2} = 2033.26 \text{ mm}$$

For a column with both end fixed

$$L = \frac{l}{2} \text{ or } l = \text{Actual length} = 2L$$

$$= 2 \times 2033.26 = 4066.53 \text{ mm}$$

$$\therefore \quad l = 4.1 \text{ m} \qquad \textbf{Ans.}$$

EXAMPLE 8.4. A solid round bar 4 m long and 50 mm in diameter was found to extend 4.6 mm under a tensile load of 50 KN. This bar is used as a strut with both ends hinged. Determine the buckling load for the bar and also the safe load taking factor of safety as 4.

Solution. Given :

Actual length of bar, l = 4 m = 4000 mm

Change in length, dl = 4.6 mm

Tensile load, P = 50 KN = 50000 N

FOS = 4

Diameter of bar, d = 50 mm

Area of cross-section of bar,

$$A = \frac{\pi}{4} \times (50)^2 = 1963.5 \text{ mm}^2$$

$$\text{Stress} = \frac{P}{A} = \frac{50000}{1963.5} = 25.46 \text{ N/mm}^2$$

$$\text{Strain} = \frac{dl}{l} = \frac{4.6}{4000} = 0.00115$$

Modulus of Elasticity,

$$E = \frac{\text{Stress}}{\text{Strain}} = \frac{25.46}{0.00115} = 2.2 \times 10^4 \text{ N/mm}^2$$

Moment of inertia, $I = \frac{\pi}{64} d^4 = \frac{\pi}{64} \times (50)^4$

$= 306796.16 \text{ mm}^4$

Effective length, $L = l = 4000$ mm

Let P_C be the buckling load

We know, $$P_c = \frac{\pi^2 EI}{L^2} = \frac{\pi^2 \times 2.2 \times 10^4 \times 306796.16}{(4000)^2}$$

$$= \frac{6.66 \times 10^{10}}{16 \times 10^6} = 4162.5 \text{ N} \quad \textbf{Ans.}$$

$$\text{Safe load} = \frac{\text{Buckling load}}{\text{FOS}}$$

$$= \frac{4162.5}{4} = 1040.62 \text{ N} \quad \textbf{Ans.}$$

EXAMPLE 8.5. Find the safe load for an elastic column made of a solid steel rod of dia 20 mm and length 1.5 mm. It is fixed at both ends. The factor of safety is 2.5 and modulus of elasticity for the rod is 210 GPa. **(Delhi, Jan. 09)**

Solution. Given :

Diameter, $d = 20$ mm

Length, $l = 1.5$ mm $= 1500$ mm

FOS $= 2.5$

$E = 210 \text{ GPa} = 210 \times 10^9 \text{ Pa} = 210 \times 10^9 \text{ N/m}^2$

$$= \frac{210 \times 10^9}{10^6} = 210 \times 10^3 \text{ N/mm}^2$$

Moment of inertia, $$I = \frac{\pi d^4}{64} = \frac{\pi \times (20)^4}{64} = 7853.98 \text{ mm}^4$$

Column is fixed at both the ends

$\therefore$ Effective length, $$L = \frac{l}{2} = \frac{1500}{2} = 750 \text{ mm}$$

Let P be the crippling load.

We know by Euler's formula

$$\text{Crippling load,} \quad P = \frac{\pi^2 EI}{L^2} = \frac{\pi^2 \times 210 \times 10^3 \times 7853.98}{(750)^2} = 28939.19 \text{ N}$$

$$\text{Safe load} = \frac{\text{Crippling load}}{\text{FOS}} = \frac{28939.19}{2.5}$$

$$= 11575.67 \text{ N} \quad \textbf{Ans.}$$

EXAMPLE 8.6. A column of timber section 150 mm × 200 mm is 6 m long both end being fixed. Find the safe load for the column. Use Euler's formula and allow a factor of safety of 3. Take E = 17500 N/mm².

Solution : Given :

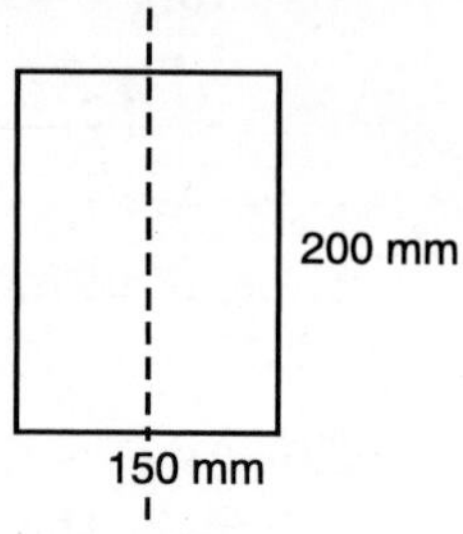

Fig. 8.9.

Width, b = 150 mm

Depth, d = 200 m

Length, l = 6 m = 6000 mm

FOS = 3

E = 17500 N/mm²

Least Moment of inertia,

$$I = \frac{db^3}{12} = \frac{200 \times 150^3}{12}$$

$$= 5.625 \times 10^7 \text{ mm}^4.$$

Effective length of column when both ends are fixed,

$$L = l/2 = \frac{6000}{2} = 3000 \text{ mm}$$

Let P be the crippling load.

By Euler's formula

$$\text{Crippling load,} \quad P = \frac{\pi^2 EI}{L^2} = \frac{\pi^2 \times 17500 \times 5.625 \times 10^7}{(3000)^2}$$

$$= 1079488 \text{ N}$$

$$\text{Safe load} = \frac{\text{Crippling load}}{\text{FOS}} = \frac{1079488}{3} = 359829.33 \text{ N}$$

$$= 360 \text{ KN} \quad \textbf{Ans.}$$

EXAMPLE 8.7. A hollow mild steel tube 5 m long and of circular cross-section is to have an outside diameter of 40 mm. Calculate the necessary internal diameter to prevent failure by buckling if the safe load is 2700 N. It is used as a strut with both ends hinged. Take factor of safety as 2.5 and E = 2.1 × 10⁵ N/mm².

Solution. Given :

Actual length, l = 5 m = 5000 mm

Outside diameter, D = 40 mm

Safe load = 2700 N

FOS = 2.5

$E = 2.1 \times 10^5$ N/mm²

Let the internal diameter be d

$$\text{Moment of inertia,} \quad I = \frac{\pi}{64}(D^4 - d^4)$$

$$= \frac{\pi}{64}(40^4 - d^4)\ \text{mm}^4 \quad ...(i)$$

Crippling load, P = Safe load × FOS

$= 2700 \times 2.5 = 6750$ N

The strut is hinged at both ends

Effective length, $L = l = 5000$ mm

As per Euler's formula

$$\text{Crippling load,} \quad P = \frac{\pi^2 EI}{L^2}$$

$$6750 = \frac{\pi^2 \times 2.1 \times 10^5 \times I}{(5000)^2}$$

$$\text{or} \quad I = \frac{6750 \times (5000)^2}{\pi^2 \times 2.1 \times 10^5}$$

$$\frac{\pi}{64}(40^4 - d^4) = \frac{6750 \times (5000)^2}{\pi^2 \times 2.1 \times 10^5} \quad \text{[using eqn. (i)]}$$

$$\frac{\pi}{64}(40^2 - d^4) = 81418.81$$

$$40^4 - d^4 = \frac{81418.81 \times 64}{\pi}$$

$$2560000 - d^4 = 1658650.4$$

$$d^4 = 2560000 - 1658650.4$$

$$d^4 = 901349.62$$

$$\therefore \quad d = \sqrt[4]{901349.62}$$

$$d = 30.81 \text{ mm} \quad \textbf{Ans.}$$

EXAMPLE 8.8. A 4 m long circular bar was found to deflect 20 mm, when it was used as a simply supported beam and subjected to a load of 100 N at its centre. If this bar is now used as a column, with both ends hinged, determined the crippling load it can carry.

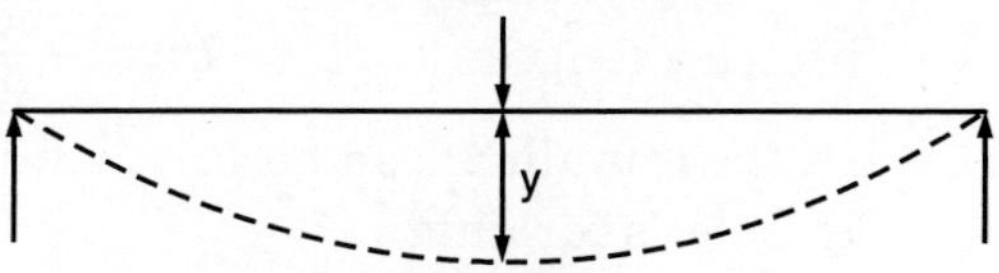

Fig. 8.10.

Solution. Given :

Actual length, $l = 4$ m $= 4000$ mm

Deflection, $y = 20$ mm

Load, $P = 100$ N

When used as beam:

Maximum deflection at the centre for simply supported beam is given by

$$y = \frac{Wl^3}{48\,EI}$$

$$20 = \frac{100 \times (4000)^3}{48 \times EI}$$

or $$EI = \frac{100 \times (4000)^3}{48 \times 20} = 6.67 \times 10^9 \text{ N mm}^2.$$

When used as column:

Effective length of column when both ends are hinged,

$$L = l = 4000 \text{ m}$$

By Euler's formula

Crippling load, $$P = \frac{\pi^2 EI}{L^2} = \frac{\pi^2 \times 6.67 \times 10^9}{(4000)^2} \quad [\because EI = 6.67 \times 10^9 \text{ N mm}^2]$$

$$= 4114.39 \text{ N} \quad \textbf{Ans.}$$

EXAMPLE 8.9. Determine Euler's crippling load for an *I*-section joist 300 mm × 150 mm × 20 mm and 5 m long which is used as a strut with both ends fixed. Take $E = 2 \times 10^5$ N/mm².

Solution. Given :

Actual length, $l = 5 \text{ m} = 5000 \text{ mm}$

$E = 2 \times 10^5 \text{ N/mm}^2$

Moment of inertia about *x-x* axis,

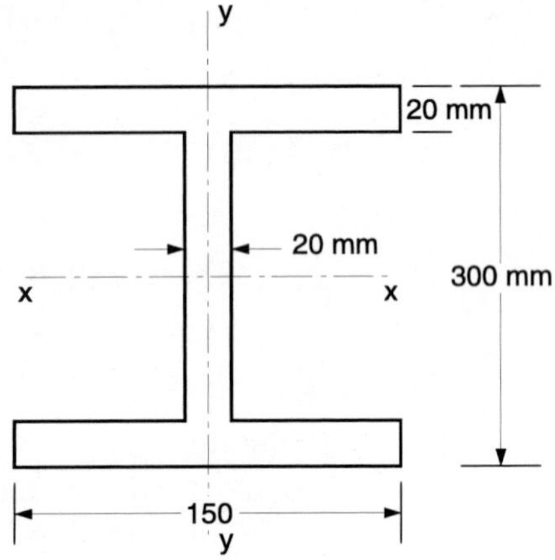

Fig. 8.11.

$$I_{xx} = \frac{1}{12} \times 150 \times 300^3 - \frac{1}{12} \times 130 \times (260)^3$$

$$= 3.375 \times 10^8 - 1.904 \times 10^8$$

$$= 1.471 \times 10^8 \text{ mm}^4$$

I_{yy} = *M.I.* about *yy* axis

$$= 2 \times \frac{1}{12} \times 20 \times 150^3 + \frac{1}{12} \times 260 \times 20^3$$

$$= 11250000 + 173333.33$$

$$= 11423333 = 1.142 \times 10^7 \text{ mm}^4$$

Least moment of inertia is I_{yy}

$$\therefore \quad I = 1.142 \times 10^7 \text{ mm}^4$$

Column with both end fixed

Effective length, $$L = l/2 = \frac{5000}{2} = 2500 \text{ mm}$$

Let the crippling load be *P*,

By Euler's formula

$$P = \frac{\pi^2 EI}{L^2} = \frac{\pi^2 \times 2 \times 10^5 \times 1.142 \times 10^7}{(2500)^2}$$

$$= 3606748.2 \text{ N} = 3606.7 \text{ KN} \quad \textbf{Ans.}$$

EXAMPLE 8.10. Compare the ratio of the strength of a solid steel column to that of a hollow of the same cross-sectional area. The internal diameter of the hollow column is 3/4 of the external diameter. Both the columns have the same length and are pinned at the ends.

Solution. Hollow Circular Column :

Let *D* and *d* be the external and internal diameter of hollow column respectively.

$$d = \frac{3}{4} D$$

Moment of inertia, $I_h = \frac{\pi}{64} (D^4 - d^4)$

$$= \frac{\pi}{64}\left\{D^4 - \left(\frac{3}{4} D\right)^4\right\} = \frac{\pi}{64} \times D^4 \left(1 - \frac{81}{256}\right)$$

$$= \frac{175\pi \times D^4}{64 \times 256}$$

Cross-sectional area,

$$A_h = \frac{\pi}{4} (D^2 - d^2) = \frac{\pi}{4}\left\{D^2 - \left(\frac{3}{4} D\right)^2\right\}$$

$$= \frac{\pi}{4} \times D^2 \left(1 - \frac{9}{16}\right) = \frac{\pi}{4} \times D^2 \times \frac{7}{16}$$

$$= \frac{7\pi D^2}{64}$$

Crippling load, $P_h = \frac{\pi^2 EI}{L^2} = \frac{\pi^2 \times E \times 175\, \pi\, D^4}{64 \times 256\, L^2}$ ...(*i*)

Solid Column:

Let d_1 be the diameter of the solid column

Cross-sectional area,

$$A_s = \frac{\pi}{4} d_1^2$$

According to question $A_s = A_h$

$$\frac{\pi}{4} d_1^2 = \frac{7\pi D^2}{64}$$

$$d_1^2 = \frac{7\pi D^2}{64} \times \frac{4}{\pi} = \frac{7}{16} D^2$$

$$d_1 = \frac{\sqrt{7}}{4} D$$

Moment of inertia, $I_s = \frac{\pi}{64} d_1^4 = \frac{\pi}{64} \times \left(\frac{\sqrt{7}}{4} D\right)^4 = \frac{\pi}{64} \times \frac{49}{16} D^4$

Crippling load, $P_s = \frac{\pi^2 EI}{L^2} = \frac{\pi^2 E \times \pi \times 49 D^4}{L^2 \times 64 \times 16}$ ...(*ii*)

Dividing (*i*) and (*ii*), we get

$$\frac{P_s}{P_h} = \frac{\pi^2 E \times \pi \times 49 D^4}{L^2 \times 64 \times 16} \times \frac{64 \times 256\, L^2}{\pi^2 E \times 175\, \pi D^4} 1 = \frac{49 \times 16}{175} = 4.48$$ **Ans.**

8.13 RANKINE'S FORMULA

Euler's formula gives correct result only for very long column which fails by buckling. Short column fails by crushing. In practice we come across struts and columns which are neither too short nor long. The failure of the column will be due to the combined effect of crushing and buckling. Rankine devised an empirical formula based on experiments for the collapse load which is applicable for both short and long columns.

Let P = Crippling load by Rankine's formula

P_c = Crushing load or compressive load

P_e = Crippling load by Euler's formula.

Then the Rankine's formula is given by

$$\frac{1}{P} = \frac{1}{P_c} + \frac{1}{P_e} \quad \text{...}(i)$$

where $P_c = \sigma_c \times A$;

σ_c = compressive stress (yield stress)

and $P_e = \dfrac{\pi^2 EI}{L^2}$ $\quad A$ = cross-sectional area of column

We have,

$$\frac{1}{P} = \frac{1}{P_c} + \frac{1}{P_e}$$

$$\frac{1}{P} = \frac{P_e + P_c}{P_c P_e}$$

$$P = \frac{P_c P_e}{P_e + P_c}$$

Dividing numerator and denominator of RHS by P_e

$$P = \frac{P_c}{1 + \dfrac{P_c}{P_e}} = \frac{\sigma_c A}{1 + \dfrac{\sigma_c A L^2}{\pi^2 EI}} \qquad \left[\because P_c = \sigma_c A \text{ and } P_e = \frac{\pi^2 EI}{L^2}\right]$$

$$= \frac{\sigma_c A}{1 + \dfrac{\sigma_c AL^2}{\pi^2 E \times AK^2}} = \frac{\sigma_c A}{1 + \dfrac{\sigma_c}{\pi^2 E} \times \left(\dfrac{L}{K}\right)^2} \qquad [\because I = AK^2]$$

$$= \frac{\sigma_c A}{1 + \alpha\left(\dfrac{L}{K}\right)^2} \qquad \text{where } \frac{L}{K} = \text{Slenderness ratio}$$

$$\alpha = \text{Rankine's constant} = \frac{\sigma_c}{\pi^2 E}.$$

8.13.1 Applicability

(*i*) In case of short column, P_e will be large as L is small. As P_e is large, $\dfrac{1}{P_e}$ will be small enough compared to $\dfrac{1}{P_c}$. Hence $\dfrac{1}{P_e}$ may be neglected and equation (*i*) becomes

$$\frac{1}{P} = \frac{1}{P_c}$$

or $$P = P_c$$

Hence, crippling load by Rankine's formulae, for a short column is approximately equal to crushing load.

***(ii)* In case of long column,** P_e will be small as L is large. As P_e is small, $\frac{1}{P_e}$ will be large enough as compared to $\frac{1}{P_c}$ · Hence $\frac{1}{P_c}$ may be neglected and equation (*i*) becomes

$$\frac{1}{P} = \frac{1}{P_e}$$

$\therefore$ $$P = P_e$$

Hence, crippling load by Rankine's formula for a long column is approximately equal to crippling load by Euler's formula.

So Rankine's formula give satisfactory result for both long and short columns.

Table 8.2. Rankine's Constants (α)

S. No.	*Material*	σ_c *N/mm*^{2}	$\alpha = \frac{\sigma_c}{\pi^2 E}$
1.	Mild steel	320	$\frac{1}{7500}$
2.	Cast iron	550	$\frac{1}{1600}$
3.	Wrought iron	250	$\frac{1}{9000}$
4.	Timber	50	$\frac{1}{750}$

Table 8.3. Rankine's Critical Stress for Mild Steel Columns

For σ_c = 320 N/mm^2 and $\sigma = \frac{1}{7500}$

S. No.	*Slenderness*	*Rankine's critical N/mm*^{2}
1.	10	316
2.	20	304
3.	30	286
4.	40	264

S. No.	*Slenderness*	*Rankine's critical N/mm²*
5.	50	240
6.	60	216
7.	70	193
8.	80	173
9.	90	153
10.	100	137
11.	110	122
12.	120	110
13.	130	98
14.	140	89
15.	150	80
16.	160	72
17.	170	66
18.	180	60
19.	190	55
20.	200	50
21.	210	47
22.	220	43
23.	230	40
24.	240	37
25.	250	34

EXAMPLE 8.11. An ISMB 250 Rolled steel joist is to be used as a column 4.0 m long with both ends fixed. Find the safe load on the column allowing a factor of safety of 3. Take σ_c = 320 N/mm² and $\alpha = \frac{1}{7500}$.

Properties of column section are:

$$A = 4755 \text{ mm}^2$$
$$I_{xx} = 5.1316 \times 10^7 \text{ mm}^4$$
$$I_{yy} = 3.345 \times 10^6 \text{ mm}^4.$$

Solution: **Given :** Actual length of column,

$$l = 4.0 \text{ m} = 4000 \text{ mm}$$
$$\text{FOS} = 3$$
$$\sigma_c = 320 \text{ N/mm}^2$$
$$\alpha = 1/7500$$
$$A = 4755 \text{ mm}^2$$

Least moment of inertia,

$$I = 3.345 \times 10^6 \text{ mm}^4$$

Effective length of column,

$$L = \frac{l}{2} = \frac{4000}{2} = 2000 \text{ mn}$$
$$I = AK^2$$

$$\therefore \qquad K = \sqrt{\frac{I}{A}} = \sqrt{\frac{3.345\times10^6}{4755}} = 26.52$$

Rankine's crippling load is given by

$$P = \frac{\sigma_c A}{1+\alpha\left(\frac{L}{K}\right)^2} = \frac{320\times4755}{1+\frac{1}{7500}\times\left(\frac{2000}{26.52}\right)^2}$$

$$= \frac{320\times4755}{1.76} = 864545.45 \text{ N}$$

$$\text{Safe load} = \frac{\text{Crippline load}}{\text{FOS}} = \frac{864545.45}{3}$$

$$= 288181.82 \text{ N} \simeq 288.18 \text{ KN} \qquad \textbf{Ans.}$$

EXAMPLE 8.12. A hollow cast iron column of external diameter 250 mm and internal diameter 200 mm is 8 m long with one end fixed and the other end hinged. Find the safe load with a factlor of safety of 5. Take σ_c = 550 N/mm² and $\alpha = \frac{1}{1600}$.

Solution. Given : Actual length of column,

l = 8 m = 8000 mm

Outer diameter, D = 250 mm

Inner diameter, d = 200 mm

FOS = 5

Compressive stress, σ_c = 550 N/mm²

Rankine's constant, $\alpha = \frac{1}{1600}$

When one end fixed and other end hinged

Effective length of column,

$$L = \frac{l}{\sqrt{2}} = \frac{8000}{\sqrt{2}} = 5656.85 \text{ mm}$$

Least moment of inertia,

$$I = \frac{\pi}{64}(D^4 - d^4) = \frac{\pi}{64}(250^4 - 200^4)$$

$$= \frac{\pi}{64} \times 2.306 \times 10^9 = 1.32 \times 10^8 \text{ mm}^4$$

Cross-sectional area,

$$A = \frac{\pi}{4}(D^2 - d^2) = \frac{\pi}{4}(250^2 - 200^2)$$

$$= \frac{\pi}{4} \times 22500 = 17671.46 \text{ mm}^2$$

$$I = AK^2$$

$$\therefore \qquad K = \sqrt{\frac{I}{A}} = \sqrt{\frac{1.132\times10^8}{17671.46}} = \sqrt{6405.81} = 80.04$$

Rankine's crippling load is given by

$$P = \frac{\sigma_c A}{1+\alpha\left(\frac{L}{K}\right)^2} = \frac{550 \times 17671.46}{1+\frac{1}{1600}\times\left(\frac{5656.85}{80.04}\right)^2}$$

$$= \frac{9719303}{4.12} = 2359054.1 \text{ N}$$

$$\text{Safe load} = \frac{\text{Crippling load}}{\text{FOS}} = \frac{2359054.1}{5}$$

$$= 471810.83 \text{ N}$$

$$\sqcup\ 471.81 \text{ KN} \quad \textbf{Ans.}$$

EXAMPLE 8.13. A hollow cylindrical cast iron column of 6 m long with both ends fixed. Determine the minimum diameter of the column if it has to carry a safe load of 300 KN with a factor of safety 4. Take the internal diameter as 0.7 times the external diameter. Take σ_c = 550 N/mm² and $\alpha = \frac{1}{1600}$.

Solution. Given :

Actual length, $l = 6 \text{ m} = 6000 \text{ mm}$

Safe load = 300 KN

FOS = 4

Internal diameter, $d = 0.7\,D$ (External diameter)

$\sigma_c = 550 \text{ N/mm}^2$

$\alpha = \frac{1}{1600}$

Effective length, $L = \frac{l}{2} = \frac{6000}{2} = 3000 \text{ mm}$

Cross-sectional area,

$$A = \frac{\pi}{4}(D^2 - d^2) = \frac{\pi}{4}\{D^2 - (0.7D)^2\}$$

$$= \frac{\pi}{4}(D^2 - 0.49\,D^2) = \frac{\pi}{4} \times 0.51\,D^2$$

$$= 0.4\,D^2 \text{ mm}^2$$

Least moment of inertia,

$$I = \frac{\pi}{64}(D^4 - d^4) = \frac{\pi}{64}\{(D^4 - (0.7D)^4\}$$

$$= \frac{\pi}{64} \times 0.76\,D^4 = 0.04D^4 \text{ mm}^4$$

$$I = AK^2$$

or $$K^2 = \frac{I}{A}$$

$\therefore$ $$K = \sqrt{\frac{I}{A}} = \sqrt{\frac{0.04D^4}{0.4D^2}} = 0.32D$$

Crippling load, P = Safe load × FOS

= 300 × 4 = 1200 KN = 1200000 N

Rankine's crippling load is given by

$$P = \frac{\sigma_c A}{1+\alpha\left(\frac{L}{K}\right)^2}$$

$$1200000 = \frac{550\times 0.4D^2}{1+\frac{1}{1600}\times\left(\frac{3000}{0.32D}\right)^2}$$

$$1200000 = \frac{220D^2}{1+\frac{1}{1600}\times\frac{9\times10^6}{0.10D^2}}$$

$$1200000 = \frac{220D^2}{1+\frac{54931.64}{D^2}}$$

$$12\times10^5 = \frac{220D^4}{D^2+54931.64}$$

$12 \times 10^5 (D^2 + 54931.64) = 220 D^4$

$5454.54 (D^2 + 54931.64) = D^4$

$D^4 - 5454.54D^2 - 3.0 \times 10^8 = 0$

$$D^2 = \frac{5454.54\pm\sqrt{(5454.54)^2+4\times3\times10^8}}{2} = \frac{5454.54\pm35067.82}{2}$$

$D^2 = 20261.18$ (neglecting the –ve value of D^2)

∴ $D = 142.34$ mm **Ans.**

$d = 0.7 \times 142.34 = 99.64$ mm. **Ans.**

EXAMPLE 8.14. Find the greatest length for which a mild steel strut of *T*-section may be used with both its ends fixed, to carry a working stress of 100 N/mm². The working load is $\frac{1}{4}$ of crippling load. Properties of *T* section are:

A = 3500 mm²

I = 230 × 10⁴ mm⁴

σ_c = 320 N/mm²; $\alpha = \frac{1}{7500}$.

Solution: Given :

Working stress = 100 N/mm²

Area of section, A = 3500 mm²

Least moment of inertia,

$I = 230 \times 10^4$ mm⁴

$$\sigma_c = 320 \text{ N/mm}^2$$

$$\alpha = \frac{1}{7500}$$

$$\text{Working load} = \frac{1}{4} \times \text{Crippling load}$$

$$\therefore \quad \text{Crippling load} = 4 \times \text{Working load}$$

$$\text{Working load} = \text{Working stress} \times A = 100 \times 3500$$

$$= 350000 \text{ N}$$

Crippling load, $P = 4 \times \text{Working load}$

$$= 4 \times 350000$$

$$= 1400000 \text{ N}$$

$$I = AK^2$$

$$\therefore \quad K^2 = \frac{I}{A} = \frac{230 \times 10^4}{3500} = 657.14$$

Rankine's crippling load is given by

$$P = \frac{\sigma_c A}{1 + \alpha \times \frac{L^2}{K^2}}$$

$$1400000 = \frac{320 \times 3500}{1 + \frac{1}{7500} \times \frac{L^2}{657.14}}$$

$$\text{or} \quad 1 + \frac{L^2}{7500 \times 657.14} = \frac{320 \times 3500}{1400000} = 0.8$$

$$\text{or} \quad 1 + \frac{L^2}{4928550} = 0.8$$

$$\frac{L^2}{4928550} = 1 - 0.8 = 0.2$$

$$L^2 = 4928500 \times 0.2 = 985710$$

$$\therefore \quad L = \sqrt{985710} = 992.83 \text{ mm}$$

Effective length, $L = \frac{l}{2}$

$\therefore$ Actual length, $l = 2L = 2 \times 992.83$

$= 1985.66 \text{ mm} \sqcup 1986 \text{ mm}$ **Ans.**

EXAMPLE 8.15. An *I*-section 300 mm × 150 mm × 10 mm is provided with a flange plate 200 mm × 12 mm for each flange. The composite member is used as a column with one end fixed and the other end hinged. Calculate the length of the member for which crippling load by Rankine's formula and Euler's formula will be the same. Take $E = 2.1 \times 10^5$ N/mm², $\sigma_c = 330$ N/mm² and $\alpha = \frac{1}{7500}$.

Solution. Given :

$$E = 2.1 \times 10^5 \text{ N/mm}^2$$

$$\sigma_c = 330 \text{ N/mm}^2$$

$$\alpha = \frac{1}{7500}$$

Cross-sectional area

$$A = 2 \times (200 \times 12 + 150 \times 10) + 280 \times 10$$
$$= 10600 \text{ mm}^2$$

$$I_{xx} = 2 \times \left[\frac{1}{12} \times 200 \times 12^3 + 200 \times 12 \times (162 - 6)^2\right]$$

$$+ 2 \times \left[\frac{1}{12} \times 150 \times 10^3 + 150 \times 10x\,(162 - 17)^2\right] + \frac{1}{12} \times 10 \times 280^3$$

$$= 2 \times [58435200] + 2(31550000) + 18293333$$
$$= 93083733 \text{ mm}^4$$

$$I_{yy} = 2 \times \left[\frac{1}{12} \times 12 \times 200^3 + \frac{1}{12} \times 10 \times 150^3\right] + \frac{1}{12} \times 280 \times 10^3$$
$$= 21625000 + 23333.33$$
$$= 21648333.33 \text{ mm}^4$$

∴ Least moment of inertia,

$$I = 21648333.33 \text{ mm}^4$$
$$I = AK^2$$

or $$K^2 = \frac{I}{A}$$

∴ $$K = \sqrt{\frac{I}{A}} = \sqrt{\frac{21648333.33}{10600}}$$
$$= 45.19$$

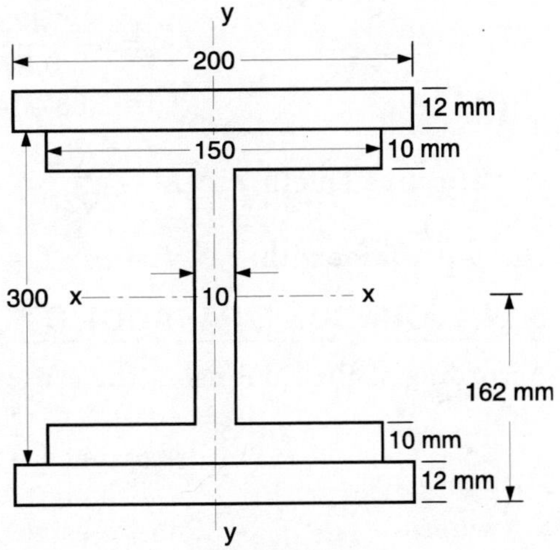

Fig. 8.12.

Rankine's crippling load

$$P = \frac{\sigma_c A}{1 + \alpha\left(\frac{L}{K}\right)^2} = \frac{330 \times 10600}{1 + \frac{1}{7500}\left(\frac{L}{45.19}\right)^2}$$

$$P = \frac{3498000}{1 + \frac{L^2}{15316021}}$$

Euler's crippling load

$$P_e = \frac{\pi^2 EI}{L^2}$$

$$= \frac{\pi^2 \times 2.1 \times 10^5 \times 21648333.33}{L^2}$$

$$= \frac{4.487 \times 10^{13}}{L^2}$$

Accroding to the question

Ranking's crippling load = Euler's crippling load

$$\frac{\dfrac{3498000}{L^2}}{1+\dfrac{L^2}{15316021}} = \frac{4.487\times10^{13}}{L^2}$$

$$\frac{3498000\times15316021}{15316021+L^2} = \frac{4.487\times10^{13}}{L^2}$$

$$\frac{5.357\times10^{13}}{15316021+L^2} = \frac{4.487\times10^{13}}{L^2}$$

$$\frac{5.357}{15316021+L^2} = \frac{4.487}{L^2}$$

$$5.357L^2 = 4.487 \times 15316021 + 4.487\ L^2$$

$$(5.357 - 4.487)L^2 = 68722986$$

$$0.87L^2 = 68722986$$

$$L^2 = \frac{68722986}{0.87} = 78991938$$

$$\therefore \quad L = 8887.74 \text{ m}$$

Effective length, $L = \dfrac{l}{\sqrt{2}}$

$\therefore$ Actual length, $l = \sqrt{2}\,L = \sqrt{2} \times 887.74 = 12569.16 \text{ mm} = 12.57 \text{ m}$ **Ans.**

8.14 JOHNSON'S PARABOLIC FORMULA

According of Prof. Johnson, the critical load is given by

$$P = A\left[f_c - r\left(\frac{L}{K}\right)^2\right]$$

where f_c = Compressive yield stress of column material

A = cross-sectional area of column

$\dfrac{L}{K}$ = Slenderness ratio

r = A constant whose value depends upon the material of the column

$= \dfrac{f_c^2}{4\pi^2 E}$.

Table 8.4. Value of *r*

S. No.	*Material*	f_c *(N/mm²)*	r
1.	Wrought iron	250	0.000039
2.	Mild steel	320	0.000057
3.	Cast iron	550	0.00016

8.15 STRAIGHT LINE FORMULA

Straight line formula is used in practical designing and is given by:

$$P = A\left[f_c - n\left(\frac{L}{K}\right)\right]$$

where P = Crippling load on the column

A = Cross-sectional area of the column

f_c = Compressive yield stress

$\frac{L}{K}$ = Slenderness ratio

n = A constant whose value depends upon the material of the column

Table 8.5. Value of *n*

S. No.	*Material*	*n*
1.	Cast iron	0.008
2.	Mild steel	0.0053
3.	Wrought iron	0.0053

8.16 INDIAN STANDARD CODE METHOD

$$\text{Safe load} = \sigma_{ac} \times A$$

where σ_{ac} = Permissible stress in axial compression

A = Cross-sectional area of column

As per IS : 800

$$\sigma_{ac} = 0.6\,\frac{f_{cc} \times f_y}{(f_{cc}^n + f_y^n)^{1/n}}$$

where σ_{ac} = Permissible stress in axial compression (N/mm^2)

f_y = Yield stress of steel (N/mm^2)

f_{cc} = Elastic critical stress in compression

$$= \frac{\pi^2 EK^2}{L^2}$$

L = Effective length of column

K = Least radius of gyration

E = Modulus of elasticity of steel = 2×10^5 N/mm^2

n = *A* factor = 1.4

Table 8.6. Safe axial compressive strees for Fe 250 grade steel

S. No.	*Slenderness ratio* $\frac{L}{K}$	*Safe axial compressive stress (N/mm^2)*
1.	10	150
2.	20	148
3.	30	145
4.	40	139

S. No.	*Slenderness ratio* $\frac{L}{K}$	*Safe axial compressive stress (N/mm²)*
5.	50	132
6.	60	122
7.	70	112
8.	80	101
9.	90	90
10.	100	80
11.	110	72
12.	120	64
13.	130	57
14.	140	51
15.	150	45
16.	160	41
17.	170	37
18.	180	33
19.	190	30
20.	200	28
21.	210	25
22.	220	23
23.	230	21
24.	240	20
25.	250	18

EXAMPLE 8.16. Determine the safe load by IS code for a hollow cylindrical mild steel tube 50 mm outside diameter and 40 mm inside diameter when the tube is used as column of length 2.0 m with both ends fixed.

Solution. Given :

Outside diameter, $D = 50$ mm

Inside diameter, $d = 40$ mm

Length of column, $l = 2.0 \text{ m} = 2000$ mm

Effective length, $L = \frac{l}{2} = \frac{2000}{2} = 1000$ mm

Cross-sectional area,

$$A = \frac{\pi}{4}(D^2 - d^2) = \frac{\pi}{4}(50^2 - 40^2)$$

$$= \frac{\pi}{4} \times (2500 - 1600) = \frac{\pi}{4} \times 900 = 706.86 \text{ mm}^2$$

Moment of inertia, $$I = \frac{\pi}{64}(D^4 - d^4) = \frac{\pi}{64}(50^4 - 40^4)$$

$$= \frac{\pi}{64} \times 3690000 = 181132.45 \text{ mm}^4$$

$$I = AK^2$$

or $$K^2 = \frac{I}{A}$$

$\therefore$ $$K = \sqrt{\frac{I}{A}} = \sqrt{\frac{181132.45}{706.86}} = 16$$

$$\frac{L}{K} = \frac{1000}{16} = 62.5$$

$$\sigma_{ac} = 122 - \frac{10 \times 2.5}{10} = 119.5 \text{ N/mm}^2$$

$$\text{Safe load} = \sigma_{ac} \times A = 119.5 \times 706.86$$
$$= 84469.77 \text{ N}$$ **Ans.**

SUMMARY

1. Any member of structure subjected to axial compressive force is known as strut. A vertical strut is known as column.
2. Slenderness ratio = $\dfrac{\text{Length of the column}}{\text{Least radius of gyration}}$.
3. Buckling load is the maximum axial compressive load at which the column starts buckling.
4. Safe load = $\dfrac{\text{Buckling load or crippling load}}{\text{Factor of safety}}$.
5. If the length of a column is less than 8 times its least lateral dimension, then the column is said to be a short column.
6. If the length of a column is more than 30 times its least lateral dimension, then it is said to a long column.
7. If the length of a column is between 8 times to 30 times its least lateral dimension, then it is said to be a medium column.
8. The failure of a column takes place due to any of the following stresses :
   (*i*) Direct compressive stress
   (*ii*) Buckling stress
   (*iii*) Combination of direct and buckling stresses.
9. Equivalent length is the length of the long column which is actually involved in bending.
10. Equivalent length of columns for different end conditions :
    Both ends fixed, $L = l/2$
    Both ends hinged, $L = l$
    One end fixed and other end free,
    $L = 2l$
    One end fixed and other hinged,
    $L = l/\sqrt{2}$.
11. Euler's crippling load for different end conditions

    Both ends fixed, $P = \dfrac{4\pi^2 EI}{l^2}$; where l = actual length of column

Both ends hinged, $P = \frac{\pi^2 EI}{l^2}$

One end fixed and other end free,

$$P = \frac{\pi^2 EI}{4l^2}$$

One end fixed and other hinged,

$$P = \frac{2\pi^2 EI}{l^2}.$$

12. Rankine's formula for crippling load is given by

$$P = \frac{\sigma_c A}{1+\alpha\left(\frac{L}{K}\right)^2}$$

where σ_c = Compressive stress (yield stress)

A = Area of cross-section

$\left(\frac{L}{K}\right)$ = Slenderness ratio

α = Rankine's constant

$$= \frac{\sigma_c}{\pi^2 E}.$$

13. Johnson's parabolic formula

$$P = A\left[f_c - r\left(\frac{L}{K}\right)^2\right]$$

where f_c = Yield stress

r = A constant = $\frac{f_c^2}{4\pi^2 E}$.

14. As per IS-800

$$\sigma_{ac} = 0.6\,\frac{f_{cc} \times f_y}{(f_{cc}^n + f_y^n)^{1/n}}$$

σ_{ac} = Permissible axial compressive stress (N/mm ^{2})

f_y = Yield stress of steel (N/mm ^{2})

f_{cc} = Elastic critical stress in compression

$$= \frac{\pi^2 EK^2}{L^2}$$

n = A factor = 1.4.

OBJECTIVE TYPE QUESTIONS

Fill in the Blanks :

(*i*) A member of structure which is subjected to axial compressive load is known as ____________ .

(*ii*) A vertical strut is known as ____________ .

(*iii*) The equivalent length (*L*) of a column of length '*l*' whose both ends are fixed is given by ____________ .

(*iv*) The ratio of length of a column to its least radius of gyration is known as ____________ .

(*v*) A column whose length is less than 8 times its diameter is called a ____________ column.

(*vi*) A column whose length is between 8 to 30 times its diameter is known as ____________ column.

(*vii*) The crippling load of a column with one end fixed and other hinged, by Euler's formula, is given by ____________ .

(*viii*) Is code-800 uses the ____________ formula for calculating safe compressive load in compressive member.

(*ix*) The maximum load at which a column buckles is known as ____________ load.

Answers

(*i*) strut, (*ii*) column, (*iii*) $L = l/2$, (*iv*) slenderness ratio, (*v*) short, (*vi*) medium, (*vii*) $P = \frac{2\pi^2 EI}{l^2}$, (*viii*) Rankine merchant formula, (*ix*) buckling load.

THEORETICAL QUESTIONS

1. What is a column ? Write the difference between strut and column.
2. Define the following :
   (*a*) Buckling load
   (*b*) Buckling factor
   (*c*) Safe load
   (*d*) Slenderness ratio.
3. What do you mean by equivalent length of a column ? Write the equivalent length of column for different end conditions.
4. Define : (*a*) Short column, (*b*) Medium column, (*c*) Long column.
5. Write Euler's formula and state the assumptions made in Euler's formula. (*Delhi, Jan. 2009*)
6. What are the limitations of Euler's formula ?
7. Derive Rankine's formula for crippling load of a column.

NUMERICALS

1. A steel column is of length 8 m and diameter 600 mm with both ends hinged. Determine the crippling load by Euler's formula. Take $E = 2.1 \times 10^5$ N/mm^2. (2.06×10^8 N)
2. A mild steel tube, 25 mm outside diameter and 20 mm inside diameter, is 3 m long and is used as a strut. Determine the safe compressive loads when this strut is used with the following end conditions:
   (*i*) Both ends are hinged.
   (*ii*) Both ends are fixed.
   (*iii*) One end is fixed and the other end is free.
   (*iv*) One end is fixed and the other end is hinged.
   Take factor of safety = 3 and $E = 2 \times 10^5$ N/mm^2. (827.64 N; 3310.6 N; 206.91 N; 1655.28 N)
3. Compare the strength of solid circular column of diameter 200 mm and hollow circular column of same cross-sectional area and thickness 30 mm. The other parameters are same for both the sections. (5.58)
4. A strut 2.50 metres long is 60 mm in diameter. One end of the strut is fixed while its other end is hinged. Find the safe compressive load for the member using Euler's formula, allowing a factor of safety of 3.5. Take $E = 2.1 \times 10^5$ N/mm^2. (120520 N)

5. A solid circular compression member 50 mm in diameter is to be replaced by a hollow circular section of the same material. Find the size of the hollow section if the internal diameter is 0.6 times the external diameter. (51.765 mm, 31.059 mm)
6. Determine the limiting length of a pin ended column of section 60 mm × 100 mm so that the critical stress is 250 N/mm^2. Assume $E = 2 \times 10^5$ N/mm^2. (1.539 m)
7. A column having a *T*-section with a flange 120 mm × 16 mm and web 150 mm × 16 mm is 3 m long. Assuming the column to be hinged at both ends, find the crippling load by using Euler's formula. Take $E = 2 \times 10^6$ kg/cm^2. (51655.32 kg)
8. Compare buckling load of a solid square pillar 150 mm × 150 mm with that of a solid circular pillar, 150 mm diameter. Both pillars are made of steel and are of the same length with same end conditions. (1.7)
9. Calculate the maximum value of the slenderness ratio of a mild steel column for which Euler's formula is valid when both the ends are hinged. (80)
10. A rolled steel joist ISMB 300 is to be used as a column of 3 metres length with both ends fixed. Find the safe axial load on the column. Take factor of safety = 3, $f_c = 320$ N/mm^2 and $\alpha = \frac{1}{7500}\cdot$ Properties of column section are :
    $A = 5626$ mm^2; $I_{xx} = 8.603 \times 10^7$ mm^4; $I_{yy} = 4.539 \times 10^7$ mm^4. (447840.79 N)
11. A hollow cast iron column of external diameter 250 mm and internal diameter 200 mm is 10 m long with both ends fixed. Find the safe load. Take factor of safety = 4, $f_c = 550$ N/mm^2 and $\alpha = \frac{1}{1600}\cdot$ (706.54 KN)
12. A column having a *T*-section with flange 180 mm × 15 mm and web 235 mm × 110 mm is 3.5 m long, with both ends fixed. Find the safe load on the column. Take factor of safety = 3, $f_c = 335$ N/mm^2 and $\alpha = \frac{1}{7500}\cdot$

❑❑❑

CHAPTER 9

Combined Direct and Bending Stress

Chapter Outline

- Ecentric Load
- Direct Stress
- Combined Direct and Bending Stress
- Limit of Ecentricity

9.1. ECENTRIC LOAD

A load whose line of action passes through the centre of gravity of the section, then the load is said to be axial load. A load whose line of action does not passes through the centre of gravity of the section, then the load is called ecentric load.

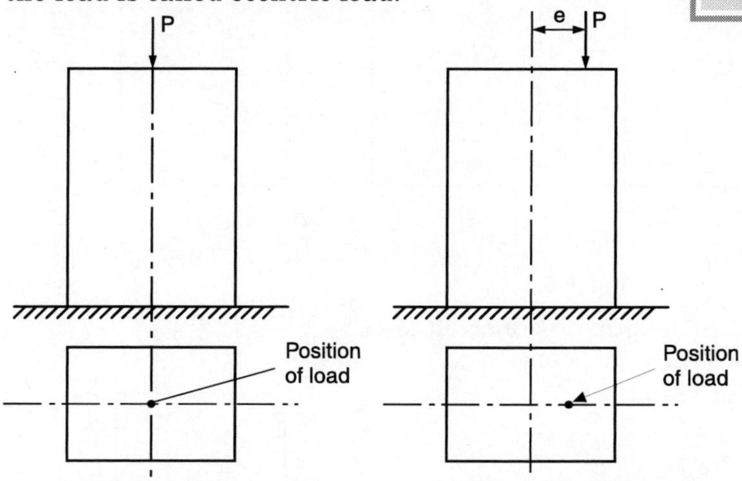

(*i*) Axial load (*ii*) Ecentric load

Fig. 9.1

Figure 9.1 (*i*) shows a section of column loaded with axial load P and figure 9.1 (*ii*) shows a section of column loaded with ecentric load. The distance between the line of action of load and axis of the column (passing through C.G.) is the ecentricity 'e' of the load. The ecentricity of a load may be about one axis or about both the axes.

9.2. DIRECT STRESS

If a column is loaded with axial load P, then the column is subjected to direct stress and is given by $\sigma_d = \dfrac{P}{A}$, where A is the cross-sectional area of the column. Figure 9.2 shows a direct stress.

9.3. COMBINED DIRECT AND BENDING STRESS

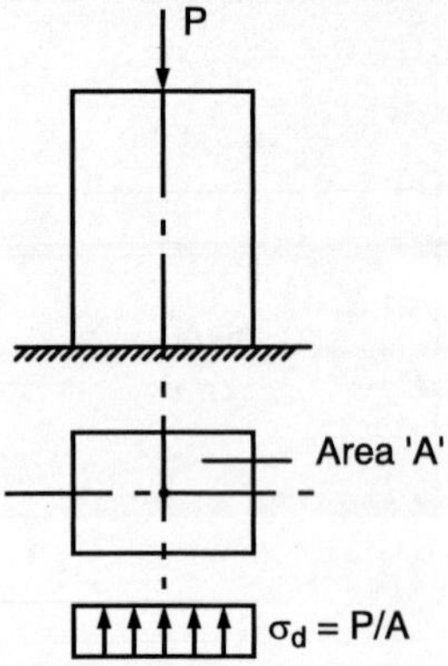

Fig. 9.2. Stress diagram.

Figure 9.3 shows a column subjected to a load '*P*' whose line of action is at a distance '*e*' from the axis of the column. Apply two equal and opposite forces along the axis of the column as shown in figure 9.3 (*ii*). Now the three forces acting on the column can be converted into two systems.

(*i*) An ixial force *P* which will produce direct stress in the column, as shown in figure 9.3 (*iii*)

(*ii*) Two equal and opposite forces forming a couple. The arm of the couple is '*e*'. The moment of the couple will be $P \times e$ and will produce bending stress (σ_b) in the column [Figure 9.3 (*iv*)].

So a column subjected to ecentric loading, is subjected to both direct and bending stress.

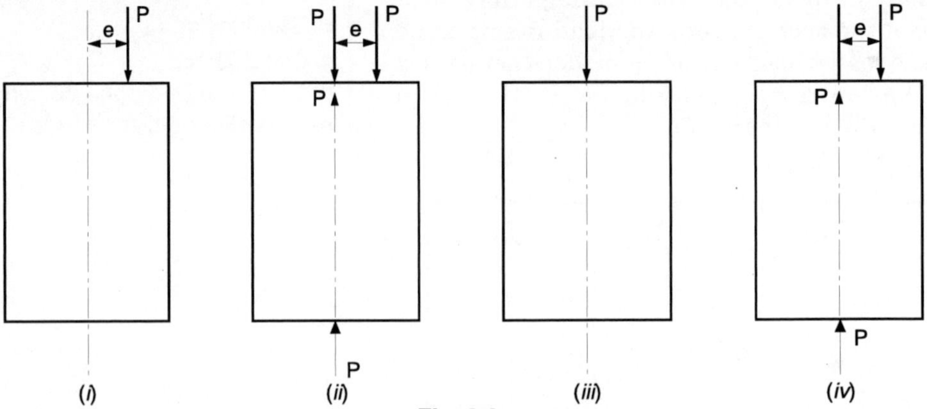

Fig. 9.3.

Consider a column of width '*b*' and depth '*d*" subjected to an ecentric load *P* as shown in figure 9.4.

Cross-sectional area, $A = b \times d$.

Direct stress, $\sigma_d = \dfrac{P}{A} = \dfrac{P}{bd}$

Due to ecentricity of load, the column is subjected to $B.M = P \times e$. This B.M. will produce bending stress in the column.

$\therefore$ Bending stress, $\sigma_b = \dfrac{M}{Z}$

where Z = section modulus $= \dfrac{I}{Y}$

$$= \frac{db^3/12}{b/2} = \frac{db^2}{6}$$

(As the ecentricity of the load is from *Y–Y* axis, $I = \dfrac{db^3}{12}$)

$$\therefore \quad \sigma_b = \frac{P \times e}{\frac{db^2}{6}} = \frac{6Pe}{Ab}$$

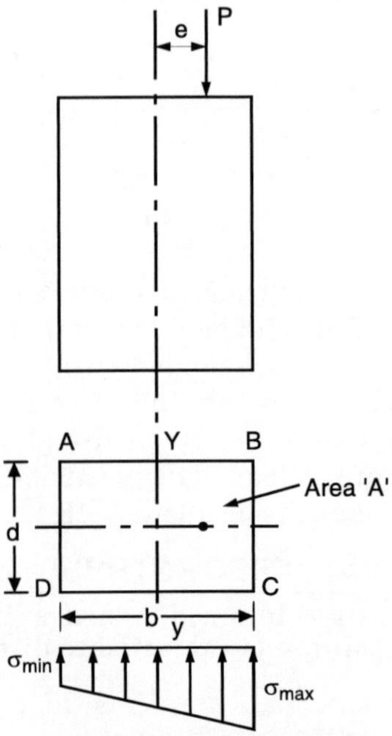

Fig 9.4. Stress diagram

The resultant stress at any point is the sum of direct and bending stress at that point.

$$\therefore \quad \text{Total stress} = \sigma_d \pm \sigma_b = \frac{P}{A} \pm \frac{M}{Z}$$

The + *ve* and −*ve* sign depends upon the position of the load. The stress will be maximum at the face *BC* as the load is near to *BC*. The stres will be minimum at the face *AD* as the load is away from *AD*.

$$\therefore \quad \sigma_{max} = \frac{P}{A} + \frac{M}{Z} = \frac{P}{A} + \frac{6Pe}{Ab} = \frac{P}{A}\left(1 + \frac{6e}{b}\right)$$

$$\text{and} \quad \sigma_{min} = \frac{P}{A} - \frac{M}{Z} = \frac{P}{A} - \frac{6Pe}{Ab} = \frac{P}{A}\left(1 - \frac{6e}{b}\right)$$

σ_{max} is always compressive. If σ_{min} is + *ve* then it is compressive and if the value of σ_{min} is − *ve*, then it is tensile.

9.4. LIMIT OF ECENTRICITY

If a column is ecentrically loaded, then both direct and bending stress are simultaneously developed. If the direct stress (σ_d) is more than bending stress (σ_b), the stress in the section all through will be compressive. If the bending stress is more than direct stress, then there will be tensile stress. As the concrete columns are weak in tension, load should be applied in such a way that there is no tensile stress in the column. To avoid tensile stress, the bending stress (σ_b) should be less than or equal to direct stress (σ_d).

Hence,

$$\sigma_b \le \sigma_d$$

$$\frac{M}{Z} \le \frac{P}{A}$$

$$\frac{Pe}{Z} \le \frac{P}{A}$$

$$\therefore \quad e \le \frac{Z}{A}$$

This shows that for tension in the column, ecentricity should be less than or equal to $\frac{Z}{A}$.

Limit of Ecentricity for a Rectangular Section

Figure 9.5 shows a rectangular section of width '*b*' and depth '*d*'. Let the section be subjected to a load at a distance '*e*', along $x - x$ axis, from $y - y$ axis. The bending will take place along $y - y$ axis.

$$I_{yy} = \frac{db^3}{12} \text{ and } y = b/2$$

$$\text{Section modulus}, \; Z = \frac{I_{yy}}{y}$$

$$= \frac{db^3}{12} \times \frac{2}{b} = \frac{db^2}{6}$$

Area of cross-section, $A = bd$.

$$\text{For no tension,} \quad e \le \frac{Z}{A}$$

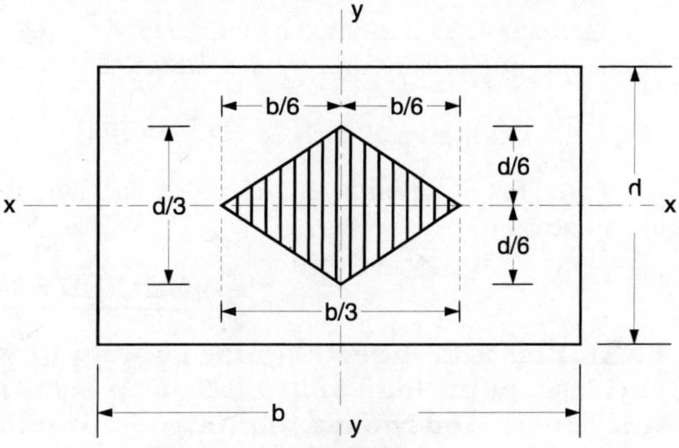

Fig 9.5.

$$e \le \frac{db^2}{6bd}$$

$$e \le \frac{b}{6}$$

This means that for no tension, load can be applied on both sides of y-y at a distance $b/6$ on x-x axis as shown in figure 9.5.

Therefore, for no tension, the load must be placed with-in $\left(\frac{b}{6}+\frac{b}{6}\right)=\frac{b}{3}$ *i.e.*, middle third of the width of section.

Similarly, for no tension, the load can be placed on y-y axis, on both sides of x-x axis, with in middle third of depth *i.e.;* $d/3$.

If we join the four points on x-x and y-y axis, we get a rhombus. This rhombus is known as **core** or **kernel** as shown in figure 9.5. Thus, **core or Kernel** of the section is the area in which any ecentric load, if placed will not produce any tension in the section.

***(ii)* Limit of ecentricity for a circular section**

Figure 9.6 shows a circular section of diameter 'd'.

Let the section be subjected to a load at a distance 'e' from centroid on x-x-axis.

$$I_{xx} = I_{yy} = \frac{\pi d^4}{6^4}$$

$$y = d/2$$

Modulus of section, $Z = \frac{I_{yy}}{y} = \frac{\pi d^4}{64} \times \frac{2}{d} = \frac{\pi d^3}{32}$

Area of cross-section, $A = \frac{\pi d^2}{4}$

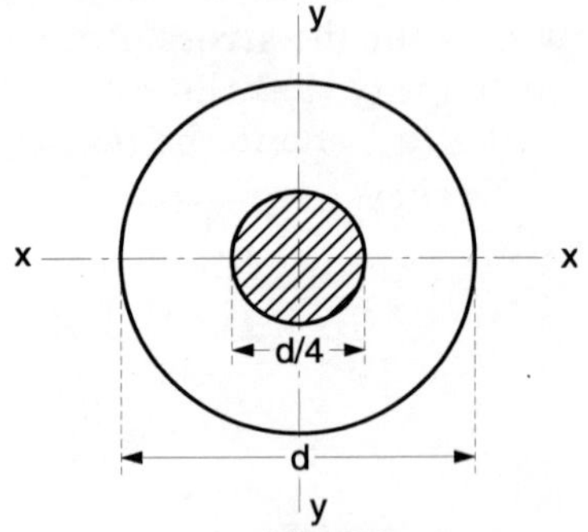

Fig. 9.6.

For no tension

$$e \le \frac{Z}{A}$$

$$e \le \frac{\pi d^3}{32} \times \frac{4}{\pi d^2}$$

$$e \le \frac{d}{8}$$

This means that for no tension the load can be ecentric, on either side of centroid by $d/8$ distance.

> **DO YOU KNOW?**
>
> If the load be applied anywhere within a concentric circle of diameter $\frac{d}{4}$, the stresses will remain wholly compressive.

$\therefore$ Diameter of core $= 2 \times \frac{d}{8} = d/4.$

Thus, for no tension in a circular section, the load must be placed within middle fourth of the section.

SOLVED EXAMPLES

EXAMPLE 9.1. A rectangular column of width 200 mm and of thickness 150 mm carries a point load of 240 KN at an ecentricity of 10 mm in a plane bisecting the thickness. Find the maximum and minimum stresses in the section.

Solution. Given :

Point load, $P = 240\ KN = 240000$ N

Ecentricity, $e = 10$ mm

Width, $b = 200$ mm

Thickness, $d = 150$ mm

Area, $A = bd = 200 \times 150 = 30000\ \text{mm}^2$

Let σ_{max} and σ_{min} use the maximum and minimum stress in the section.

Using the relation

$$\sigma_{max} = \frac{P}{A}\left(1+\frac{6e}{b}\right) = \frac{240000}{30000}\left(1+\frac{6\times10}{200}\right)$$

$$= 8\ (1 + 0.3) = 10.4\ \text{N/mm}^2\ \text{(comp)}$$

$$\sigma_{min} = \frac{P}{A}\left(1-\frac{6e}{b}\right) = \frac{240000}{30000}\left(1-\frac{6\times10}{200}\right)$$

$$= 8\ (1 - 0.3)$$

$= 5.6\ \text{N/mm}^2$ (comp) **Ans.**

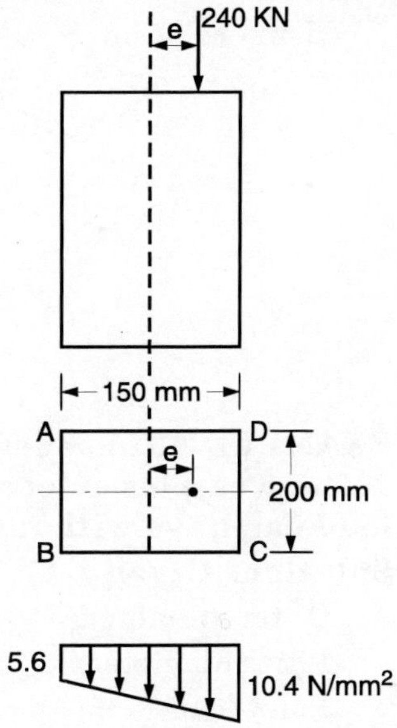

Fig. 9.7.

EXAMPLE 9.2. A short column of hollow cylindrical section 250 mm outside diameter and 150 mm inside diameter carries a vertical load of 400 KN along one of the planes 100 mm away from the axis of the column. Find the extreme stress intensities and state their nature.

Solution: Given :

Inner diameter, $d_i = 150$ mm

Outer diameter, $d_o = 250$ mm

Point load, $P = 400$ KN $= 400000$ N

Ecentricity, $e = 100$ mm

Distance of extreme fiber,

$$y = \frac{250}{2} = 125\ \text{mm}$$

$$\text{Area, } A = \frac{\pi}{4}(d_o^2 - d_i^2) = \frac{\pi}{4}\left(250^2 - 150^2\right)$$

$$= 31415.93\ \text{mm}^2$$

$$\text{Moment of inertia, } I = \frac{\pi}{64}\left(250^4 - 150^4\right)$$

$$= 1.67 \times 10^8\ \text{mm}^4$$

Moment about the axis of the column,

$$M = P \times e = 400000 \times 100$$

$$= 4 \times 10^7\ \text{N-mm}$$

$$\text{Section modules, } Z = \frac{I}{y} = \frac{1.67\times10^8}{125} = 1336000\ \text{mm}^3$$

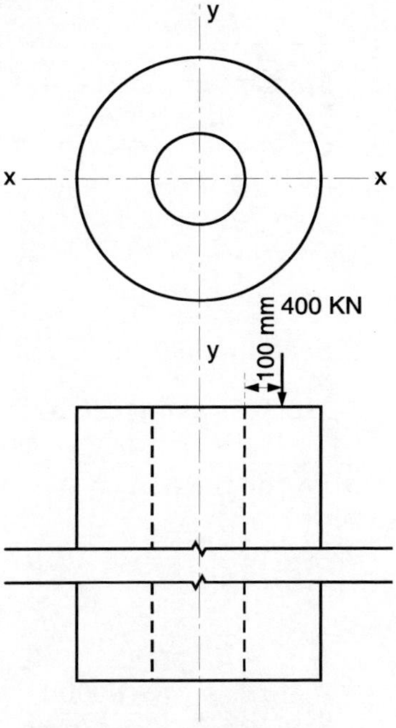

Fig. 9.8.

Using relation

$$\sigma_{max} = \frac{P}{A}+\frac{M}{Z} = \frac{400000}{31415.93}+\frac{4\times10^7}{1336000} = 12.73 + 29.94$$

$$= 42.67 \text{ N/mm}^2 \text{ (compressive)}$$ **Ans.**

$$\sigma_{min} = \frac{P}{A}-\frac{M}{Z} = \frac{400000}{31415.93}-\frac{4\times10^7}{1336000}$$

$$= 12.73 - 29.94 = -17.21 \text{ N/mm}^2$$

$$= 17.21 \text{ N/mm}^2 \text{ (Tensile)}$$ **Ans.**

EXAMPLE 9.3. A short column of external diameter 400 mm and internal diameter 200 mm carries an ecentric load of 80 KN. Find the greatest ecentricity which the load can have without producing tension on the cross-section.

Solution. Given :

Outer diameter, $d_o = 400$ mm

Inner diameter, $d_i = 200$ mm

Point load, $W = 80\ KN = 80 \times 1000$

$= 80000$ N

Let the greatest ecentricity be x mm which the load can have without producing tension.

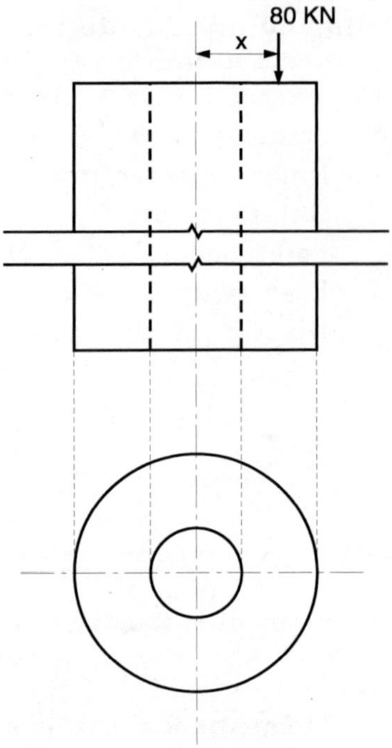

Fig. 9.9.

Area, $$A = \frac{\pi}{4}\left(d_o^2 - d_i^2\right) = \frac{\pi}{4}\left(400^2 - 200^2\right)$$

$$= 94247.78 \text{ mm}^2$$

Moment of inertia, $$I = \frac{\pi}{64}\left(d_o^4 - d_i^4\right) = \frac{\pi}{64}\left(400^4 - 200^4\right)$$

$$= \frac{\pi}{64} \times 2.4 \times 10^{10} = 1.18 \times 10^9 \text{ mm}^4$$

$$y = \frac{d_o}{2} = \frac{400}{2} = 200 \text{ mm}$$

Section modulus, $$Z = \frac{I}{y} = \frac{1.18\times10^9}{200} = 5.9 \times 10^6 \text{ mm}^3$$

Moment about the axis of the column,

$$M = P \times e = 80000\,x \text{ N-mm}$$

For no tension

$$\sigma_{min} = \frac{P}{A}-\frac{M}{Z} = 0$$

$$\therefore \quad \frac{P}{A} = \frac{M}{Z}$$

or, $$\frac{80000}{94247.78} = \frac{80000x}{5.9\times10^6}$$

or, $$\frac{80000\times5.9\times10^6}{94247.78\times80000} = x$$

$$\therefore \quad x = \frac{5.9\times10^6}{94247.78} = 62.6 \text{ mm}$$ **Ans.**

EXAMPLE 9.4. In a compreassion testing specimen 13 mm is diameter the line of thrust is parallel to the axis of the specimen but is displaced from it. Determine the distance of the line of thrust from the axis, when the maximum stress is 15 %. greater than the mean stress on a section normal to the axis.

Solution. Given : Diameter of specimen,

$$d = 13 \text{ mm}$$

$$\sigma_{max} = 15\% \text{ greater than mean stress}$$

Area, $$A = \frac{\pi}{4}d^2 = \frac{\pi}{4}\times(13)^2 = 132.73 \text{ mm}^2$$

Moment of inertia, $$I = \frac{\pi}{64}d^4 = \frac{\pi}{64}\times(13)^4$$

$$= 1401.98 \text{ mm}^4.$$

$$y = \frac{13}{2} = 6.5 \text{ mm}$$

Section modulus, $$Z = \frac{I}{y} = \frac{1401.98}{6.5} = 215.69 \text{ mm}^3$$

Let the ecentricity be x mm.

∴ Moment, $$M = Px$$

$$\text{Mean stress} = \frac{P}{A} = \frac{P}{132.73} \text{ N/mm}^2$$

$$\text{Maximum stress} = \text{Mean stress} + 15\% \text{ of mean stress}$$

$$\sigma_{max} = \frac{115}{100} \times \text{Mean stress}$$

$$\sigma_{max} = \frac{115}{100}\times\frac{P}{132.73} \text{ N/mm}^2 \qquad ...(i)$$

Using relation

$$\sigma_{max} = \frac{P}{A}+\frac{M}{Z}$$

$$\frac{115}{100\times132.73} = \frac{P}{132.73}+\frac{Px}{215.69} \qquad \text{[from } (i)\text{]}$$

or, $$\frac{115}{13273} = \frac{1}{132.73}+\frac{x}{215.69}$$

or, $$\frac{115}{13273} - \frac{1}{132.73} = \frac{x}{215.69}$$

or, $$0.0087 - 0.0075 = \frac{x}{215.69}$$

$$0.0012 \times 215.69 = x$$

∴ $$x = 0.2588 \simeq 0.26 \text{ mm}$$ **Ans.**

EXAMPLE 9.5. A hollow rectangular masonry pier is 1.2 mm × 0.80 m, overall, the wall thickness being 0.15 m. A vertical load of 100 KN is transmitted in the vertical plane bisecting 1.2 m side at an ecentricity of 0.1 m from the geometric axis of the section. Calculate the maximum and minimum stress intensities is the section.

Solution. Given :

Point load, $P = 100 \text{ KN} = 100000 \text{ N}$

Ecentricity, $e = 0.1 \text{ m} = 100 \text{ mm}$

$b_o = 1200 \text{ mm}$

Outer diameter, $d_o = 800 \text{ mm}$

$b_i = 1200 - 150 - 150 = 900 \text{ mm}$

Inner diameter, $d_i = 800 - 150 - 150 = 500 \text{ mm}$

Cross-sectional Area,

$$A = 1200 \times 800 - 900 \times 500$$
$$= 960000 - 450000$$
$$= 510000 \text{ mm}^2$$

Moment of inertia,

$$I = \frac{1}{12} \times 1200 \times 800^3 - \frac{1}{12} \times 900 \times 500^3$$
$$= 5.12 \times 10^{10} - 9.37 \times 10^9$$
$$= 4.18 \times 10^{10} \text{ mm}^4.$$

Moment, $M = P \times e = 100000 \times 100$

$$= 10 \times 10^6 \text{ N-mm}$$

$$y = \frac{800}{2} = 400 \text{ mm}$$

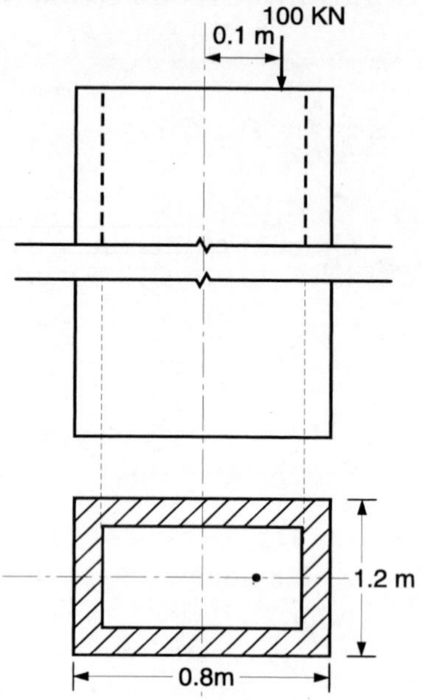

Fig. 9.10.

$$\text{Section modulus, } Z = \frac{I}{y} = \frac{4.18 \times 10^{10}}{400} = 10.45 \times 10^7 \text{ mm}^3.$$

Using relation

$$\sigma_{max} = \frac{P}{A} + \frac{M}{Z} = \frac{100000}{510000} + \frac{10 \times 10^6}{10.45 \times 10^7}$$
$$= 0.196 + 0.096 = 0.292 \text{ N/mm}^2 \quad \textbf{Ans.}$$

$$\sigma_{min} = \frac{P}{A} - \frac{M}{Z} = \frac{100000}{510000} - \frac{10 \times 10^6}{10.45 \times 10^7}$$
$$= 0.196 - 0.096 = 0.1 \text{ N/mm}^2. \quad \textbf{Ans.}$$

EXAMPLE 9.6. A short hollow pier 1.5 m square outside and 1 m square inside, supports a vertical point load of 7 KN located on a diagonal 0.8 m from the vertical axis of the pier. Neglecting the self weight of the pier. Calculate the normal stresses at the four outside corners on a horizontal section of the pier.

Solution: Given :

Outer area, $A_o = 1.5 \times 1.5$

$$= 2.25 \text{ m}^2$$

Inner area, $A_i = 1 \times 1 = 1 \text{ m}^2$

Point load, $P = 7 \text{ KN}$

Ecentricity, $e = 0.8$ m

Area of cross-section of pier $= A - A_i$

$= 2.25 - 1 = 1.25\ \text{m}^2$

Moment of inertia about the diagonal

$$I = 2\times\frac{1}{12}\left(2.12\times1.06^3 - 1.41\times0.70^3\right)$$

$$= \frac{1}{6}\times(2.52-0.48) = 0.34\ \text{m}^4.$$

$$y = \frac{2.12}{2} = 1.06\ \text{m}$$

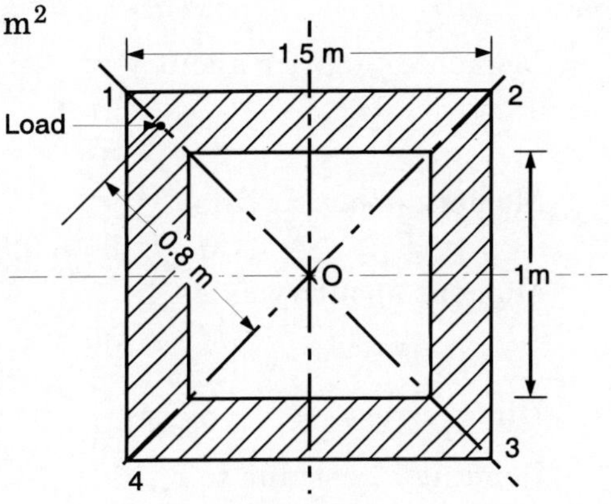

Fig. 9.11.

Section modulus, $z = \dfrac{I}{y} = \dfrac{0.34}{1.06} = 0.32\ \text{m}^3$

Moment, $M = p \times e$

$= 7 \times 0.8 = 5.6$ KN-m

Direct stress, $\sigma_d = \dfrac{P}{A} = \dfrac{7}{1.25} = 5.6\ \text{KN/m}^2$

Bending stress, $\sigma_b = \dfrac{M}{Z} = \dfrac{5.6}{0.32} = 17.5\ \text{KN/m}^2$

At the corners 2 and 4 there will be no bending stress

$\therefore$ Stress at corner 1 $= \sigma_d + \sigma_b = 5.6 + 17.5$

$= 23.1\ \text{KN/m}^2$ (Compressive)

Stress at corner 2 $= \sigma_d = 5.6\ \text{KN/m}^2$ (Comp.)

Stress at corner 3 $= \sigma_d - \sigma_b = 5.6 - 17.5$

$= -11.9\ \text{KN/m}^2$

$= 11.9\ \text{KN/m}^2$ (Tensile)

Stress at corner 4 $= \sigma_d = 5.6\ \text{KN/m}^2$ (Comp.)

EXAMPLE 9.7. A masonary pies of 3 m × 4 m supports a vertical load of 80 KN as shown in Figure 9.12.

(*a*) Find the stress developed at each corner of the pier.

(*b*) What additional load should be placed at the centre of the pier, so that there is no tension anywhere in the pier section ?

(*c*) What are the stresses at the corners with the additional load in the centre ?

Solution. Given :

Width, $b = 4$ m

Depth, $d = 3$ m

Point load, $P = 80$ KN

Ecentricity from x-x axis

$e_x = 0.5$ m

Ecentricity from y-y axis is

$e_y = 1$ m

Area, $A = 4 \times 3 = 12\ \text{m}^2$

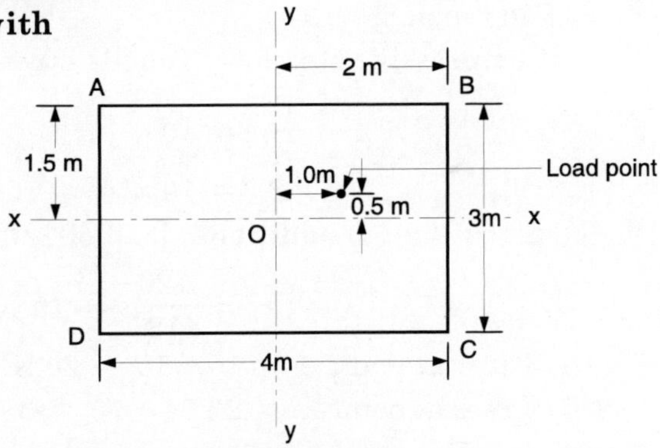

Fig. 9.12.

Moment of inertia about x-x,

$$I_{xx} = \frac{1}{12} \times 4 \times 3^3 = 9 \text{ m}^4$$

Moment of inertia about y-y,

$$I_{yy} = \frac{1}{12} \times 3 \times 4^3 = 16 \text{ m}^4$$

Moment about xx axis,

$$M_x = 80 \times 0.5 = 40 \text{ KN-m}$$

Moment about yy axis,

$$M_y = 80 \times 1 = 80 \text{ KN-m}$$

Direct stress, $$\sigma_d = \frac{P}{A} = \frac{80}{12} = 6.67 \text{ KN/m}^2$$

Bending stress due to e_x,

$$\sigma_{bx} = \frac{M_x \times y}{I_{xx}} = \frac{40 \times 1.5}{9} = 6.67 \text{ KN/m}^2$$

Bending Stress due to e_y,

$$\sigma_{by} = \frac{M_y \times x}{I_{yy}} = \frac{80 \times 2}{16} = 10 \text{ KN/m}^2$$

(*a*) (*i*) Resultant stress at point *A*,

$$\sigma_A = \sigma_d - \sigma_{bx} + \sigma_{by} \qquad [\because \text{Point A is in 2nd quadrant}]$$

$$\sigma_A = 6.67 - 6.67 + 10 = 10 \text{ KN/m}^2 \text{ (Compressive)}$$

(*ii*) Resultant stress at point *B*,

$$\sigma_B = \sigma_d + \sigma_{bx} + \sigma_{by} = 6.67 + 6.67 + 10$$
$$= 23.34 \text{ KN/m}^2 \text{ (Compressive)}$$

(*iii*) Resultant stress at point *C*,

$$\sigma_C = \sigma_d + \sigma_{bx} - \sigma_{by} = 6.67 + 6.67 - 10$$
$$= 3.34 \text{ KN/m}^2 \text{ (Compressive)}$$

(*iv*) Resultant stress at point *D*,

$$\sigma_D = \sigma_d - \sigma_{bx} - \sigma_{by} = 6.67 - 6.67 - 10$$
$$= -10 \text{ KN/m}^2 = 10 \text{ KN/m}^2 \text{ (Tensile)}$$

(*b*) Let *P*, be the additional load that should be placed in the pier section for no tension.
For no tension,

Compressive stress = Tensile stress

$$\frac{P_1}{A} = 10$$

$$P_1 = 10 \times A = 10 \times 12 = 120 \text{ KN}$$

(c) Stress due to additional load of 120 KN at the centre

$$= \frac{120}{12} = 10 \text{ KN/m}^2$$

(*i*) Stress at point *A* = 10 + 10 = 20 KN/m^2

(*ii*) Stress at point *B* = 23.34 + 10 = 33.34 KN/m^2

(*iii*) Stress at point *C* = 3.34 + 10 = 13.34 KN/m^2

(*iv*) Stress at point *D* = –10 + 10 = 0.

EXAMPLE 9.8. A short column of rectangular cross-section 80 mm × 60 mm carries a load of 40 KN at a point 20 mm from the longer side and 35 mm from the shorter side. Determine the maximum compressive and tensile stresses in the section.

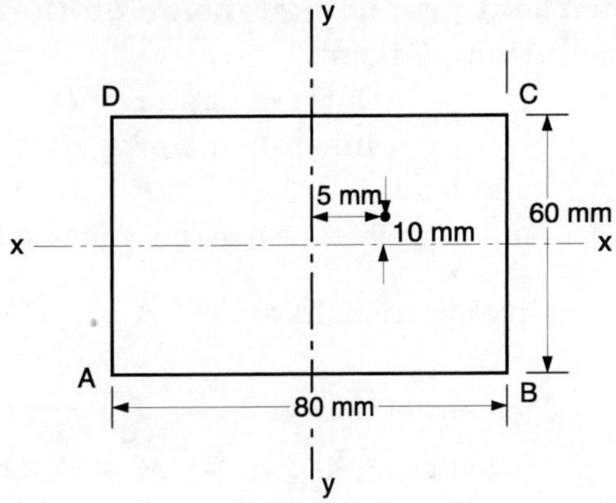

Fig. 9.13.

Solution: Given :

Width, $b = 80$ mm

Depth, $d = 60$ mm

Point load, $P = 40$ KN $= 40000$ N

$e_x = 10$ mm

$e_y = 5$ mm

Area, $A = b \times d = 80 \times 60 = 4800 \text{ mm}^2$

Moment of inertia about x-x axis,

$$I_{xx} = \frac{1}{12}bd^3 = \frac{1}{12} \times 80 \times (60)^3 = 1.44 \times 10^6 \text{ mm}^4$$

Moment of inertia about y-y axis,

$$I_{yy} = \frac{1}{12}bd^3 = \frac{1}{12} \times 60 \times (80)^3 = 2.56 \times 10^6 \text{ mm}^4$$

$$y = \frac{60}{2} = 30 \text{ mm}$$

$$x = \frac{80}{2} = 40 \text{ mm}$$

Moment about x-x axis,

$$M_x = P \times e_x = 40000 \times 10 = 400000 \text{ N-mm}$$

Moment about y-y axis,

$$M_y = P \times e_y = 40000 \times 5 = 200000 \text{ N-mm}$$

Direct stress, $\sigma_d = \frac{P}{A} = \frac{40000}{4800} = 8.33 \text{ N/mm}^2$

Bending stress due to e_x, $\sigma_{bx} = \frac{M_x \times y}{I_{xx}} = \frac{400000 \times 30}{1.44 \times 10^6}$

$= 8.33 \text{ N/mm}^2$

Bending stress due to c_y,

$$\sigma_{by} = \frac{M_y \times x}{I_{yy}} = \frac{200000 \times 40}{2.56 \times 10^6}$$

$= 3.125 \text{ N/mm}^2$

Maximum stress $= \sigma_d + \sigma_{bx} + \sigma_{by} = 8.33 + 8.33 + 3.125$

$= 19.78 \text{ N/mm}^2$ (Compressive)

Minimum stress $= \sigma_d - \sigma_{bx} - \sigma_{by} = 8.33 - 8.33 - 3.125$

$= -3.125 \text{ N/mm}^2$

$= 3.125 \text{ N/mm}^2$ (Tensile)

EXAMPLE 9.9. A short column of external diameter '*D*' and internal diameter '*d*' carries an ecentric load '*W*'. Find the greatest ecentricity which the load can have without producing tension on the cross-section of the column.

Solution: Given :

Outer diameter $= D$

Inner diameter $= d$

Point load, $P = W$

Let the greatest ecentricity for no tension be e.

Cross-sectional area, $A = \frac{\pi}{4}(D^2 - d^2)$

Moment of inertia, $I = \frac{\pi}{64}(D^4 - d^4)$

Moment, $M = W \times e$

$$y = \frac{D}{2}$$

Direct stress, $\sigma_d = \frac{P}{A} = \frac{W}{\frac{\pi}{4}(D^2 - d^2)}$

Bending stress, $\sigma_b = \frac{My}{I} = \frac{W \times e \times D/2}{\frac{\pi}{64}(D^4 - d^4)}$

For no tension

Direct stress = Bending stress

$$\frac{W}{\frac{\pi}{4}(D^2 - d^2)} = \frac{We \times D}{\frac{\pi}{64}(D^4 - d^4) \times 2}$$

$$\frac{4W}{\pi(D^2 - d^2)} = \frac{32\ We \times D}{\pi(D^2 + d^2)(D^2 - d^2)}$$

Dividing both sides by $\frac{4W}{\pi(D^2 - d^2)}$, we get $1 = \frac{8eD}{D^2 + d^2}$

or, $\frac{D^2 + d^2}{8D} = e$

$\therefore$ Greatest ecentricity,

$$e = \frac{D^2 + d^2}{8D}$$ **Ans.**

EXAMPLE 9.10. A column section 300 mm external diameter and 150 mm internal diameter supports an axial load of 2600 KN and an ecentric load of P KN at an ecentricity of 400 mm. If the compressive and tensile stresses are not to exceed 140 N/mm² and 60 N/mm² respectively, find the magnitude of the load P.

Solution: Given :

Outer diameter, $d_0 = 300$ mm

Inner diameter, $d_i = 150$ mm

Axial load, $W = 2600 \text{ KN} = 2600 \times 1000 \text{ N}$

Ecentric load $= P$ KN

Ecentricity, $e = 400$ mm

Cross-sectional area,

$$A = \frac{\pi}{4}(d_0^2 - d_i^2) = \frac{\pi}{4}(300^2 - 150^2)$$

$$= 53014.38 \text{ mm}^2$$

Moment of inertia,

$$I = \frac{\pi}{64}(d_0^4 - d_i^4) = \frac{\pi}{64}(300^4 - 150^4)$$

$$= \frac{\pi}{64} \times 7.59 \times 10^9 = 37.26 \times 10^7 \text{ mm}^4.$$

$$y = \frac{300}{2} = 150 \text{ mm}$$

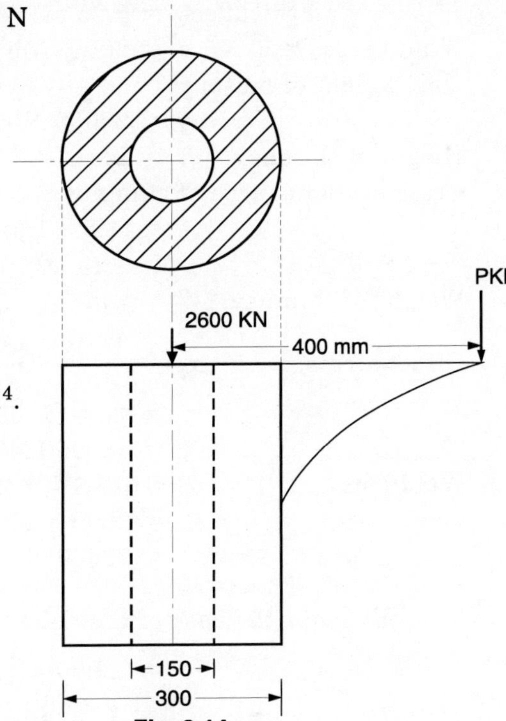

Fig. 9.14

Sectional modulus, $Z = \frac{I}{y} = \frac{37.26 \times 10^7}{150}$

$$= 2484000 \text{ mm}^3$$

Moment, $M = P \times e = P \times 400$

$$= 400\text{P KN-mm}$$

$$= 400000 \text{ PN-mm}$$

Direct stress, $\sigma_d = \frac{(2600 + P) \times 1000}{53014.38} = 0.019\ (2600 + P)$

Bending stress, $\sigma_b = \frac{400000P}{2484000} = 0.16\ P \text{ N/mm}^2$

$$\sigma_{max} = \sigma_d + \sigma_b$$

$$140 = 0.019\ (2600 + P) + 0.16\ P$$

or $140 = 49.4 + 0.019P + 0.16P$

$$140 - 49.4 = 0.179\ P$$

$\therefore$ $90.6 = 0.179$ P

$\therefore$ $P = \frac{90.6}{0.179} = 506.14$ KN ...(*i*)

$$\sigma_{min} = \sigma_b - \sigma_d$$

$$60 = 0.16P - 0.019\ (2600 + P)$$

$$= 0.16P - 49.4 - 0.019\ P$$

$$60 + 49.4 = 0.141P$$

$\therefore$ $P = 775.89$ KN ...(*ii*)

The magnitude of P will be minimum of (*i*) and (*ii*)

i.e., $P = 506.14$ KN **Ans.**

EXAMPLE 9.11. A square chimney 24 m high has an opening of 1 m × 1 m and wall thickness 0.25 m. Calculate the maximum stress in the masonry if the horizontal wind pressure is 2000 N/m² and masonry weight 20000 N/m³.

Solution: Given :

Area of opening $= 1 \text{ m} \times 1 \text{ m}$

Wall thickness $= 0.25$ m

Wind pressure, $p = 2000$ N/m^2

Unit weight of masonry,

$\gamma = 20000$ N/m^3

Height of chimney, $h = 24$ m

Cross-sectional area of chimney,

$$A = (1.5 \times 1.5) - (1 \times 1) = 1.25 \text{ m}^2$$

Weight of chimney, $W = \gamma A h$

$$\text{Direct stress, } \sigma_d = \frac{W}{A} = \frac{\gamma A h}{A} = \gamma h = 20000 \times 24 = 480000 \text{ N/m}^2$$

$$\text{Wind load, } P = \text{p} \times 1.5 \times 24 = 2000 \times 1.5 \times 24 = 72000 \text{ N}$$

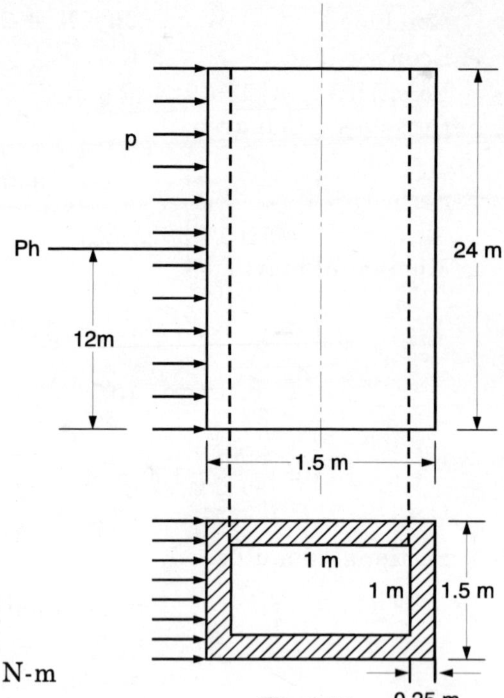

Fig. 9.15.

$$\text{Moment about the base of the chimney} = P \times \frac{h}{2} = 72000 \times 12 = 864000 \text{ N-m}$$

$$\text{Moment of inertia, } I = \frac{1}{12}(1.5^4 - 1^4) = \frac{1}{12} \times 4.06 = 0.338 \text{ m}^4.$$

$$y = \frac{1.5}{2} = 0.75 \text{ m}$$

$$\text{Bending stress, } \sigma_b = \frac{My}{I} = \frac{864000 \times 0.75}{0.338} = 1917159.8 \text{ N/m}^2$$

$$\sigma_{max} = \sigma_b + \sigma_d = 480000 + 1917159.8 = 2397159.8 \text{ N/m}^2 \text{ (Compressive)}$$ **Ans.**

EXAMPLE 9.12. A masonry chimney of a hollow circular section has external and internal diameters of 2.5 m and 2 m respectively. It is subjected to uniform horizontal wind pressure of 1800 N/m². Determine the maximum height of chimney for developing no tension at the base. Density of masonry is 20000 N/m³.

Solution: Given :

$d_0 = 2.5$ m

$d_i = 2.0$ m

$p = 1800$ N/m^2

$\gamma = 20000$ N/m^3

Let the height of chimney be h m.

Weight of chimney $= \gamma A h$

$$\text{Direct stress, } \sigma_d = \frac{\gamma A h}{A} = \gamma h = 20000 \text{ h N/m}^2$$

$$\text{Wind load, } P = p \times 2.5 \times h = 1800 \times 2.5 \times h = 4500\, h \text{ N}$$

Moment due to P about the base

$$M = P \times e = 4500\, h \times \frac{h}{2}$$

$$= 2250\, h^2 \text{ N-m}$$

Moment of inertia, $I = \dfrac{\pi}{64} \times (2.5^4 - 2^4) = 1.132 \text{ m}^4$

$$y = \frac{2.5}{2} = 1.25 \text{ m}$$

Section modulus, $Z = \dfrac{I}{y} = \dfrac{1.132}{1.25} = 0.906$

Bending stress, $\sigma_b = \dfrac{M}{Z} = \dfrac{2250h^2}{0.906}$

$$= 2483.4\, h^2 \text{ N/m}^2$$

For no tension $\sigma_d = \sigma_b$

$$20000\, h = 2483.4\, h^2$$

$$\frac{20000}{2483.4} = h = 8.05 \text{ m} \qquad \textbf{Ans.}$$

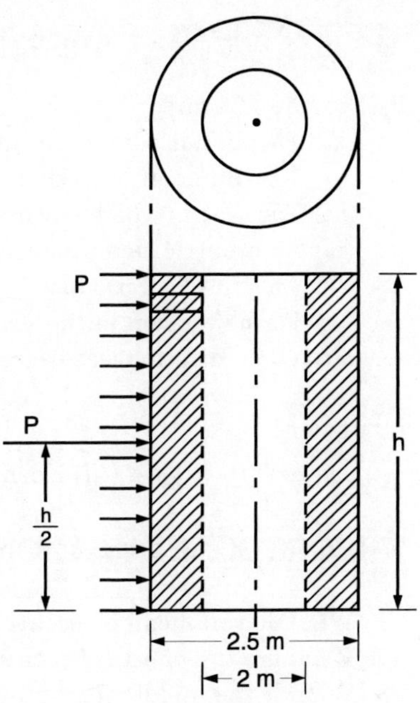

Fig. 9.16.

SUMMARY

1. The load whose line of action coincides with the axis of the column is known as axial load.
2. The load whose line of action does not coincide with the axis of column is known as ecentric load.
3. Axial load produces direct stress in a column whereas ecentric load produces both direct and bending stress in the column.
4. Ecentricity of a load is the distance between the line of action of the load and the C.G. of the section.
5. If a column is subjected to an ecentric load P, then

$$\sigma_{max} = \frac{P}{A} + \frac{M}{Z}$$

$$\sigma_{min} = \frac{P}{A} - \frac{M}{Z}$$

where A = Area of cross-section

Z = Modulus of section

M = Bending moment.

6. If direct stress is equal to bending stress, then tensile stress will be zero.
7. If bending stress is more than direct stress, than there will be tensile stress across the section.
8. Core or Kernel of the section is the area in which any ecentric load if placed will not produce tension in the section.
9. In a rectangular section, for no tension, the load should be placed within middle third of the section *i.e.*, $e \leq \dfrac{b}{6}.1$
10. For no tension in a circular section, $e \leq \dfrac{d}{8}$ where d = diameter of the section.

OBJECTIVE TYPE QUESTIONS

Fill in the Blanks

(*i*) The distance between line of action of a load and the *C.G.* of the section is known as ________.

(*ii*) The diameter of Kernel of a circular section is ________.

(*iii*) The area in which any ecentric load may be placed without causing tension, is known as ________.

(*iv*) An ecentric load produces direct as well as ________ stress.

(*v*) An axial load produces ________ stress in a column.

(*vi*) For no tension in the section, the ecentricity must not exceed ________.

(*vii*) When direct stress is greater than bending stress, there will be no ________.

Answers

(*i*) ecentricity (*ii*) *d*/4 (*iii*) Kernel (*iv*) bending (*v*) direct (*vi*) $\frac{Z}{A}$ (*vii*) tension.

THEORETICAL QUESTIONS

1. What do you mean by ecentric load ?
2. What are the effects of ecentric loads on a short column ?
3. What is the middle third rule ?
4. Draw the stress distribution diagram in case of an ecentrically loaded column.
5. Give the limits of ecentricity in the following cases:
   (*i*) Rectangular section
   (*ii*) Circular section.

NUMERICALS

1. A short hollow cylindrical column is 300 mm in external diameter and 25 mm thick. It carries a vertical load of 1000 KN at an ecentricity of 20 mm. Calculate the maximum and minimum normal stress intensities induced in the section.
   (σ_{max} = 60.87 MN/m^2 (comp) ; σ_{min} = 31.73 MN/m^2 (comp))
2. A rectangular column of width 120 mm and of thickness 100 mm carries a point load of 120 KN at an ecentricity of 10 mm. Determine the maximum and minimum stresses at the base of the column. Also draw the stress distribution diagram.
   (σ_{max} = 15 N/mm^2 (comp) ; σ_{min} = 5 N/mm^2 (comp))
3. A square column 250 mm × 250 mm carries a vertical load 120 KN at a distance 600 mm from the axis as shown in figure 9.17. Find the maximum and minimum stresses induced in the section.
   (σ_{max} = 412.86 N/cm^2(comp) ; σ_{min} = –28.86 N/cm^2(tensile))
4. A steel pillar of circular section is of outside diameter 40 cm and metal thickness 4 cm. A projecting bracket attached to it carries a load of 10 tones. The ecentricity of load is 25 cm. Compute the maximum and minimum intensities of stress in the section.
   (σ_{max} = 0.103 *t*/cm^2 (comp) ; σ_{min} = 0.059 *t*/cm^2 (tensile))

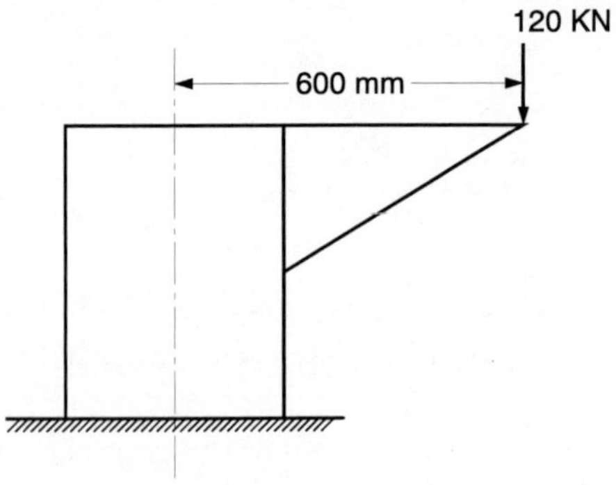

Fig. 9.17.

5. A load of 75 KN is carried by a column made of cast iron. The external and internal diameters are 200 mm and 180 mm respectively. If the ecentricity of the load is 35 mm, find :
   (*i*) The maximum and minimum stress intensities

(*ii*) upto what ecentricity there is no tensile stress in the column ?

(σ_{max} = 22.28 MN/m^2 (comp) ; σ_{min} = 2.84 MN/m^2 (comp) ; 45.2 mm)

6. A hollow rectangular masonry pies is 1.2 m × 0.8 m, overall. The wall thickness being 0.15 m. A vertical load of 100 KN is transmitted in the vertical plane bisecting 1.2 m side at an ecentricity of 0.1 m from the geometric stress intensities in the section. Calculate the maximum and minimum stress intensities in the section. (σ_{max} = 291.6 KN/m^2 (comp) ; σ_{min} = 100.4 KN/m^2 (comp))
7. A short column of *I*-section 25 cm × 20 cm has a cross-sectional area 52 cm^2 and maximum radius of gyration 10.7 cm. A vertical load W KN acts through the centroid of the section together with a parallel load of *W*/4 KN acting through a point on the centre line of the web distant 6 cm from the centroid. Calculate the greatest allowable value of *W* if the maximum stress is not to exceed 65 MN/m^2. What is then the minimum stress. (239.4 KN ; 49.4 MN/m^2)
8. A column section 200 mm external diameter and 160 mm internal diameter supports an axial load of 200 KN and centric load of *P* KN (on the bracket) at an ecentricity of 300 mm. If the compressive and tensile stresses are not to exceed 140 N/mm^2 and 105 N/mm^2 respectively, find magnitude of the load *P*. (165.68 KN)

❑❑❑

CHAPTER 10 Torsion

10.1. TORSION

Torsion is the twisting of a structural member when it is loaded by couples that produce rotation about its longitudinal axis.

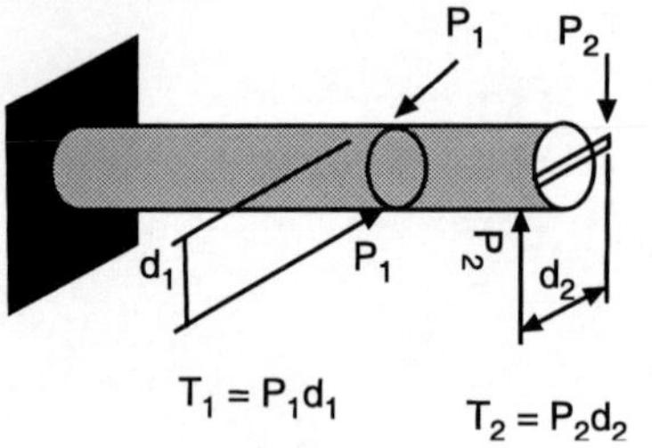

Fig. 10.1

Figure 10.1 shows a circular straight bar fixed at one end and loaded by two pairs of forces. Each pair of forces $P_1 - P_1$ and $P_2 - P_2$ forms a couple that tends to twist the bar about its longitudinal axis.

The moment of couple of forces $P_1 - P_1$,

$$T_1 = P_1 d_1$$

and moment of couple of forces $P_2 - P_2$,

$$T_2 = P_2 d_2$$

Moment of couples T_1 and T_2 are shown in figure 10.2. These couples T_1 and T_2 are called **torques**, **twisting couples** or **twisting moments**.

Units of Torque is N-m or N-mm

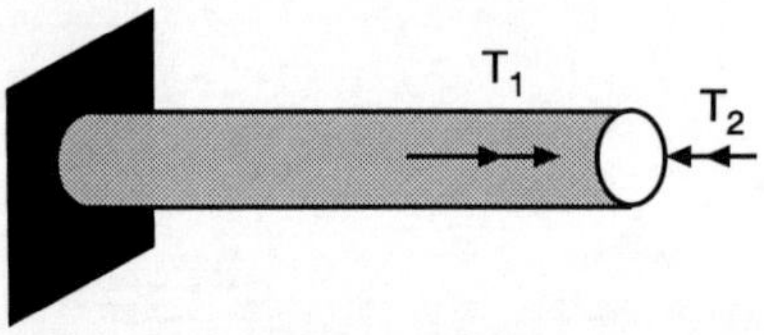

Fig. 10.2

10.2. PURE TORSION

A member is said to be under *pure torsion* when it is subjected to two equal and opposite torques.

Chapter Outline

- Torsion
- Pure Torsion
- Angle of Twist
- Assumption in the Theory of Pure Corsion
- Theory of Pure Torsion
- Power Transmitted by a Shaft
- Strength of a Shaft
- Strength of a Hollow Circular Shaft
- Torsional Rigidity (or Stiffness)
- Comparison of Hollow and Solid Shafts

DO YOU KNOW?

If a moment is applied in a vertical plane containing the longitudinal axis of the beam, it will set bending stresses in the beam.

Figure 10.3 shows pure torsion in a circular shaft. T is the torque or twisting couple acting in the shaft.

While a beam bends due to bending moment, a shaft twists due to torsion. When a shaft is under the action of pure torsion, its cross-sections are under pure shear stresses. By the principle of complementary shear stresses, the shear stresses acting on the plane of the cross-section are accompanied by shear stresses of the same magnitude acting on the longitudinal planes as shown in figure 10.4.

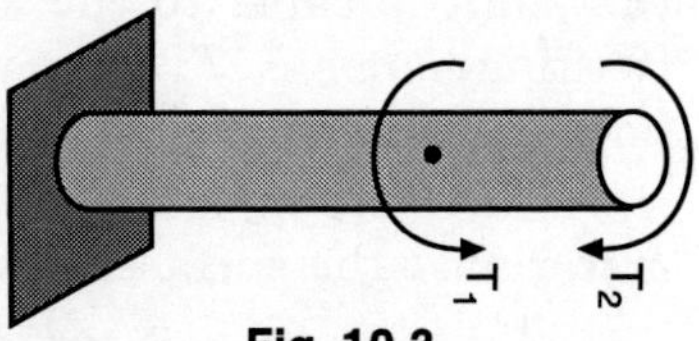

Fig. 10.3

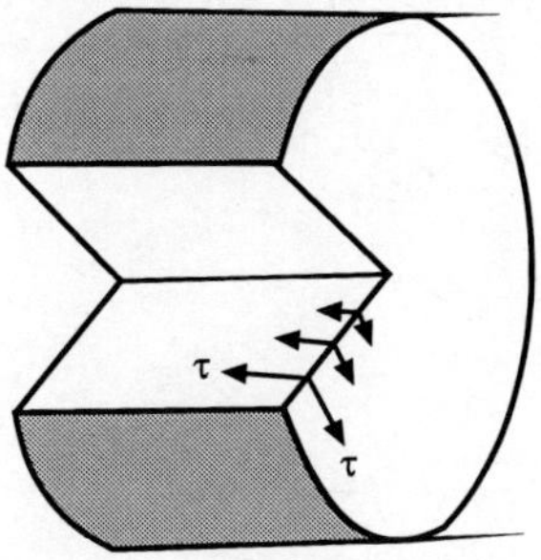

Fig. 10.4

DO YOU KNOW?

If a moment is applied in a vertial plane perpendicular to the longitudinal axis of the beam it will be subjected to torque causing torsion in the member.

10.3. ANGLE OF TWIST

During twisting of shaft, there will be a rotation about the longitudinal axis of one end w.r.t to the other. This angle of rotation is known as **angle of twist (θ)**. Figure 10.5 shows angle of twist.

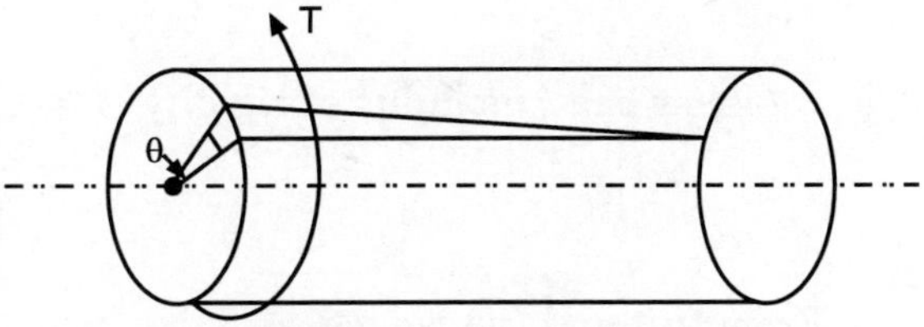

Fig. 10.5. Angle of twist.

10.4. ASSUMPTION IN THE THEORY OF PURE TORSION

The torsion equation is based on the following assumptions :

(*i*) The material of the shaft is homogeneous and isotropic.

(*ii*) The shaft is of uniform circular section throughout its length.

(*iii*) The twist is uniform along the length of the shaft.

(*iv*) Cross-section of the shaft, which are plane before twist remain plane after twist.

(*v*) Stresses induced in the shaft due to torsion do not exceed the proportional limit.

(*vi*) The distance between any two cross-sections remains the same after the application of torque.

10.5. THEORY OF PURE TORSION

10.5.1 For Solid Shaft

Consider a solid circular shaft of length 'l' and radius 'R' subjected to a twisting moment 'T' at one end and the other end is fixed as shown in figure 10.6.

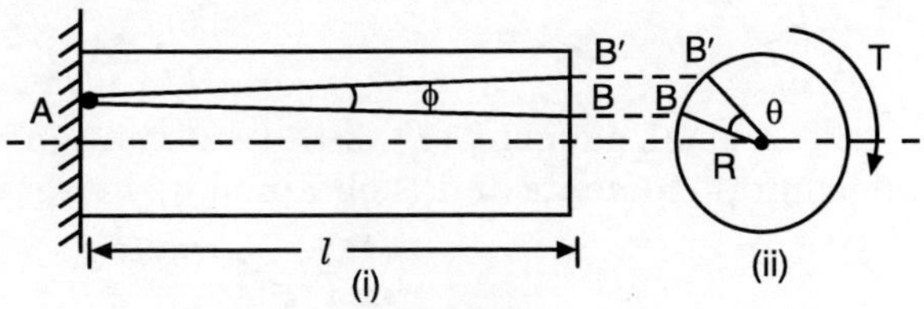

Fig. 10.6

Note: The mathematical treatment of stress and deformations induced by torsion is too complicated, except in the cases of solid and hollow circular shaft.

Let us consider an imaginary line AB on the surface of the shaft parallel to the axis of the shaft. Due to this twisting moment 'T', the point B will be shifted to B' and line AB will be deformed to AB'.

$\angle BAB'$ represents the shear strain (ϕ) of the shaft material at the surface.

DO YOU KNOW ?

Shear stress is maximum at the surface of a shaft and zero at its longitudinal axis. Thus the material near the centre contributes little to the shaft's torque resisting capacity.

In Δ BAB' $\qquad \tan\phi = \dfrac{BB'}{AB}$

or $\qquad \phi = \dfrac{BB'}{l}$ $\qquad$ [$\because$ $\tan\phi = \phi$, as ϕ is very small]

$\therefore \qquad BB' = l\phi \qquad$ ...(*i*)

From Fig. 10.6 (*ii*), $BB' = R\theta$ ...(*ii*)

From (*i*) and (*ii*), we get

$$l\phi = R\theta$$

$\therefore \qquad \phi = \dfrac{R\theta}{l} \qquad$ ...(*iii*)

we know that, modulus of rigidity (C) $= \dfrac{\text{Shear stress } (\tau)}{\text{Shear strain } (\phi)}$

$\therefore \qquad \phi = \dfrac{\tau}{c} \qquad$ ...(*iv*)

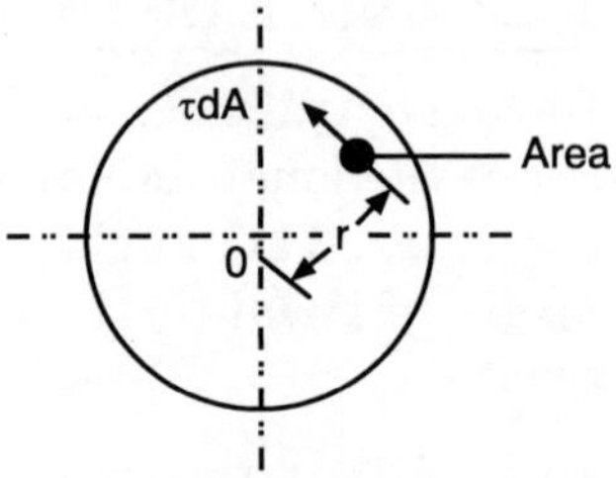

Fig. 10.7

From (*iii*) and (*iv*) we get

$$\frac{\tau}{c} = \frac{R\theta}{l}$$

or $\qquad \dfrac{\tau}{R} = \dfrac{C\theta}{l} \qquad$ ...(*v*)

Torsional Moment of Resistance : Figure 10.7 shows a cross-section of a circular shaft of radius 'R'. Consider an elementary area 'dA' in the cross-section at a distance 'r' from centre o. The shear force acting on the elementary area 'dA' $= \tau dA$.

where τ = shear stress for the elementary area $= \dfrac{C\theta r}{l}$ $\qquad$ [using (*v*)]

Moment of this force about the centre o $= (\tau dA) \times r$

$$= \left(\frac{C\theta r}{l} dA\right) \times r = \frac{C\theta}{l} r^2 dA$$

Torsional moment (T) is the summation over the entire cross-sectional area 'A' of such elemental moments and is obtained by integrating this elemental moment.

$$T = \int \frac{C\theta}{l} r^2 dA = \frac{C\theta}{l} \int r^2 dA$$

$$= \frac{C\theta}{l} I_p \text{ where } I_p = \text{ Polar moment of inertia} = \int r^2 dA$$

$\therefore \qquad \dfrac{T}{I_p} = \dfrac{C\theta}{l} \qquad$ ...(*vi*)

From (v) and (vi) we get,

$$\boxed{\frac{\tau}{R} = \frac{T}{I_p} = \frac{C\theta}{l}} \quad \ldots(vii)$$

Equation (vii) is called the *torsional equation for circular shaft.*

where τ = Shear stress
T = Torsion in the shaft
R = Radius of circular shaft
l = Length of the shaft
θ = Angle of twist
C = Modulus of rigidity
I_p = Polar moment of inertia

10.5.2 Hollow Circular Shafts

Consider a hollow circular shaft subjected to a twisting moment (torque) 'T'.

Take an elementary circular ring of thickness 'dx' at a distance 'x' from the centre.

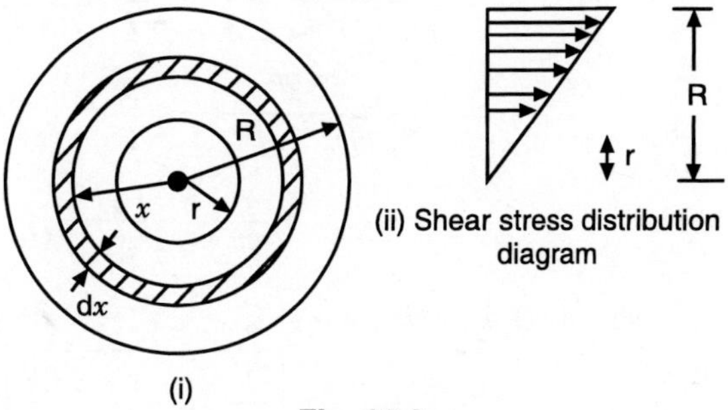

Fig. 10.8

Area of elementary ring, $A = 2\pi x \times dx$

Shear stress at radius 'x' $= \frac{\tau}{R} \times x$

Turning force on the elementary ring

$$= \text{Shear stress} \times \text{A}$$

$$= \frac{\tau}{R} \times x \times 2\pi x dx = \frac{\tau}{R} \times 2\pi\, x^2 dx$$

Turning moment on the elementary ring, dT

$$= \text{Turning force} \times \text{distance}$$

$$dT = \frac{\tau}{R} \times 2\pi\, x^2 dx \times x$$

$$dT = \frac{\tau}{R} \times 2\pi\, x^3 dx$$

DO YOU KNOW?

Intensity of shear stress at any point in the cross section of a shaft subjected to pure torsion is proportional to its distance from the centre.

Total turning moment on the shaft,

$$T = \int_r^R \frac{\tau}{R} \times 2\pi\, x^3 dx = \frac{\tau}{R} \times 2\pi \int_r^R x^3 dx$$

$$= \frac{\tau}{R} \times 2\pi \left[\frac{x^4}{4}\right]_r^R = \frac{2\pi}{4} \times \frac{\tau}{R} [x^4]_r^R = \frac{\pi}{2} \times \frac{\tau}{R} [R^4 - r^4]$$

$$= \frac{\pi}{2} \times \frac{\tau}{\frac{D}{2}} \left[\left(\frac{D}{2}\right)^4 - \left(\frac{d}{2}\right)^4\right] = \frac{\pi\tau}{D}\left[\frac{D^4}{16} - \frac{d^4}{16}\right]$$

$$\therefore \quad T = \frac{\pi\tau}{16\, D}(D^4 - d^4)$$

$$\therefore \qquad \tau = \frac{16\,T}{\pi}\left(\frac{D}{D^4 - d^4}\right)$$

Using relation

$$\frac{\tau}{R} = \frac{C\theta}{l} \qquad ...(i)$$

$$\frac{\tau \times 2}{D} = \frac{C\theta}{l} \qquad \left[\because R = \frac{D}{2}\right]$$

$$\frac{16\,T \times 2 \times D}{D\pi\,(D^4 - d^4)} = \frac{C\theta}{l} \qquad \left[\because \tau = \frac{16T}{\pi}\left(\frac{D}{D^4 - d^4}\right)\right]$$

$$\frac{32\,T}{\pi(D^4 - d^4)} = \frac{C\theta}{l}$$

$$\frac{T}{I_p} = \frac{C\theta}{l} \quad ...(ii) \quad \left[\because \frac{\pi}{32}(D^4 - d^4) = I_p = \text{Polar moment of inertia}\right]$$

From (*i*) and (*ii*), we get,

$$\boxed{\frac{T}{I_p} = \frac{\tau}{R} = \frac{C\theta}{l}}$$

10.6. POWER TRANSMITTED BY A SHAFT

The power transmitted by a shaft is given by;

Power, $$P = \frac{2\pi N T_a}{60}\text{ watts} = \frac{2\pi N T_a}{60 \times 1000}\text{ KW}$$

where T_a = Average or mean torque transmitted in Nm

N = Number of revolution per minute (r.p.m)

Note:

1. The torque varies from zero to maximum w.r.t radius of shaft. It is zero at the centre and maximum at the circumference. Hence, average or mean torque is used to calculate power transmitted by a shaft.
2. Care should be taken while solving a numerical as we get maximum torque, T from torsion equation and for power calculation we need average torque, T_a. The maximum torque is always 30 to 40% more than the average torque.

10.7. STRENGTH OF A SHAFT

The maximum torque or maximum power which a shaft can transmit is known as **strength of a shaft**.

Strength of a Solid Circular Shaft

Let, τ = Shear stress induced in the shaft

T = Maximum torque transmitted by the shaft

D = Diameter of the shaft

I_p = Polar moment of inertia = $\frac{\pi}{32} D^4$

DO YOU KNOW?

The resistance torque of the shaft must atleast be equal to mean torque, so that maximum shear stress is not exceeded.

Using relation

$$\frac{T}{I_p} = \frac{\tau}{R}$$

$$T = \frac{I_p}{R} \times \tau = \frac{\pi \times 2}{32 \times D} \times D^4 \times \tau$$

$$\therefore \quad \boxed{T = \frac{\pi}{16} D^3 \times \tau}$$

10.8. STRENGTH OF A HOLLOW CIRCULAR SHAFT

Let D = Outer diameter of the shaft

d = Inner diameter of the shaft

τ = Shear stress induced in the shaft

T = Maximum torque transmitted by the shaft

I_p = Polar moment of inertia = $\frac{\pi}{32}(D^4 - d^4)$

Using relation

$$\frac{T}{I_p} = \frac{\tau}{R}$$

$$T = \frac{I_p}{R} \times \tau = \frac{\frac{\pi}{32}(D^4 - d^4)}{\frac{D}{2}} \times \tau$$

$$\therefore \quad \boxed{T = \frac{\pi}{16} \frac{(D^4 - d^4)}{D} \times \tau}$$

10.9. TORSIONAL RIGIDITY (OR STIFFNESS)

The torque required to produce a twist of one radian per unit length of a shaft is known as *torsional rigidity*.

Let a torque 'T' produces a twist of θ radian in a shaft of length 'l'.

Then by using relation

$$\frac{T}{I_p} = \frac{C\theta}{l}$$

$$T = \frac{C\theta}{l} \times I_p \qquad ...(i)$$

If $l = 1$ unit and $\theta = 1$ radian, then equation (*i*) becomes

$$T = C \times I_p$$

$\therefore$ Torsional rigidity = Modulus of rigidity × Polar moment of inertia

10.10. COMPARISON OF HOLLOW AND SOLID SHAFTS

(A) Comparison by strength

Note: For non circular sections, an exact study of stresses and strains produced by torsion is rather complicated and is beyond the scope of this book.

Figure 10.9 shows cross-section of a solid and hollow shaft. Both the shafts have same length, material and weight and have same maximum shear stress.

Let, D_s = Diameter of the solid shaft
D = Outer diameter of hollow shaft = nd
d = Inner diameter of hollow shaft
A_s = Cross-sectional area of solid shaft
A_h = Cross-sectional area of hollow shaft
T_h = Torque transmitted by hollow shaft
T_s = Torque transmitted by solid shaft
τ = Shear stress in both the shafts

(i) Solid shaft (ii) Hollow shaft

Fig. 10.9

We know

$$T_s = \frac{\pi}{16} D_s^3 \times \tau$$

$$T_h = \frac{\pi}{16} \frac{(D^4 - d^4)}{D} \times \tau$$

$$\frac{\text{Strength of hollow shaft, } T_h}{\text{Strength of solid shaft, } T_s} = \frac{\frac{\pi}{16} \frac{(D^4 - d^4)}{D} \times \tau}{\frac{\pi}{16} \times D_s^3 \times \tau}$$

or,
$$\frac{T_h}{T_s} = \frac{D^4 - d^4}{D \times D_s^3} \qquad ...(i)$$

Putting $D = nd$ in (i), we get

$$\frac{T_h}{T_s} = \frac{(nd)^4 - d^4}{nd \times D_s^3} = \frac{(n^4 - 1)\, d^4}{n \times d \times D_s^3}$$

$$\frac{T_h}{T_s} = \frac{(n^4 - 1)\, d^3}{n \times D_s^3} \qquad ...(ii)$$

As length, material and weight of both the shafts are same,

$$\therefore \quad A_h = A_s$$

$$\frac{\pi}{4}(D^2 - d^2) = \frac{\pi}{4} D_s^2$$

or,
$$D_s^2 = D^2 - d^2$$

$$\therefore \quad D_s = \sqrt{D^2 - d^2}$$

Now
$$D_s^3 = D_s^2 \times D_s = (D^2 - d^2)\sqrt{D^2 - d^2}$$

$$= (n^2 d^2 - d^2)\sqrt{n^2 d^2 - d^2} \qquad [\because D = nd]$$

$$= d^2 (n^2 - 1) \times d\sqrt{n^2 - 1}$$

$$\therefore \quad D_s^3 = d^3 (n^2 - 1)\sqrt{n^2 - 1}$$

Putting the value of D_s^3 in (*ii*), we get

$$\frac{T_h}{T_s} = \frac{(n^4 - 1)\,d^3}{n \times d^3\,(n^2 - 1)\sqrt{n^2 - 1}}$$

$$= \frac{(n^2 - 1)(n^2 + 1)}{n\,(n^2 - 1)\sqrt{n^2 - 1}} \qquad [\because n^4 - 1 = (n^2 - 1)(n^2 + 1)]$$

$$\therefore \qquad \frac{T_h}{T_s} = \frac{(n^2 + 1)}{n\sqrt{n^2 - 1}}$$

Let $\qquad n = 1.5$

Then, $\qquad \dfrac{T_h}{T_s} = \dfrac{(1.5)^2 + 1}{1.5\sqrt{(1.5)^2 - 1}} = \dfrac{3.25}{1.68} = 1.93$

This shows that the torque transmitted by hollow shaft is more than that of solid shaft. Hence, the strength of hollow shaft is more than the solid shaft.

***(B)* Comparison by Weight :** While comparing a solid and hollow shaft by weight it is assumed that both the shafts have same length and material. If the torque applied to both shafts is same, then the maximum shear stress will also be same in both the shafts.

Let D_s = Diameter of the solid shaft
D = Outer diameter of hollow shaft = nd
d = Inner diameter of hollow shaft
A_s = Cross-sectional area of solid shaft
A_h = Cross-sectional area of hollow shaft
T_h = Torque transmitted by hollow shaft
T_s = Torque transmitted by solid shaft
τ = Shear stress in both the shafts

$$\frac{\text{Weight of hollow shaft, } W_h}{\text{Weight of solid shaft, } W_s} = \frac{A_h \times l \times \rho_h}{A_s \times l \times \rho_s} = \frac{A_h}{A_s}$$

[$\because \rho_h = \rho_s$ as the material of both the shafts are same. Length of shafts are also same.]

$$\frac{W_h}{W_s} = \frac{A_h}{A_s} = \frac{\frac{\pi}{4}(D^2 - d^2)}{\frac{\pi}{4}D_s^2}$$

$$\therefore \qquad \frac{W_h}{W_s} = \frac{D^2 - d^2}{D_s^2} \qquad ...(i)$$

Putting $D = nd$ in (*i*), we get

$$\frac{W_h}{W_s} = \frac{n^2 d^2 - d^2}{D_s^2} = \frac{d^2(n^2 - 1)}{D_s^2} \qquad ...(ii)$$

Torque applied in both the shafts are same

$$\therefore \quad T_s = T_h$$

$$\text{or,} \quad \frac{\pi}{16} D_s^3 \times \tau = \frac{\pi}{16} \frac{(D^4 - d^4)}{D} \times \tau$$

$$\text{or,} \quad D_s^3 = \frac{D^4 - d^4}{D} = \frac{n^4 d^4 - d^4}{nd} \qquad [\because D = nd]$$

$$\therefore \quad D_s^3 = \frac{d^3(n^4 - 1)}{n}$$

$$\text{and so} \quad D_s = \frac{d(n^4 - 1)^{\frac{1}{3}}}{n^{\frac{1}{3}}}$$

Putting the value of D_s in (*ii*), we get

$$\left[\because D_s = \frac{d(n^4 - 1)}{n^{\frac{1}{3}}} \qquad \because D_s = \frac{d^2(n^4 - 1)^{\frac{2}{3}}}{n^{\frac{2}{3}}}\right]$$

$$\frac{W_h}{W_s} = \frac{d^2(n^2 - 1) \times n^{\frac{2}{3}}}{d^2(n^4 - 1)^{\frac{2}{3}}}$$

$$\frac{W_h}{W_s} = \frac{(n^2 - 1) \times n^{\frac{2}{3}}}{(n^4 - 1)^{\frac{2}{3}}}$$

Let $n = 1.5$

$$\therefore \quad \frac{W_h}{W_s} = \frac{[(1.5)^2 - 1] \times (1.5)^{\frac{2}{3}}}{[(1.5)^4 - 1]^{\frac{2}{3}}} = \frac{1.638}{2.546} = 0.64$$

This shows that the weight of a hollow shaft of same length, material and torque to that of a solid shaft is less than the solid shaft. Hence, hollow shafts are economical than the solid shafts.

SOLVED EXAMPLES

EXAMPLE 10.1. A solid steel shaft is to transmit a torque of 20,000 Nm. If the shear stress is not to exceed 45 N/mm², find the minimum diameter of the shaft.

Solution : Given,

Torque, $T = 20{,}000 \text{ Nm} = 20{,}000 \times 1000 = 2 \times 10^7 \text{ N mm}$

Shear stress, $\tau = 45 \text{ N/mm}^2$

REMEMBER

$$\frac{T}{I_P} = \frac{\tau}{R}$$

we know,

$$T = \frac{\pi}{16} D^3 \times \tau$$

Note:

$$I_p = \frac{\pi}{32} D^4 \text{ and } R = \frac{D}{2}$$

$$\therefore \quad \frac{I_p}{R} = \frac{\frac{\pi}{32} D^4}{\frac{D}{2}} = \frac{\pi}{16} D^3$$

$$2 \times 10^7 = \frac{\pi}{16} \times D^3 \times 45$$

$$\frac{2 \times 10^7 \times 16}{\pi \times 45} = D^3$$

$$D^3 = 2263537$$

$$\therefore \quad D = 131.3 \text{ mm} \quad \textbf{Ans.}$$

EXAMPLE 10.2. A solid shaft of 150 mm diameter is used to transmit torque. Find the maximum torque transmitted by the shaft if the maximum shear stress induced in the shaft is 45 N/mm².

Solution : Given,

Diameter of shaft, $D = 150$ mm

Shear stress, $\tau = 45$ N/mm²

Using relation

$$\frac{T}{I_p} = \frac{\tau}{R}$$

$$T = \frac{I_p}{R} \times \tau = \frac{\frac{\pi}{32} D^4}{\frac{D}{2}} \times \tau$$

$$T = \frac{\pi D^3}{16} \times \tau = \frac{\pi \times (150)^3}{16} \times 45$$

$$= 29820587 \text{ N-mm} = 29820.6 \text{ Nm} \quad \textbf{Ans.}$$

EXAMPLE 10.3. Find the length of a 10 mm diameter wire so that it can be twisted through three-fourth of a revolution without exceeding a shear stress of 50 N/mm². Take C = 8 × 10⁴ N/mm².

Solution : Given,

Diameter, $D = 10$ mm $\Rightarrow$ Radius, $R = 5$ mm

Angle of twist, $\theta = \frac{3}{4}$ of a revolution

$$= \frac{3}{4} \times 2\pi = 4.712$$

Shear stress, $\tau = 50$ N/mm²

$$C = 8 \times 10^4 \text{ N/mm}^2$$

REMEMBER
$\frac{T}{I_p} = \frac{\tau}{R} = \frac{C\theta}{l}$

Using relation

$$\frac{\tau}{R} = \frac{C\theta}{l}$$

$$\frac{50}{5} = \frac{8 \times 10^4 \times 4.712}{l}$$

$$\therefore \quad l = \frac{8 \times 10^4 \times 4.712 \times 5}{50} = 37696 \text{ mm} = 37.7 \text{ m} \quad \textbf{Ans.}$$

EXAMPLE 10.4. Find the maximum torque transmitted by a hollow circular shaft of outer and inner diameters 200 mm and 100 mm respectively if the shear stress is not to exceed 40 N/mm².

Solution : Given,

Outer diameter, $D = 200$ mm

Inner diameter, $d = 100$ mm

Shear stress, $\tau = 40$ N/mm²

Using relation,

$$\frac{T}{I_p} = \frac{\tau}{R}$$

$$T = \frac{I_p}{R} \times \tau = \frac{\frac{\pi}{32}(D^4 - d^4)}{\frac{D}{2}} \times \tau = \frac{\pi}{16}\frac{(D^4 - d^4)}{D} \times \tau$$

$$= \frac{\pi}{16}\frac{[(200)^4 - (100)^4]}{200} \times 40 = \frac{\pi}{16}\frac{[1.6 \times 10^9 - 1 \times 10^8]}{200} \times 40$$

$$= \frac{\pi}{16} \times \frac{1.5 \times 10^9}{200} \times 40 = 58904862 \text{ N-mm}$$

$T = 58904.86$ Nm. **Ans.**

EXAMPLE 10.5. Find the maximum shear stress induced in a solid shaft of diameter 250 mm when the shaft is to transmit 200 KW at 250 rpm.

Solution : Given,

Diameter of shaft, $D = 250$ mm

Power, $P = 200$ KW = 200000 W

Speed of shaft, $N = 250$ rpm

Let τ be the maximum shear stress induced in the shaft

We know that

Power, $$P = \frac{2\pi NT_a}{60}$$

$$2,00,000 = \frac{2\pi \times 250 \times T_a}{60}$$

REMEMBER

Power transmitted by a shaft,

$$P = \frac{2\pi NT}{60} \text{ Watt}$$

$$\therefore \quad T_a = \frac{2,00,000 \times 60}{2\pi \times 250} = 7639.437 \text{ Nm} = 7639437 \text{ Nmm.}$$

Again, $$T = \frac{\pi}{16} D^3 \times \tau$$

$$7639437 = \frac{\pi}{16} \times (250)^3 \times \tau$$

$$\frac{7639437 \times 16}{\pi \times (250)^3} = \tau$$

$$\therefore \quad \tau = \frac{1.2223 \times 10^8}{49087385} = 2.49 \text{ N/mm}^2$$ **Ans.**

EXAMPLE 10.6. Find the maximum torque that can be applied safely to a shaft of 300 mm diameter. The permissible angle of twist is 1.5 degree in a length of 7.5 m and the shear stress is not to exceed 42 N/mm². Take c = 84.4 KN/mm².

Solution : Given,

Diameter of shaft, $D = 300$ mm

Angle of twist, $\theta = 1.5° = \frac{1.5 \times \pi}{180}$ radian $= 0.026$ radian

Length of shaft, $l = 7.5$ m $= 7500$ mm

Shear stress, $\tau = 42$ N/mm^2

$C = 84.4$ KN/mm^2 $= 84,400$ N/mm^2

Case I: Torque based on shear stress

Using relation

$$\frac{T}{I_p} = \frac{\tau}{R}$$

$$\therefore \quad T = \frac{I_p}{R} \times \tau = \frac{\frac{\pi}{32} D^4}{\frac{D}{2}} \times \tau = \frac{\pi D^3}{16} \times \tau = \frac{\pi \times (300)^3}{16} \times 42 = 2.23 \times 10^8 \text{ Nmm}$$

Case II: Torque based on angle of twist

Using relation

$$\frac{T}{I_p} = \frac{C\theta}{l}$$

$$T = \frac{C\theta}{l} \times I_p = \frac{C\theta}{l} \times \frac{\pi}{32} D^4 = \frac{84400 \times 0.026}{7500} \times \frac{\pi}{32} \times (300)^4 = 2.32 \times 10^8 \text{ Nmm}$$

Maximum torque that can be applied safely is 2.23×10^8 N-mm (*i.e.*, Minimum of two values of T). **Ans.**

EXAMPLE 10.7. A hollow shaft 400 mm outer diameter and 200 mm inner diameter is subjected to a torque of 300 KNm. Find the shear stresses at the outer and inner surfaces of the shaft. Also draw the stress distribution diagram.

Solution : Given,

Outer diameter, $D = 400$ mm

Inner diameter, $d = 200$ mm

Torque, $T = 300$ KNm $= 300 \times 10^6$ Nmm

$$\text{Polar moment of inertia, } I_p = \frac{\pi}{32}(D^4 - d^4) = \frac{\pi}{32}[(400)^4 - (200)^4]$$

$$= \frac{\pi}{32} \times 2.4 \times 10^{10} = 2.356 \times 10^9 \text{ mm}^4.$$

Let τ and τ_1 be the shear stresses induced at outer and inner surfaces of shaft.

Shear stress at the outer surface of shaft

Radius, $R = \frac{400}{2} = 200$ mm

We know

$$\frac{T}{I_p} = \frac{\tau}{R}$$

$$\frac{300 \times 10^6}{2.356 \times 10^9} = \frac{\tau}{200}$$

$$\frac{300 \times 10^6 \times 200}{2.356 \times 10^9} = \tau$$

$$\therefore \quad \tau = 25.47 \text{ N/mm}^2 \quad \textbf{Ans.}$$

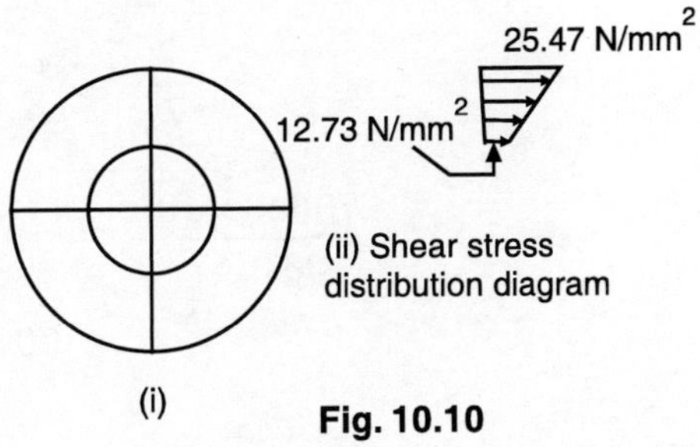

Fig. 10.10

Shear stress at the inner surface of shaft

$$\frac{\tau}{R} = \frac{\tau_1}{r} \text{ where } r = \frac{d}{2} = 100 \text{ mm} \qquad \left[\because \frac{\tau}{R} = \text{const}\right]$$

$$\frac{25.47}{200} = \frac{\tau_1}{100}$$

$$\frac{25.47 \times 100}{200} = \tau_1$$

$$\therefore \quad \tau_1 = 12.73 \text{ N/mm}^2 \quad \textbf{Ans.}$$

EXAMPLE 10.8. A solid circular shaft of 100 mm diameter is to transmit a maximum torque of 70620 Nm. Find the shear stress at :

(*i*) Centre of shaft

(*ii*) Outermost fibre of shaft

Solution : Given,

Diameter of shaft, $D = 100$ mm

Maximum torque, $T = 70620$ Nm $= 70620 \times 10^3$ Nmm

Polar moment of inertia,

$$I_p = \frac{\pi}{32} D^4$$

$$= \frac{\pi}{32} \times (100)^4 = 9817477 \text{ mm}^4$$

(*i*) Shear stress at the centre of the shaft

We know $\dfrac{T}{I_p} = \dfrac{\tau}{R}$ ...(*i*)

at the centre $R = 0$

$\therefore$ Equation (*i*) becomes

$$\frac{T}{I_p} = \frac{\tau}{0}$$

$$\therefore \quad \tau = 0 \quad \textbf{Ans.}$$

(*ii*) Shear stress at the outermost fibre of shaft

At the outer fibre, $R = \dfrac{D}{2} = \dfrac{100}{2} = 50$ mm

$$\frac{T}{I_p} = \frac{\tau}{R}$$

$$\frac{70620 \times 10^3}{9817477} = \frac{\tau}{50}$$

$$\therefore \quad \tau = \frac{70620 \times 10^3 \times 50}{9817477} = 359.66 \text{ N/mm}^2.$$

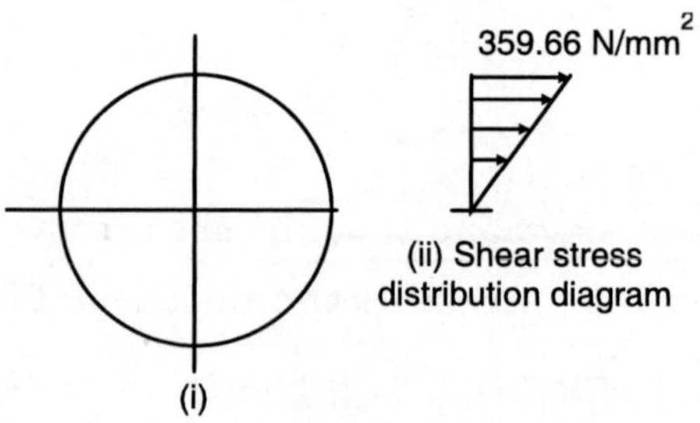

Fig. 10.11

EXAMPLE 10.9. A Solid shaft is to transmit a torque of 80 KNm. If the shearing stress is not to exceed 45 MN/m², find the maximum diameter of the shaft.

Solution : Given,

Torque, $T = 80$ KNm

Shear stress, $\tau = 45$ MN/m^2 = 45×10^3 KN/m^2

Let D be the diameter of the solid shaft

Polar moment of inertia, $I_p = \frac{\pi}{32} D^4$

Using relation

$$\frac{T}{I_p} = \frac{\tau}{R}$$

$$\frac{80}{\frac{\pi}{32} D^4} = \frac{45 \times 10^3}{\frac{D}{2}}$$

$$\frac{80 \times 32}{\pi D^4} = \frac{2 \times 45 \times 10^3}{D}$$

$$\frac{80 \times 32}{\pi \times 2 \times 45 \times 10^3} = \frac{D^4}{D}$$

$$\frac{2560}{282743.34} = D^3$$

$$\therefore \quad D = 0.208 \text{ m} = 208 \text{ mm} \quad \textbf{Ans.}$$

EXAMPLE 10.10. A solid circular shaft of 100 mm diameter is transmitting 110 KW at 150 r.p.m. Find the intensity of induced shear stress in the shaft.

Solution : Given,

Diameter, $D = 100$ mm $= 0.1$ m

Power, $P = 110$ kW

Speed, $N = 150$ rpm

Using relation

$$P = \frac{2\pi NT}{60 \times 1000} \text{ kW}$$

$$110 = \frac{2\pi \times 150\, T}{60 \times 1000}$$

$$\frac{110 \times 60 \times 1000}{2\pi \times 150} = T$$

$$\therefore \quad T = 7002.82 \text{ Nm}$$

$$I_p = \frac{\pi}{32} D^4$$

$$R = \frac{D}{2}$$

$$\therefore \quad \frac{I_p}{R} = \frac{\frac{\pi}{32} D^4}{\frac{D}{2}} = \frac{\pi D^3}{16} \quad \ldots(i)$$

We have the relationship

$$\frac{T}{I_p} = \frac{\tau}{R}$$

$$T \times \frac{R}{I_p} = \tau \qquad \text{...(ii)}$$

Putting the value of $\frac{R}{I_p}$ in (*ii*) from (*i*)

$$\frac{T \times 16}{\pi D^3} = \tau$$

$$\frac{7002.82 \times 16}{\pi \times (0.1)^3} = \tau$$

$$\therefore \quad \tau = 35665069 \text{ N/m}^2$$

$$= 35.66 \text{ MN/m}^2 \quad \textbf{Ans.}$$

EXAMPLE 10.11. A solid shaft of 60 mm diameter is running at 160 r.p.m. Find power in KW which the shaft can transmit, if the permissible shear stress is 80 MN/m² and maximum torque is likely to exceed the mean by 20%.

Solution : Given,

Diameter, $D = 60 \text{ mm} = 0.060 \text{ m}$

Speed, $N = 160 \text{ r.p.m}$

Shear stress, $\tau = 80 \text{ MN/m}^2 = 80 \times 10^3 \text{ KN/m}^2$

$$T_{max} = 1.2\ T_{mean}$$

$$\frac{I_p}{R} = \frac{\frac{\pi}{32} \times D^4}{\frac{D}{2}} = \frac{\pi}{16} D^3$$

Using relation $\frac{T}{I_p} = \frac{\tau}{R}$

$$T = \frac{I_p}{R} \times \tau = \frac{\pi}{16} D^3 \times \tau = \frac{\pi}{16} \times (0.060)^3 \times 80 \times 10^3$$

$$\therefore \quad T_{max} = \text{T} = 3.39 \text{ KNm}$$

$$T_{max} = 1.2 \times T_{mean}$$

or $$3.39 = 1.2 \times T_{mean}$$

$$\therefore \quad T_{mean} = \frac{3.39}{1.2} = 2.825 \text{ KNm} = 2825 \text{ Nm}$$

We know $$P = \frac{2\pi NT}{60 \times 1000} = \frac{2\pi \times 160 \times 2825}{60 \times 1000} = 47.33 \text{ KW} \quad \textbf{Ans.}$$

EXAMPLE 10.12. A solid shaft of 100 mm diameter is to transmit 117.6 KW at 100 r.p.m. Find the maximum intensity of shear induced and the angle of twist for a length of 6 meters. Take C = 80 GN/m².

Solution : Given,

Diameter of shaft, $D = 100 \text{ mm} = 0.1 \text{ m}$

Power, $P = 117.6 \text{ kW}$

Speed, $N = 100 \text{ r.p.m}$

Length, $L = 6 \text{ m}$

$$C = 80 \text{ GN/m}^2 = 80 \times 10^9 \text{ N/m}^2$$

Using the relation $P = \dfrac{2\pi NT}{60 \times 1000}$

$$117.6 = \frac{2\pi \times 100 \times T}{60 \times 1000}$$

$$\frac{117.6 \times 60 \times 1000}{2\pi \times 100} = T$$

$$\therefore \quad T = 11230 \text{ N-m}$$

We know, $\dfrac{T}{I_p} = \dfrac{\tau}{R}$

$$\frac{R}{I_p} \times T = \tau$$

$$\frac{D/2}{\frac{\pi}{32} D^4} \times T = \tau$$

$$\frac{16\, T}{\pi D^3} = \tau$$

$$\therefore \quad \tau = \frac{16 \times 11230}{\pi \times (0.1)^3} = 57193920 \text{ N/m}^2$$

$$= 57.19 \text{ MN/m}^2$$

We know, $\dfrac{T}{I_p} = \dfrac{C\theta}{l}$

$$\frac{T \times l}{I_p \times C} = \theta$$

$$\frac{T \times l}{\frac{\pi}{32} D^4 \times C} = \theta$$

$$\frac{32\, Tl}{\pi D^4 \times C} = \theta$$

$$\frac{32 \times 11230 \times 6}{\pi \times (0.1)^4 \times 80 \times 10^9} = \theta$$

$$\frac{2156160}{25132741} = \theta$$

$$\therefore \quad \theta = 0.086 \text{ rad}$$

$$= \frac{0.086 \times 180}{\pi} = 4.91° \qquad \textbf{Ans.}$$

EXAMPLE 10.13. Design a suitable diameter for a circular shaft required to transmit 80.2 kW at 180 r.p.m. The shear stress in the shaft is not to exceed 70 MN/m² and the maximum torque exceeds the mean by 40%. Also calculate the angle of twist in a length of 2 m. Take C = 90 GN/m².

Solution : Given,

Power, $P = 80.2$ KW

Speed, $N = 180$ r.p.m

Shear stress, $\tau = 70\ \text{MN/m}^2 = 70 \times 10^6\ \text{N/m}^2$

$T_{max} = 1.4\ T_{mean}$

length, $l = 2\text{m}$

$C = 90\ \text{GN/m}^2 = 90 \times 10^9\ \text{N/m}^2$

Using relation

$$P = \frac{2\pi\ NT}{60 \times 1000}$$

$$80.2 = \frac{2\pi \times 180 \times T}{60 \times 1000}$$

$$\frac{80.2 \times 60 \times 1000}{2\pi \times 180} = T$$

$$\therefore \quad T = \frac{48,12,000}{1131} = 4254.64\ \text{Nm}$$

$$T_{max} = 1.4\ T_{mean} = 1.4 \times 4254.64 = 5956.5\ \text{Nm}$$

We know,

$$\frac{T}{I_p} = \frac{\tau}{R}$$

$$T = \frac{I_p}{R} \times \tau = \frac{\pi}{16} D^3 \times \tau$$

$$\left[\because \frac{I_p}{R} = \frac{\pi}{16} D^3 \text{ where } D \text{ is the diameter of the shaft}\right]$$

$$5956.5 = \frac{\pi}{16} \times D^3 \times 70 \times 10^6$$

$$\frac{5956.5 \times 16}{\pi \times 70 \times 10^6} = D^3$$

$$\therefore \quad D = 0.076\ \text{m} = 76\ \text{mm}$$

$$I_p = \frac{\pi}{32} \times D^4 = \frac{\pi}{32} \times (0.076)^4$$

$$= 3.27 \times 10^{-6}\ \text{mm}^4$$

We know

$$\frac{T}{I_p} = \frac{C\theta}{l}$$

$$\Rightarrow \quad \frac{T \times l}{I_p \times C} = \theta$$

$$\theta = \frac{5956.5 \times 2}{3.27 \times 10^{-6} \times 90 \times 10^{9}}$$

$$\therefore \quad \theta = 0.0404 \text{ rad}$$

$$= \frac{0.0404 \times 180}{\pi} = 2.31° \quad \textbf{Ans.}$$

EXAMPLE 10.14. A hollow shaft of diameter ratio $\frac{3}{8}$ is required to transmit 588 KW at 110 r.p.m, the maximum torque being 20% greater than the mean. The shear stress is not to exceed 63 MN/m^2 and the twist in a length of 3m not to exceed 1.4 degree. Calculate the external diameter of the shaft which would satisfy these conditions. Take C = 84 GN/m^2.

Solution : Given,

Power, $P = 588$ KW

Speed, $N = 110$ r.p.m

$T_{max} = 1.2\ T_{mean}$

Shear stress, $\tau = 63$ MN/m^2 $= 63 \times 10^6$ N/m^2

Length of shaft, $l = 3$m

Angle of twist, $\theta = 1.4°$

$C = 84$ GN/m^2 $= 84 \times 10^9$ KN/m^2

Let D and d be the outer and inner diameter of the hollow shaft.

$$\frac{d}{D} = \frac{3}{8} = 0.375 \text{ (given)}$$

$$\therefore \quad d = 0.375\, D$$

We know that

$$P = \frac{2\pi\, NT}{60 \times 1000}$$

$$588 = \frac{2\pi \times 110 \times T}{60 \times 1000}$$

$$\frac{588 \times 60 \times 100}{2\,\pi \times 110} = T$$

$$\therefore \quad T = 51045.33 \text{ N-m}$$

$$T_{max} = 1.2 \times 51045.33 = 61254.4 \text{ N-m}$$

Case I : When shear stress is not to exceed 63 MN/m^2

Using relation $\quad \dfrac{T}{I_p} = \dfrac{\tau}{R}$

or, $$T = \frac{I_p}{R} \times \tau = \frac{\pi}{32}\frac{(D^4 - d^4)}{D} 2\tau \qquad \left[\because I_p = \frac{\pi}{32}(D^4 - d^4)\ R = \frac{D}{2}\right]$$

$$61254.4 = \frac{\pi}{16}\left[\frac{D^4 - (0.375D)^4}{D}\right] \times \tau$$

$$61254.4 = \frac{\pi}{16} \times \frac{D^4}{D}[1 - (0.375)^4] \times \tau$$

$$61254.4 = \frac{\pi}{16} \times D^3 \times 0.98 \times 63 \times 10^6$$

$$61254.4 = 12122621\ D^3$$

$$\frac{61254.4}{12122621} = D^3$$

$$\therefore \qquad D = 0.1715 \text{ m} = 172 \text{ mm} \qquad \ldots(i)$$

Case II : When angle of twist in a length of 3 m not to exceed 1.4 degree

Using the relation

$$\frac{T}{I_p} = \frac{C\theta}{l}$$

$$\frac{T \times 32}{\pi(D^4 - d^4)} = \frac{C\theta}{l}$$

$$\frac{T \times 32}{\pi[D^4 - (0.375\ D)^4]} = \frac{C\theta}{l}$$

$$\frac{T \times 32}{\pi \times D^4\ (1 - 0.0198)} = \frac{C\theta}{l}$$

$$\frac{T \times 32}{\pi \times D^4 (1 - 0.0198)} = \frac{C\theta}{l}$$

$$\frac{61254.4 \times 32}{\pi \times D^4 \times 0.98} = \frac{84 \times 10^9 \times \pi \times 1.4}{3 \times 180°}$$

$$\frac{1960140.8}{3.08\ D^4} = \frac{3.69 \times 10^{11}}{540°}$$

$$\frac{1961040.8 \times 540}{3.08 \times 3.69 \times 10^{11}} = D^4$$

$$\therefore \qquad D = 0.1747 \text{ m} = 175 \text{ mm} \qquad \ldots(ii)$$

Greater of the two values *i.e.*, 175 mm will be the external diameter of the shaft. **Ans.**

EXAMPLE 10.15. A hollow shaft of internal diameter 400 mm and of metal thickness 30 mm is required to transmit power at 180 r.p.m. Determine the power it can transmit, if the shear stress in the shaft is not to exceed 60 MN/m² and the maximum torque exceed the mean by 30%.

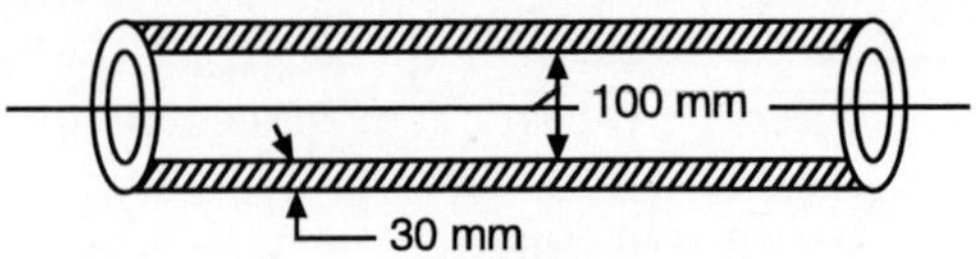

Fig. 10.12

Solution : Given,

Internal diameter $\qquad d = 400 \text{ mm} = 0.4 \text{ m}$

Thickness, $\qquad t = 30 \text{ mm}$

Outer diameter, $\qquad D = 400 + 30 + 30 = 460 \text{ mm} = 0.46 \text{ m}$

Speed, $\qquad N = 180 \text{ r.p.m.}$

Shear stress, $\qquad \tau = 60 \text{ MN/m}^2 = 60 \times 10^6 \text{ N/m}^2$

$$T_{max} = 1.3\ T_{mean}$$

$$I_p = \frac{\pi}{32}(D^4 - d^4)$$

$$R = \frac{D}{2}$$

$$\therefore \quad \frac{I_p}{R} = \frac{\pi}{32}(D^4 - d^4) \times \frac{2}{D} = \frac{\pi (D^4 - d^4)}{16\, D}$$

Using relation

$$\frac{T}{I_p} = \frac{\tau}{R}$$

$$T = \frac{I_p}{R} \times \tau = \frac{\pi (D^4 - d^4)}{16 D} \times \tau$$

$$= \frac{\pi [(0.46)^4 - (0.4)^4]}{16 \times 0.46} \times 60 \times 10^6$$

$$= \frac{\pi \times 0.019}{7.36} \times 60 \times 10^6$$

$$\therefore \quad T_{\max} = 486605.38 \text{ N-m}$$

$$T_{\text{mean}} = \frac{T_{\max}}{1.3} = \frac{486605.38}{1.3} = 374311.83 \text{ N-m}$$

Power,

$$P = \frac{2\pi\, NT}{60 \times 1000} = \frac{2\pi \times 180 \times 374311.83}{60 \times 1000}$$

$$= 7055.61 \text{ KW}$$ **Ans.**

EXAMPLE 10.16. A solid aluminium shaft 1 m long and 50 mm diameter is to be replaced by a tubular steel shaft of the same length and same outside diameter (*i.e.*, 50 mm) such that each of the two shafts could have the same angle of twist per unit torsional moment over the total length. What must the inner diameter of the tubular steel shaft be? Modulus of rigidity of steel is three times that of aluminium.

Solution : Given,

Length of shaft, $l = 1$m

Outer diameter of aluminium shaft, $D_a = 50 \text{ mm} = 0.050 \text{ m}$

$$C_s = 3\, C_a$$

$$\left(\frac{\theta}{T}\right)_{\text{aluminium shaft}} = \left(\frac{\theta}{T}\right)_{\text{steel shaft}}$$

For aluminium shaft

Let 'd' be the inner diameter of the shaft

$$I_p = \frac{\pi}{32}\left[(0.050)^4\right] = \frac{\pi}{32}\left[6.25 \times 10^{-6}\right]$$

Using relation

$$\frac{T_a}{I_p} = \frac{C_a\, \theta_a}{l}$$ ['a' denots aluminium]

or

$$\frac{l}{I_p C_a} = \frac{\theta_a}{T_a}$$

or

$$\frac{1}{\frac{\pi}{32}[6.25 \times 10^{-6}]} \times \frac{1}{C_a} = \frac{\theta_a}{T_a}$$

For steel shaft

$$\frac{T_s}{I_p} = \frac{C_s\,\theta_s}{l} \qquad \text{[s denotes steel]}$$

$$\frac{l}{I_p\,C_s} = \frac{C_s}{T_s}$$

$$\frac{1}{\frac{\pi}{32}[(0.05)^4 - d^4]} \times \frac{1}{C_s} = \frac{\theta_s}{T_s}$$

A.T.Q

$$\frac{\theta_a}{T_a} = \frac{\theta_s}{T_s}$$

$$\frac{1}{\frac{\pi}{32}[6.25 \times 10^{-6}]} \times \frac{1}{C_a} = \frac{1}{\frac{\pi}{32}(6.25 \times 10^{-6} - d^4)} \times \frac{1}{C_s}$$

$$\frac{1}{(6.25 \times 10^{-6})} \times \frac{1}{C_a} = \frac{1}{(6.25 \times 10^{-6} - d^4)} \times \frac{1}{3C_a}$$

$$\frac{1}{6.25 \times 10^{-6}} = \frac{1}{18.75 \times 10^{-6} - 3d^4}$$

or $\quad 6.25 \times 10^{-6} = 18.75 \times 10^{-6} - 3d^4$

or $\quad 3d^4 = 18.75 \times 10^{-6} - 6.25 \times 10^{-6}$

$3d^4 = 12.5 \times 10^{-6}$

$$d^4 = \frac{12.5 \times 10^{-6}}{3} = 4.17 \times 10^{-6}$$

$\therefore \quad d = 0.0452 \simeq 45.2$ mm **Ans.**

EXAMPLE 10.17. A hollow steel shaft of 150 mm external diameter and 100 mm bore twists through an angle of 0.8 degree in a length of 2m when subjected to an axial torque. What are the values of shearing stresses at the inner and outer surfaces of the shaft?

Calculate the power that is transmitted by the shaft at a speed of 250 r.p.m., when the above torque is applied. C = 90 GN/m².

Solution : Given,

Outer diameter, $\quad D = 150 \text{ mm} = 0.15 \text{ m}$

Inner diameter, $\quad d = 100\text{mm} = 0.10 \text{ m}$

Angle of twist, $\quad \theta = 0.8 = \dfrac{0.8 \times \pi}{180}$ radian

Length of shaft, $\quad l = 2\text{m}$

Speed, $\quad N = 250$ r.p.m.

$C = 90 \text{ GN/m}^2 = 90 \times 10^9 \text{ N/m}^2$

Outer radius, $\quad R = \dfrac{D}{2} = \dfrac{0.15}{2} = 0.075 \text{ m}$

Inner radius, $r = \frac{d}{2} = \frac{0.10}{2} = 0.05$ m

Let τ_i and τ_o be the shearing stresses at the inner and outer surfaces of the shaft.
Using relation

$$\frac{C\theta}{l} = \frac{\tau}{R}$$

or $$\frac{C\theta}{l} = \frac{\tau_i}{r}$$

or $$\tau_i = \frac{C\theta r}{l} = \frac{90 \times 10^9 \times 0.014 \times 0.05}{2} = 31.5 \text{ MN/m}^2 \quad \textbf{Ans}$$

$$\frac{C\theta}{l} = \frac{\tau_0}{R}$$

or $$\tau_0 = \frac{C\theta \times R}{l} = \frac{90 \times 10^9 \times 0.014 \times 0.075}{2}$$

$$= 47.25 \text{ MN/m}^2 \quad \textbf{Ans.}$$

$$I_p = \frac{\pi}{32}(D^4 - d^4) = \frac{\pi}{32}\{(0.15)^4 - (0.10)^4\}$$

$$= \frac{\pi}{32} \times 4.06 \times 10^{-4} = 3.988 \times 10^{-5} \text{ m}^4$$

We know $$\frac{T}{I_p} = \frac{\tau_0}{R}$$

$$T = \frac{I_p}{R} \times \tau_0 = \frac{3.988 \times 10^{-5}}{0.075} \times 47.25 \times 10^6$$

$$= 25124.4 \text{ N-m}$$

$$P = \frac{2\pi NT}{60 \times 1000} = \frac{2\pi \times 250 \times 25124.4}{60 \times 1000} = 657.75 \text{ KW} \quad \textbf{Ans.}$$

EXAMPLE 10.18. Calculate the maximum torque and the mean power being transmitted in the case of a hollow shaft of which outer diameter is 200 mm and the inner 100 mm if the shear stress is not to exceed 60 MN/m² and the shaft-speed is 300 r.p.m. Assume maximum torque to be 25% more than mean.

Solution : Given,

Outer diameter, $D = 200$ mm

Inner diameter, $d = 100$ mm

Shear stress, $\tau = 60 \text{ MN/m}^2 = 60 \times 10^6 \text{ N/m}^2$

Speed, $N = 300$ r.p.m.

$$T_{max} = 1.25\, T_{mean}$$

$$I_p = \frac{\pi}{32}[(0.2)^2 - (0.1)^4] = 1.47 \times 10^{-4} \text{ m}^4$$

$$R = \frac{200}{2} = 100 \text{ mm} = 0.1 \text{ m}$$

Using relation $\dfrac{T}{I_p} = \dfrac{\tau}{R}$

$$T = \frac{I_p}{R}\tau = \frac{1.47 \times 10^{-4}}{0.1} \times 60 \times 10^6$$

$$T_{max} = 88375.293 \text{ N-m} = 88.37 \text{ KN-m} \quad \textbf{Ans}$$

$$T_{mean} = \frac{T_{max}}{1.25} = \frac{88375.29}{1.25} = 70700.23 \text{ N-m}$$

We know $P = \dfrac{2\pi NT}{60 \times 1000} = \dfrac{2\pi \times 300 \times 70700.23}{60{,}000}$

$$= 2221.11 \text{ KW} \quad \textbf{Ans.}$$

EXAMPLE 10.19. A hollow shaft has an external diameter of 300 mm and internal diameter of 150 mm. Compare its strength with that of a solid shaft of the same weight per unit length.

Solution : Given

For Hollow shaft $D_o = 300$ mm

$D_i = 150$ mm

Let d be the diameter of the solid shaft and τ be the shear stress.

Using relation

$$\frac{T_h}{I_p} = \frac{\tau}{R}$$

$$T_h = \frac{I_p}{R}\tau = \frac{\pi}{16}\frac{(0.3)^4 - (0.15)^4}{(0.3)} \times \tau \qquad \left[\because I_P = \frac{\pi}{32}(D^4 - d^4) \text{ and } R = \frac{D}{2}\right]$$

$$= \frac{\pi}{16} \times \frac{7.59 \times 10^{-3}}{0.3} \times \tau = \frac{\pi}{16} \times 0.0253 \times \tau \qquad ...(i)$$

For solid shaft

$$T_s = \frac{I_p}{R}\tau = \frac{\pi}{16} \times d^3\tau \qquad ...(ii)$$

Dividing (*i*) and (*ii*), we get

$$\frac{T_h}{T_s} = \frac{\frac{\pi}{16} \times 0.0253 \times \tau}{\frac{\pi}{16} \times d^3\tau} = \frac{0.0253}{d^3} \qquad ...(iii)$$

As the weight per unit length of both the shafts are equal

$\therefore$ $A_h \times 1 \times \rho = A_s \times 1 \times \rho$

or $A_h = A_s$

$$\frac{\pi}{4}[D_0^2 - D_i^2] = \frac{\pi}{4}d^2$$

$$(0.3)^2 - (0.15)^2 = d^2$$

$\therefore$ $d = \sqrt{0.0675} = 0.26$ m

Putting the value of d in (*iii*) we get

$$\frac{T_h}{T_s} = \frac{0.0253}{(0.26)^3} = \frac{0.0253}{0.017} = 1.49$$

∴ The strengths of hollow shaft is 1.49 times the strength solid shaft. **Ans.**

EXAMPLE 10.20. A solid shaft of 200 mm diameter is to be replaced by a hollow steel shaft with internal diameter equal to 0.5 D, where D is the external diameter. Design the hollow shaft and find out the saving in material. The value of maximum shear stress may be assumed as same for both the shafts.

Solution : Given,

Diameter of solid shaft, d = 200 mm = 0.2 m

Internal diameter of hollow shaft, $D_i = 0.5\ D$

Where, D is the external diameter of the hollow shaft.

Twisting moment of solid shaft

Using relation

$$\frac{T_s}{I_p} = \frac{\tau}{R}$$

$$T_s = \frac{I_p}{R} \times \tau$$

$$= \frac{\frac{\pi}{32}[D_0^4 - D_i^4]}{\frac{D_o}{2}} \times \tau$$

$$= \frac{\frac{\pi}{16}[D^4 - (0.5D)^4]}{D} \times \tau$$

$$= \frac{\pi}{16} D^3 \times 0.9375\tau \text{ N-m}$$

Twisting moment of hollow shaft

Using relation $\frac{T_h}{I_p} = \frac{\tau}{R} \Rightarrow T_h = \frac{I_p}{R}\tau$

$$= \frac{\frac{\pi}{32} d^4}{\frac{d}{2}} \tau = \frac{\pi}{16} d^3 \tau = \frac{\pi}{16} \times (0.2)^3\ \tau = \frac{\pi}{16} \times 0.008\ \tau$$

The twisting moment of both the shaft should be equal.

$$T_s = T_h$$

$$\frac{\pi}{16} D^3 \times 0.9375\ \tau = \frac{\pi}{16} d^3 \tau = \frac{\pi}{16} \times 0.008\ \tau \qquad [\because d^3 = 0.008]$$

$$D^3 = \frac{0.008}{0.9375}$$

∴ $D = 0.204 \text{ m} = 204 \text{ mm}$

⇒ $D_o = 200 \text{ mm and } D_i = 102 \text{ mm}$

Percentage saving in material

$$\text{Percentage saving} = \frac{W_s - W_h}{W_s} \times 100$$

$$= \left(1 - \frac{W_h}{W_s}\right) \times 100 \qquad \ldots(i)$$

$$\frac{W_h}{W_s} = \frac{\frac{\pi}{4}(D_h^2 - D_i^2) \times l \times \rho}{\frac{\pi}{4} d^2 \times l \times \rho}$$

[Assuming that the length and material of both the shafts are same]

$$= \frac{D_h^2 - D_i^2}{d^2} = \frac{(0.204)^2 - (0.102)^2}{(0.2)^2}$$

$$= \frac{0.312}{0.04} = 0.78$$

Putting the value of $\frac{W_h}{W_s}$ in (*i*) we get

Percentage saving = (1 – 0.78) × 100 = 0.22 × 100 = 22 %. **Ans.**

EXAMPLE 10.21. Two shafts of the same material and same length are subjected to the same torque. If the first shaft is of a solid circular section and the second shaft is of hollow circular section whose internal diameter is 2/3 of the outside and in each shaft, maximum shear is same, then compare the weight of two shafts.

Solution : Given,

$T_s = T_h$ ($\because$ Torque are same)

$\rho_s = \rho_h = \rho$ ($\because$ Shafts are of same material)

$l_s = l_h = l$ ($\because$ Both the shaft have same length)

$D_i = \frac{2}{3} D_o$ of hollow shaft

Let d be the diameter of the solid shaft.

$$\frac{T_s}{I_p} = \frac{\tau}{R}$$

$$T_s = \frac{I_p}{R}\tau = \frac{\frac{\pi}{32}d^4}{\frac{d}{2}}\tau = \frac{\pi}{16}d^3\tau$$

$$\frac{T_h}{I_p} = \frac{\tau}{R}$$

$$T_h = \frac{I_p}{R}\tau = \frac{\frac{\pi}{32}[D_o^4 - D_i^4]}{D_o}\tau = \frac{\pi}{16}\frac{\left[D_o^4 - \left(\frac{2}{3}D_o\right)^4\right]^2}{D_o}\tau$$

$$= \frac{\pi}{16} \frac{D_o^4 \left(1 - \frac{16}{81}\right)}{D_o} \tau = \frac{\pi}{16} \times 0.802\, D_o^3\, \tau$$

$$T_s = T_h \text{ (given)}$$

$$\frac{\pi}{16} d^3 \tau = \frac{\pi}{16} \times 0.802\, D_o^3\, \tau$$

$$\therefore \qquad d^3 = 0.802\, D_o^3 \Rightarrow d = 0.93\, D_o$$

Weight of solid shaft, $W_s = \frac{\pi}{4} d^2 \times l \times \rho$

Weight of hollow shaft, $W_h = \frac{\pi}{4}\left(D_o^2 - D_i^2\right) \times l \times \rho$

Ratio of weights, $$\frac{W_h}{W_s} = \frac{\frac{\pi}{4}(D_o^2 - D_i^2)\,\rho l}{\frac{\pi}{4} d^2 \rho l} = \frac{D_o^2 - D_i^2}{d^2} = \frac{D_o^2 - \left(\frac{2}{3} D_o\right)^2}{(0.93\, D_o)^2}$$

$$= \frac{D_o^2\left(1 - \frac{4}{9}\right)}{0.865\, D_o^2} = \frac{0.55}{0.865} = 0.636 \quad \textbf{Ans.}$$

EXAMPLE 10.22. What power can be transmitted at 300 r.p.m by a hollow steel shaft of 7.5 cm external and 5 cm internal diameters, When permissible shear stress for the steel is 70 MN/m² and maximum torque is 1.3 times the mean?

Compare the strength of this hollow shaft with that of a solid shaft of the same material, weight and length.

Solution : Given,

Speed, $N = 300$ r.p.m

Outer diameter, $D = 7.5 \text{ cm} = 0.075 \text{ m}$

Internal diameter, $d = 5.0 \text{ cm} = 0.05 \text{ m}$

Shear stress, $\tau = 70 \text{ MN/m}^2$

$= 70 \times 10^6 \text{ N/m}^2$

$T_{max} = 1.3\, T_{mean}$

Using relation

$$\frac{T}{I_p} = \frac{\tau}{R}$$

or, $$T = \tau \times \frac{I_p}{R} = \frac{\tau \times \frac{\pi}{32}(D^4 - d^4)}{\frac{D}{2}}$$

$$T = \pi \times \frac{\pi}{16} \frac{(D^4 - d^4)}{D} = 70 \times 10^6 \times \frac{\pi}{16} \frac{[(0.075)^4 - (0.05)^4]}{0.075}$$

$$= \frac{70 \times 10^6 \times \pi \times 2.54 \times 10^{-5}}{16 \times 0.075}$$

$$\therefore \quad T_{max} = \frac{5585.75}{1.2} = 4654.79 \text{ N-m}$$

$$T_{mean} = \frac{T_{max}}{1.3} = \frac{4654.79}{1.3} = 3580.61 \text{ N-m}$$

Power, $$P = \frac{2\pi NT}{60 \times 1000} = \frac{2\pi \times 300 \times 3580.61}{60 \times 1000} = 112.49 \text{ KW}$$

Let d_s be the diameter of the solid shaft.

Area of solid shaft (A_s) = Area of hollow shaft (A_h)

$$\frac{\pi}{4} \times d_s^2 = \frac{\pi}{4}(D^2 - d^2)$$

$$d_s^2 = [(0.075)^2 - (0.05)^2]$$

$$\therefore \quad d_s = 0.056 \text{ m}$$

Let T_s and T_H be the torque transmitted by solid and hollow shaft respectively.

$$\frac{T_s}{I_P} = \frac{\tau}{R}$$

$$\therefore \quad T_s = \frac{I_p}{R}\tau = \frac{\frac{\pi}{32} d_s^4}{\frac{d_s}{2}} \times \tau$$

$$= \frac{\pi}{16} d_s^3 \tau = \frac{\pi}{16} \times (0.056)^3 \tau$$

Again,

$$\frac{T_H}{R} = \frac{\tau}{I_p}$$

$$T_H = \frac{I_p}{R}\tau = \frac{\frac{\pi}{32}(D^4 - d^4)}{\frac{D}{2}} \times \tau$$

$$= \frac{\pi}{16} \frac{(D^4 - d^4)}{D} \tau = \frac{\pi}{16} \frac{[(0.075)^4 - (0.050)^4]}{0.075} \tau$$

$$\frac{T_H}{T_S} = \frac{\frac{\pi}{16}[(0.075)^4 - (0.050)^4]\tau}{0.075 \times \frac{\pi}{16} \times (0.056)^3 \tau}$$

$$= \frac{[(0.075)^4 - (0.050)^4]}{0.075 \times (0.056)^3}$$

$$= \frac{2.54 \times 10^{-5}}{1.32 \times 10^{-5}} = 1.924 \quad \textbf{Ans.}$$

SUMMARY

1. Torsion is the twisting of a structural member when it is loaded by couples that produce rotation about its longitudinal axis.
2. Units of torque is Nm or N mm.
3. A member is said to be under pure torsion when it is subjected to two equal and opposite torques.
4. The shear stresses produced in a shaft, subjected to pure torsion, are maximum on the surfaces of shafts and is zero at the axis of shafts.
5. Torsion equation is given by;
$$\frac{T}{I_p} = \frac{\tau}{R} = \frac{C\theta}{l}$$
6. The torsion transmitted by a solid shaft is given by;
$$T = \frac{\pi}{16} D^3 \times \tau$$
7. The torque transmitted by a hollow shaft is given by;
$$T = \frac{\pi}{16} \times \frac{(D^4 - d^4)}{D} \times \tau$$
8. Power transmitted by a shaft is given by;
$$P = \frac{2\pi N\, T_a}{60} \text{ watts}$$
9. The maximum torque or maximum power which a shaft can transmit is known as strength of a shaft.
10. Torsional rigidity is defined as the torque required to produce a twist of one radian per unit length of a shaft.

OBJECTIVE TYPE QUESTIONS

Fill in the Blanks :

(*i*) ________ stress is produced in a shaft-when it is subjected to pure torsion.

(*ii*) When equal and opposite torques are applied at the two ends of a shaft, then the shaft is in ________.

(*iii*) The torsion equation is $\frac{T}{\_\_} = \frac{\tau}{R} = \frac{c\theta}{\_\_}$.

(*iv*) The shear stress is ________ on the surface of the shaft and is ________ at the centre.

(*v*) The unit of torque is ________.

(*vi*) The maximum torque transmitted by a shaft is known as ________ of a shaft.

(*vii*) The power transmitted by a shaft is given by
$P =$ ________ watts.

(*viii*) In the formula for power of a shaft, $P = \frac{2\pi\, NT}{60}$ watt. T is the ________ torque.

(*ix*) The product of modulus of rigidity and polar moment of inertia is known as ________ .

(*x*) Shafts are designed on the basis of ________ and ________ .

(*xi*) A hollow shaft is ________ than the solid shaft of same length, material and weight.

Answers

(*i*) Shear	(*ii*) Pure torsion	(*iii*) I_p, l	(*iv*) maximum; zero
(*v*) N-m	(*vi*) strength	(*vii*) $\frac{2\pi\, NT_a}{60}$	(*viii*) Average
(*ix*) torsional rigidity	(*x*) strength, rigidity	(*xi*) stronger	

THEORETICAL QUESTIONS

1. What do you mean by torsion of a shaft? What is effect of torsion on a shaft?
2. What assumptions are made in the theory of pure torsion?
3. Derive the torsion equation : $\frac{I}{I_p} = \frac{\tau}{R} = \frac{C\theta}{l}$
4. Write short notes on :
   (*i*) Power transmitted by a shaft
   (*ii*) Strength of a shaft
   (*iii*) Torsional rigidity of a shaft.
5. What do you mean by strength of a shaft? How will you calculate it for a hollow shaft?
6. Show that a hollow shaft is lighter than a solid shaft if the shafts have same length, material and are subjected to the same torque.
7. Show that a hollow shaft is stronger than a solid shaft if it has same length, material and same weight.

NUMERICALS QUESTIONS

1. Find the length of a 10 mm diameter wire so that it can be twisted through three-fourth of a revolution without exceeding a shear stress of 40 N-mm^2. Take $C = 2.5 \times 10^4$ N-mm^2. **[14.72 m]**
2. Find the maximum torque transmitted by solid shaft of 250 mm diameter if the maximum shear stress induced in the shaft is 50 N/mm^2. **[153398 Nm]**
3. A solid shaft has to transmit a torque of 40,000 Nm. Find the minimum diameter of the shaft if the shear stress is not to exceed 45 N/mm^2. **[165.4 mm]**
4. Find the maximum torque transmitted by a hollow shaft of outer diameter 350 mm and internal diameters 150 mm. The shear stress in the shaft should not exceed 40 N/mm^2. **[325379 Nm]**
5. A solid steel shaft has to transmit 75 kW at 200 r.p.m. Find suitable diameter for a shaft, if the maximum torque transmitted on each revolution exceeds the mean by 30%. The allowable shear stress in the shaft is 80 MN/m^2. **[69.7 Nmm]**
6. A solid shaft of 100 mm diameter is to transmit 160 kW at 100 r.p.m. If the maximum torque exceeds the mean by 40%. Find the angle of twist for a length of 6 m. Take $C = 8 \times 10^4$ N/mm^2. **[9.36°]**
7. A solid shaft of 200 mm diameter is to transmit 187.5 kW at 200 r.p.m. Find the maximum shear stress induced in the shaft. **[5.7 N/mm^2]**
8. A solid shaft transmits 75 kW at 200 r.p.m. Calculate the shaft diameter, if the twist in the shaft is not to exceed 1° in 2 metres length of shaft and the shear stress is not to exceed 50 MN/m^2. Take $C = 100$ GN/m^2. **[80 mm]**
9. A solid circular shaft of 120 mm diameter is to transmit a maximum torque of 71.62 kNM. Find the shear stress at
   (*i*) Centre of the shaft
   (*ii*) Outermost fibre of the shaft **[0; 211.1 N/mm^2]**
10. A hollow shaft is to transmit 300 kW at 80 r.p.m. If the shear stress is not to exceed 60 MN/m^2 and internal diameter of 0.6 of the external diameter, find the external and internal diameters assuming that the maximum torque is 1.4 times mean. **[170 mm ; 102 mm]**
11. A hollow shaft of outer diameter of 450 mm and inner diameter of 250 mm is subjected to a torque of 350 KNm. Find the shear stresses of the outer and inner surfaces of the shaft. Draw the shear stress distribution diagram. **[21.62 N/mm^2 ; 12.01 N/mm^2]**

❏❏❏

CHAPTER 11

Springs

11.1 SPRINGS

Springs are elastic bodies which are used to absorb energy and to release it as and when required. These are used to absorb energy due to resilience. Any material can be used to construct a spring, so long as the material has the required combination of rigidity and elasticity.

Chapter Outline

11.2 TYPES OF SPRINGS

Springs may be broadly divided into two categories:

(*i*) Torsion springs

(*ii*) Bending springs

11.3 TORSION SPRINGS

A torsion spring is a spring which is subjected to torsion (twisting moment) and resilience occurs due to torsion. Helical spring is a type of torsion spring.

Helical springs are of two types :

(*i*) Open coil helical springs

(*ii*) Closed coil helical springs

DO YOU KNOW?

The best form of spring is that which absorbs greatest amount of energy for a given stress.

11.4 BENDING SPRING

A bending spring is a spring which is subjected to bending only and resilience occurs due to bending. Leaf spring or laminated spring is a type of bending spring.

Laminated springs are of two types :

(*i*) Semi-elliptical type

(*ii*) A quarter-elliptical type or cantilever type

11.5 STIFFNESS OF A SPRING

It is defined as load required to produce unit deflection.

Mathematically,

$$\text{Stiffness} = \frac{\text{Load}}{\text{Deflection}}$$

$$\therefore \quad K = \frac{W}{\delta}$$

> **DO YOU KNOW ?**
>
> Stiffness of a spring is also known as spring constant.

11.6 HELIX ANGLE

Helix angle is the angle which the axis of the spring wire makes with horizontal line.

11.7 ANGLE OF TWIST

Angle of twist is the angle through which cross-section of a bar is twisted due to twisting moment.

11.8 CLOSE COILED HELICAL SPRINGS

The helical spring in which helix angle is very small is known as close coiled helical spring. In other words, if the pitch between two adjacent turns is small, then the spring is known as close coiled helical spring.

11.9 OPEN COILED HELICAL SPRINGS

The helical spring in which the pitch between two adjacent turns is large, is known as open coiled helical spring.

11.10 CLOSE COILED HELICAL SPRINGS

(*a*) Close coiled helical spring under axial load '*W*'

Figure 11.1 shows a close coiled helical spring subjected to axial load '*W*'

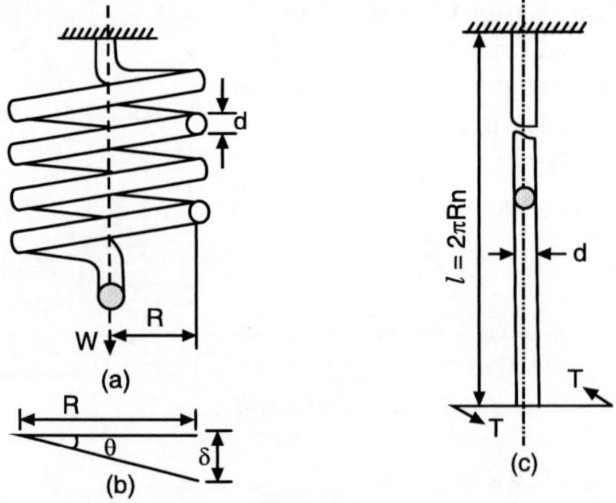

Fig. 11.1

Let,

R = Mean radius of the coil

d = Diameter of spring wire

Note: A close coiled helical spring is so closely coiled that each turn is practically a plane at right angle to the axis of the helix and the stresses upon the material are almost of pure torsion.

n = Number of coils
L = Total length of the wire
θ = Angle of twist of the wire
C = Modulus of rigidity of the spring material
δ = Deflection of the spring due to axial load
τ = Maximum shear stress induced in the spring due to twisting moment
I_p = Polar moment of inertia of the wire

$$= \frac{\pi}{32} d^4$$

Total length of coil, L = circumference of the spring × number of coils = 2π Rn

As the angle of the helix is small, each section of the coil is under torsion. The bending and shearing stresses may be neglected being small.

$$\therefore \quad T = WR$$

Shear stress

From the relation, $\frac{T}{I_p} = \frac{C\theta}{l} = \frac{\tau}{R}$

$$\frac{T}{I_p} = \frac{\tau}{r} \quad \text{[where } r = \text{radius of spring wire} = d/2]$$

$$\frac{T}{I_p} \times r = \tau$$

$$\frac{T \times r}{\frac{\pi}{32} d^4} = \tau$$

$$\tau = \frac{32\, T \times d}{\pi d^4 \times 2}$$

$$\therefore \quad \tau = \frac{16\, T}{\pi d^3} = \frac{16\, WR}{\pi d^3} \quad [\because T = WR]$$

Deflection δ

Again, $\frac{T}{I_p} = \frac{C\theta}{l}$

$$\theta = \frac{Tl}{I_p C} = \frac{T \times L}{I_p\, C} \quad [\because l = L \text{ length of coil}]$$

$$\theta = \frac{WR \times 2\pi\, Rn}{\frac{\pi}{32} d^4\, C} = \frac{32 \times 2\, WR^2 n}{d^4 C}$$

$$\therefore \quad \theta = \frac{64\, WR^2 n}{d^4 C}$$

Note: Shear stress is induced at any section of a spring due to both T and W. If shear stress due to W is not neglected then.

Total maximum shear stress = $\tau_{max} = \frac{16\, WR}{\pi d^3} + \frac{4W}{\pi d^2}$

The free end will twist through angle θ.

$\therefore$ Deflection, $\delta = R\theta$

$$= R \times \frac{64\,WR^2 n}{d^4\,C}$$

$$\therefore \qquad \boxed{\delta = \frac{64\,WR^3 n}{d^4\,C}}$$

Stiffness of spring, *K*

Load per unit deflection is known as stiffness of the spring.

$$K = \frac{W}{\delta} = \frac{W}{\dfrac{64\,WR^3 n}{d^4 C}}$$

$$\therefore \qquad \boxed{K = \frac{Cd^4}{64\,R^3 n}}$$

DO YOU KNOW?

If a spring is made up of square section wire, then the stiffness is given by

$$K = \frac{Cb^4}{44.7 n R^3}$$

where b = side of the square

Energy stored in the spring, *U*

Energy stored, $$U = \frac{1}{2} T\theta = \frac{1}{2} WR \times \frac{64\,WR^2 n}{d^4 C}$$

$$= \frac{1}{2C} \times \frac{16\,WR}{\pi\,d^3} \times \frac{4\,WR^2 n}{d} \times \frac{\pi^2}{\pi}$$

$$= \frac{1}{2C} \times \tau \times \frac{16\,WR}{\pi\,d^3} \times \frac{Rnd^2 \pi^2}{4}$$

$$= \frac{1}{4C}\tau^2 \times \frac{Rd^2 n\,\pi^2}{2} = \frac{1}{4C}\tau^2 \times \text{volume of wire}$$

($\because$ Volume of wire of spring = Length of wire × Area of cross-section)

$$= 2\pi\,Rn \times \frac{\pi}{4} d^2 = \frac{\pi^2 R d^2 n}{2}$$

Again, energy stored, $$U = \frac{1}{2} T\theta = \frac{1}{2} WR \times \frac{\delta}{R}$$

$$\therefore \qquad U = \frac{1}{2} W\delta.$$

11.11 LEAF SPRINGS

Leaf springs are also known as *Laminated springs*. This type of springs is commonly used in carriages such as trucks, trains, etc. It consists of a number of leaves of equal width and thickness, but of varying length, held together at the centre with clamps. These springs are of two types :

(*i*) Semi-elliptical springs

(*ii*) Quarter - elliptical or cantilever springs

11.11.1 Semi-elliptical Springs

These springs are simply supported at its ends and centrally loaded. Fig. 11.2 shows a laminated spring which consists of number of parallel plates of metal of same width, but of varying length, placed one above the other. The spring rest on the axle of the vehicle and are pin jointed to the chassis. The plates are bent to the same radius so that they contact only at their edges, when the spring is loaded at the centre, the change of curvature of all the plates is uniform and become flat. The central deflection of the spring disappears. The load at which the plates become flat is called *proof load.*

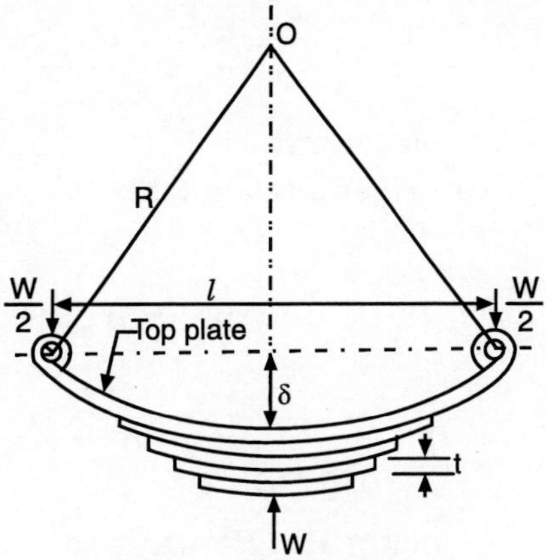

Fig. 11.2. Laminated spring

Consider the laminated spring which is carrying a central load '*W*'. The loaded spring acts as a simply supported beam. The load '*W*' will be equally distributed at the two ends and hence load at both ends of the spring shall be $\frac{W}{2}$ as shown in figure 11.2.

> **DO YOU KNOW?**
>
> An open coiled helical spring falls under both the catagories of spring *i.e.*, torsion spring and bending spring.

Let,

l = Span of the spring
t = Thickness of plates
b = Width of plates
n = Number of plates
R = Radius of curvature of plates
δ = Original central deflection of the top plate
σ_b = Maximum bending stress developed in the plates

Maximum B.M at the centre due to load '*W*', $M = \frac{W}{2} \times \frac{l}{2} = \frac{Wl}{4}$

M. I of each strip, $I = \frac{bt^3}{12}$

We know, $\frac{M}{I} = \frac{\sigma_b}{y}$

$$M = \frac{I}{y} \times \sigma_b = \frac{\frac{bt^3}{12}}{\frac{t}{2}} \times \sigma_b = \frac{bt^2}{6} \times \sigma_b \qquad \left[\because y = \frac{t}{2}\right]$$

Total resisting moment of n strips

$$= n \times M = \frac{n\,\sigma_b\,bt^2}{6}$$

Since, maximum bending moment due to load = Total resisting moment

$$\frac{Wl}{4} = \frac{n\,\sigma_b bt^2}{6}$$

or, $$\frac{Wl \times 6}{4 \times n \times bt^2} = \sigma_b$$

$$\therefore \quad \sigma_b = \frac{3\,Wl}{2n\,bt^2} \quad \ldots(i)$$

Refer figure 11.3

In right angle $\Delta\ OSP$

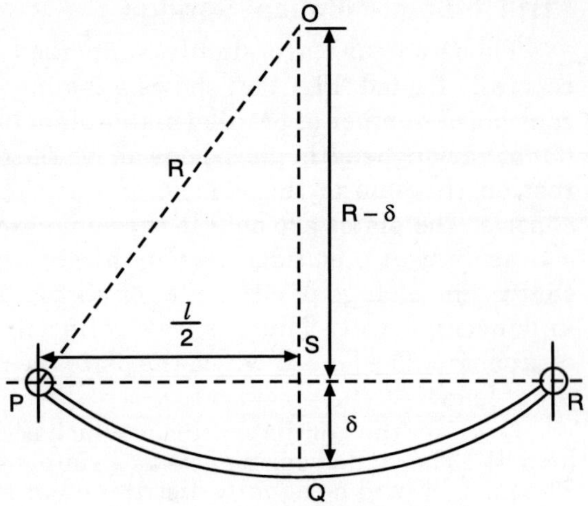

Fig. 11.3

$$OP^2 = OS^2 + PS^2$$

$$R^2 = (R-\delta)^2 + \left(\frac{l}{2}\right)^2$$

$$R^2 = R^2 + \delta^2 - 2\,R\delta + \frac{l^2}{4}$$

$$R^2 - R^2 = \delta^2 - 2\,R\delta + \frac{l^2}{4}$$

$$0 = \delta^2 - 2\,R\delta + \frac{l^2}{4}$$

$$0 = -2\,R\delta + \frac{l^2}{4} \quad \text{[Neglecting } \delta^2 \text{ being very small]}$$

$$2\,R\delta = \frac{l^2}{4}$$

$$\therefore \quad \delta = \frac{l^2}{8R} \quad \ldots(ii)$$

we know,

$$\frac{M}{I} = \frac{\sigma_b}{y} = \frac{E}{R}$$

i.e.,

$$\frac{\sigma_b}{y} = \frac{E}{R}$$

or

$$R = \frac{E \times y}{\sigma_b}$$

$$R = \frac{E \times t}{2\,\sigma_b} \quad \left[\because y = \frac{t}{2}\right]$$

DO YOU KNOW?

Resilience due to bending

$$= \frac{\sigma_b^2}{6E}$$

where σ_b = bending stress

Putting the value of R in (ii), we get

$$\delta = \frac{l^2 \times 2\,\sigma_b}{8\,E \times t} = \frac{\sigma_b\,l^2}{4\,Et} \quad \ldots(iii)$$

Putting the value of σ_b from (i) into (iii), we get

$$\delta = \frac{3\,Wl^3}{8\,nEbt^2}$$

11.11.2 Quarter Elliptic Springs

Quarter Elliptic springs are also known as cantilever laminated springs. Figure 11.4 shows a quarter elliptic spring subjected to a load 'W' at its free end.

Let, l = Span of the spring

t = Thickness of plates

b = Width of plates
n = Number of plates
R = Radius of curvature of plates
δ = Deflection at the free end

B. M at the fixed end due to load 'W', $M = W \times l$

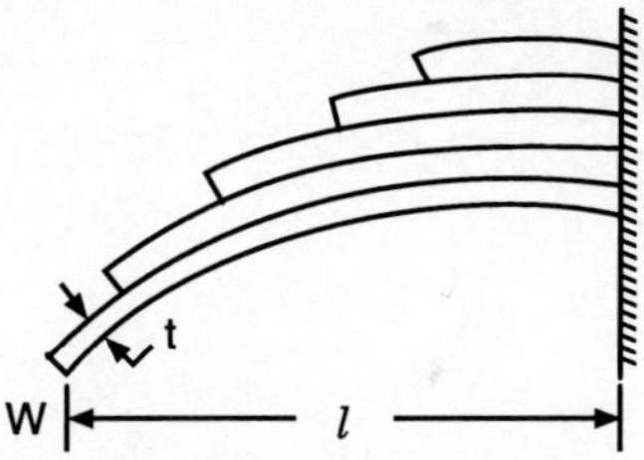

Fig. 11.4. Quarter elliptical spring

M. I of each strip, $= \dfrac{bt^3}{12}$

We know, $\dfrac{M}{I} = \dfrac{\sigma_b}{y}$ [$\because$ σ_b = Maximum bending stress]

$$M = \frac{I \times \sigma_b}{y} = \frac{\frac{bt^3}{12} \times \sigma_b}{\frac{t}{2}} = \frac{bt^2}{6} \times \sigma_b$$

Total resisting moment of n strips = $n \times M$

$$= \frac{nbt^2\ \sigma_b}{6}$$

Maximum bending moment = Total resisting moment

$$W \times l = \frac{nbt^2\ \sigma_b}{6}$$

$$\therefore \quad \sigma_b = \frac{6\ Wl}{nbt^2} \quad ...(i)$$

Again from properties of a circle

$$(2R - \delta)\ \delta = L \times L$$

or $$2\ R\delta = L^2$$ (Neglecting δ^2, being very small)

$$\therefore \quad \delta = \frac{L^2}{2R} \quad ...(ii)$$

We know, $\dfrac{\sigma_b}{y} = \dfrac{E}{R}$

$$\therefore \quad R = \frac{E \times y}{\sigma_b} = \frac{E \times t}{2\ \sigma_b} \quad \left[\because y = \frac{t}{2}\right] \quad ...(iii)$$

Putting the value of σ_b from (*i*) into (*iii*), we get

$$R = \frac{E \times t \times nbt^2}{2 \times 6\ WL} = \frac{n\,Ebt^3}{12\ WL} \quad ...(iv)$$

Putting the value of R from (*iv*) into (*ii*), we get

$$\delta = \frac{L^2 \times 12\ WL}{2\ nEbt^3} = \frac{6\ WL^3}{nEbt^3}$$

11.12 SPRINGS IN SERIES

Two springs of stiffness K_1 and K_2 are connected in series and carrying a load 'W'. The total extension of the springs will be the sum of extension of two springs.

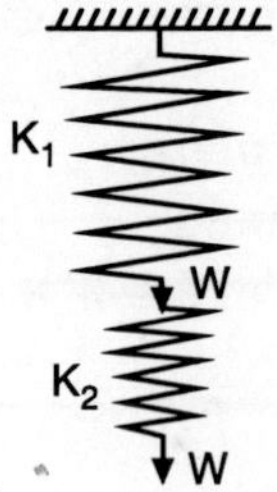

Fig. 11.5 Springs in series

Let, δ_1 = Extension of spring, K_1

δ_2 = Extension of spring, K_2

δ = Total extension of spring

K = Stiffness of composite spring

$\therefore$ Total extension, $\delta = \delta_1 + \delta_2$

$$\frac{W}{K} = \frac{W}{K_1} + \frac{W}{K_2} \qquad \left[\because \text{Stiffness} = \frac{\text{load}}{\text{extension}}\right]$$

$$\therefore \quad \frac{1}{K} = \frac{1}{K_1} + \frac{1}{K_2}$$

Note:

When two springs are in series, both the spring will carry equal load.

11.13 SPRINGS IN PARALLEL

When the springs are connected in parallel, the extension of each spring will be the same and the total load will be shared by two springs.

Let,

δ = Extension of springs I and II of the composite spring

W = Total load carried by composite spring

K = Stiffness of composite spring

K_1 = Stiffness of spring I

K_2 = Stiffness of spring II

W_1 = Load carried by spring K_1

W_2 = Load carried by spring K_2

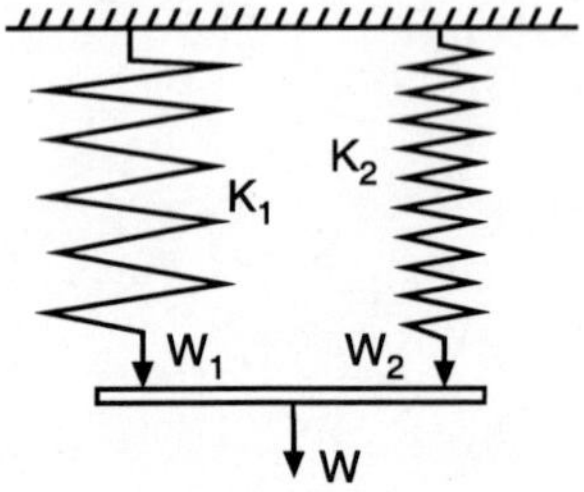

Fig. 11.6

$\therefore$ Total load, $W = W_1 + W_2$

$$\delta K = \delta K_1 + \delta K_2$$

$$K = K_1 + K_2$$

SOLVED EXAMPLES

EXAMPLE 11.1. A close coiled helical spring is to carry a load of 110 N and the mean coil diameter is to be 8 times that of the wire diameter. Calculate these diameter if the maximum shear stress is to be 100 N/mm².

Solution : Given, Load, $W = 110\ N$

Maximum shear stress, $\tau = 100\ \text{N/mm}^2$

Let d = diameter of wire

$\therefore$ Mean coil diameter, $D = 8\,d \Rightarrow R = 4d$

We know,

Shear stress, $$\tau = \frac{16\,WR}{\pi\,d^3}$$

$$100 = \frac{16 \times 110 \times 4d}{\pi \times d^3}$$

$$100 = \frac{7040\,d}{\pi\,d^3}$$

or $$d^2 = \frac{7040}{\pi \times 100} = 22.41$$

∴ $d = 4.73$ mm **Ans.**

and $D = 8 \times 4.73 = 37.84$ mm **Ans.**

EXAMPLE 11.2. A close coiled helical spring of round steel wire 5 mm in diameter having 12 complete coils of 50 mm mean diameter is subjected to an axial load of 100 N. Find the deflection of the spring and the maximum shear stress in the material. Take C = 8 × 10⁴ N/mm².

Solution : Given,

Spring wire diameter, $d = 5$ mm

Number of coils, $n = 12$

Mean diameter of coil, $D = 50$ mm ⇒ $R = 25$ mm

Axial load, $W = 100\,N$

$C = 8 \times 10^4$ N/mm²

> **REMEMBER**
>
> Deflection of spring,
> $$\delta = \frac{64\,WR^3 n}{d^4 C}$$

We know,

Deflection of spring, $$\delta = \frac{64\,WR^3 n}{d^4 C} = \frac{64 \times 100 \times (25)^3 \times 12}{(5)^4 \times 8 \times 10^4}$$

∴ $$\delta = \frac{1.2 \times 10^9}{5.0 \times 10^7} = 24 \text{ mm} \quad \textbf{Ans.}$$

Maximum shear stress, $$\tau_{max} = \frac{16\,WR}{\pi d^3} = \frac{16 \times 100 \times 25}{\pi \times (5)^3}$$

$$= \frac{40{,}000}{392.7} = 101.86 \text{ N/mm}^2 \quad \textbf{Ans.}$$

EXAMPLE 11.3. A close-coiled helical spring of round steel wire is to have a stiffness of 20 N/mm. It is required to carry a maximum load of 500 N. The maximum permissible shear stress is 350 N/mm². Take C = 8.40 × 10⁴ N/mm². Find the diameter of the wire and the number of turns if the mean diameter of the coil is 40 mm.

Solution : Given,

Stiffness of spring, $K = 20$ N/mm

Load, $W = 500$ N

Maximum shear stress, $\tau_{max} = 350$ N/mm²

$C = 8.4 \times 10^4$ N/mm²

Mean diameter of coil, $D = 40$ mm ⇒ $R = 20$ mm

Let the diameter of wire of spring be 'd'.

We know, $$\tau_{max} = \frac{16\,WR}{\pi\,d^3}$$

$$350 = \frac{16 \times 500 \times 20}{\pi \times d^3}$$

$$d^3 = \frac{16 \times 500 \times 20}{\pi \times 350} = \frac{1,60,000}{1099.56}$$

$$d^3 = 145.51$$

$\therefore$ $$d = 5.26 \text{ mm}$$

REMEMBER

Stiffness, $K = \dfrac{cd^4}{64\,R^3 n}$

Stiffness, $$K = \frac{cd^4}{64\,R^3 n}$$

$$20 = \frac{8.4 \times 10^4 \times (5.26)^4}{64 \times (20)^3 \times n}$$

$$n = \frac{8.4 \times (5.26)^4 \times 10^4}{64 \times 20 \times (20)^3} = \frac{64301672}{10,240,000} = 6.28 \approx 7 \quad \textbf{Ans.}$$

EXAMPLE 11.4. A close-coiled helical spring, with coil diameter as 100 mm and wire diameter as 12 mm consists of 16 coils. If it is subjected to an axial tension of 400 N, find the maximum stress induced in the coil, the extension suffered by the spring and the energy stored in it. Modulus of rigidity, C = 84 GN/m².

Solution : Given,

Coil diameter, $D = 100$ mm $\quad\therefore$ Radius of coil, $R = 50$ mm

Wire diameter, $d = 12$ mm

Number of coils, $n = 16$

Axial tension, $W = 400$ N

Modulus of rigidity, $$C = 84 \text{ GN/m}^2 = \frac{84 \times 10^9}{10^6} = 84 \times 10^3 \text{ N/mm}^2$$

Stress induced in coil, $$\tau = \frac{16\,WR}{\pi\,d^3} = \frac{16 \times 400 \times 50}{\pi \times (12)^3}$$

$$= \frac{32,00,00}{5428.67} = 58.95 \text{ N/mm}^2 \quad \textbf{Ans.}$$

Extension of spring, $$\delta = \frac{64\,WR^3 n}{Cd^4} = \frac{64 \times 400 \times (50)^3 \times 16}{84 \times 10^3 \times (12)^4}$$

$$= \frac{5.12 \times 10^{10}}{1.74 \times 10^9} = 29.39 \quad \textbf{Ans.}$$

Volume of wire $$= \frac{\pi}{4} d^2 \times 2\pi\,Rn = \frac{\pi}{4}(12)^2 \times 2\pi \times 50 \times 16$$

$$\left(\because \text{Area of cross-section of wire} = \frac{\pi}{4} d^2\right)$$

(length of wire in the spring = circumference of spring × number of coil = 2 π Rn)

$$= 568489.21 \text{ mm}^3$$

Energy stored $= \dfrac{\tau^2}{4C} \times \text{volume of wire}$

$$= \frac{(58.95)^2}{4 \times 84 \times 10^3} \times 568489.21 = \frac{1.975 \times 10^9}{336 \times 10^3}$$

$$= 5877.98 \text{ Nmm} = 5.88 \text{ N-m} \quad \textbf{Ans.}$$

Alternate method for finding out stored energy

Energy stored, $U = \dfrac{1}{2} W\delta = \dfrac{1}{2} \times 400 \times 29.39$

$$= 5878 \text{ Nmm} = 5.88 \text{ Nm} \quad \textbf{Ans}$$

EXAMPLE 11.5. A close-coiled helical spring having 100 mm mean diameter is made of 20 turns of 10 mm diameter steel rod. The spring carries an axial load of 100 N. Find the shearing stress developed in the spring and the deflection of the load. C = 84 GN/m².

Solution : Given,

Mean diameter of spring, $D = 100$ mm

∴ Radius of spring, $R = \dfrac{D}{2} = \dfrac{100}{2} = 50$ mm

Number of turns, $n = 20$

Diameter of steel rod, $d = 10$ mm

Axial load, $W = 100\ N$

Modulus of rigidity, $C = 84 \text{ GN/m}^2$

$$= \frac{84 \times 10^9}{10^3 \times 10^3} = 84 \times 10^3 \text{ N/mm}^2$$

Shearing stress, $\tau = \dfrac{16\,WR}{\pi\, d^3} = \dfrac{16 \times 100 \times 50}{\pi \times (10)^3}$

$$= \frac{80{,}000}{3141.59} = 25.46 \text{ N/mm}^2 \quad \textbf{Ans}$$

Deflection, $\delta = \dfrac{64\,WR^3 n}{Cd^4} = \dfrac{64 \times 100 \times (50)^3 \times 20}{84 \times 10^3 \times (10)^4}$

$$= \frac{1.6 \times 10^{10}}{84 \times 10^7} = 19.05 \text{ mm} \quad \textbf{Ans}$$

EXAMPLE 11.6. A helical closed-coiled spring is made of round steel wire 6.35 mm diameter. The mean diameter of coil is 63.50 mm, number of turns = 12. If C = 8.4 × 10⁴ N/mm². Find:

(*i*) the pull required to extend the spring 25.4 mm, and

(*ii*) the stress in the spring wire

Solution : Given,

Diameter of wire, $d = 6.35$ mm

Mean diameter of coil, $D = 63.50$ mm $\Rightarrow R = 31.75$ mm

Number of coils, $n = 12$

$C = 8.4 \times 10^4$ N/mm^2

Deflection, $\delta = 25.4$ mm

(*i*) Pull required for deflection of 25.4 mm

Deflection, $$\delta = \frac{64\,WR^3n}{d^4C}$$

or, $$W = \frac{\delta \times d^4 \times C}{64 \times R^3 \times n} = \frac{25.4 \times (6.35)^4 \times 8.4 \times 10^4}{64 \times (31.75)^3 \times 12}$$

$$= \frac{3.469 \times 10^9}{24580596} = 141.13 \text{ N} \quad \textbf{Ans}$$

(*ii*) Stress in the spring

Shear stress, $$\tau = \frac{16\,WR}{\pi\,d^3}$$

$$\tau = \frac{16 \times 141.13 \times 31.75}{\pi \times (6.35)^3} = \frac{71694.04}{804.398} = 89.13 \text{ N/mm}^2 \quad \textbf{Ans}$$

EXAMPLE 11.7. A railway wagon weighing 40 KN and moving with a speed of 8 Km/hr is stopped by a buffer of 4 springs whose allowable maximum compression is 150 mm. Find out the number of turns in each spring; if the diameter of the spring wire is 14 mm and the diameter of the coil is 80 mm. Take C = 8.4 × 10⁴ N/mm².

Solution : Given,

Speed, $$v = 80 \text{ Km/hr} = \frac{8 \times 1{,}000}{60 \times 60} = 2.22 \text{ m/s}$$

Deflection, $\delta = 150$ mm

Diameter of spring wire, $d = 14$ mm

Diameter of coil, $D = 80$ mm $\Rightarrow R = 40$ mm

$C = 8.4 \times 10^4$ N/mm^2

Weight of railway wagon, 40 KN = 40,000 *N*

Mass of railway wagon, $$m = \frac{W}{g} = \frac{40{,}000}{9.81} = 4077.47 \text{ Kg}$$

K.E of the wagon $$= \frac{1}{2}mv^2 = \frac{1}{2} \times 4077.47 \times (2.22)^2$$

$$= 10047.702 \text{ Nm} = 10047702 \text{ Nmm}$$

Let *W* be the gradually applied load on one spring so as to compress it by 150 mm.

Deflection, $$\delta = \frac{64\,WR^3n}{d^4C}$$

$$150 = \frac{64 \times W \times (40)^3 \times n}{(14)^4 \times 8.4 \times 10^4}$$

$$\therefore \quad W = \frac{150 \times (14)^4 \times 8.4 \times 10^4}{64 \times (40)^3 \times n} = \frac{118164.06}{n} N$$

Strain energy stored by one spring $= \frac{1}{2} W\delta$

$$U = \frac{1}{2} \times \frac{118164.06}{n} \times 150 = \frac{8862304.7}{n}$$

Strain energy stored by four spring $= \frac{8862304.7 \times 4}{n}$

Strain energy of springs = K.E of wagon

$$\frac{8862304.7 \times 4}{n} = 10047702$$

$$= \frac{8862304.7 \times 4}{10047702} = n$$

$$\therefore \quad n = 3.53 = 4 \text{ coils} \quad \textbf{Ans.}$$

EXAMPLE 11.8. A close-coiled helical spring is made out of a 10 mm diameter steel rod. The coils have 12 complete turns. The mean diameter of the coil is 100 mm. It carries an axial pull of 300 N. Calculate the shear stress induced in the spring. If $C = 0.8 \times 10^5$ N/mm², calculate the deflection under the pull and the amount of energy stored in the spring during the extension.

Solution : Given,

Diameter of steel rod, $d = 10$ mm

Number of turns, $n = 12$

Mean diameter of the coil, $D = 100$ mm $\Rightarrow R = 50$ mm

Axial pull, $W = 300$ N

$C = 0.8 \times 10^5$ N/mm²

We know, shear stress, $\tau = \frac{16\,WR}{\pi\, d^3} = \frac{16 \times 300 \times 50}{\pi \times (10)^3} = \frac{2,40,000}{3141.59} = 76.39$ N/mm²

Deflection, $\delta = \frac{64\, WR^3 n}{d^4 C} = \frac{64 \times 300 \times (50)^3 \times 12}{(10)^4 \times 0.8 \times 10^5} = 36$ mm **Ans**

Amount of energy stored in the spring, $u = \frac{1}{2} W\delta$

$$= \frac{1}{2} \times 300 \times 36 = 5400 \text{ Nmm} \quad \textbf{Ans}$$

EXAMPLE 11.9. A weight of 3000 N is dropped on to a close-coiled compression helical spring with 15 coils. Calculate the height of the drop before striking so that the spring is compressed by 200 mm. The diameter of the rod is 25 mm and the mean radius of the coil is 100 mm. Take C = 8 × 10⁴ N/mm².

Solution : Given,

Number of coils, $n = 15$

Compression/deflection of spring, $\delta = 200$ mm

Diameter of rod, $d = 25$ mm

Mean radians of coil, $R = 100$ mm

$$C = 8 \times 10^4 \text{ N/mm}^2$$

Falling weight, $W_1 = 3000\ N$

Let W be the gradually applied load which produces the compression of spring equal to 200 mm.

We know,

Deflection $$\delta = \frac{64\, WR^3 n}{d^4 C}$$

$$200 = \frac{64 \times W \times (100)^3 \times 15}{(25)^4 \times 8 \times 10^4}$$

$$\frac{200 \times (25)^4 \times 8 \times 10^4}{64 \times (100)^3 \times 15} = W$$

$\therefore$ $$W = 6510.4\ N$$

Energy stored in the spring, $U = \frac{1}{2} W\delta$

$$U = \frac{1}{2} \times 6510.4 \times 200 = 651040 \text{ Nmm} \quad ...(i)$$

Let the weight 3000 N be dropped from a height of h mm.

Workdone by falling weight on spring, (W.D) = $W_1(h + \delta)$

$\therefore$ $$\text{W.D} = 3000 \times (h + 200) \quad ...(ii)$$

Work done on spring = Energy stored in the spring

$$3000\ (h + 200) = 651040$$

$$h + 200 = \frac{651040}{3000} = 217.01$$

$$217.01 - 200 = 17.01 \text{ mm}$$ **Ans.**

EXAMPLE 11.10. A close-coiled helical spring has a stiffness of 2 N/mm under a compressive load of 60 N. The shear stress in the spring is 130 N/mm². The solid length of the spring is 45 mm. Find

(*i*) the diameter of the wire

(*ii*) the mean diameter of the coils

(*iii*) the number of coils

Take C = 0.42 × 10⁵ N/mm²

REMEMBER
Solid length of spring = Number of coils × diameter of wire

Solution : Given,

Stiffness of spring, $K = 2$ N/mm

Compressive load, $W = 60$ N

Shear stress, $\tau = 130$ N/mm²

Solid length of spring = 45 mm

Solid length of spring = Number of coils × diameter of wire

$$45 = n \times d$$

$\therefore$ $$n = \frac{45}{d}$$

We know,

Shear stress, $\tau = \dfrac{16\,WR}{\pi\,d^3}$

$$130 = \frac{16 \times 60 \times R}{\pi\,d^3}$$

$$\frac{130 \times \pi}{16 \times 60} = \frac{R}{d^3}$$

$$\frac{408.4}{960} = \frac{R}{d^3}$$

$$0.425 = \frac{R}{d^3}$$

$\therefore$ $R = 0.425\,d^3$

Stiffness, $K = \dfrac{W}{\delta}$

$$2 = \frac{60}{\delta}$$

$\therefore$ $\delta = 30$ mm

Deflection, $\delta = \dfrac{64\,WR^3 n}{Cd^4}$

$$30 = \frac{64 \times 60 \times (0.425\,d^3)^3 \times 45}{0.42 \times 10^5 \times d^4 \times d} \qquad \left[\because n = \frac{45}{d} \text{ and } R = 0.425\,d^3\right]$$

$$30 = \frac{13265.1 \times d^9}{42000 \times d^5} = 0.316 \times d^4$$

or $d^4 = \dfrac{30}{0.316} = 94.94$

$\therefore$ $d = 3.12$ mm **Ans**

Radius of coil, $R = 0.425\,d^3$

$= 0.425 \times (3.12)^3 = 12.9$ mm

Diameter of coil, $D = 2 \times R$

$= 2 \times 12.9 = 25.8$ mm **Ans**

Number of coil, $n = \dfrac{45}{3.12} = 14.42 \approx 15$ **Ans**

EXAMPLE 11.11. A weight of 250 N is dropped through a height of 750 mm to a close-coiled compression helical spring which instantaneously compresses by 150 mm under the impact. If the diameter of the spring rod is 25 mm and the mean radius of the coil is 100 mm, find the maximum instantaneous stress produced by the impact and the number of coils in the spring. Take C = 8 × 10⁴ N/mm².

Solution : Given,

Falling weight, $W_1 = 250$ N

Height of fall, $h = 750$ mm

Compression of spring, $\delta = 150$ mm

Diameter of spring rod, $d = 25$ mm

Mean radius of coil, $R = 100$ mm

$C = 8 \times 10^4$ N/mm²

Work done by falling weight on spring (W.D) $= W_1(h + \delta)$

$= 250\ (750 + 150)$

$= 250 \times 900 = 225000$ N mm

Let W be the gradually applied load

Energy stored in the spring $= \frac{1}{2} W\delta$

Energy stored = W.D

$$\frac{1}{2} W \times 150 = 225000$$

$$W = \frac{225000 \times 2}{150} = 3000 \text{ N}$$

Maximum stress, $\tau = \frac{16\,WR}{\pi\, d^3} = \frac{16 \times 3000 \times 100}{\pi \times (25)^3}$

$$= \frac{4800000}{49087.38} = 97.78 \text{ N/mm}^2 \quad \textbf{Ans}$$

We know,

Deflection, $\delta = \frac{64 W R^3 n}{d^4 C}$

$$150 = \frac{64 \times 3000 \times (100)^3 \times n}{(25)^4 \times 8 \times 10^4}$$

$$\frac{150 \times (25)^4 \times 8 \times 10^4}{64 \times 3000 \times 10^6} = n$$

$$\frac{4.6875 \times 10^{12}}{1.92 \times 10^{11}} = n$$

$\therefore$ $n = 24.41 \approx 25$ **Ans.**

EXAMPLE 11.12. Two springs are connected together in series to form a composite spring of stiffness 1 N/mm. Each spring has a mean diameter 8 times the wire diameter. One of the spring has 24 coils and wire diameter 2 mm. Determine the diameter of the other spring which has 32 coils. Find the maximum load that can be applied in tension to the composite spring and calculate the extension. The permissible shear stress is 200 N/mm² and C = 8 × 10⁴ N/mm².

Solution : Given,

Stiffness of composite spring, $K = 1$ N/mm

$C = 8 \times 10^4$ N/mm²

Shear stress, $\tau = 200$ N/mm²

1st spring

Number of coils, $n_1 = 24$

Diameter of wire, $d_1 = 2$ mm

Diameter of coil, $D_1 = 8 \times d_1$

$= 8 \times 2 = 16$ mm $\Rightarrow R_1 = 8$ mm

Stiffness, $$K_1 = \frac{Cd_1^4}{64\, R_1^3\, n_1} = \frac{8 \times 10^4 \times (2)^4}{64 \times (8)^3 \times 24} = \frac{128 \times 10^4}{786432} = 1.63 \text{ N/mm}$$

2nd spring

Number of coils, $n_2 = 32$

Let diameter of wire $= d_2$

Diameter of coil, $D_2 = 8\, d_2 \Rightarrow R = 4\, d_2$

Stiffness, $$K_2 = \frac{C\, {d_2}^4}{64\, R_2^3\, n_2} = \frac{8 \times 10^4 \times d_2^4}{64 \times (4d_2)^3 \times 32}$$

$$\therefore \quad K_2 = \frac{8 \times 10^4 \times d_2}{131072} = 0.61\, d_2$$

As the springs are connected in series

$$\therefore \quad \frac{1}{K} = \frac{1}{K_1} + \frac{1}{K_2}$$

$$\frac{1}{1} = \frac{1}{1.63} + \frac{1}{0.61\, d_2}$$

$$1 - \frac{1}{1.63} = \frac{1}{0.61\, d_2}$$

$$\frac{0.63}{1.63} = \frac{1}{0.61\, d_2}$$

$$d_2 = \frac{1}{0.61 \times 0.386}$$

$$\therefore \quad d_2 = 4.25 \text{ mm}$$

$$\therefore \quad K_2 = 0.61 \times 4.25 = 2.59 \text{ N/mm}$$

We know,

Shear stress, $$\tau = \frac{16\, WR}{\pi\, d^3}$$

$$200 = \frac{16 \times W \times R_1}{\pi\, d_1^3}$$

[$\because d_1 < d_2$, the greatest load is carried by smaller diameter]

Note :

The greatest load that can be carried by the spring will correspond to the smaller diameter.

$$200 = \frac{16 \times W \times 8}{\pi \times (2)^3}$$

$$200 = \frac{128 W}{8\pi}$$

$$\frac{200 \times 8\pi}{128} = W$$

$\therefore$ $W = 39.27$ N **Ans**

Deflection, $\delta_1 = \frac{W}{K_1} = \frac{39.27}{1.63} = 24.09$

$$\delta_2 = \frac{W}{K_2} = \frac{39.27}{2.59} = 15.16 \text{ mm}$$

Total deflection, $\delta = \delta_1 + \delta_2 = 24.09 + 15.16 = 39.25$ mm **Ans.**

EXAMPLE 11.13. A helical spring B is placed inside the coils of a second helical spring 'A', having the same number of coils, length and of the same material. The two springs are compressed by an axial load of 2300 N, which is shared between them. The mean coil diameters of A and B are 100 mm and 70 mm and the wire diameters are 13 mm and 8 mm respectively. Calculate the load taken and the maximum stress in each spring.

Solution : Given axial load, $W = 2300$ N

Spring A

Mean coil diameter, $D_A = 100$ mm $\Rightarrow R_A = 50$ mm

Wire diameter, $d_A = 13$ mm

Let the load shared by spring 'A' be W_A.

Spring B

Mean coil diameter, $D_B = 70$ mm $\Rightarrow R_B = 35$ mm

Wire diameter, $d_B = 8$mm

Let the load shared by spring 'B' be W_B.

$\therefore$ $W_B = W - W_A = (2300 - W_A)$

As the spring are connected in parallel

Deflection of spring 'A', δ_A = Deflection of spring 'B', δ_B

$$\frac{64\, W_A\, R_A^3 n}{C d_A^4} = \frac{64\, W_B\, R_B^3 n}{C d_B^4}$$

$$\frac{W_A\, R_A^3}{d_A^4} = \frac{W_B\, R_B^3}{d_B^4}$$

$$\frac{W_A \times (50)^3}{(13)^4} = \frac{(2300 - W_A) \times (35)^3}{(8)^4}$$

$$\frac{W_A \times 125000}{28561} = \frac{(2300 - W_A) \times 42875}{4096}$$

$$4.38\, W_A = 10.47\,(2300 - W_A)$$

$$4.38\, W_A = 24081 - 10.47\, W_A$$

$$14.85\, W_A = 24081$$

$\therefore$ $W_A = \frac{24081}{14.85} = 1621.62\, N$ **Ans**

$\therefore$ Load shared by second spring, $W_B = 2300 - 1621.62 = 678.38\, N$ **Ans.**

Shear stress in spring 'A', $\tau_A = \frac{16\, W_A\, R_A}{\pi\, d_A^3} = \frac{16 \times 1621.62 \times 50}{\pi \times (13)^3}$

$$= \frac{1297296}{6902.08} = 187.96 \text{ N/mm}^2 \quad \textbf{Ans}$$

Shear stress in spring '*B*',

$$\tau_B = \frac{16\, W_B R_B}{\pi\, d_B^3} = \frac{16 \times 678.38 \times 35}{\pi \times 8^3}$$

$$= \frac{379892.8}{1608.49} = 236.18 \text{ N/mm}^2 \quad \textbf{Ans.}$$

EXAMPLE 11.14. A leaf spring is to be made of seven steel plates 65 mm wide and 6.5 mm thick. Calculate the length of the spring, so that it may carry a central load of 2.75 KN, the bending stress being limited to 160 N/mm^2.

Solution : Given,

Width of plate, $b = 65$ mm

Thickness of plate, $t = 6.5$ mm

Bending stress, $\sigma_b = 160$ N/mm^2

Central load on spring, $W = 2.75$ KN $= 2750$ N

Let l be the length of the spring

Number of plates, $n = 7$

We know,

$$\text{Bending stress, } \sigma_b = \frac{3\,Wl}{2\,nbt^2}$$

$$160 = \frac{3 \times 2750 \times l}{2 \times 7 \times 65 \times (6.5)^2}$$

$$160 = \frac{8250\, l}{38447.5}$$

$$\frac{160 \times 38447.5}{8250} = l$$

$$\therefore \quad l = 745.65 \text{ mm} \quad \textbf{Ans.}$$

EXAMPLE 11.15. A laminated spring 1m long is made up of plates each 50 mm wide and 10 mm thick. If the bending stress in the plates is limited to 100 N/mm^2, how many plates are required to enable the spring to carry a central load of 2000 N? If $E = 2.1 \times 10^5$ N/mm^2, what is deflection under the given load?

Solution : Given,

Width of plate, $b = 50$ mm

Thickness of plate, $t = 10$ mm

Bending stress, $\sigma_b = 100$ N/mm^2

Load, $W = 2000\ N$

$E = 2.1 \times 10^5$ N/mm^2

Length of spring, $l = 1\text{m} = 1000$ mm

Let the number of plates required be n.

We know,

$$\text{bending stress, } \sigma_b = \frac{3Wl}{2n\, bt^2}$$

$$100 = \frac{3 \times 2000 \times 1000}{2 \times n \times 50 \times (10)^2}$$

$$n = \frac{6 \times 10^6}{2 \times 100 \times 50 \times 100} = \frac{6 \times 10^6}{1 \times 10^6} = 6 \quad \textbf{Ans}$$

Deflection, $$\delta = \frac{\sigma_b\, l^2}{4\,Et} = \frac{100 \times (1000)^2}{4 \times 2.1 \times 10^5 \times 10} = 11.90 \text{ mm} \quad \textbf{Ans.}$$

EXAMPLE 11.16. A carriage spring is to be 600 mm long and made of 9.5 mm thick steel plates and 50 mm wide. How many plates are required to carry a load of 4500 N, without the stress exceeding 230 N/mm²? What would be central deflection and initial radius of curvature, if the plates straighten under the load? Take E = 2 × 10⁵ N/mm².

Solution : Given,

Length of spring, $l = 600$ mm

Thickness of plate, $t = 9.5$ mm

Width of plate, $b = 50$ mm

Load, $W = 4500$ N

Bending stress, $\sigma_b = 230$ N/mm^2

$E = 2 \times 10^5$ N/mm^2

Let number of plates required be n

We know, bending stress,

$$\sigma_b = \frac{3\,Wl}{2\,nbt^2}$$

$$230 = \frac{3 \times 4500 \times 600}{2 \times n \times 50 \times (9.5)^2}$$

$$n = \frac{3 \times 4500 \times 600}{2 \times 230 \times 50 \times (9.5)^2} = \frac{81{,}00{,}000}{20{,}75{,}750} = 3.9 \approx 4$$

Deflection, $$\delta = \frac{\sigma_b l^2}{4\,Et} = \frac{230 \times (600)^2}{4 \times 2 \times 10^5 \times 9.5} = \frac{8{,}28{,}00{,}000}{76{,}00{,}000} = 10.89 \text{ mm} \quad \textbf{Ans}$$

Using the relation

Deflection, $$\delta = \frac{l^2}{8R}$$

or, $$R = \frac{l^2}{8\delta}$$

$\therefore$ $$R = \frac{(600)^2}{8 \times 10.89} = 4132.23 \text{ mm} \quad \textbf{Ans.}$$

SUMMARY

1. *Springs* are elastic bodies which are used to absorb energy and to release it as and when required.
2. A spring which is subjected to torsion and resilience occurs due to torsion is known as **torsion spring**.

3. A spring which is subjected to bending only and resilience occurs due to bending is known as *bending spring.*
4. The helical spring in which the pitch between two adjacent turns is large, is known as *open coil helical spring.*
5. The helical spring in which helix angle is very small is known as *close-coiled helical spring.*
6. Stiffness of a spring is defined as load required to produce unit deflection.

   Stiffness, $$K = \frac{\text{load } (W)}{\text{deflection } (\delta)}$$
7. The maximum shear stress developed in a close-coiled helical spring subjected to axial load '*W*' is given by;
   $$\tau = \frac{16\,WR}{\pi d^3}$$
8. Deflection of close-coiled helical spring is given by;
   $$\delta = \frac{64\,WR^3 n}{d^4 c}$$
9. Strain energy in a close-coiled helical spring is given by;
   $$U = \frac{1}{4c} \times \tau^2 \times \text{volume of wire of spring}$$
10. The maximum bending stress developed in the laminated (semi-elliptical) spring is given by;
    $$\sigma_b = \frac{3\,Wl}{2\,nbt^2}$$
11. Central deflection of the laminated spring is given by;
    $$\delta = \frac{\sigma_b\, l^2}{4\,Et} \text{ or } \frac{3\,Wl^3}{8nEbt^2}$$
12. The maximum bending stress developed in a quarter elliptic spring is given by;
    $$\sigma_b = \frac{6\,Wl}{nbt^2}$$
13. Deflection of a quarter elliptic laminated spring is given by
    $$\delta = \frac{6\,WL^3}{nEbt^3}$$
14. When two springs are connected in series, then
    $$\frac{1}{K} = \frac{1}{K_1} + \frac{1}{K_2}$$
    where, K = Stiffness of composite spring

    K_1 = Stiffness of spring - I

    K_2 = Stiffness of spring - II
15. When two springs are connected in parallel, then
    $$K = K_1 + K_2$$
16. For springs in series, total extension, $\delta = \delta_1 + \delta_2$
17. For springs in parallel, total load, $W = W_1 + W_2$.

OBJECTIVE TYPE QUESTIONS

Fill in the Blanks :

(*i*) A ________ spring is a length of wire wound into a helix.

(*ii*) Stiffness of spring is the load per unit ________ .

(*iii*) If two springs are connected in parallel, then the stiffness of composite spring is ________ of the stiffness of two springs.

(*iv*) ________ springs are also known as semi-elliptical springs.

(*v*) Shear stress is developed in ________ springs.

(*vi*) Bending stress is developed in ________ springs.

(*vii*) ________ load is the load at which the plates of a laminated spring become straight.

(*viii*) $\frac{1}{K} = \frac{1}{K_1} + \frac{1}{K_2}$ is the formula used when the springs are connected in ________ .

(*ix*) ________ angle is the angle which the axis of the spring wire makes with a horizontal line.

(*x*) Close-coiled helical springs are the springs in which, helix angle is very ________ .

(*xi*) Solid length of a spring = Number of coils × ________ .

(*xii*) The elastic members which distort under load and regain their original shape on removal to load is known as ________ .

Answers

(*i*) Helical	(*ii*) Deflection	(*iii*) Sum	(*iv*) Laminated
(*v*) Helical	(*vi*) Laminated	(*vii*) Proof	(*viii*) Parallel
(*ix*) Helix	(*x*) Small	(*xi*) diameter of wire	(*xii*) springs

THEORETICAL QUESTIONS

1. What is a spring?
2. What are the different types of springs? Write their applications.
3. What are the functions of a spring?
4. Define the followings in respect of a spring :

   (*a*) Stiffness (*b*) Solid length

   (*c*) Angle of twist (*d*) Helix angle
5. What is a difference between close-coiled spring and open coiled spring?
6. Derive the formula for stiffness of a close-coiled helical spring.
7. What is a laminated spring? Derive the formula for deflection of a laminated spring.

NUMERICALS

1. A close-coiled helical spring has a radius of 50 mm and 20 turns. It is made of wire of 10 mm diameter. Find the maximum shear stress in the wire and the deflection of the spring if it is subjected to a load of 200 *N*. Assume $C = 0.725 \times 10^5$ N/mm^2. **[50.93 N/mm^2; 1.38 mm]**
2. A close-coiled helical spring is to carry a load of 100 *N* and the mean coil diameter is to be 8 times that of the wire diameter. Determine the diameter of the wire if the maximum stress is to be 75 N/mm^2. **[5.2 mm]**
3. A close-coiled helical spring of round steel wire 10 mm in diameter having 14 complete coils is subjected

to an axial load of 250 *N*. Calculate the deflection of the spring and the maximum shear stress in the material. Take $C = 80$ kN/mm^2 and spring index = 10. **[35 mm; 63.66 N/mm^2]**

4. Design a close-coiled helical spring of stiffness 8 N/mm. The maximum shear stress in the spring material is not to exceed 400 MN/m^2 under a load of 800 *N*. The diameter of the coil is to be 10 times the diameter of the wire. Take $C = 80 \times 10^9$ N/m^2. **[d = 7.14 mm; D = 71.4 mm; n = 9]**

5. A close coiled helical spring is made of 10 mm diameter steel wire by winding it on a 150 mm diameter mandrel. If the spring has 10 coils, then what is the spring constant? What force must be applied to the spring to elongate it by 60 mm? Take $C = 82000$ N/mm^2. **[2.5 N/mm; 150 N]**

6. A close-coiled helical spring having 10 coils of mean diameter 80 mm, is made of steel wire of 8 mm diameter. In unloaded condition, pitch of the coil is 15 mm. Determine the axial load which will cause the coils to close. Take $C = 84000$ N/mm^2. **[588 *N*]**

❑❑❑

EXPERIMENTS

EXPERIMENT NO. 1

OBJECT

To conduct a tensile test on mild steel specimen and determine the yield stress, ultimate stress, percentage elongation and Young's modulus of elasticity.

APPARATUS

Universal Testing Machine (UTM), Extensometer, mild steel specimen etc.

PROCEDURE

Same as tension test of ch - I

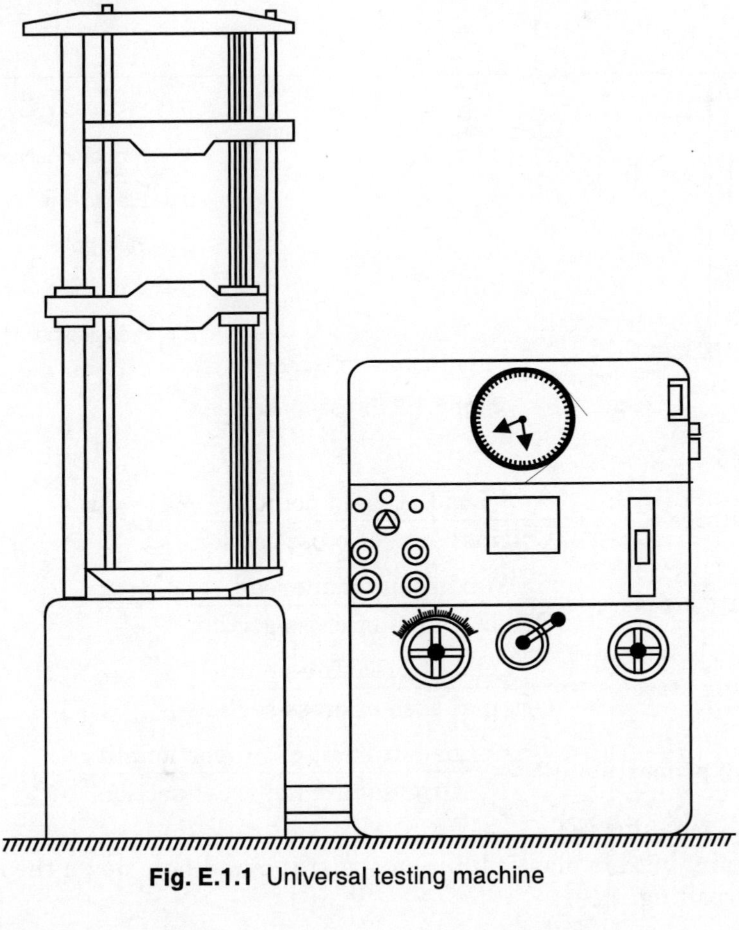

Fig. E.1.1 Universal testing machine

OBSERVATIONS

Original dimensions of the specimen

Length (l) = ...

Diameters (d) = ...

Cross-sectional area (A) = $\frac{\pi}{4} d^2$ = ...

Final dimension of the specimen after the test

Length (l) = ...

Diameter (d) =

Cros-sectional area (A) = $\frac{\pi}{4} d^2$ = ...

S. No.	Load (P)	Extension (δl)	Stress = $\frac{P}{A}$	Strains = $\frac{\delta l}{l}$

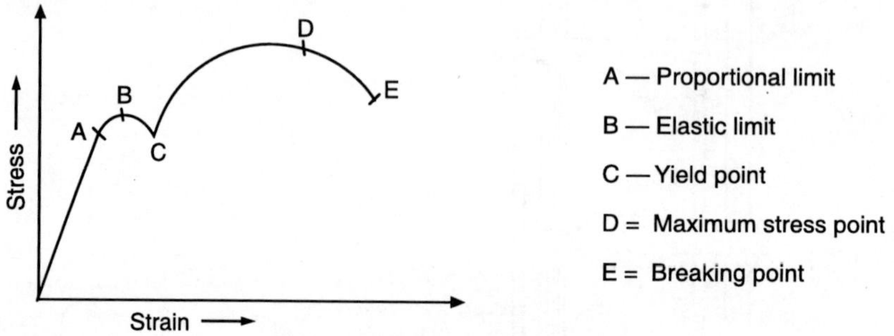

Fig. E.1.2 Stress-strain curve

CALCULATIONS

(*i*) Yield stress $= \dfrac{\text{Load at yield point}}{\text{Original area of cross-section}}$

(*ii*) Ultimate stress $= \dfrac{\text{Maximum tensile load}}{\text{Original area of cross-section}}$

(*iii*) Breaking stress $= \dfrac{\text{Load at breaking point}}{\text{Original area of cross-section}}$

(*iv*) Limit of proportionality $= \dfrac{\text{Load at limit of proportionality}}{\text{Original area of cross-section}}$

(*v*) Young's modulus (*E*)

Take any point with in elastic limit on the graph and note down the corresponding load and extension reading.

$$\text{Stress} = \frac{\text{Load}}{\text{Original area}}$$

$$\text{Strain} = \frac{\text{Change in length}}{\text{Original length}}$$

$$\therefore \text{ Young's modulus, } E = \frac{\text{Stress}}{\text{Strain}}$$

(*vi*) Percentage elongation = $\dfrac{\text{Final length} - \text{original length}}{\text{Original length}} \times 100$

PRECAUTIONS

(*i*) Specimen should be fixed tightly in the machine :

(*ii*) Reading of load and extension should be noted carefully.

(*iii*) Before necking, the extensiometer should be removed.

EXPERIMENT NO. 2

OBJECT

To determine young's modulus of elasticity for steel wire with searl's apparatus.

APPARATUS

Searle's apparatus, Micrometer, meter rod, two identical wires, weights etc.

THEORY

Searle's apparatus is used for finding young's modulus of elasticity of a wire. It consists of two rectangular frames *P* and *Q* of equal sizes connected together at the centre by a cross link '*R*'. Both the frames are suspended by identical wires *S* and *T*. Wire '*S*' is the reference wire and wire '*T*' is the test wire. The wires are held by chucks (torsion heads) at each end. A spirit level is connected to the frames as shown in figure 2*E*. The wire '*S*' is kept tight by a constant weight and slotted weights are added to the test wire '*T*' to get the extension. The young's modulus is determined by using Hooke's law *i.e.*,

$$E = \frac{\text{Stress}}{\text{Strain}} \quad \text{within elastic limit}$$

$$\text{Stess} = \frac{W}{A} \text{ and strain} = \frac{\delta l}{l}$$

$$\therefore \quad E = \frac{W/A}{\frac{\delta l}{l}} = \frac{W \times l}{\delta l \times A}$$

where W = load

l = original length of wire

δl = increase in length of wire

A = cross-sectional area of wire

PROCEDURES

(*i*) Fix the two wires with the frames *P* and *Q* with the help of chucks at each end of the wires.

(*ii*) Measure accurately the lengths of the wires '*s*' and '*T*'.

(*iii*) Measure the diameter of the wires with the micrometer at different places and find the mean diameter of the wire.

(*iv*) Place a constant weight on frame '*P*' to keep the wire '*S*' tight.

(*v*) By operating the micrometer screw bring the bubble of the spirit level at the centre.

(*vi*) Note down the initial reading of micrometer. Let it be x_1.

(*vii*) Place a known slotted weight in the frame *Q*. Due to this weight, the wire '*T*' elongates and the frame *Q* comes down. The bubble tube becomes inclined and the bubble goes out of centre.

(*viii*) By bringing the bubble at the centre, note down the final reading of micrometer. Let it be x_2.

(*ix*) The difference between the final and initial reading of the micrometers gives the elongation 'δl' of the wire.

(*x*) Repeat the experiment four or five times with different loads.

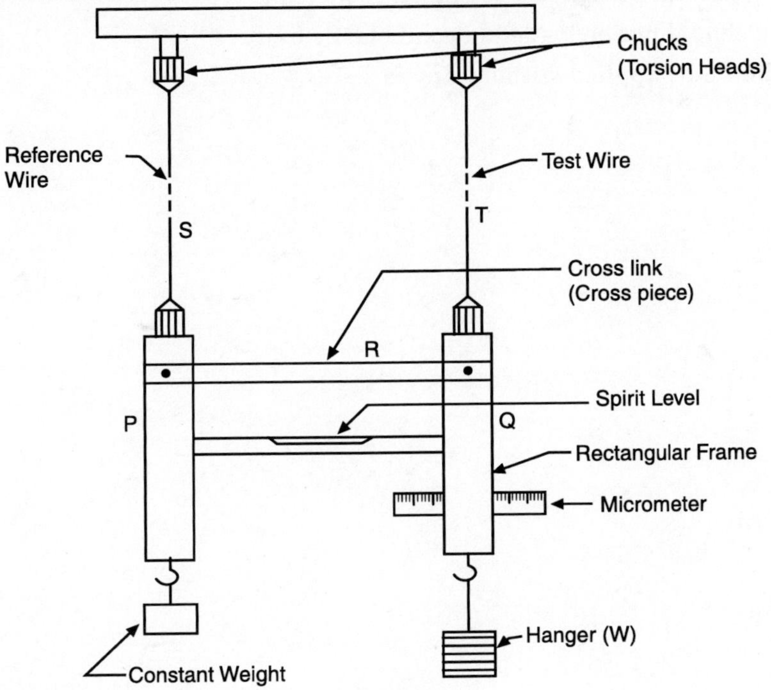

Fig. E2.1

OBSERVATIONS

Diameter of wire :

(*i*) ...
(*ii*) ...
(*iii*) ...
(*iv*) ...
(*v*) ...

Mean diameter $= \dfrac{i + ii + iii + iv + v}{5}$

S. No.	Length of of wire (l)	Area of cross section (A)	Load (*W*)	Stress $\sigma = \dfrac{W}{A}$	Micrometer reading		Increase in length $\delta l = x_2 - x_1$	Strain $e = \dfrac{\delta l}{t}$	Young's modulus of elasticity $E = \dfrac{\sigma}{e}$
					x	x_2			

PRECAUTIONS

(*i*) Wire should be smooth and free from any kinks.

(*ii*) The wires should be tight before placing the load.

(*iii*) The wire should not be loaded beyond elastic limit.

(*iv*) Loads should be applied without any jerk.

EXPERIMENT NO. 3

OBJECT

To determine the modulus of rupture (maximum binding stress) of a timber beam.

APPARATUS

Transverse bending testing machine, steel ruler, specimen etc.

THEORY

The bending equation is given by

$$\frac{M}{I} = \frac{\sigma_b}{y}$$

The maximum bending stress f_{max} occurs at the greatest distance y_{max} from neutral axis of the beam.

$$\therefore \quad \sigma_{b\ max} = \frac{M}{I} \times y_{max}$$

where, M = Maximum bending moment = $\frac{Wl}{A}$

I = Moment of inertia about N.A = $\frac{bd^3}{12}$

y_{max} = Distance of outermost fibre from N.A = $\frac{d}{2}$

l = Effective length of the beam

W = Load at which the beam fracture

$$\therefore \quad \sigma_{b\ max} = \frac{\frac{Wl}{4} \times \frac{d}{2}}{\frac{bd^3}{12}}$$

$$\sigma_{b\ max} = \frac{Wl \times 12}{4\ bd^3} \times \frac{d}{2} = \frac{3\ Wl}{2\ bd^2}$$

PROCEDURES

(*i*) Measure the dimenssions of the beam *i.e.*, length, breadth and depth. The depth of beam should be greater than the width of the beam.

(*ii*) Fix the beam depthwise on the rollers.

(*iii*) Measure centre to centre distance between the roller supports.

(*iv*) Close the release value and start the machine. The loading frame slowly moves upward due to pumping and the beam will come in contact to the upper roller.

(*v*) Continue pumping till the fracture of the specimen.

(*vi*) Open the release value and note down the maximum load at which the specimen fracture.

OBSERVATIONS

Length of beam, l =
Width of beam, b =
Depth of beam, d =
Load at fracture, N =

$$y = \frac{d}{2}$$

CALCULATION

Moment of inertia, $I = \frac{bd^3}{12}$

Maximum, bending movement $M = \frac{Wl}{4}$

Maximum bending stress, $\sigma_{b\ max} = \frac{M}{I} \times y_{max}$

PRECAUTIONS

(*i*) The dimensions of the beam should be measured accurately.
(*ii*) The maximum load should be noted accurately.
(*iii*) The calculation of bending moment and moment of inertia should be done correctly.

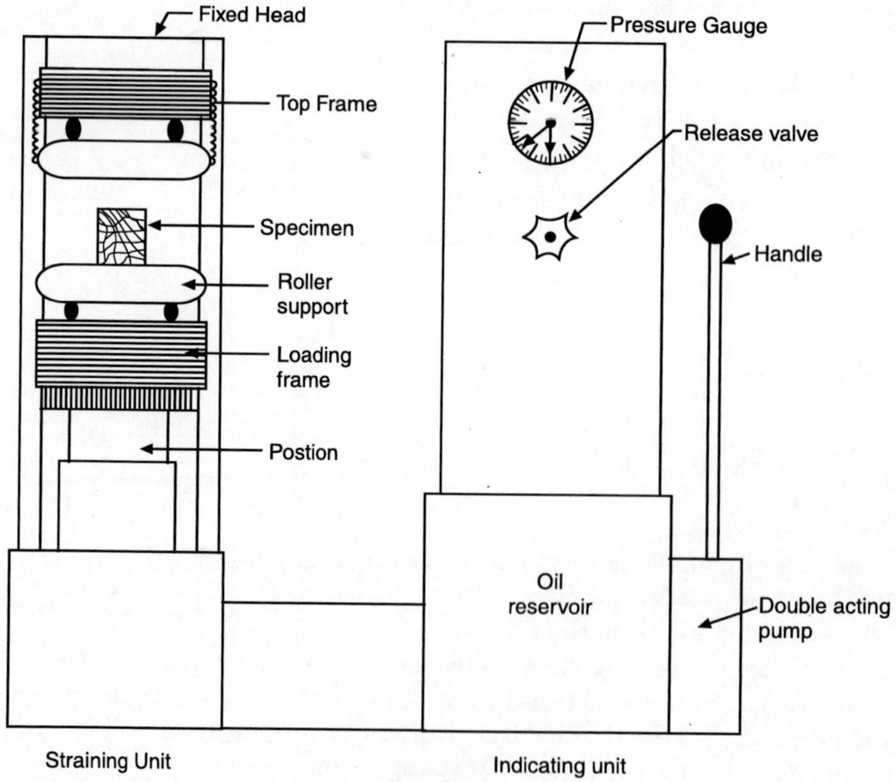

Fig. E3.1

EXPERIMENT NO. 4

OBJECT

Veryfication of forces in a framed structure.

APPARATUS

A simple roof truss apparatus, weights, ruler.

THEORY

A structure made up of several members connected together is known as frame. Frames are of two types ; (*i*) perfect frame (*ii*) imperfect frame, the frame is said to be perfect if it satisfies the equation $n = 2j - 3$, where, n = number of members in a frame and j = number of joints. The member of a frame which is under tension is known as 'tie' member and member under compression is known as 'strut'.

PROCEDURES

(*i*) Note down the initial readings of the spring balance, F_1, F_2 and F_3.

(*ii*) Place the load 'W' on the hook at the crown.

(*iii*) Note down the readings of the spring balances F_1, F_2 and F_3 again.

(*iv*) Measure the lengths l_1, l_2 and l_3 with the help of ruler.

(*v*) Repeat the experiment with different loads and note down the different readings in the observation table.

(*vi*) In order to find the forces graphically draw the vector diagram as shown in figure 4E.1

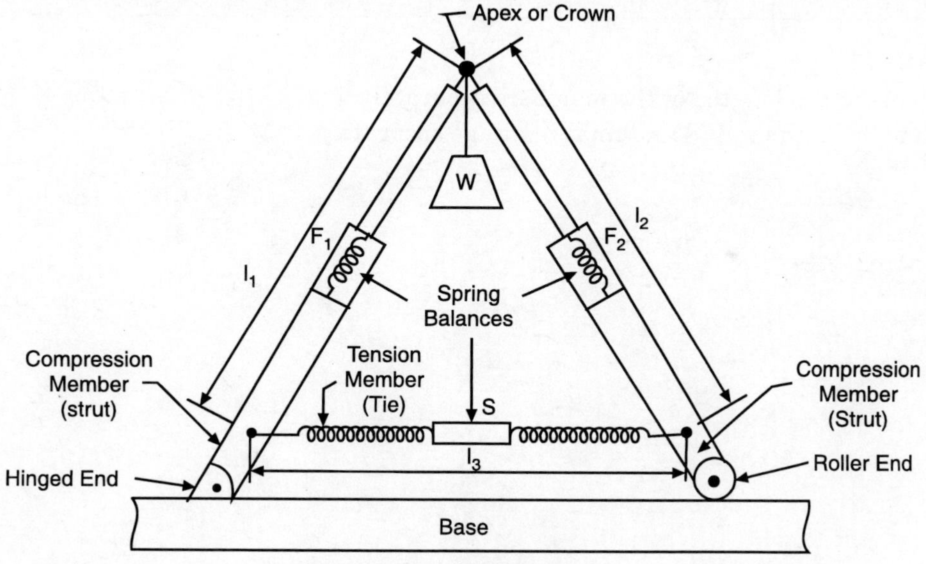

Simple Roof Truss Apparatus

Fig. E4.1

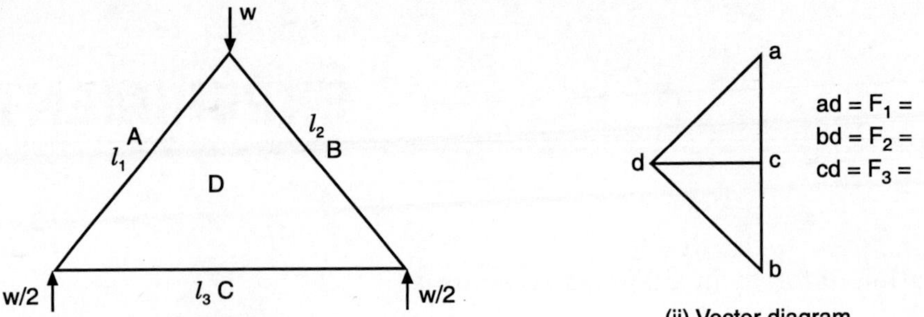

(i) Space diagram

(ii) Vector diagram

Fig. E.4.2

OBSERVATION TABLE

S. No.	Load	Length of members			Initial readings of spring balance (1)			Final reading (2)			Corrected reading (2)-(1)			Forces determined graphically		
	W	l_1	l_2	l_3	F_1	F_2	F_3	F_1	F_2	F_3	F_1	F_2	F_3	F_1	F_2	F_3

PRECAUTIONS

1. Measure the lengths of the members accurately.
2. Note down the initial and final readings accurately.

❑❑❑

INDEX

P

R

S

T

U

V

W

Y

❑❑❑

OTHER DETAILS

(i) Who recommended you this book ? (Please tick (✓) in the box near the option relevant to you.

Teacher ☐ Friends ☐ Bookseller ☐

....

....

....

....

(ii) Name of the recommending teacher, his designation and address :

....

....

....

....

(iii) Name and address of the bookseller you purchased the book from :

....

....

....

....

(iv) Name and address of your institution (Please mention the University or Board as the case may be)

....

....

....

....

(v) Your name and complete postal address :

....

....

....

....

(vi) Course you are enrolled in :

....

....

....

....

The best assessment will be awarded each month: The award will be in the form of our publications as decided by the Editorial Board.

Please mail the filled up coupon at your earliest to :

S.K. KATARIA & SONS®
4885/109, Prakash Mahal, Dr. Subhash Bhargav Lane,
Opposite Delhi Medical Association, Daryaganj, New Delhi–110002 (INDIA)
Phone: +91-11-43551243, *Mobile:* +91-9871775858, 7678425209
e-mail: katariabook@gmail.com, katariabooks@yahoo.com
Website: www.skkatariaandsons.com